Plumbing: Installation and Design

Plumbing: Installation and Design

James A. Sullivan
Southern Illinois University
Carbondale, Illinois

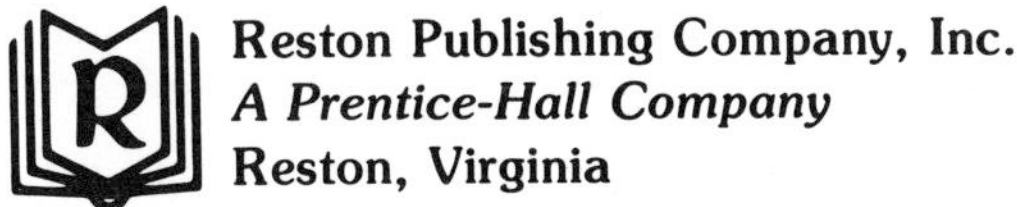

Reston Publishing Company, Inc.
A Prentice-Hall Company
Reston, Virginia

Library of Congress Cataloging in Publication Data

Sullivan, James A.
Plumbing, installation and design.

Includes index.
1. Plumbing. I. Title.
TH6123.S94 696′.1 79-27284
ISBN 0-8359-5552-4
ISBN 0-8359-5553-2 pbk.

Reston Publishing Company, Inc.
A Prentice-Hall Company
Reston, Virginia 22090

10 9 8 7 6 5 4 3 2 1

Printed in the United States of America

To Emily and Eileen

Contents

Preface

This book promotes the job and image of the plumber as a craftsman and professional who uses the combination of knowledge and skills in the practice of a trade. The knowledge base consists of the materials and practices used in plumbing, the flow of fluids, basic mathematics, and plumbing codes. The plumber must have hand and machine tool skills to rough in systems, work with pipe, install fixtures, and make basic repairs. The combination of knowledge and skills, together with a mature attitude and safety mindedness are the substance of a professional person. Assisting in the development of this type of individual is the underlying purpose of this book.

The twelve chapters in *Plumbing: Installation and Design* may be completed in sequence, or the order may be changed to suit the reader's purpose. For example, chapters three, six, and seven, which deal with mathematics and the behavior of fluids, may be used when they are appropriate or omitted entirely if they are not needed, without any loss of continuity. The same is true of chapters eleven and twelve which describe systems components and solar plumbing.

Both English and SI metric units are used throughout the text. English units are given first with SI units following in brackets. While conversion to metric units is still in its infancy in some places, facility with metric units is encouraged. They are used together in this text to develop a familiarity with common comparisons which are made in such applications as threaded pipe and tube sizes.

Many persons and companies have assisted the author in the preparation of this work. Special thanks is extended to the several reviewers for their advice and constructive criticism, particularly to Patrick J. Higgins (National Association of Plumbing-Heating-Cooling Contractors), T. C. Brown (Cast Iron Soil Pipe Institute), George Brazil (Brazil Plumbing), Robert C. Carmody (Copper Development Association Inc.), and Professor Donald Bergeson (University of Illinois).

James A. Sullivan

1 Introduction

1-1 History of Plumbing

Plumbing is a craft and a plumber is a craftsman. A craft in the modern sense of the word is an occupation which requires special skills that are peculiar to a trade. A plumber works with tools to install and repair pipes, plumbing fixtures, appliances, and many components such as water pumps in buildings. There are two sets of pipes in a plumbing system. One set of pipes brings in the fresh water supply called "potable" water, and the other removes the waste water to the sewer outside the building.

No one is sure how far plumbing dates back in history, but it is known that during the Dark Ages (400 to 1400 A.D.) the Egyptians and Romans constructed stone- and tile-lined aqueducts to transport water to cities, thus applying the principle that water flows downhill. Lead was also formed with a hammer over wooden mandrels, and the seams were welded to make pipes. Some of these relics have been tested to pressures of 250 $lbf/in.^2$ without failure. Pipes of this sort, leading from a higher pond, could operate a fountain, demonstrating the principle that water seeks its own level. The Romans are generally credited with constructing plumbing systems.

A plumber in ancient Rome was called a *plumbarius,* which is derived from the Latin *plumbum,* meaning lead. Until recently, plumbers still worked with lead extensively. Lead is still used in plumbing, particularly to solder copper water supply tubing and replace lead caulk joints in waste piping.

The history of plumbing can be dated to 5000 B.C. with the tile-lined bathtubs fashioned for the ancient Greeks and Egyptians, and the cisterns used by the Babylonians to collect rainwater for public use. About 4000 years ago King Minos of Crete is credited with having toilets in the royal palace which consisted of holes in the floor covered by a lid. They were also protected with a venting system. The principle is not much different from that used to construct the *outhouses* in this country, which were replaced in some rural areas by inside plumbing as late as 1950. Some historians include the waterholes and dished-out places in rocks that were used by prehistoric man as plumbing fixtures and appurtenances.

Not much happened in the plumbing trade during the Dark Ages. Rome was overrun by heathens who held bathing in low esteem, thinking its effects were adverse. Perhaps it was the high lead content in the pipes and drinking vessels, which was unhealthy to the Romans, that gave rise to this belief. Current practice keeps exposure to lead to a minimum in the interest of public health.

During the Renaissance there was also a resurgence in plumbing practices. Promising young plumbers were indentured to a master craftsman under a system that bound the apprentice to work for the master for a number of years. Apprenticeships were often arranged for by fathers, who paid the master to accept the responsibility for raising and teaching the young son in the trade. It followed the old Jewish proverb, "He who does not teach his son a trade teaches him to become a thief." By the early 1700's, England had apprenticeship laws establishing the term of service under the master plumber to seven years. The current term for a plumbing apprenticeship is four to six years.

A number of inventions helped the trade as well. The modern bathtub was credited to Lord Russell in England about 1830. His American counterpart was Adam Thompson, who improved the system. Thomas Crapper invented the siphon flush valve for water closets in the 1870's and for his contribution was dubbed Sir Thomas Crapper by the Lord Chamberlain. Some time after, Sir Thomas Crapper was elevated to the position of Royal Plumber and is also credited with invention of a precision-fitting manhole cover. The slang use of the term "crapper" is generally attributed to American servicemen returning from England during World War I.

Modern plumbing practice is guided by a set of 22 principles[1] that have been developed over time. They were first set forth by the National Plumbing Code (American Standards Association A40.8–1955). When there are no local plumbing codes or unforeseen circumstances arise, these principles can be used to define the intent of accepted practice. Although they sound a bit stilted because of the language usage, they have tremendous practical importance, as well as historical significance to the trade.

1. *Provide a safe potable water supply.* All premises intended for human habitation, occupancy, or use shall be provided with a supply of pure and wholesome water, neither connected with unsafe water supplies nor subject to the hazards of backflow or back-siphonage.

2. *Provide an adequate water supply.* Plumbing fixtures, devices, and appurtenances shall be supplied with (hot and cold) water in sufficient volume and at pressures adequate to enable them to function satisfactorily and without undue noise under normal conditions of use.

3. *Conserve water.* Plumbing shall be designed and adjusted to use the minimum quantity of water consistent with proper performance and cleaning.

4. *Safety devices are required.* Devices for heating and storing water shall

be so designed and installed as to prevent dangers from explosion through overheating.

5. *Connect to a public sewer where available.* Every building having plumbing fixtures installed and intended for human habitation, occupancy, or use on premises abutting on a street, alley, or easement in which there is a public sewer shall have a connection with the sewer.
6. *Minimum number of fixtures.* Each family dwelling unit on premises abutting on a sewer or with a private sewage-disposal system shall have, at least, one water closet and one kitchen-type sink. It is further recommended that a lavatory and bathtub or shower shall be installed to meet the basic requirements of sanitation and personal hygiene. All other structures for human occupancy or use on premises abutting on a sewer or with a private sewage-disposal system shall have adequate sanitary facilities, but in no case less than one water closet and one other fixture for cleansing purposes.
7. *Quality plumbing fixtures.* Plumbing fixtures shall be made of smooth nonabsorbent material, shall be free from concealed fouling surfaces, and shall be located in ventilated enclosures.
8. *Nonfouling drainage system.* The drainage system shall be designed, constructed, and maintained so as to guard against fouling, deposit of solids, and clogging, and with adequate cleanouts so arranged that the pipes may be readily cleaned.
9. *Quality piping materials.* The piping of the plumbing system shall be of durable material, free from defective workmanship and so designed and constructed as to give satisfactory service for its reasonable expected life.
10. *Water-seal fixture traps.* Each fixture directly connected to the drainage system shall be equipped with a water-seal trap.
11. *Trap seal protection.* The drainage system shall be designed to provide an adequate circulation of air in all pipes with no danger of siphonage, aspiration, or forcing of trap seals under conditions of ordinary use.
12. *Outside vent.* Each vent terminal shall extend to the outer air and be so installed as to minimize the possibilities of clogging and the return of foul air to the building.
13. *Test the system.* The plumbing system shall be subjected to such tests as will effectively disclose all leaks and defects in the work.
14. *Exclude foreign substances from the system.* No substance which will clog the pipes, produce explosive mixtures, destroy the pipes or their joints, or interfere unduly with the sewage-disposal process shall be allowed to enter the building drainage system.
15. *Prevent contamination.* Proper protection shall be provided to prevent contamination of food, water, sterile goods, and similar materials by

backflow of sewage. When necessary, the fixture, device, or appliance shall be connected indirectly with the building drainage system.

16. *Adequate light and ventilation required.* No water closet shall be located in a room or compartment that is not properly lighted and ventilated.
17. *Alternate sewage disposal systems.* If water closets or other plumbing fixtures are installed in buildings where there is no sewer within a reasonable distance, suitable provision shall be made for disposing of the building sewage by some accepted method of sewage treatment and disposal.
18. *Sewer flooding prevention.* Where a plumbing drainage system may be subject to backflow of sewage, suitable provision shall be made to prevent its overflow in the building.
19. *System maintenance.* Plumbing systems shall be maintained in a sanitary and serviceable condition. (Also see the definition of plumbing in Appendix A.)
20. *Fixture spacing.* All plumbing fixtures shall be so installed with regard to spacing as to be reasonably accessible for their intended use.
21. *Structural integrity.* Plumbing shall be installed with due regard to preservation of the strength of structural members and prevention of damage to walls and other surfaces through fixture usage.
22. *Ground and surface water protection.* Sewage or other waste from a plumbing system which may be deleterious to surface or subsurface waters shall not be discharged into the ground or into any waterway unless it has first been rendered innocuous through subjection to some acceptable form of treatment.

1-2 Structure of the Trade and Apprenticeship System

The work force in the plumbing trade is divided into several levels. Starting at the bottom is the *plumber's helper,* who assists a licensed plumber as a skilled laborer. Next is the *apprentice plumber,* who is *indentured;* that is, he or she has been accepted into the trade and has signed an agreement to work and study for four or five years under a licensed journeyman or master plumber. At the next level are the *journeyman plumbers,* who have served an apprenticeship and have passed the journeyman's license test. The journeyman is a free agent and may work for a contractor or as an independent plumber. The *master plumber* has the highest skill level and may have under his supervision either journeymen or indentured apprentices. The master plumber has served a number of years as a journeyman and has passed the master plumber's license examination. Two other levels of plumbers who have supervisory authority are the *job foreman,* who is at least a journeyman plumber, and the *superintendent,* who oversees all the plumbing jobs run by the general contractor. In addition to supervising, the foreman lays out the work for a particular job and sees that the tools and materials are available when needed to complete the work. The superintendent supervises all the job foremen for the general contractor and has wide responsibility for seeing that the work at the various jobs gets done on schedule within the time and cost estimates that have been established.

Large plumbing firms employ an *estimator,* who plans and assembles bids for jobs. This often includes making isometric drawings from blueprints and specifications, making "take-offs," which consist of listing the materials from the drawings, and formulating the final bid for the contractor. The *contractor* is the person who is responsible for running the business. The contractor combines business experience and financial capital to secure jobs and provide work for the skilled craftsmen.

The apprenticeship system used to train plumbers grew out of the craft guilds of the Middle Ages (A.D. 1100 to 1500). It is a means whereby the trade replenishes the supply of craftsmen in keeping with certain standards and work demands. In this country approximately one hundred trades offer apprenticeships in 300 or more skilled occupations. Benjamin Franklin, for example, was apprenticed to his older brother, who was a master printer, to learn the printing trade. The system has been an effective means of training for more than two hundred years. Training for the apprentice includes a specified number of years "on the job" doing prescribed work, as well as classroom instruction held "after hours" during the same time to learn related trade information in blueprint reading, welding, and plumbing theory. Four to five years is the normal term of the apprenticeship, with 144 to 220 hours of related classroom instruction held at night each year during that time.

Apprenticeships are regulated by the Bureau of Apprenticeship and Training, which was established in the Department of Labor in 1937. Its function is to oversee the programs in the various states, to set standards for the training of skilled workers in the industry, and to protect the welfare of the apprenticeship as an employed worker. Standards include minimum age for an apprentice, ratio of apprenticeships to licensed plumbers (journeymen), length of apprenticeships, education requirements, hours of work, and the wages the apprentice will receive during the training period.

In the various states, apprenticeships are controlled by a local Joint Apprenticeship and Training Committee, composed of equal representation of labor and management. The JATC establishes local standards for apprenticeships which govern what will be learned on the job site and in the related classroom instruction to pass state or municipal plumbing tests for licensure. State and local governments usually permit plumbing to be done only by a licensed plumber. Apprentices are licensed but are not allowed to complete work alone at the job site. Rather, they must be under the supervision of a journeyman or master plumber. At the end of the apprenticeship, the apprentice receives a certificate of completion from the JATC, which is registered with the Bureau of Apprenticeship and Training, and qualifies to take the examination for the journeyman's license. When the apprentice passes the examination given by the state or local board of examiners, he or she is issued a journeyman's license to do plumbing within the jurisdiction. Larger cities often require a city license. The rules governing licensure in the states are enforced by the state board of public health. A sample plumber's license is shown in Fig. 1-1. In Illinois, for example, the apprentice license is the same as the journeyman's, but includes the condition that to be valid the apprentice must be working at the time with the sponsor designated on the license. This assures that the apprentice is complying with the conditions of the apprenticeship agreement.

The approved program of work experience and related instruction is developed by the local JATC to incorporate the skills necessary to become

Figure 1-1

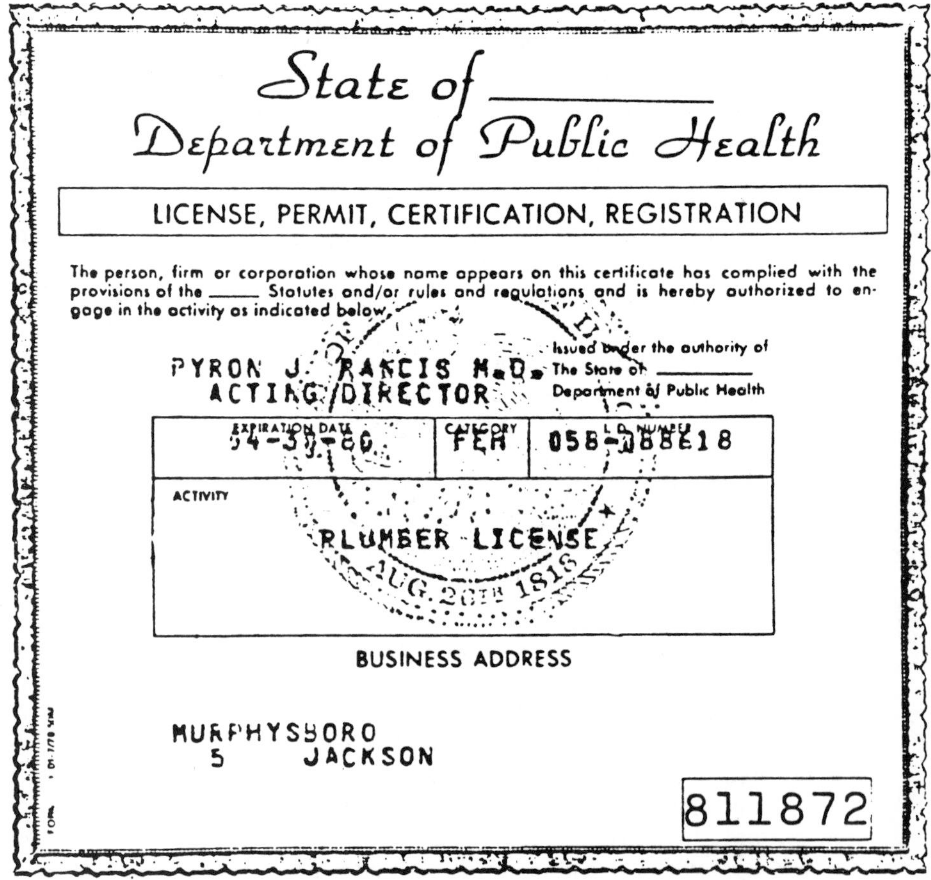

State of ______

Department of Public Health

LICENSE, PERMIT, CERTIFICATION, REGISTRATION

The person, firm or corporation whose name appears on this certificate has complied with the provisions of the ______ Statutes and/or rules and regulations and is hereby authorized to engage in the activity as indicated below.

PYRON J. FRANCIS M.D.
ACTING DIRECTOR

Issued under the authority of
The State of ______
Department of Public Health

EXPIRATION DATE	CATEGORY	I.D. NUMBER
04-30-80	FER	058-088618

ACTIVITY

PLUMBER LICENSE

BUSINESS ADDRESS

MURPHYSBORO
5 JACKSON

811872

Sample Plumber's License

a successful plumber and to pass the state, city, or local plumbing license examinations. Figure 1-2 illustrates a sample apprenticeship training program. The qualifications are common for a four-year program.[2] Some programs last five years or more and may extend the related instruction. The sample plan proposes work that has priority to the JATC, and related instruction topics that describe a knowledge of plumbing work and materials, mathematics, blueprint reading, hydraulic principles, and related safety. In some states, for example, Illinois, the state Plumbing Code specifies that the authority for constructing the examination and issuing licenses rests with the state Board of Plumbing Examiners, or with municipalities over 500,000 to which this power is granted. They also have the authority to

Figure 1-2

ENTRY QUALIFICATIONS

Minimum age - 17
Education - high school or equivalent
Health - physical examination required
Citizenship - citizen or declared intention of becoming a naturalized American citizen
Term of apprenticeship — 4 years minimum
Work experience - 7400 Hours (1850 hours per year)
Related instruction - 864 hours (216 hours per year)
Sponsor - must have a sponsoring employer

PROGRAM OF WORK EXPERIENCE	HOURS
Installation of hot- and cold-water distribution piping	700
Installation of drain, waste, and vent piping	2000
Installation of fixtures and appliances	500
Maintenance and repair	2000
Installation and maintenance of plumbing systems components (pumps, valves, meters, and gauges)	700
Other plumbing work	1500
	7400

RELATED INSTRUCTION	
The origin and development of the plumbing trade	20
Tools and plumbing work	20
Plumbing mathematics	80
Pipes and fittings	40
Cutting, joining, and supporting pipes	40
Science of fluids	40
Flow in pipes	40
Cold- and hot-water distribution systems	80
Drain, waste, and vent systems	80
Installing, and testing systems	160
Plumbing system components	80
Solar plumbing	40
Safety and first aid	40
Drafting and blueprint reading	40
Welding	64
	864

Sample Apprenticeship Training Program

prescribe approved training courses of instruction in vocational schools and colleges. These established rules assure that an approved course of instruction will adequately teach the design, planning, installation, replacement, extension, alteration, and repair of plumbing systems. Where the formal school training rather than the apprenticeship system is used to prepare for the plumbing trade, it is common to require an approved two-year program of study followed by at least five years of work experience as an apprentice under a licensed plumber before the examination can be taken.

The related instruction portion of the apprenticeship can be completed in a number of ways by the JATC who has this responsibility: by instruction using the designated instructor and course work approved by the JATC, by approved courses at local public and private schools, or by a combination of the two. The latter is the usual method, since the combination of existing school courses taught by persons with specializations in such areas as blueprint reading, drafting, and welding, and plumbing instructors who are expert mechanics and conversant with plumbing codes gives the student the best available experts from whom to learn. A number of resources are available to assist the instructor and student learner in the guided plan of study.[3,4,5,6,7,8,9]

1-3 Plumbing Codes

Early efforts to establish plumbing codes in this country date back to the early nineteen-hundreds. The need for minimum standards for plumbing to protect the health of the public had been recognized by then for some time and various organizations had published local plumbing codes for that purpose. At the national level the early work to coordinate these codes into a comprehensive code can be traced to the Subcommittee on Plumbing of the Building Code Committee of the U.S. Department of Commerce, which published BH2, *Recommended Minimum Requirements for Plumbing in Dwellings and Similar Buildings,* at the National Bureau of Standards in 1924. Revisions, based in part upon research conducted by the National Bureau of Standards, resulted in BH13 by the same title, which was widely known as the "Hoover Code," published in 1928 and further revised in 1932. In 1933, the National Association of Master Plumbers published its *Standard Plumbing Code,* and the latest revision of this was made in 1942. Western states plumbing officials released the first draft of the Uniform Plumbing Code in 1938. The current edition was released in 1979.

A major code milestone was reached with the revision of the BH Recommendations in 1936 and 1940 by the Central Housing Committee for Plumbing, after which it was given the designation BMS66, *Plumbing Manual.* It was important in that it formed the basis for a wartime plumbing code which was to stress the conservation of materials. This was developed and designated *Emergency Plumbing Standards,* which remained in effect through the war years. It is also considered to be the first plumbing code to have been followed as a standard on a nationwide basis.

The *National Plumbing Code* (A40.8–1955), which was a standard designated by what is now the American National Standards Institute, Inc. until 1972, has been in existence since 1955. Its sponsors were the American Public Health Association and the American Society of Mechanical Engineers. Its history can be traced to efforts after World War II to establish a peacetime plumbing code based upon successful experience with the "Emergency Standards" used during the war years. It was the culmination of efforts by a joint committee known as the Uniform Plumbing Code Committee formed in 1946 by the National Association of Plumbing Contractors, the United Association of Journeymen and Apprentices, and other interested government agencies. Basic research to fill information voids in the behavior of fluids, particularly in the hydraulics and pneumatics of plumbing systems, was conducted by the National Bureau of Standards, universities, and laboratories across the country. These included the University of Illinois, the University of Iowa, and the U.S. Public Health Service Environmental Center, Cincinnati, Ohio. A number of plumbing references have been written to parallel and teach the rudiments of the National Plumbing Code.[10,11,12]

Withdrawal of A40 as a National Standard in 1972 occurred when consensus on the revisions could not be reached. Since then the *National Plumbing Code* has been made available from the American Society of Mechanical Engineers, which will continue its publication until such time as an American National Standard Plumbing Code is available.

One current effort to arrive at a uniform standard for plumbing is being given leadership by the National Association of Plumbing-Heating-Cooling Contractors, which published the *National Standard Plumbing Code.* This organization traces its origin to the National Association of Master Plumbers, which was founded in 1883, and published and revised the *Standard Plumbing Code* until 1942. The NAPHCC participated in the development of the wartime "Emergency Standards," and was also represented on the coordinating committee for the National Plumbing Code. The sequence and format of the *National Standard Plumbing Code* follow the A40.8 code, which permits a convenient cross-reference between the two codes. The present edition is cosponsored and endorsed by the American Society of Plumbing Engineers, which also participates on the Code Committee with responsibility for future revisions of the standard.[13]

1-4 Plumbing Systems

There are three separate plumbing systems that work together in a dwelling. They are the fresh water distribution system, or "potable" water supply as it is called, the sanitary drainage and vent system, and the storm drain system. The water distribution system brings the fresh water under pressure into the dwelling and pipes it to bathrooms, kitchens, and other places where it is needed. The sanitary drainage and vent system removes this water from the building in larger pipes after it has been used, allows fresh air to circulate through the system, and permits gases to escape, preventing a pressure buildup in the pipes. The storm drainage system takes care of rain water from the roof and areas on the ground where there is water runoff. The sanitary drainage and storm water systems are kept separate.

In corporations and other communities with water districts, the water supply originates for practical purposes at the corporation main and extends into and throughout the building, supplying fresh water to sinks, lavatories, water closets, bathtubs, and outside hose bibbs or sill cocks to which hoses can be connected.

The term *potable* (pronounced pō′ tə bəl) *water* means water free from impurities in sufficient amounts to cause disease or harmful physical effects. There is a standard for the amounts of impurities published by the state health department. The water is tested from time to time, and it must pass the test before it can be distributed to the community as potable water. Protecting the potable water supply is one of the very important reasons for having qualified and licensed plumbers in the trade. Private well systems draw water from shallow and deep wells, as well as cisterns and lakes. The integrity of this water supply is the responsibility of the owner and frequently requires treatment including chlorination, filtering, and softening to bring it to a safe and usable standard. The local health department will assist in this by testing the water supply and supplying a report of its findings.

The components of a fresh-water distribution system are shown in Fig. 1–3. Beginning at the corporation water main, there are:

Corporation cock or corporation stop. The valve at the corporation main.

Water service pipe. The pipe that runs from the corporation main to the curb stop, and from there to the water meter and into the dwelling.

Curb cock or *curb stop.* The valve at the curb or near the sidewalk in a stop box or valve box, which is an adjustable cast iron enclosure for the valve. The adjustment allows the top to be brought up to ground level from varying depths.

Meter valve. The valve is in the water meter setting, which is enclosed in a round, square, oval, or rectangular compartment in the ground with a metal lid over it, on the corporation side. If the meter is in a building, it is next to it but still on the corporation side to allow removal of the water meter.

Water meter. Measures the number of gallons (liters) or cubic feet (cubic meters) of water used in the building.

Building main valve. Located just inside the building. It turns the water off. If the meter is inside the building, then it commonly is located just after the meter.

Building main. The main water artery in the building. It is also the largest fresh-water pipe in the building and extends from the meter to

Figure 1-3

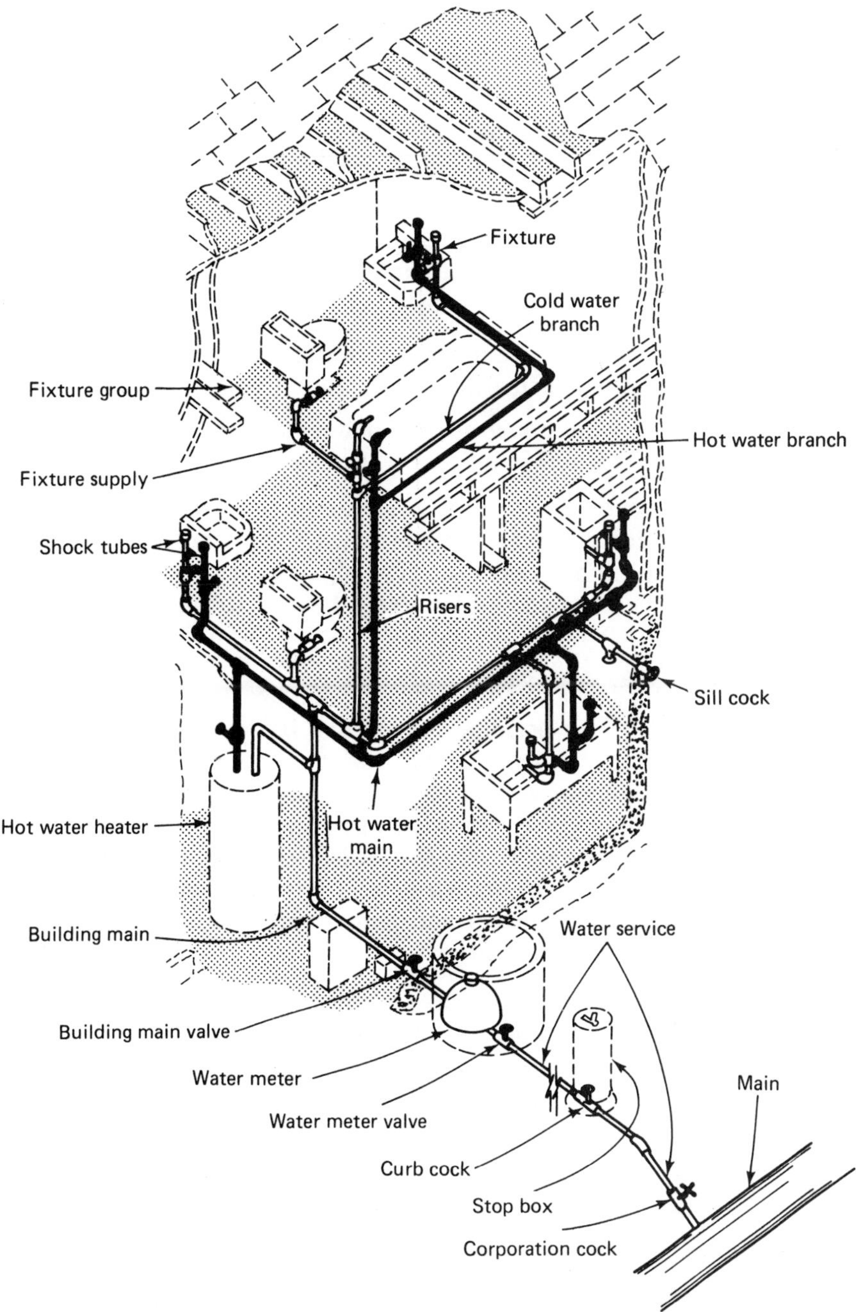

Fresh-Water Distribution System

the principal branches, which distribute water to the various fixture groups. In family dwellings, typically, the building main takes the shortest route to the water heater.

Principal branches. Convey fresh water from the building main to fixture groups in various locations, such as bathrooms and kitchens.

Riser. A vertical pipe that extends upward one full story or eight feet, whichever is less, to convey water from story to story to the various fixture groups.

Fixture group. A cluster of fixtures in one or more rooms close together; for example, a bathroom fixture group would consist of the water closet, lavatory, and bathtub–shower combination.

Fixture supply. The individual water pipe that conveys water to a fixture from a principal branch or an individual fixture branch.

Fixture. A plumbing device such as a water closet, lavatory, bathtub, or kitchen sink.

The second of the three systems is the sanitary drainage and vent system, which removes the water with the water-borne waste from the dwelling after it has been used. The pipes are larger than the water distribution pipes because they must drain by gravity flow. Even the horizontal pipes are pitched slightly so that the water will float the wastes downstream toward the sewer. The vent part of the system consists of air pipes which allow air and other gases that are generated by the waste to circulate freely. This permits the gravity flow and allows noxious odors to escape from the system through the roof, where the vents terminate.

A drainage and vent system is shown in Fig. 1–4. The necessary components include the corporation sanitary sewer, building sewer, building drain, main cleanout, main soil stack and stack vent, waste stack and secondary stack vent, branch drains, fixture drains, fixture traps, fixture vents, horizontal branches, and vent terminal jacket and flashing. The purpose of each of these follows:

Corporation sanitary sewer. Carries the raw sewage and waste to the treatment plant. Storm, drain, and ground water are excluded.

Building sewer. Carries the raw sewage from the dwelling to the corporation sewer.

Building drain. Carries waste water and soil under the floor of the building. It is the lowest horizontal run and connects to the soil and waste stacks.

Main cleanout. A place in a fitting with access to clean the drain. It is located at the base of the stacks.

Main soil stack. Carries soil and waste from the fixture groups. It extends from the base of the stack through the roof.

Figure 1-4

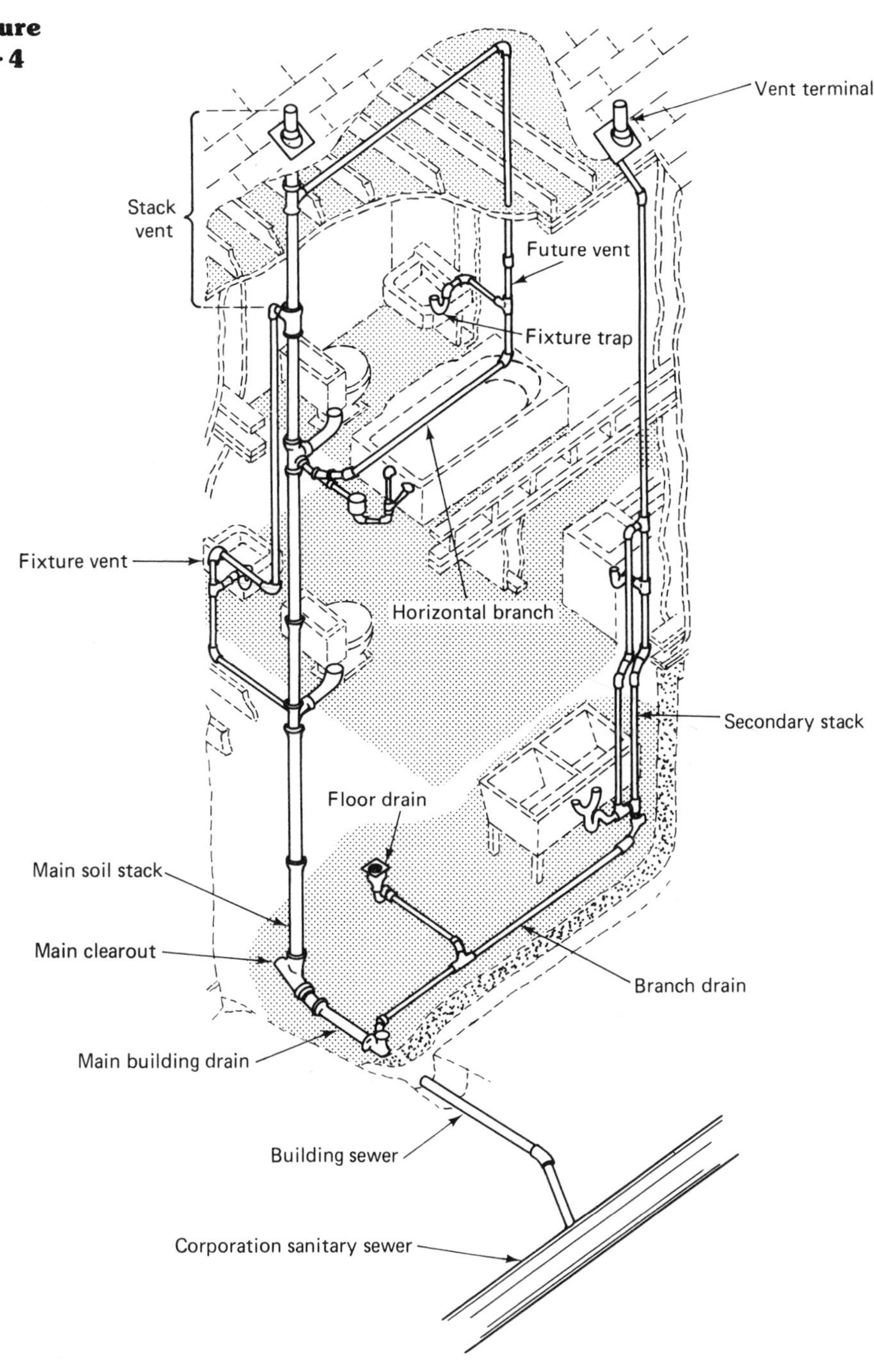

Sanitary Drainage System

Stack vent. The part of the main stack above the highest fixture that extends through the roof. It lets air in and out of the system.

Vent stack. Main air vent stack which connects near the base of the soil stack and extends through the roof, connecting back vents, loop vents, and circuit vents to the outside air.

Waste stack and *secondary stack vent.* Carries waste water from fixtures other than water closets. The upper portion above the highest fixture is the secondary stack vent. It extends through the roof and allows air to enter and leave the system.

Branch drains. The pipes that carry soil and waste from the fixtures to the main or secondary stacks.

Fixture drains. The pipes that lead from each fixture trap to the branch drains.

Fixture traps. Provide a liquid seal at each fixture.

Fixture vent. Each fixture trap seal is protected by a vent.

Horizontal branches. Pipes that run horizontally and carry soil and waste from one or more branch or fixture drains to the stack.

Vent terminal jacket and *flashing.* The end of the stack vent or vent stack that comes through the roof. It has a jacket over it, which is sealed to the roof with a metal flashing. The jacket keeps the roof from leaking around the vent terminal. It can be used to trap warm air around the vent to prevent stoppage at the terminal by frost in colder climates.

The third of the three systems is the storm drain system. It carries water from the roof and ground surface areas such as patios to the storm drain. The parts of the system include the roof drains or gutters, the leaders or downspouts, and the building storm drain, which carries the water to the storm sewer. These components are shown in Fig. 1–5, and serve the following purposes:

Roof drains. Catch the water on the roof. They are protected by a grill to prevent the entry of debris that would plug the drain.

Gutters. Mounted to the facia of the building. The water runs off the roof directly into the gutter.

Leaders. Run vertically from the roof drain to the building storm drain; they are attached inside the building.

Downspouts. Run vertically from the roof drain to the building storm drain; they are attached to the outside of the building.

Storm sewers. Horizontal conductors that carry rain and runoff water only. Storm sewers and sanitary sewers are not connected except in special circumstances; when they are, the storm system must be trapped to prevent the circulation of sewer gases in the storm drain system.

Figure 1-5

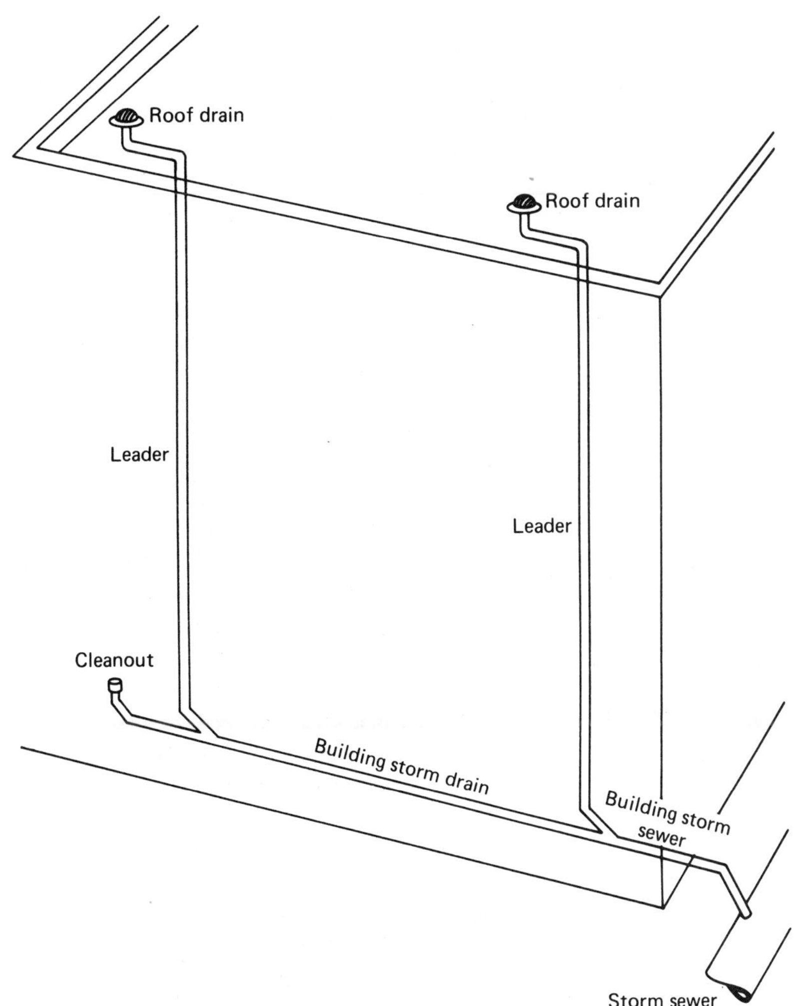

Storm Drain System

1-5 Safety on the Job

Safety prevents accidents. It is an attitude about the way you work. It causes you to be careful when otherwise you would not be. It causes you to consider the consequences before you act. To be effective at the job site, safety has to be practiced by everyone. Development of a safe attitude takes training, good work habits, and the foresight to analyze a situation before a careless act results in an accident that injures someone who now must lose time from work to recuperate. Many accidents permanently injure people, resulting in loss of limbs, eyesight, and hearing and in other damage to the body.

A trained craftsman works nearly forty years in the trade and will earn over $800,000 in today's dollars. That is a lifetime at work and investment. When a person considers that a whole lifetime of work is something that can be ended by one careless practice or work habit, it makes one stop and think. Is an unsafe act worth it, just to get today's work done a little sooner? The answer is always no, it is not.

Under the Occupational Health and Safety Act of 1971, known as OSHA, the employer has the responsibility to provide a safe and healthful workplace for employees. Employees have the responsibility to follow the safety rules that will prevent many of the common accidents, such as wearing hard hats, safety shoes, gloves, and eye protection. OSHA is enforced by compliance inspectors from the Department of Labor. Violations by employers are punishable by fines and in severe cases by imprisonment. A copy of the Act is available from your local congressman.

Under OSHA a safety program is required under the leadership of a designated person. This means that safety for your specialty will be taught, probably as part of the related instruction of the apprenticeship program, but this could occur at work. The instruction should consist of safe personal work habits; the safe way of performing such tasks as lifting, erecting a scaffold, and bracing a trench; and performing first aid procedures to treat an injury or accident victim. First aid courses are offered by instructors who are certified by the Red Cross, and you receive a certificate as a credential that you have completed the course and can perform as required. Taking safety courses and keeping certificates current is important to being a professional in the trade.

Safe work habits can be divided into a general code of conduct on the job, and specific habits that relate to plumbing work itself. The purpose of developing a general code of conduct is to protect you in your life space at work. It is like defensive driving on the highway. A certain amount of area around you, free from hazards, is necessary if you are to remain safe. You do not want to get hurt, so you must evaluate the area where you will be working and the people who work there for the possibility that you are in danger. This cannot be left to chance. Here are some of the hazards and their causes which can affect the safety of your life space at work:

Hazard	*Cause*
1. Other workers	Horseplay and a careless attitude about your safety. (You will be fired for horseplay, probably the first time.) Expecting you to work unsafely.
2. Falling or bumping into things	Cluttered and dimly lit areas. Improper stacking of pipe. Materials stored where they can fall. Unsafe

scaffolds, lifting rigging, unbraced trenches, and oil and water on the floor.

3. Imminent dangers — Unguarded machines, unsafe equipment, ungrounded electrical service or equipment, moving and motorized equipment transporting materials, no marked walkways or handrails in danger zones.
4. Inadequate safety and prevention plan — No safety zones or signs, no plan or fire protection equipment in case of fire, no first aid facilities or first aid kit. Who is notified in case of accident has not been identified.

Now let us be more specific and refer to work that each plumber does within that life space that he occupies, and how it affects his personal safety and that of others. The purpose is to protect yourself from you rather than others. That is, you are trying to avoid hurting yourself; in this case, getting hurt will be your own fault. You will need safety equipment and a few skills to protect yourself. Consider the equipment first. To be safe at work in the plumbing shop and at the work site, you need to protect parts of the body with special equipment. It is your responsibility to secure it, carry it, and maintain it.

1. For the eyes: Goggles or safety glasses. (Have a face mask available for grinding.)
2. For the hands: Gloves to prevent blisters, burning, cuts, and abrasions.
3. For the head: A hard hat that fits properly and is comfortable.
4. For the feet: Hard-toe shoes with thick, puncture-resistant soles.
5. For the hearing: Ear protection by caps or approved plugs. Cotton does not protect. (Noise is perhaps the most insidious of all hazards, since its deafening effects are cumulative. After a few years of exposure, you lose appreciation for the stereo, and then you begin to lose conversations. During this time, no one is aware, not even you, that you are suffering more and more permanent hearing damage.)
6. For the lungs: Protection consists of a breathing mask or respirator that filters out dust, gases, and abrasive materials that can poison or damage the lungs. Do not work alone when there is the potential for poisonous gas, such as in manholes, pits, and sewers. A safety harness and rope may also be required.
7. For the body: Clothing of good quality, hard-surfaced and close-fitting shirts and trousers without cuffs. Flimsy or loose-fitting garments are absolutely the best way to get caught and hurt by sharp objects and machinery such as grinders that can "grab."

Consider also the hazards of long hair and beards. It is true that it is not the hair but what is under it that is important. That is just the point. If "what is under it" may get hurt, then there is a potential problem. Long hair or a beard has the potential for being caught, for being set afire, or for just plain being in the way when you need to see what is going on. If these difficulties can be resolved, then there is no problem. If they cannot, then you are a safety hazard—to yourself.

By its very nature, jewelry is not safe. A ring can be smashed into a finger, or can even pull the finger off should the hand become wedged or pulled by a machine or object that causes entrapment. Necklaces can be caught and can pull you into machinery. If they are metal, they are also conductors and can transmit heat and electric shocks to the body very near the heart. Metal watch bands have the same potential. A rule of thumb is to be sure these items can "break away" without entrapping you. If they will not, you are not safe.

Now that you own all this safety equipment, how do you fit it all in and work, too? That is, what do you do with the goggles, gloves, ear protection device, and respirator? It is a good idea to have a special place for this equipment so that it will be available when it is needed. The gloves and eye protection should be carried on your person. They will be worn a good part of the time. The ear plugs can be carried in your lunch box or car. So can the respirator, if the work schedule indicates that you will need it. The important thing is that you own it and that it is accessible when you need it.

Working safely with hand and machine tools is as much a matter of craftsmanship and personal pride as it is of safety. Craftsmen just do not use cheap and broken tools. And their good tools are not used carelessly or hurriedly in a manner that will damage the tool and result in personal injury. Craftsmanship generates a high level of pride and self-esteem for oneself, and the notion of performing work skills in a sloppy manner is foreign to that self-image. The technique the craftsman develops is to know the proper use of specific tools and the approved sequences in which they are used. The manufacturer of high-quality tools insures both by first maintaining quality standards in their production, and then by supplying directions for their proper use and maintenance. These skills are first learned and then followed scrupulously. If there are also improper uses that are to be avoided, or safety precautions that must be observed, these too are noted in the directions. The directions are first read and saved in a file. For example, pounding on a screwdriver or using it for a chisel is to be avoided. As another example, when chisels are used, gloves and eye protection must be worn. Finally, when both chisels and screw drivers become worn, there are specific procedures that are followed to maintain them. You may notice that your fellow craftsman will not loan you his or her hand tools. You should not take offense. It is just a matter of personal pride.

Operating machine tools is a different matter. Most often this equipment belongs to the contractor and many people use it. It is something like the company truck; everybody uses it, but no one is responsible for it.

Someone should be. Before using a piece of machine equipment such as a drill, power saw, threading machine, or chop saw, look it over very carefully. Ask these questions: Is the area around it clean, safe, and well lighted? Does it have the proper guards? Is the electrical service safe and grounded? Have I had a demonstration on how to use it correctly? If you are in doubt about this last point, see if you can find the directions and read them. Finally, do I have the proper protective equipment to operate the machine? Here a note of caution is important. Many plumbers have a common characteristic. They always carry a rag. Watch out for rags around rotating machinery. They are as dangerous as gloves. Be sure to read the directions to know whether gloves are to be used with a machine or not. Remember the machine that grabs your rag or glove also has your hand. There are other areas and ways in which you work that affect your personal safety. These include the way you lift and your work in trenches, on ladders, on scaffolds, around electricity, and with torches and lead-melting equipment.

Trenching and working in trenches are dangerous. Trenching may bring you or the machine in contact with electric utilities, phone lines, water pipes, gas pipes, and other obstructions. Contact the utility companies *before* you dig. Anyone can do it afterward. There are obvious dangers that can be avoided, such as a pile of dirt too close to the trench, where it might slide back in, but others are less obvious. For example, trenches more than five feet deep should be braced or made wide enough to ensure that they will not fall in on you. A collapsing trench can bury and suffocate a workman. When you are in the trench, be very much aware of others who are lowering in pipe, backfilling, pouring concrete, or doing a number of other jobs that may cause you to become injured. A heavy machine run close to an otherwise safe trench can cause it to collapse. A concrete truck can collapse the shoring and trap the person in the trench. A simple rule of thumb is to be sure that either someone who is competent is assigned to insure your safety, such as the person who directs the machinery near the trench, or that you stay clear when there is activity over or near the trench. For example, do not stand in the trench when machinery approaches, and do not stand under the boom of a lift lowering pipe. Above all, do not ride the load. The requirements for trenching and working in trenches are well established. The terms are given and illustrated in Fig. 1–6. There have been more than enough accidents for you to know when you are in danger. Do not take a chance. Two of many approved shoring systems for trenches are shown in Fig. 1–7. The minimum sizes for shoring lumber are given in Table 1–1 (page 22).

Lifting is a simple skill. Plumbers do it a lot. If you get hurt lifting, you nearly always lose work. Few other injuries are more painful or bothersome. First, you should know how much weight you are attempting to lift. If it is more than half your weight, think about getting help. When you lift, plant both feet firmly near the object. Get set up for the lift, but do not jerk. Bend your knees and squat, keeping your back as straight as possible. Lift

Figure 1-6

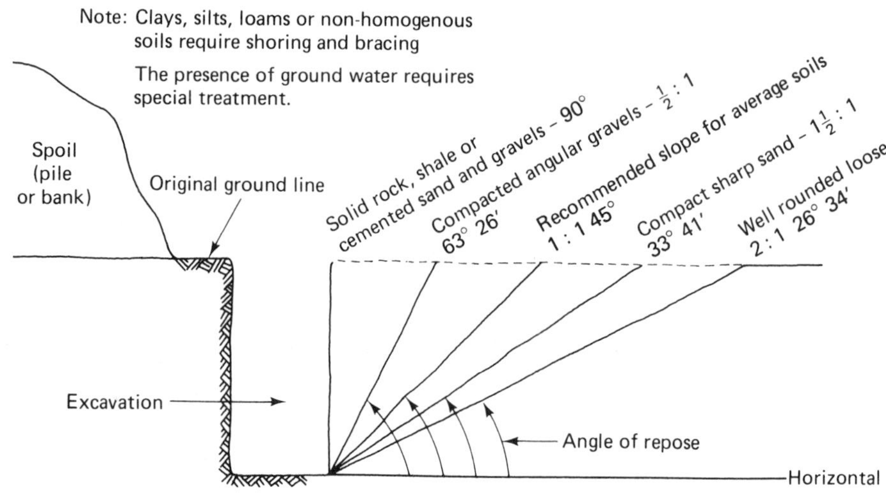

Trenching and Shoring Terms

with your legs, not your back. If the object cannot be handled easily, get help. Remember, work is an endurance contest of many years, not a show-off exercise of strength for just today.

Work on ladders and scaffolds is dangerous because you can fall. There is a correct way to set a ladder, and several correct methods to erect scaffolds. Look the ladder and scaffold (or material for the scaffold) over carefully before you use it. Do not ask yourself, "Will it hold me?" but "Is it safe?". Metal ladders around electricity are dangerous. Painted wooden ladders are also dangerous, because paint covers flaws and cracks in the wood. The ladder and scaffolding material should come from storage where it is off the floor and dry. Select a ladder of the correct length; too long or too short will not do. When the ladder is placed, be sure that the bottom is approximately one-fourth of its length away from the wall. That is, a ladder leaning against a support 12 ft. (3.7 m) up its length should be 3 ft out at the bottom (0.91 m). If the ladder is secured to climb from one floor to another, or to a scaffold above, leave at least three feet minimum extending above the next floor or scaffold level. If it is to be used from time to time in the same place, tie it at the top. If it is to be at the construction site over an extended period, also attach it at the bottom by ties or stakes. If it is in a passageway used by vehicles, machinery, and equipment, establish a safety zone around the ladder, even protecting it with a barricade when necessary.

There will usually be electrical power at the construction site, and this is where the potential for accidents is greatest. Defective extension lights, power cords, and ungrounded equipment are the worst offenders. They can deliver the electricity to you rather than for the intended use. Look each

Figure 1-7

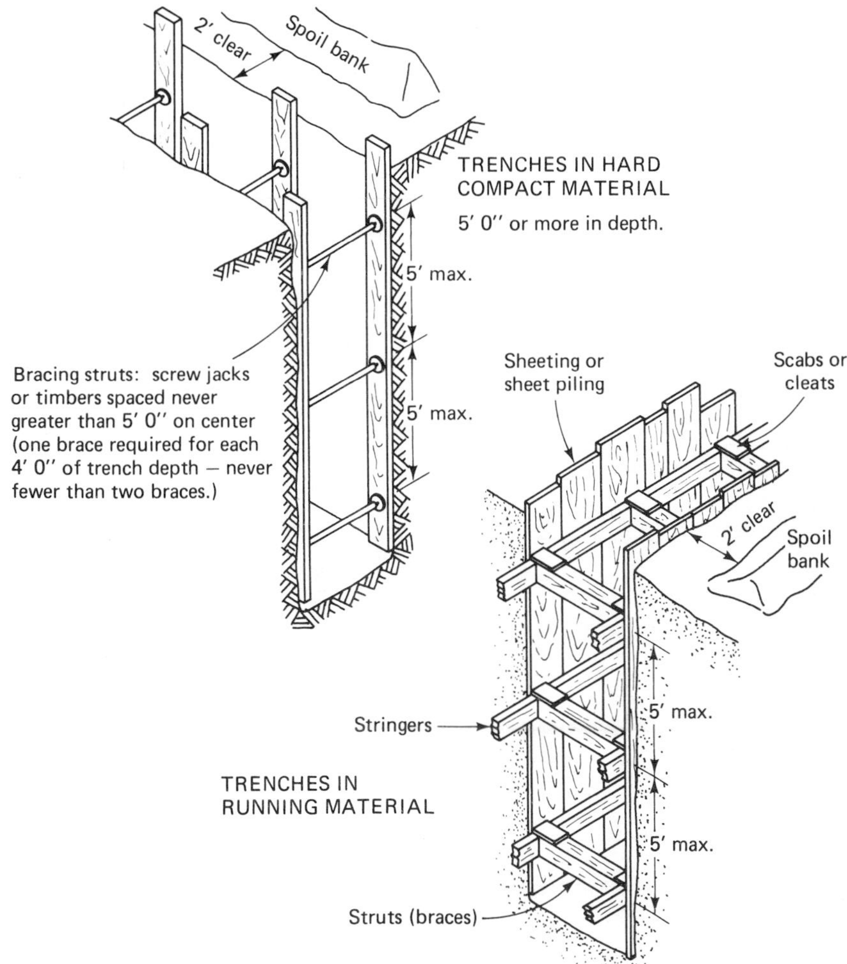

Trench Shoring Systems

one over carefully before you use it. If the cords have become frayed or have loose plugs on the end, do not use them until they are repaired. Do not use such equipment as electric drills, hammers, and saws if the grounding wire has been tampered with. When you do, you are the ground. Another basic rule to remember is that electricity and water do not mix with safety. If you are in a trench that is damp or has water in the bottom, you are taking a chance on becoming grounded if you use electrical power cords, drop lights, and tools that are not specifically designed for these conditions. Most are not. Submersible pumps, of course, are.

Minimum Sizes for Shoring Lumber (National Safety Council) **Table 1-1**

Depth of Trench	Kind or Condition of Earth	Size and Spacing of Members									
		Uprights		Stringers		Cross Braces* (Width of Trench)				Max. Spacing	
		Min. Dim.	Max. Spac.	Min. Dim.	Max. Spac.	Up to 3 ft	3 to 6 ft	6 to 9 ft	9 to 12 ft	Vert.	Horiz.
Feet		Inches	Feet	Inches	Feet	Inches	Inches	Inches	Inches	Feet	Feet
4 to 10	Hard, compact	3 × 4 or 2 × 6	6	—	—	2 × 6	4 × 4	4 × 6	6 × 6	4	6
	Likely to crack	”	3	4 × 6	4	”	”	”	”	”	”
	Soft, sandy, or filled	”	Close sheeting	4 × 6	”	4 × 4	4 × 6	6 × 6	6 × 8	”	”
	Hydrostatic pressure	”	”	6 × 8	”	”	”	”	”	”	”
10 to 15	Hard	”	4	4 × 6	”	”	”	”	”	”	”
	Likely to crack	”		4 × 6	4	”	”	”	”	”	”
	Soft, sandy, or filled	”	Close sheeting	4 × 6	”	4 × 6	6 × 6	6 × 8	8 × 8	”	”
	Hydrostatic pressure	3 × 6	”	8 × 10	”	”	”	”	”	”	”
15 to 20	All kinds or conditions	”	Close sheeting	4 × 12	”	4 × 12	6 × 8	8 × 8	8 × 10	”	”
Over 20	All kinds of conditions	”	”	6 × 8	”	”	8 × 8	8 × 10	10 × 10	”	”

* Trench jacks may be used in lieu of, or in combination with, cross braces.

Fire is another one of those hazards that is peculiar to plumbing. The plumber uses a lead-melting furnace mostly (for repair jobs), and torches for making solder joints. It is a mistake to underestimate the risks involved. Do not use a lead-melting furnace in flammable surroundings or leave it unattended. A portable fire extinguisher should be at the site when soldering is done with torches. The critical time to stop a fire is in the first minute. After that it could be out of control. Some of the newer propane-air and acetylene-air torches have safety valves in the handles that close automatically if the torch is dropped. All things being considered, a wet rag and fire extinguisher are the best insurance against having a fire in a frame dwelling.

1-6 Summary of Practice

Plumbing and plumbers have been with us for a long time. The status of the trade is acknowledged in society, and craftsmanship is an established tradition. The pay received by plumbers is one of the highest in the construction trades. Repairs and retrofitting, the part of the trade which requires expanding or replacing plumbing, accounts for a large part of the market.

Plumbing safeguards the health of the public. Because of the importance of this charge, plumbers are licensed by the state and local governments with enforcement overseen by the state health department. Local municipalities and authority require the license as proof that the plumber is competent, and have established ordinances that require a plumbing permit to start a job. Upon completion, the plumbing job is certified by an inspector from the code enforcement department that the work is in compliance with the state and local plumbing codes, and completed by a licensed plumber.

Craftsmen are developed for the trade through the apprenticeship system. Apprentices learn through formal course work and on-the-job experiences in a structured program established by the local joint apprenticeship and training committee. Half the members of the committee are from labor, while the other half come from the contractors who hire and sponsor them. Other representation on the committee comes from local schools and the state bureau of apprenticeship. A typical apprenticeship training program in plumbing lasts four or five years.

Plumbing codes are established from principles that have been developed over time from accepted practice. They put together the basics every plumber should know. National plumbing codes, sometimes called "model codes" because they are nonbinding, have come into being to standardize practices and make them uniform. State and city codes are developed from them. They are binding. Accepted practices and the principles listed in the various codes can be considered to be the same as they affect the three basic systems: potable water supply and distribution system; sanitary drainage and vent system; and storm drain system.

Safety at work protects against injury and loss of property. Job safety can be divided into two parts: the work station and surrounding area (including other workmen), and the safety practices that each person incorporates in his or her own work. To be safe, the workplace must be free from recognizable hazards and hazardous people, for example, a person inclined toward horseplay. The safe practices that each person incorporates in his or her work reflect the personal knowledge, pride, and craftsmanship of each. Only a safe person can be considered a craftsman, since the care and proper use of quality tools is the very basis of the craft.

REVIEW QUESTIONS AND PROBLEMS

1. Trace the history of plumbing in your area.
2. What is the origin of the term "potable water"?
3. How did plumbers and plumbing receive their names?
4. What is the purpose of the plumbing principles?
5. How does one become an apprentice in the plumbing trade in your area? List other defined jobs in the plumbing trade.
6. How are national plumbing codes established?
7. What is the difference between model codes and state codes?
8. Describe the purpose of the three basic plumbing systems.
9. What is the purpose of the joint apprenticeship committee and who has membership?
10. List the requirements and program of study for completing the apprenticeship program in your area.
11. What is a safe work site?
12. What is an unsafe work practice?
13. Explain two ways to prevent a personal injury at the work site.
14. What safety equipment should every plumber own, and what safety equipment should be provided by the contractor at the work site?
15. Interview an OSHA inspector and have him or her explain the most frequent causes of injury around construction sites, particularly as they affect plumbers.

REFERENCES

[1] Used with the permission of, and available from, The American Society of Mechanical Engineers, United Engineering Center, 345 East 47th

Street, New York, N.Y. 10017. Designated ASA A40.8–1955, the ASME will continue to make this publication available until an American National Standard Plumbing Code is available.

[2] A model (voluntary) standard is published by the U.S. Department of Labor, Employment and Training Administration, and has the approval of the United Association of Journeymen and Apprentices of the Plumbing and Pipefitting Industry of the United States and Canada, and of the National Association of Plumbing-Heating-Cooling Contractors. *National Apprenticeship Standards for Plumbing and Steam-fitting-Pipefitting,* 1977.

[3] Mario J. Fala, *Uniform Plumbing Code Study Guide* (Los Angeles: The International Association of Plumbing and Mechanical Officials), 1977.

[4] Jules Oravetz, *Questions and Answers for Plumbers Examinations* (Indianapolis, Indiana: Howard W. Sams), 1977.

[5] Apprentice *Training Lesson Plans* (First and Second Year), National Association of Plumbing-Heating-Cooling Contractors, 1978.

[6] National Joint Plumbing Apprentice and Journeyman Training Committee, United Association Building, 901 Massachusetts Avenue, N.W., Washington, D.C. 20001.

[7] National Joint Steamfitter-Pipefitter Apprenticeship Committee, United Association Building, 901 Massachusetts Avenue, N.W., Washington, D.C. 20001.

[8] United Association Training Department for Apprentices and Journeymen, United Association of Journeymen and Apprentices of the Plumbing and Pipe Fitting Industry of the United States and Canada, 901 Massachusetts Avenue, N.W., Washington, D.C. 20001.

[9] Bureau of Apprenticeship and Training, U.S. Department of Labor, Washington, D.C. 20213.

[10] Vincent T. Manas, *National Plumbing Code Handbook* (New York: McGraw-Hill Publishing Company), 1957.

[11] Vincent T. Manas, *National Plumbing Code Illustrated* (St. Petersburg, Florida: Manas Publications), 1968.

[12] Harold E. Babbitt, *Plumbing* (New York: McGraw-Hill Publishing Company), 1960.

[13] *National Standard Plumbing Code* (Washington, D.C.: National Association of Plumbing-Heating-Cooling Contractors), 1968.

2 Tools and Plumbing Work

2-1 Introduction

The plumber is a craftsman who uses tools to install pipe and systems connected by pipe. Craftsmen, both men and women, are known by the tools that they use and the skill with which they use them. Craftsmanship requires a high degree of skill in performing the many tasks and operations of the plumbing trade, coupled with pride and satisfaction from the work and finished project. Thus, satisfactory performance as a plumber requires not only a substantial investment in tools and equipment, but concerted effort and experience in developing hand and machine tool skills that are safe, orderly, and effective in the shop and at the work site.

Current practice in the plumbing trade is characterized by the plumbing shop from which supplies and subassemblies are brought to the work site by the business truck. The uniformed craftsman on the truck is supported by at least one person or an answering service at the shop office. Frequently, trucks are radio-dispatched to increase efficiency as the craftsman makes repair or installation calls. Such other items as billing time can be reported by using the truck radio, as well as calls made to request supplies and assistance from the shop. Emergency calls coming into the shop can also be relayed to the truck. A modern plumbing truck is shown in Fig. 2-1.

While final assembly and testing of plumbing systems is accomplished at the work site, much of the work can be prepared at the shop in advance. In some regions during the winter months and inclement weather, this shop capability is particularly important in that it increases both the productivity and comfort of the plumber, which would otherwise be reduced at the work site by rain, cold, and darkness. Figure 2-2 illustrates a complete fabricated assembly with hot and cold water piping positioned and braced, ready for delivery by the truck to the installation. Notice that a simple three-piece channel assembly with only two welds is used to support and position the entire unit. The rigid support frame is connected to the assembly at six places to brace the entire unit and provides support for the hot- and cold-water lines with eight clamps. The main brace on the starter fitting (bottom center) assures proper installation and sets fixture height in

Figure 2-1

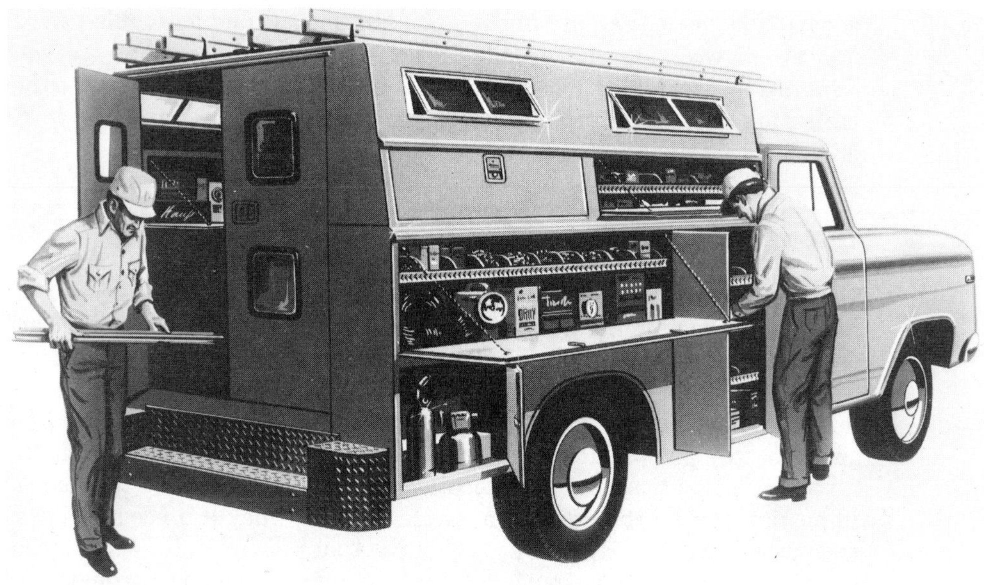

Modern Plumbing Truck
(Reading Body Works, Inc.)

Figure 2-2

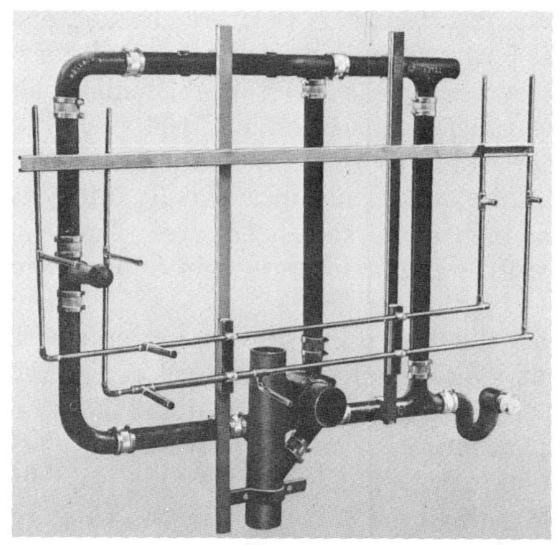

Shop Fabricated No-Hub Waste and Vent Assembly with Copper Tube Assembly Attached, Including Hot and Cold Piping
(Courtesy of Tyler Pipe Company)

the chase, the space between the walls. Notice that the main waste and vent stacks are not a part of the unit, but are cut and assembled with the unit at the work site during installation. The inset photo shows a No-Hub joint being assembled with an electric-powered Jenny Wrench, further illustrating time-saver techniques employed to ease assembly and increase efficiency.

2-2 Hand Tools

Hand tools are the most basic requirement of the craftsman. Some of the hand tools the plumber uses are also used in other trades, including carpentry and sheet metal work, whereas others are peculiar only to plumbing. The responsibility for having and maintaining hand tools is nearly always left to the individual. That is, they are personal property, and by virtue of their use, are personalized. For this reason, care and safe storage should be given attention first, because keeping one's tools secure and together is an important part of having them available when they are needed. Perhaps, then, the most important first tool is the tool box itself, for it provides the means to store and transport the tools to and from work, or from the truck to the work site.

Tools of high quality are the only best buy. Not only are they more accurate and perform the work better, but they are safer to use and last longer. Quality hand tools are guaranteed against defects in material and workmanship, usually for the life of the tool; power tools are guaranteed for one year from the date of purchase. Quality in a tool means that it is made from the best materials, the best practices being used, and that it conforms in size, balance, accuracy, and finish to standards that are checked and adhered to by the manufacturer. Hammers, wrenches, screwdrivers, chisels, rulers, and pliers are just a few examples of tools in which materials and manufacturing standards have a marked effect on both the quality of work that can be produced and the safety of the craftsman. Fragments from inferior hammers and chisels, for example, can cause bodily injury to the person using the tools as well as to other persons in the area. The most reliable signs of quality in tools are the manufacturers' names and their written guarantees.

Measuring and marking tools are illustrated in Fig. 2–3, together with their names and special designations. Folding rules are made in regular and heavy-duty models. The extension folding rule has a graduated 6-in. brass lid fitted into one section to permit accurate inside measurements. Rulers are typically marked to the nearest sixteenth of an inch (1.59 mm). English/metric rulers are also available. Markings are imbedded in the wood or fiberglass, which is then covered with a clear epoxy coating. The brass spring joints lock to prevent end play, and the ends are covered with a durable brass end cap, which is also graduated. The inside measurements on a *plumber's rule* are used to figure 45-degree angle offsets. Automatic locking measuring tapes are available in a variety of lengths from 6 to 20 ft.

Figure 2-3

Heavy-Duty Extension Rule

Metric/English Fiberglass Folding Rule

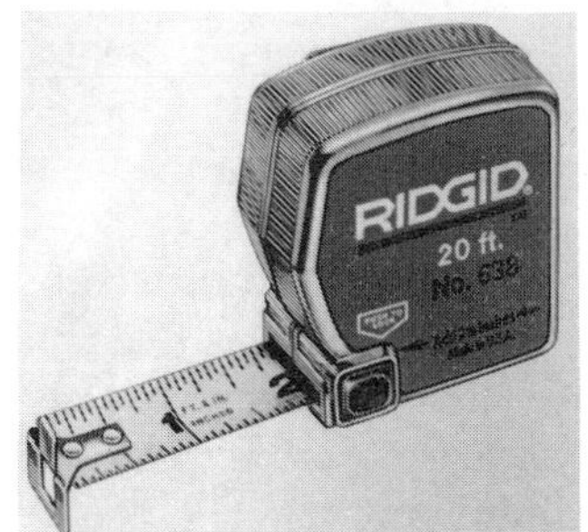

Automatic Locking Steel Tape

General Purpose Steel Tape

Torpedo Level

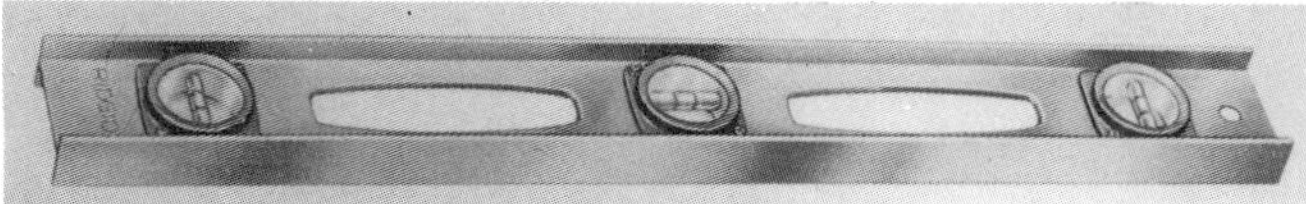

Magnesium Level

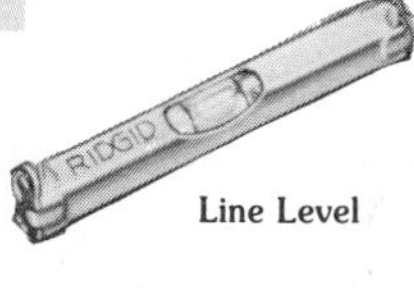

Line Level

Aluminum Level

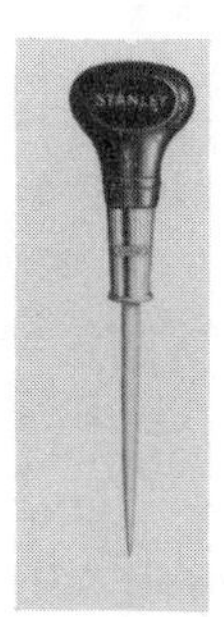

Scratch Awl

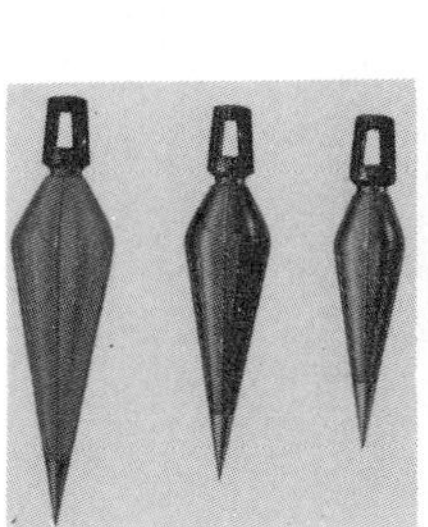

Plumb Bob

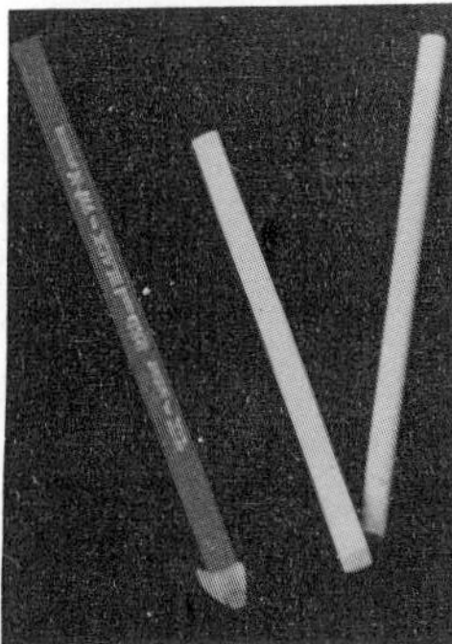

Pencil and Soapstone

Chalk line and Chalk

Measuring and Marking Tools
(Ridge Tool Company and Stanley Tools)

Steel tapes with fold-in rewind mechanisms are used for longer measurements of 50 ft, 100 ft, and farther. The plumb bob is balanced about its vertical axis and pointed at the bottom. When suspended from a line, it accurately indicates the exact vertical drop from the point where it is suspended. It is invaluable for centering openings for pipes dropping from one floor to the next, as well as for designating the center point of the pipe on the floor below, from which center-to-center measurements can be taken. The chalk line is used to lay out the path of pipe on floors, ceilings, and walls, including drain lines, when the pitch is established with a string level. When the walls have not yet been erected, this layout can be marked on the floor with the chalk line so that plumbing subassemblies can be installed without them. The two most common marking tools are the pencil and soapstone, used to mark soil pipe. The scratch awl is also used to mark wood, metal, and other surfaces where the scribe should penetrate the material surface slightly. Levels of aluminum and magnesium are used to check the level plumb, grade, and angle of pipes. The torpedo level is ideal for general use and work in tight places. The three vials indicate plumb, level, and 45 deg. Also notice that the edge is grooved for pipe and conduit.

Figure 2-4

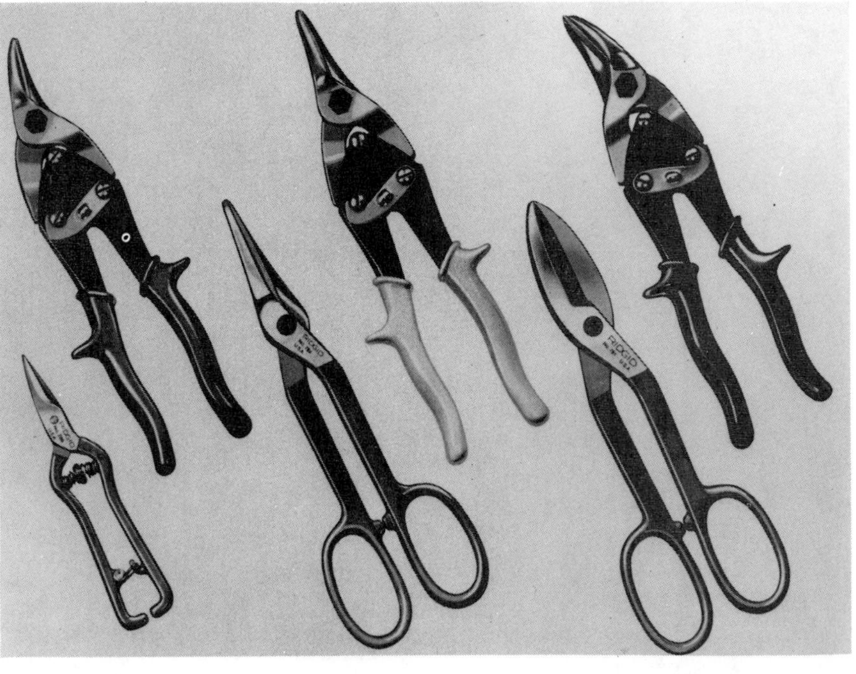

Light Metal Cutting Shear | Cuts Left | Duckbill Snip (circular) | Cuts Straight | Straight Snip | Cuts Right

Metal Cutting Snips (Ridge Tool Company)

Some models are magnetized for pipe work and easy storage. Longer levels, 18 in. and 48 in., are more accurate because of their length and are useful for layout as well as for checking longer plumbing assemblies.

Metal cutting snips are illustrated in Fig. 2–4. Straight snips are used for cutting sheet metal, screening, wire, leather, cloth, gasket material, and roofing. The handles are designed for either right- or left-hand use. The duckbill snip is for cutting circles. The light metal cutting shear is a self-opening utility snip for light materials. Aviation snips are for cutting 18-gauge or lighter cold-rolled sheet steel. Compound action at the joint results in maximum jaw power with minimum hand effort. Aviation snips are designated for cutting straight, left hand and right hand, which is particularly useful when one is cutting passage holes in sheet metal.

Pliers and screwdrivers are available in a variety of types and sizes (Fig. 2–5). The most common is the slip joint utility type. Tongue and

Figure 2-5

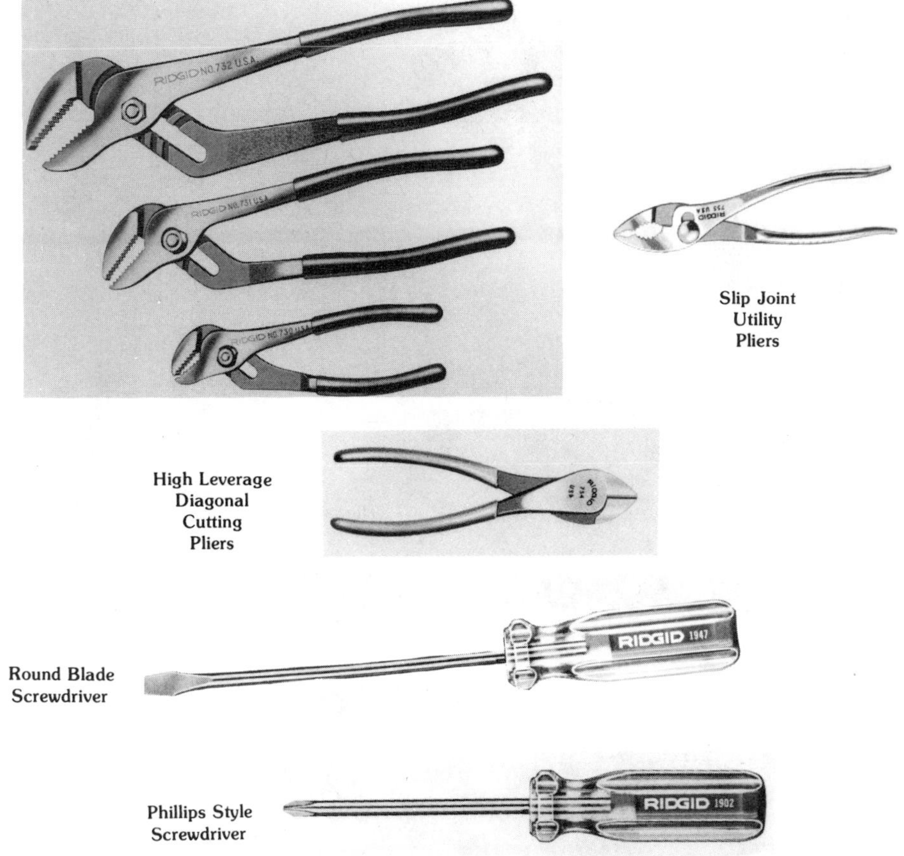

Pliers and Screwdrivers (Ridge Tool Company)

groove pliers are available with both toothed and smooth jaws. Only *smooth jaws* are appropriate to assemble finished fittings. Diagonal cutting pliers are used to cut wire and other soft materials. Screwdrivers have the phillips blade, or flat blade, and round or square shanks. Square shanks can be given additional twisting effort with a wrench. The handle should give a comfortable grip as well as resist breakage.

Hand saws are available for cutting pipe, steel, iron, nonferrous metals, plastic, and wood. For plumbing, the hacksaw is the most common of these and uses blades from 10-in. to 12-in., with a tooth pitch of 18, 24, and 32 teeth per inch (Fig. 2–6). The jab-type saw is a modified hacksaw for hard-to-get-at places. The compass saw is used for both metal and wood and uses blades with 8, 10, and 24 teeth per inch. The universal saw is a modified miter box saw with teeth on both sides. It is used for cutting plastic pipe,

Figure 2-6

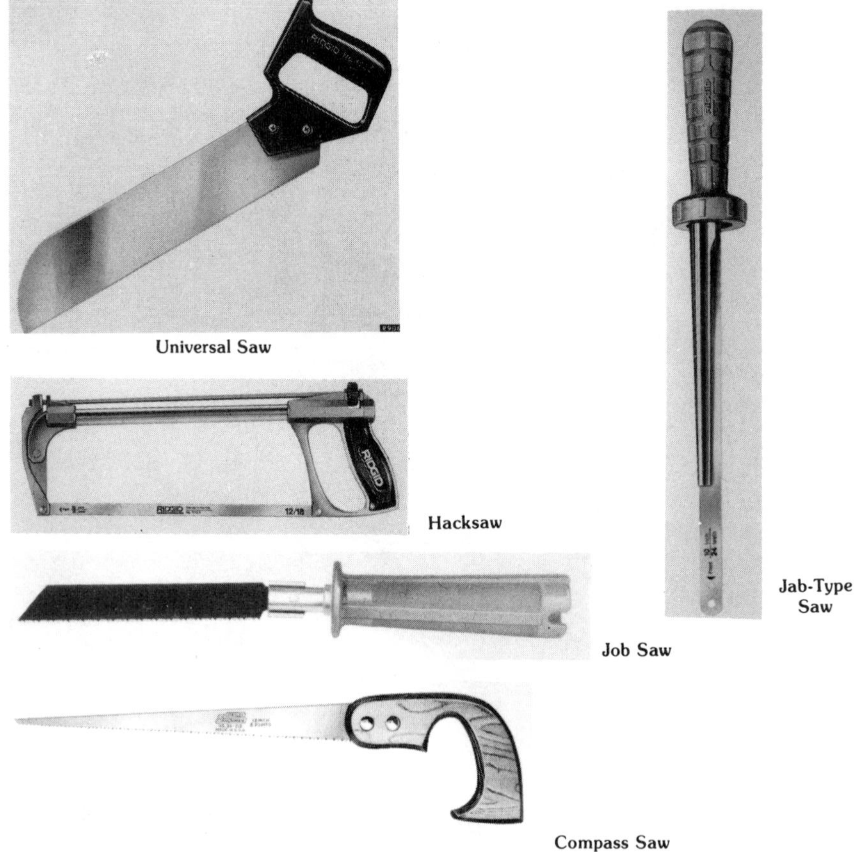

Hand Saws
(Ridge Tool Company)

laminates, plywood, and veneers. The special toothing on the curved back edge cuts slots. Saw blade terminology is illustrated in Fig. 2-7.

Bolt threading in low volume is done with hand threaders. Threading dies cut threads on the outside; threading taps cut threads on the inside. Bolt threads are straight as compared with pipe threads, which are tapered. Both threaders are available in sizes to 1-in. The ratchet head die shown in Fig. 2-8 permits threading with only a partial turning and arc of the handle. This is particularly important for ease of operation when the bolt is held in a vise or threading is done in close quarters. One-piece button dies can be reversed to thread down to the shoulder of the bolt. The tapered side is for starting the thread. Notice the slot in the die, which is used to make

Figure 2-7

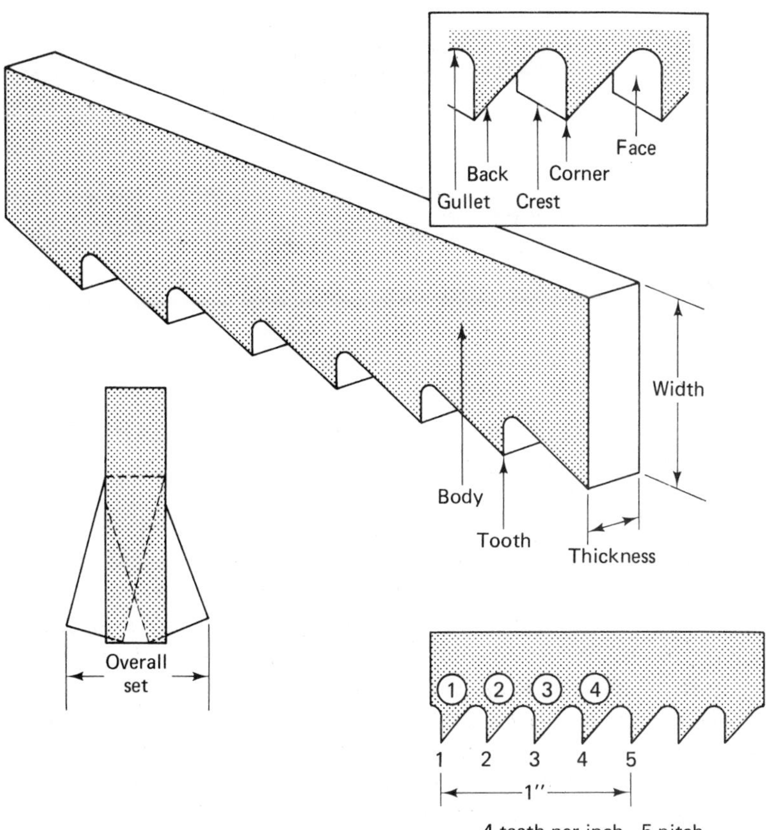

Saw Tooth Terminology

Figure 2-8

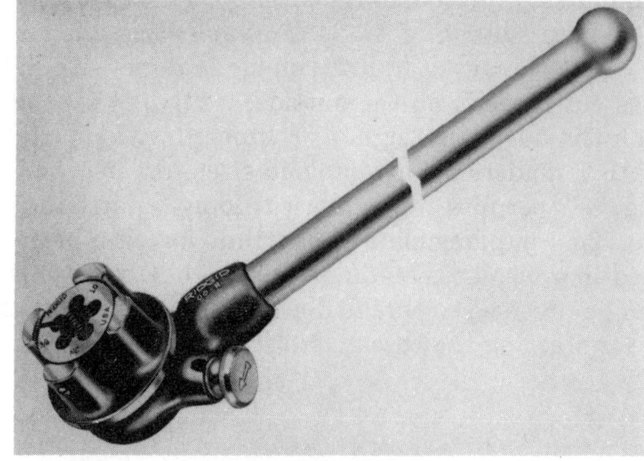

Ratchet Bolt Threader

3-Way Bolt Threaders

Bolt Threaders
(Ridge Tool Company)

adjustments for over- or undersizing threads. The three-way bolt threader is employed for threading common sizes, holding three such adjustable button dies which do not have to be removed. A generous amount of cutting oil must be supplied to both taps and dies when one is cutting threads.

Figure 2–9 illustrates a number of wrenches used by the plumber. Straight pipe wrenches are available in sizes from 6-in. through 60-in. (15-cm–150-cm) with pipe size capacity from $\frac{3}{4}$-in. to 8-in. Replacement jaws are made of hardened alloy steel. The full-floating jaw assures an instant grip and release as the wrench is worked. The I-beam handle is made of malleable iron or aluminum. Small wrenches are stored in the tool box. Larger wrenches are typically hung on a hook placed through the hang-up hole in the handle. Chain wrenches are also used for turning pipes in close quarters. Strap wrenches are used with polished pipe. The sturdy nylon strap gives a tight grip but will not mar the finish. Smooth-jaw wrenches are used for assembling nuts and fittings. The spud wrench is a utility monkey wrench. The smooth jaws are made of heat-treated steel. The offset hex wrench has a thin jaw with an extra-wide opening, and the offset jaw gives an easy-on secure grip on sink and tub nuts (inset figure) and other hard-to-get-at nuts and fittings. Adjustable wrenches have smooth jaws and are

Figure 2-9

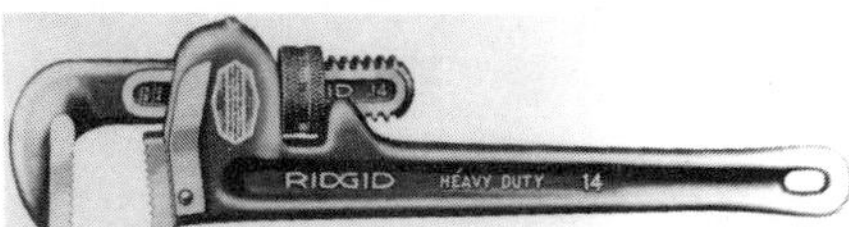

Straight Pipe Wrench

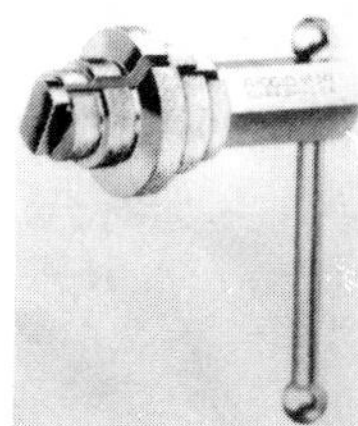

Internal Wrench

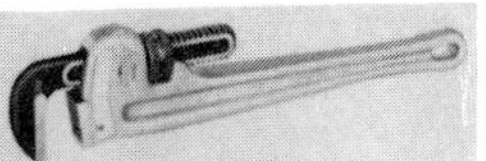

Aluminum Handle Straight Pipe Wrench

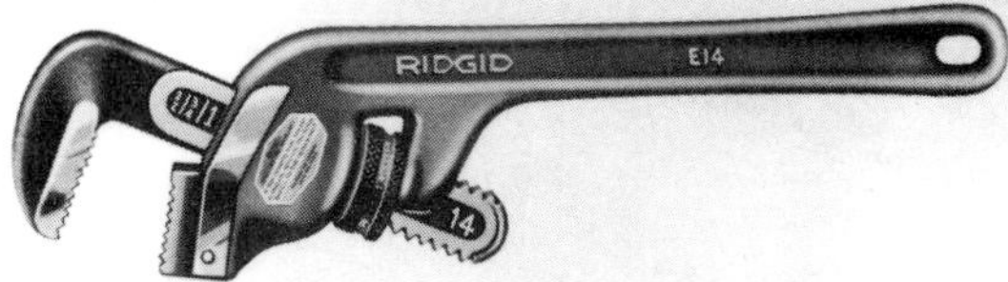

End Pipe Wrench

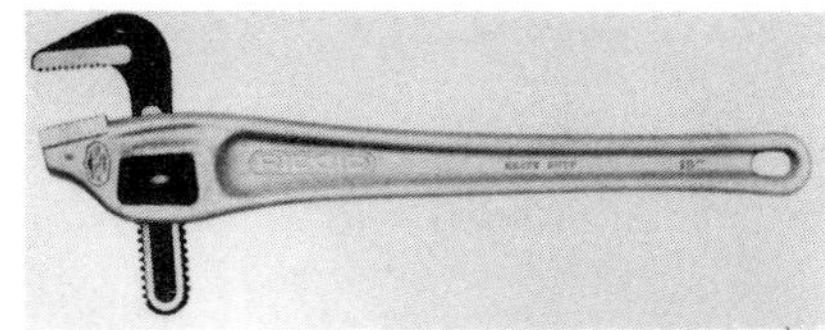

Aluminum Handle Offset Pipe Wrench

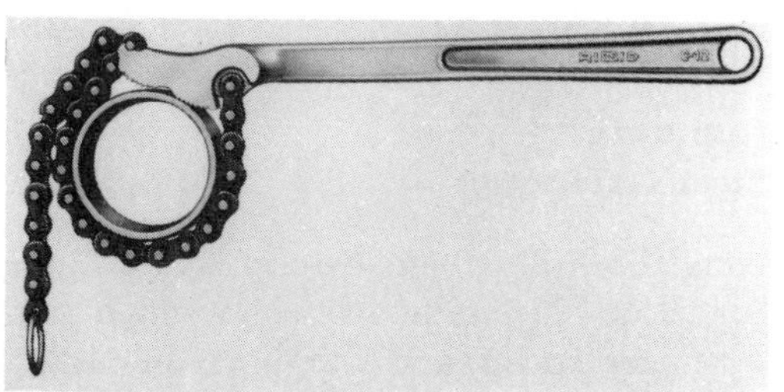

Light-Duty Chain Wrench

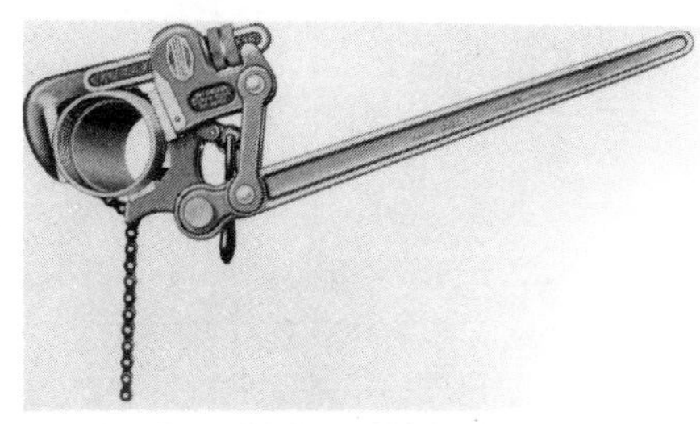

Compound Leverage Pipe Wrench

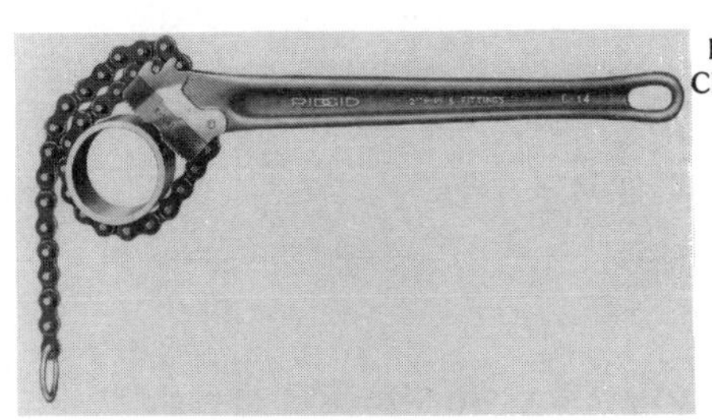

Heavy-Duty Chain Wrench

Offset Hex Wrench

Spud Wrench 12 "

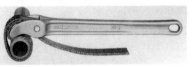

Strap Wrench

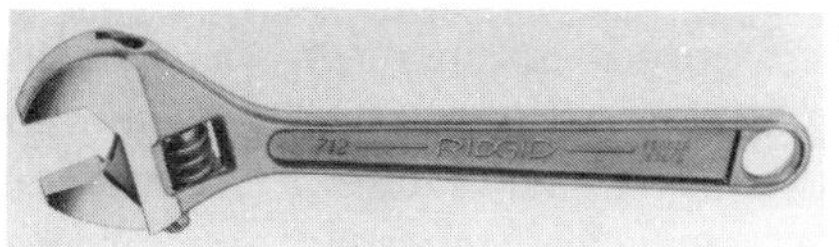

Adjustable Wrench

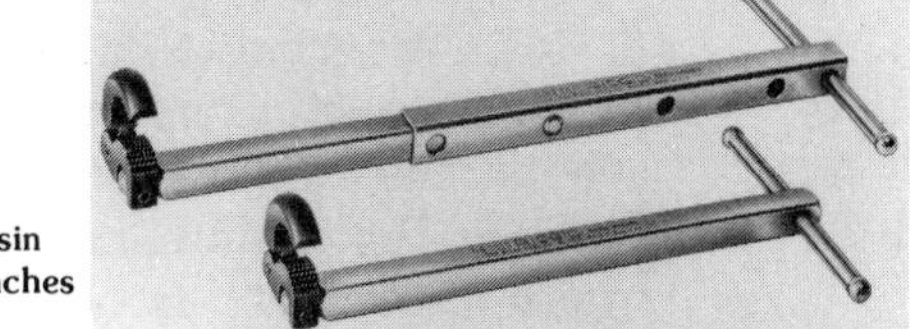

Basin Wrenches

Plumber's Wrenches (Ridge Tool Company)

Hammers
(Ridge Tool Company)

available in sizes from 4-in. through 24-in. (10-cm–61-cm). Adjustable wrenches particularly must be of high quality to serve the intended purpose. Inferior-quality adjustable wrenches spring and slip, both damaging the work and endangering the plumber.

The most common hammer the plumber uses is the ball peen hammer, available in sizes from 8-oz. to 32-oz. (0.23-kg to 0.91-kg). Occasionally the plumber may need a claw hammer or hatchet. Sledgehammers with hickory handles are available in sizes from $2\frac{1}{2}$-lb to 10-lb (1.13-kg to 4.54-kg) and larger.

2-3 Cast Iron Pipe Tools

Cast iron soil pipe is available in hub and spigot, and no-hub types. Hub and spigot pipe is joined with a caulked joint of oakum and lead or with a compression gasket, whereas no-hub pipe is joined by using special no-hub gaskets and clamps like the assembly in Fig. 2–2.

The tools in Fig. 2–10 are used to cut cast iron pipe and assemble it. Cutting can be accomplished with hammer and chisel, hand or power saw, or with a soil pipe cutter. Saw cuts are the most predictable and accurate. Cutting with a hammer and chisel is time consuming, however, and drastically increases labor costs. The heavy-duty ratchet cutter shown is more labor efficient and brings chain-mounted cutting wheels to bear around the circumference of the pipe. Cutting occurs when the ratchet tightens the chain squeezing the pipe along the line, causing the pipe to break. Cutters of this type are also used to sever clay, asbestos, and cement pipe.

Figure 2-10

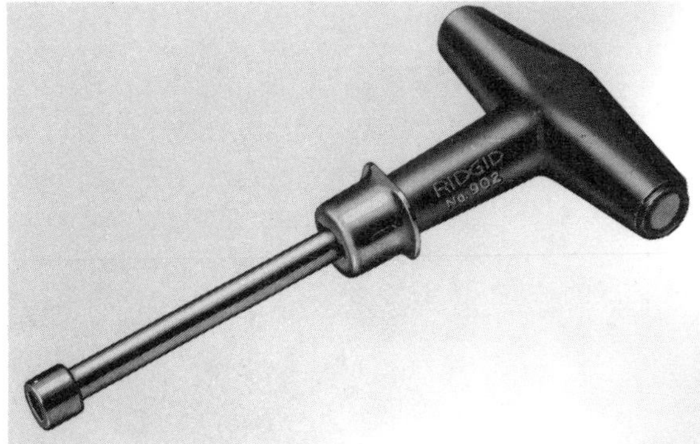

Torque Wrench for Cast Iron No-Hub®
Soil Pipe

Abrasive
Cut-Off Machine

Soil Pipe Cutter

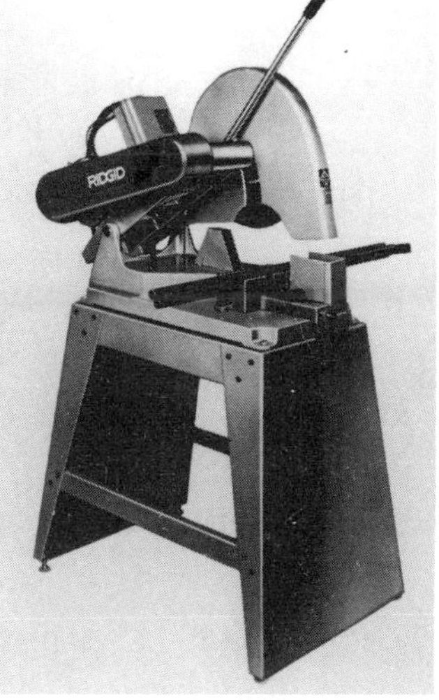

**Cast Iron Pipe Tools
(Ridge Tool Company)**

Caulked joints are made of oakum and lead. The oakum is yarned (stuffed) into the joint with the yarning iron and then is packed with a light hammer and packing iron. Lead is melted in the lead pot on the furnace and taken to the joint and poured with the ladle. The joint runner, an asbestos rope, contains the lead in the joint until it hardens. Caulking irons and a 2-lb hammer are then used to caulk the joint, after which the excess lead is removed with the cut-off chisel.

Compression gasket soil pipe joints may be assembled by striking the fitting or pipe with a lead maul. They may also be brought together by using an assembly tool such as that shown in Fig. 2-11. The compression

Figure 2-11

Compression Gasket Assembly Tool (Ridge Tool Company)

gasket that seals the space between the hub and spigot is inserted in the hub end and lubricated, after which the male end of the fitting or plain end of the pipe is forced into the gasket until it seats.

No-hub soil pipe connection requires that the clean assembly consisting of the neoprene gasket, the stainless steel sleeve, and the clamps are tightened together to the required torque of 60 in.-lb. This can be done by hand with a torque wrench or by an electric or air driver, which overrides at the preset torque.

2-4 Threaded Pipe Tools

Threaded steel and galvanized pipe comes in joint lengths of 21 ft (6.4 m). Most pipe fitting work requires shorter lengths, and the pipe must be measured, cut to length, reamed, and threaded before assembly.

In addition to pipe wrenches, tools used with threaded pipe include vises and stands for holding, pipe cutters for sizing the length, reamers for resizing the inside diameter of the cut end, and threading equipment to cut the threads.

Holding the pipe securely for cutting, reaming, and threading is very important, and a number of special vises are available for this purpose. Figure 2-12 illustrates several of them. Notice that the holding action is accomplished by using either two hardened jaws, or an iron base jaw with a tightening chain. The portable three-legged pipe vise is particularly useful in that it is not only movable to the work site, but also provides a work station at the correct height where other related tools can be stored for ready use. The adjustable pipe support locates the pipe at the same height as the vise and lets the pipe move forward from the complete joint as pieces are cut, reamed, and threaded.

Figure 2-12

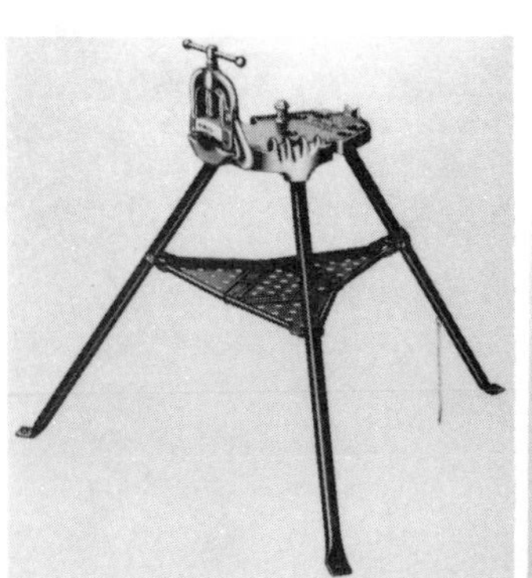

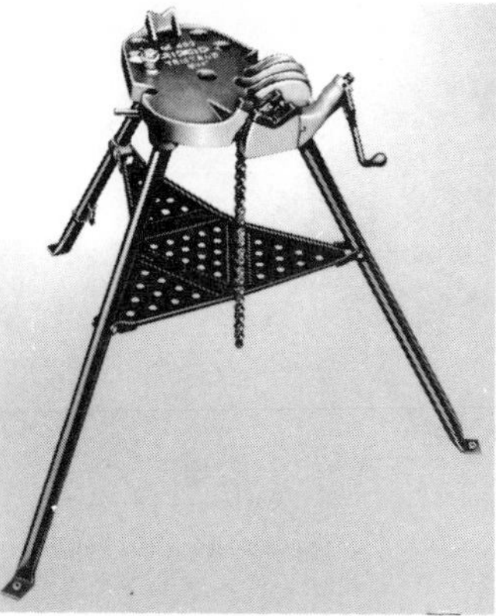

Portable Stand Vises with Built-in Folding Tool Tray

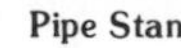

Pipe Stand

Top Screw Post Chain Vise

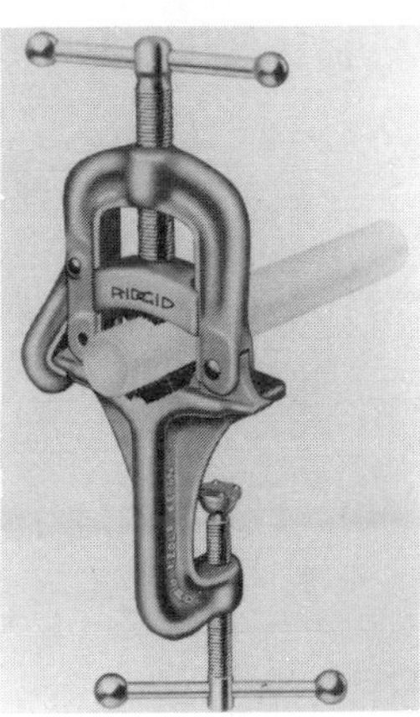

Portable Kit Yoke Vise

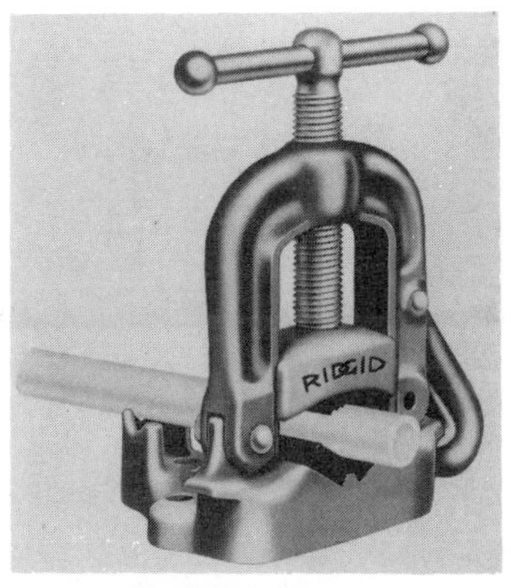

Bench Yoke Vise

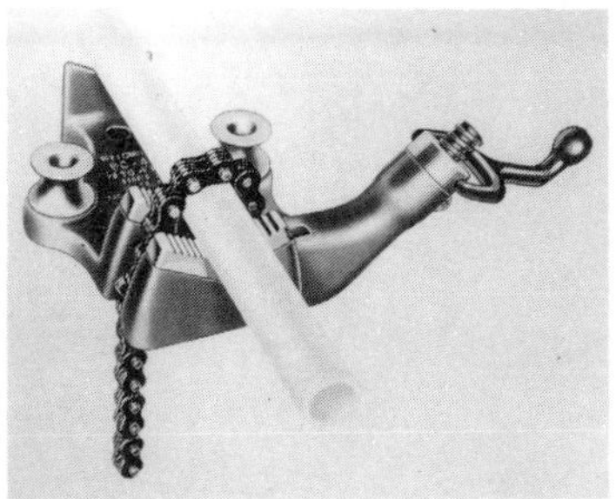

Top Screw Bench Chain Vise

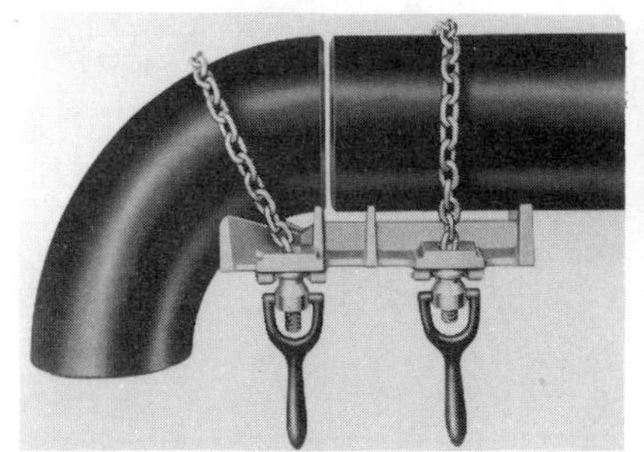

Pipe Welding Vises

Plumber's Vises (Ridge Tool Company)

Figure 2-13

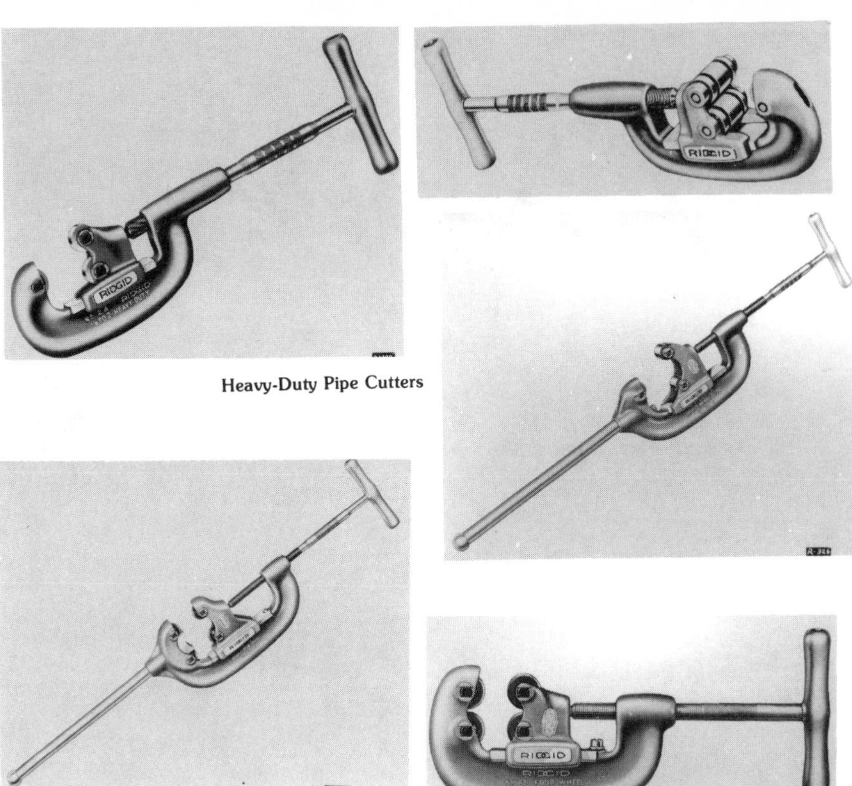

Heavy-Duty Pipe Cutters

Heavy-Duty 4-Wheel Pipe Cutters

Axial Pipe Cutter

Threaded Pipe Cutters (Ridge Tool Company)

After the pipe is secured in a vise, it is cut to length with a wheeled cutter that leaves the end square. This is important because the threading die must start squarely on the end if the threads are to be straight. Wheeled cutters such as those shown in Fig. 2–13 cut the pipe in sizes from $\frac{1}{8}$ in. through 12 in. by squeezing the material apart as the rollers go around the pipe, moved inward with a wrist turning action given to the handle. Single-wheel cutters must rotate completely around the pipe, whereas multiple-wheel cutters need traverse only a portion of the circumference to accomplish the cutting task.

Spreading of the material by the cutter wheel to accomplish the cut reduces the size of the pipe, and before it is threaded it *must be reamed.* The burr left after cutting constitutes a significant obstruction and causes not only flow losses, but provides a sharp edge where foreign material will catch and obstruct flow. Thus the common three rules are given for threaded pipe: ream, ream, and ream. The pipe is reamed after cutting but before threading, because the reaming action not only removes material but bell-shapes the end of the pipe. The size is restored by the threading operation. Reaming after threading would also bell-shape the end, preventing the threads from starting. Three common designs in ratchet pipe reamers are shown in Fig. 2–14, together with their respective pipe size applications.

Pipe threaders are illustrated in Fig. 2–15. Threading action can be accomplished either by turning the die by hand, or by turning the pipe with a power drive. The three-way threader holds three commonly used dies in sizes through 1 in., whereas the ratchet drop head threader uses interchangeable dies through $1\frac{1}{4}$ in. which "drop out" when the ratchet knob is released. Dies can be removed and reversed in both threaders for close-to-wall threading. The *Jam-Proof* ratchet threader is adjustable for threading

Figure 2-14

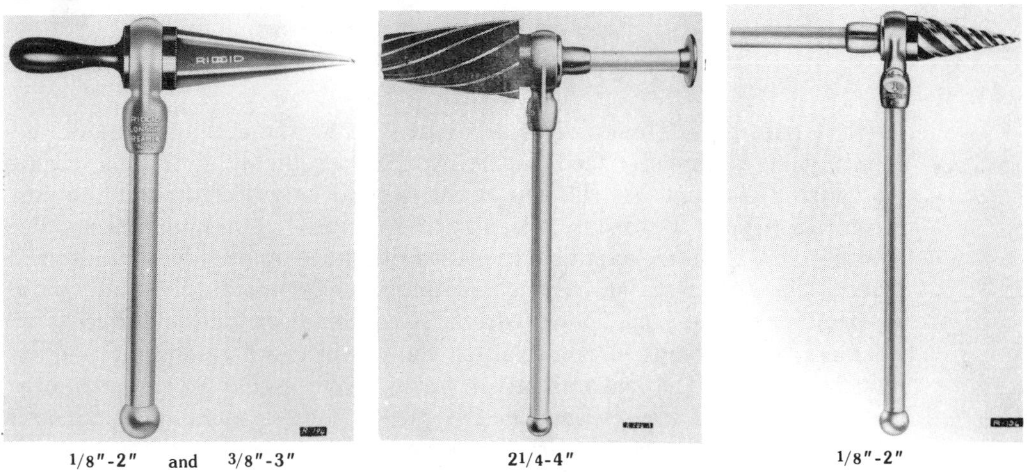

1/8"-2" and 3/8"-3" 21/4-4" 1/8"-2"

Ratchet Pipe Reamers (Ridge Tool Company)

Figure 2-15

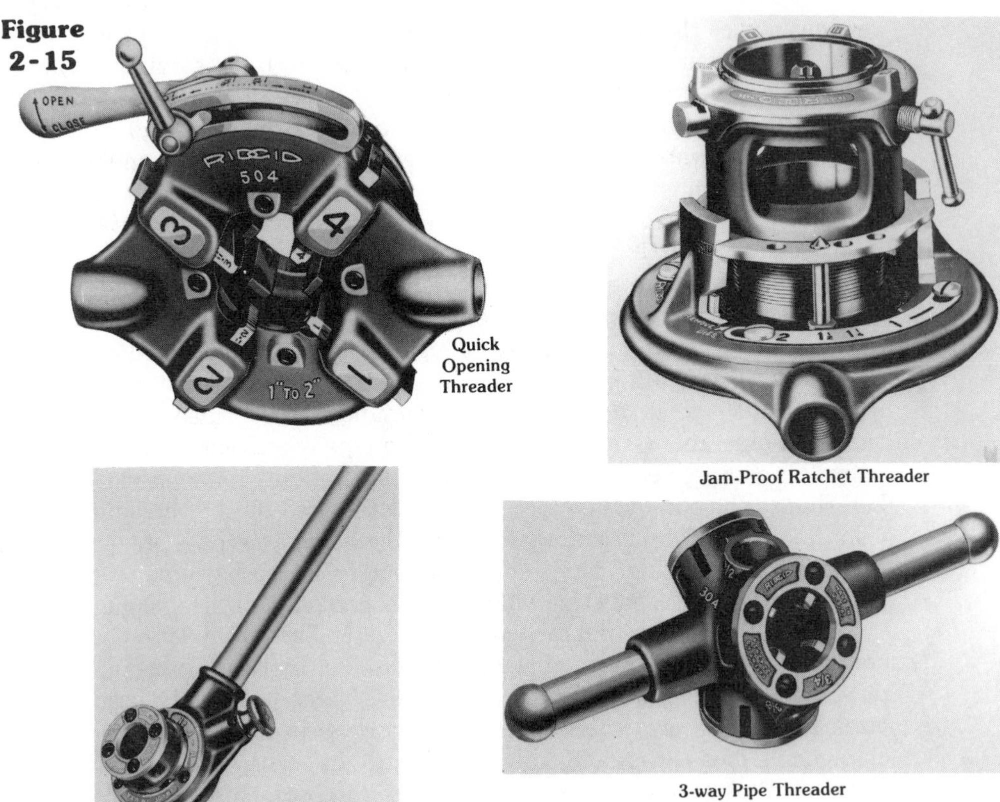

Quick Opening Threader

Jam-Proof Ratchet Threader

3-way Pipe Threader

Enclosed Ratchet Drop Head Threader

Pipe and Conduit Threaders (Ridge Tool Company)

1-in. to 2-in. pipe. Threading is accomplished by the action of the die advancing on the threaded feed mechanism for the required distance as it cuts threads on the pipe. At the end of the desired length of thread, the jam-proof drive pawl kicks out, releasing the threader. Power drives such as those in Fig. 2–16 are available to rotate either the pipe or the threader and speed the operation of cutting, reaming, and threading. Small shops typically use the basic power drive in combination with hand cutters, reamers, and threaders. Some larger shops will use a threading machine which combines the features of the power drive with matching mounted cutters, reamers, and threaders. The portable hand-operated continuous oiler in Fig. 2–17 is used to provide oil to the pipe dies. Flooding of the die aids the cutting action and chip removal, and prevents overheating and galling. The double screen pan catches waste oil for recycling, thereby eliminating the potential for a messy floor, and strains the chips from the

Figure 2-16

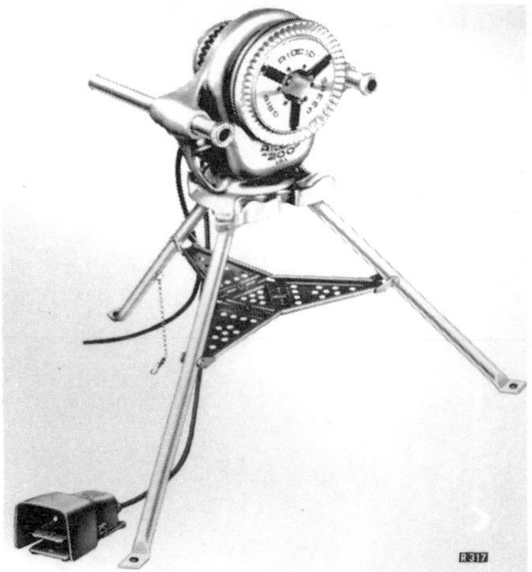

Power Drive

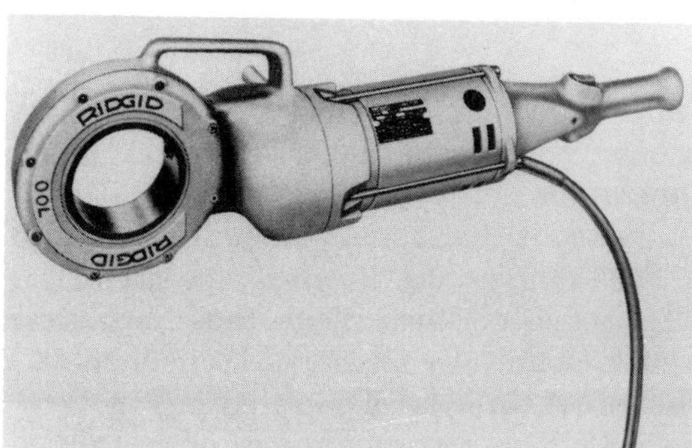

Portable Power Drive with
Geared Threader

Power Drives
(Ridge Tool Company)

Figure 2-17

Recirculating
Oiler

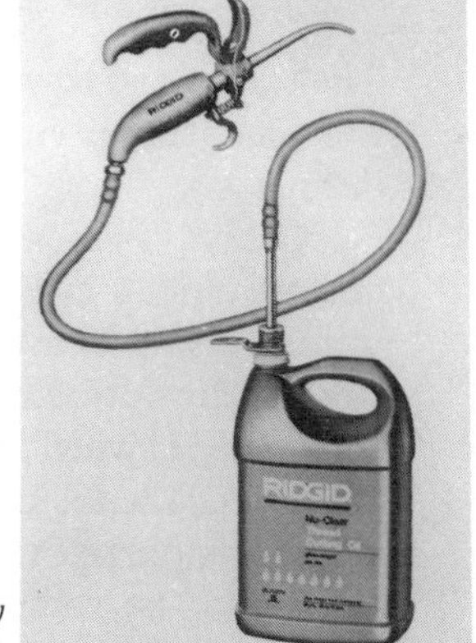

Utility
Oiler

Oilers
(Ridge Tool Company)

Table 2-1 **Pipe Sizes and Threads per Inch**

Pipe Size in Inches	*Threads per Inch*	*Pipe Size in Inches*	*Threads per Inch*
$\frac{1}{8}$	27	$2\frac{1}{2}$	8
$\frac{1}{4}$	18	3	8
$\frac{3}{8}$	18	$3\frac{1}{2}$	8
$\frac{1}{2}$	14	4	8
$\frac{3}{4}$	14	5	8
1	$11\frac{1}{2}$	6	8
$1\frac{1}{4}$	$11\frac{1}{2}$	8	8
$1\frac{1}{2}$	$11\frac{1}{2}$	10	8
2	$11\frac{1}{2}$	12	8

oil to prevent damage to the pump. Threading machines commonly incorporate a motorized recirculating oiler that directs the stream of cutting oil to the die during the threading operation.

The reason pipe dies can be adjusted is that usually more than one size has the same number of threads. For example, the threaded pipe sizes 1 in., $1\frac{1}{4}$ in., $1\frac{1}{2}$ in. and 2 in., all have $11\frac{1}{4}$ threads per inch. Thus, these pipe sizes can be threaded with the same dies, simply by changing their diameter, which is accomplished by the jam-proof threader illustrated in Fig. 2–17 and the geared threader in Fig. 2–18. The correct number of threads is assured not only by the cutting dies but by the threaded feed mechanism, which also has a thread with the same pitch (complete threads per inch). Table 2–1 lists pipe sizes from $\frac{1}{8}$ in. to 12-in. and their respective thread pitch.

2-5 Metal Tubing Tools

The tools used to work with copper, brass, aluminum, and thin wall perform the operations: cutting and reaming, bending, flaring, cleaning, and soldering.

Cutting is most often accomplished with one of several available tubing cutters shown in Fig. 2–18 with capacities from $\frac{1}{8}$ in. to 6 $\frac{5}{8}$ in. (3 mm to 168 mm). Cutting is accomplished by the action of the cutter wheel against the tube around its circumference. Tightening the handle each rotation forces the wheel toward the center of the tube while rotating it around the circumference, causing the cutter wheel to part the tubing. When the cutter wheel has parted the tubing sufficiently, it separates, leaving a burr inside that must be removed with the reamer on the cutter or with a separate reamer, such as those shown in Fig. 2–19. Internal tubing cutters are used for trimming extended ends off stub outs and installed closed bowl or shower waste lines below the flange level. Grooved roller tubing cutters are

Figure 2-18

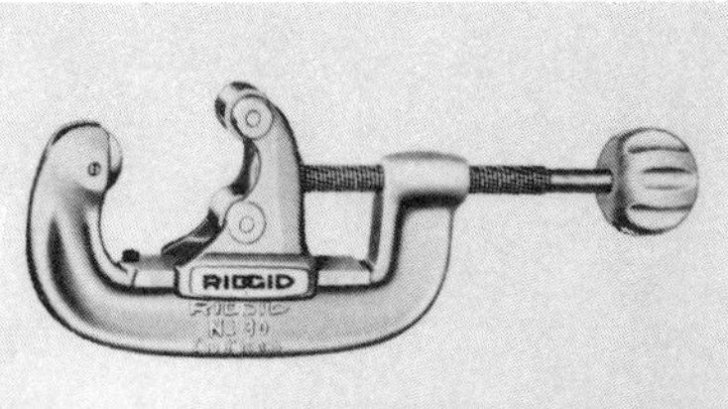

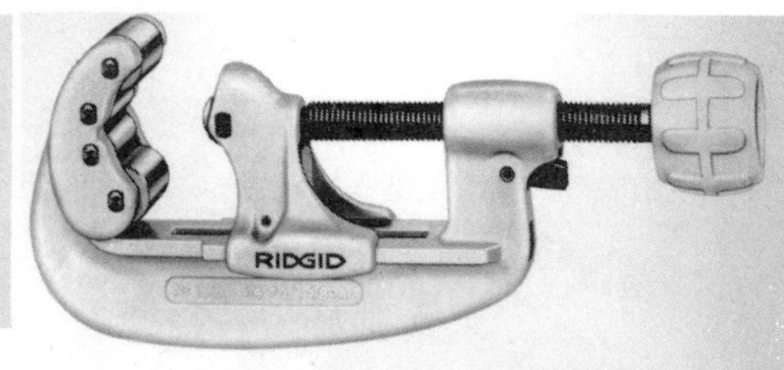

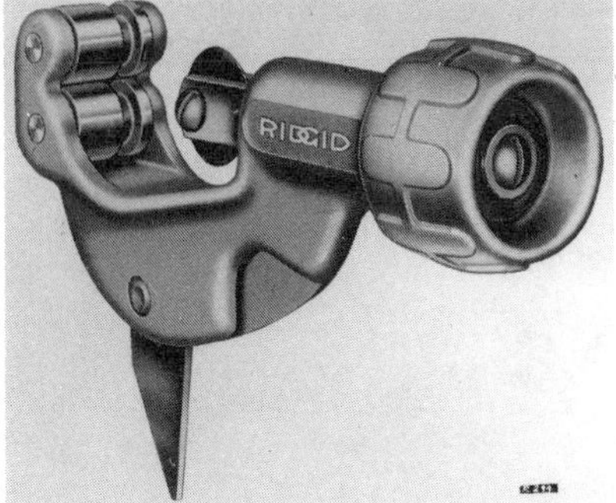

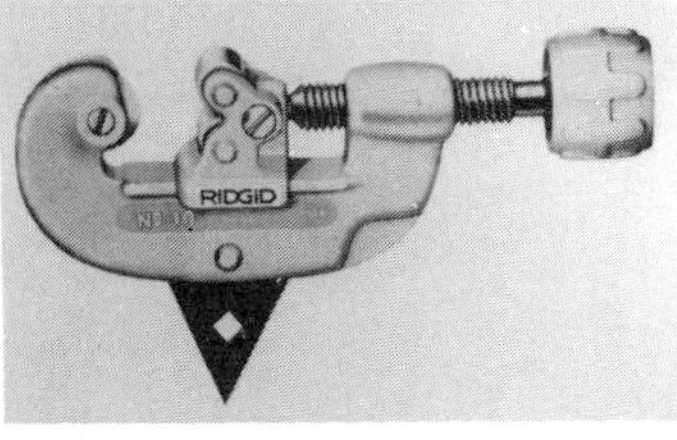

External Cutters

Internal Cutters

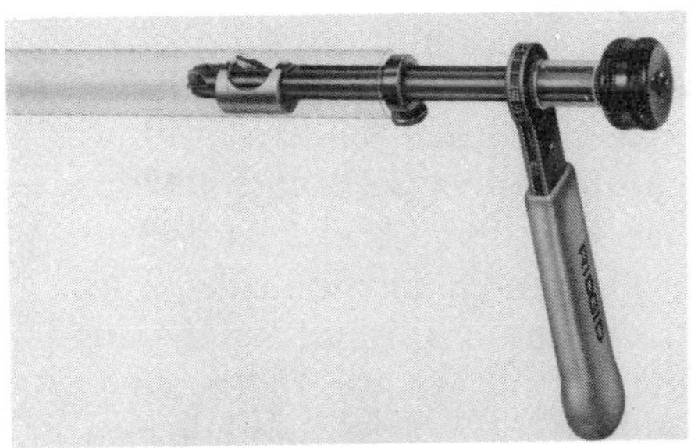

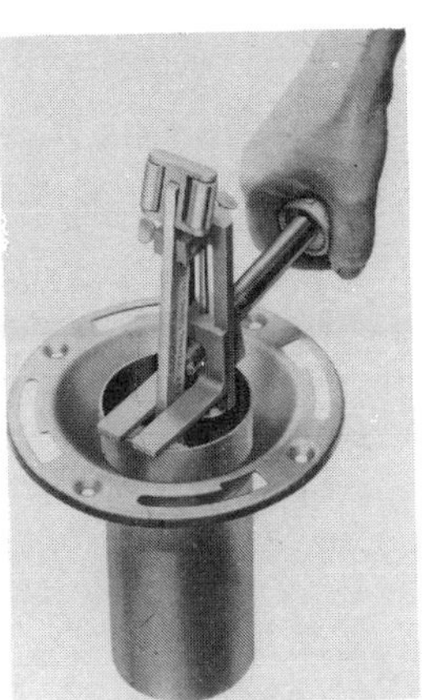

Tubing Cutters (Ridge Tool Company)

Figure 2-19

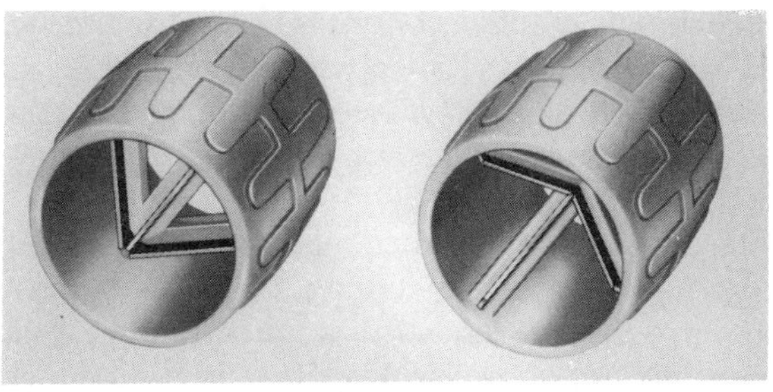

Tubing Reamer (Ridge Tool Company)

Figure 2-20

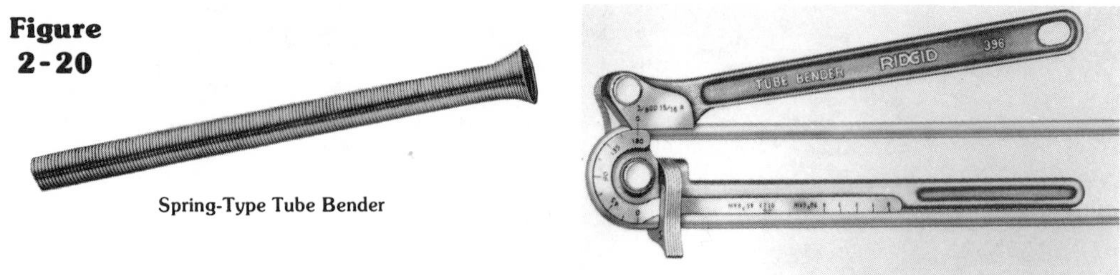

Spring-Type Tube Bender

Level-Type Tube Bender

Geared Ratchet Lever-Type Tube Bender

Tubing Benders (Ridge Tool Company)

used for close-to-flare cuts. This often eliminates tube replacement, because a solder fitting can be used on the existing tube to make the joint. Tubing benders (Fig. 2–20) are used to produce uniform bends without distorting the cross section of the tube. Spring-type benders are the quickest and are used for free-form bending of soft temper copper and aluminum tube. They are available in sizes from $\frac{1}{4}$ in. through $\frac{7}{8}$ in., and are useful for making change-of-direction and offset bends of single runs. The bend is made by slipping the spring over the outside of the tube to the place where the bend is to be made, making the bend by hand, and then removing the spring. The lever bender is a more exact tool, and forms bends on soft and tempered copper, brass, aluminum, steel, and stainless tubing from $\frac{3}{16}$ in. through $\frac{7}{8}$ in. O.D. (outside diameter). Bends are made by degree designations from 0 deg through 180 deg, and account for the linear dimension of the arc to produce bends accurate to $\frac{1}{32}$ in. in length. The ratchet bender reduces bending effort by providing a handle travel to bending arm ratio of more than 10 to 1. Hickey and conduit benders are used to bend heavy-duty and thin-wall conduit.

Flaring tools are used to make up assembled fittings. They have traditionally been used with copper, brass, and aluminum tube, but more recently have been extended to applications with plastic pipe. The hammer-type flaring tool shown in Fig. 2–21 is available in sizes from $\frac{3}{8}$ in. through 2 in. and spreads the end of the metal tube into a flare by being inserted into the end of the tube and tapped with a hammer. A positive diameter

Figure 2-21

Flaring Tool

Hammer-Type Flaring Tool

Flaring Tools (Ridge Tool Company)

stop reduces the possibility of over- or undersize flares and assures a tight seal with the fitting. The screw-type flaring tool accurately spreads the tube end to form either $37\frac{1}{2}$-deg or 45-deg SAE flares. The flaring cone that spreads the tube end is mounted in needle bearings to reduce friction and produce a rolling action for even metal flow, rather than galling, which would occur should the cone turn during the flaring operation. Although most flaring is done to soft copper and aluminum, precision flaring tools are also recommended for steel, stainless steel, hard copper, and brass. Sizes are available from $\frac{1}{8}$ in. to 1 in. in English sizes, and 6, 8, 10, 12, 14, 16, and 18 mm metric. Fittings with the $37\frac{1}{2}$-deg flare require less force to effect the seal, but tend to stick when taken apart. Fittings with the 45-deg flare require greater force to attain the seal but diassemble easily for reuse.

Copper tube cleaning tools are shown in Fig. 2–22. They brighten the tube ends and fittings inside so that solder can be sweated in the space between them to make a watertight and mechanically strong joint. Brightening the tube ends and fittings can be accomplished with steel wool brushes, internal and external cleaning abrasive cloth, special tools made to fit and/or resize the tube and fittings, and powered cleaning machines.

Soldering tools include a heat torch, flux, and solder. The air-propane torch may be hand-held with the tank, valve, regulator, and nozzle tip being assembled together, or the tank and regulator can be one assembly and the torch handle and valve another, connected by a hose. Air-acetelene torches are used for soldering and light brazing and deliver nearly twice the heat of air-propane torches. Having an adequate heat source is required when one is making *hard* (high temperature) solder joints for solar plumbing installations, since system temperatures at the collectors can approach the melting point of *soft* solder joints.

Figure 2-22

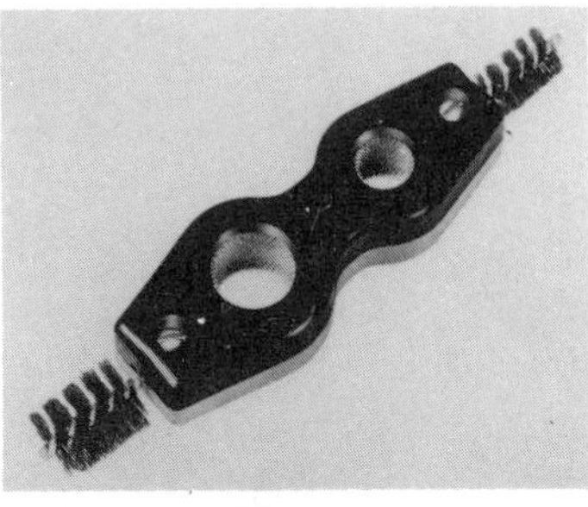

Wire Brush

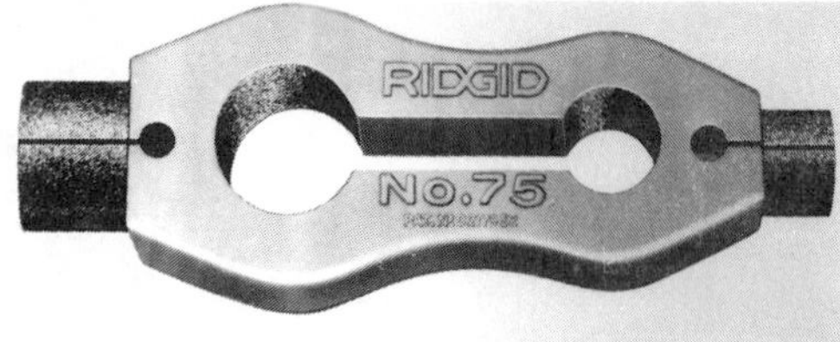

Abrasive Impregnated Rubber

Copper Cleaning Tools
(Ridge Tool Company & Mill-Rose Company)

2-6 Plastic Pipe Tools

Tools are used to cut, ream, flare, clean, and join plastic pipe. Cutting can be accomplished with a hand or power hacksaw, plastic hand saw, special blade-type tubing cutter, or power "chop saw." Sawing leaves little material that cannot be removed by hand. When plastic tubing is cut with a roller cutter, the material and cutting burr are removed with a reamer, such as that shown in Fig. 2-23, or a knife.

Figure 2-23

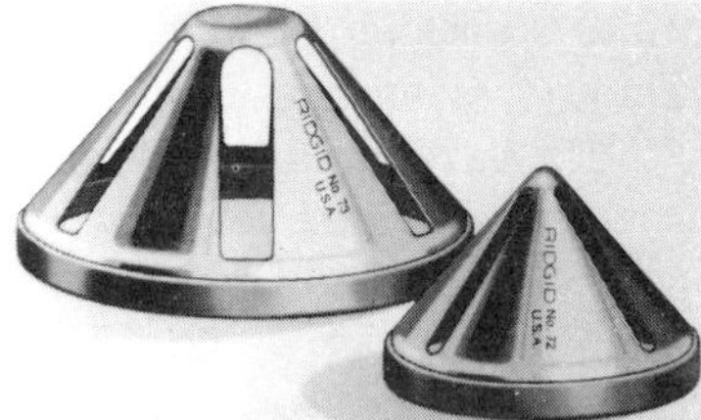

Deburring Tools

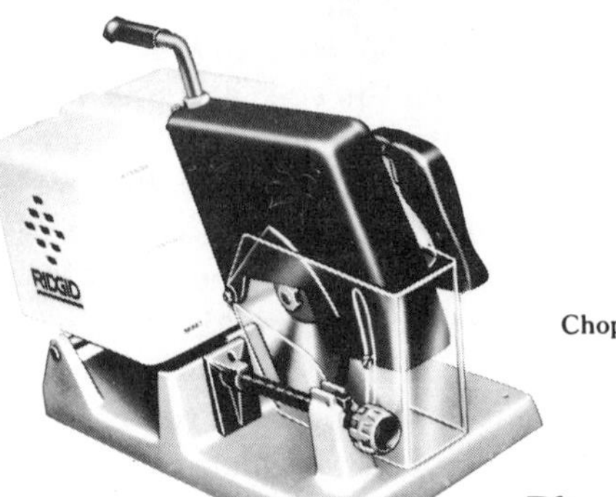

Chop Saw

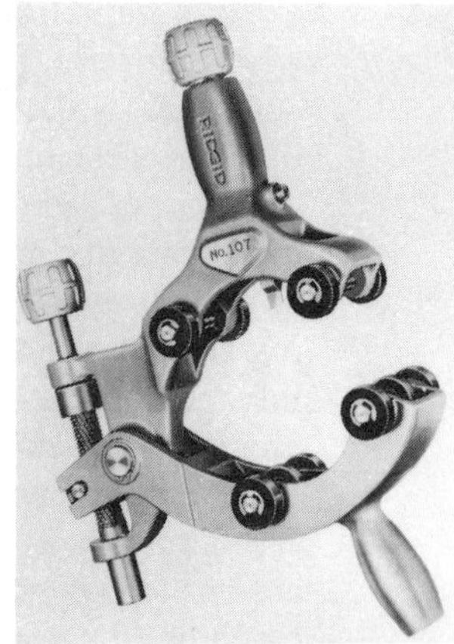

Blade-Type Cutter

Plastic Pipe Tools
(Ridge Tool Company)

2-7 Roughing-in Tools

The roughing-in tools shown in Fig. 2-24 are used to make room for plumbing installations and include saber saws, hole saws, hammer drills, hole drills, and boring machines. Hammer drills are used on concrete as are boring machines. Hole drills are used on wood to gain access for pipes. The hole saw is a general-purpose tool useful for roughing-in pipe in wood, plastic, formica, and plywood.

Figure 2-24

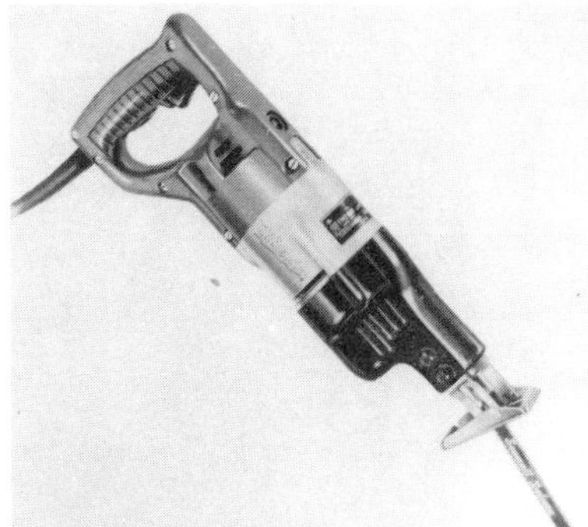

Two-Speed Reciprocating Saw
(Milwaukee Electric Tool Company)

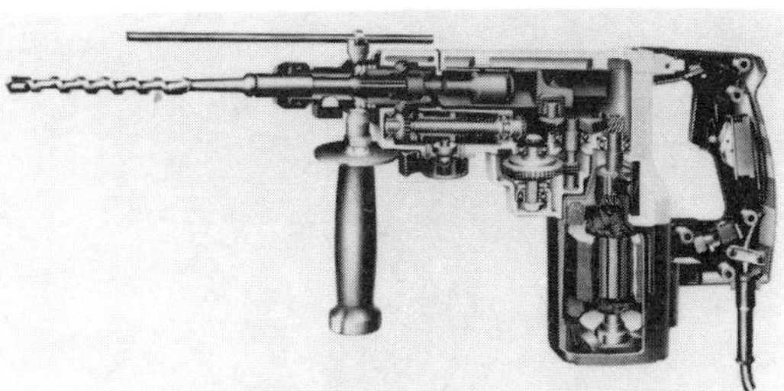

Hammer Drill
(Black & Decker - U.S. - Inc.)

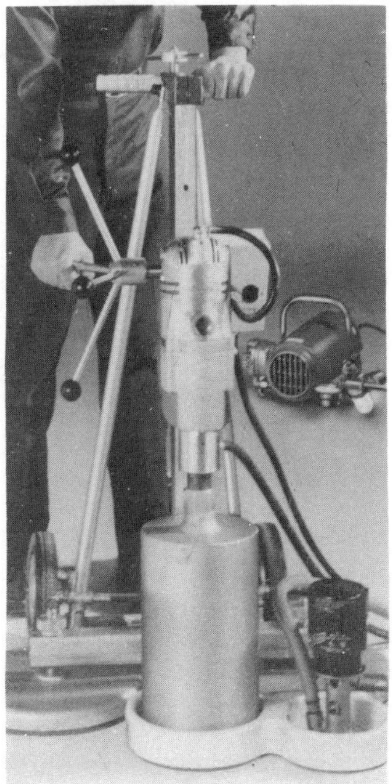

Concrete Hole Saw
(Milwaukee Electric Tool Company)

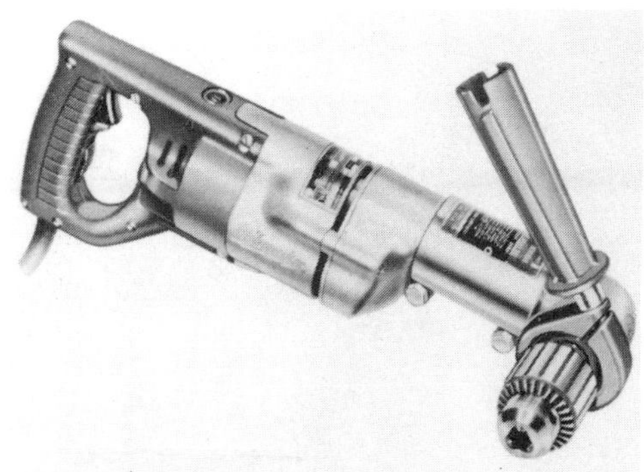

Right Angle Hole Drill
(Milwaukee Electric Drill Company)

Sabre Saw
(Millers Falls)

Roughing-In Tools

The bits shown in Fig. 2-25 are accessories to the tools. Masonry bits for hammer drills use carbide tips brazed to a fluted shank to remove dust. The general-purpose bits shown are for drilling holes in metal, and other materials to $\frac{1}{2}$-in. diameter. The plumber's bits shown drill holes from $\frac{7}{8}$ in. to $2\frac{9}{16}$ in. for pipe access. The hole saws are for larger sizes, and use a pilot drill to align the saw. Hole saws are available in sizes from $\frac{9}{16}$ in. to 6 in. (14.3 mm to 152.4 mm).

Figure 2-25

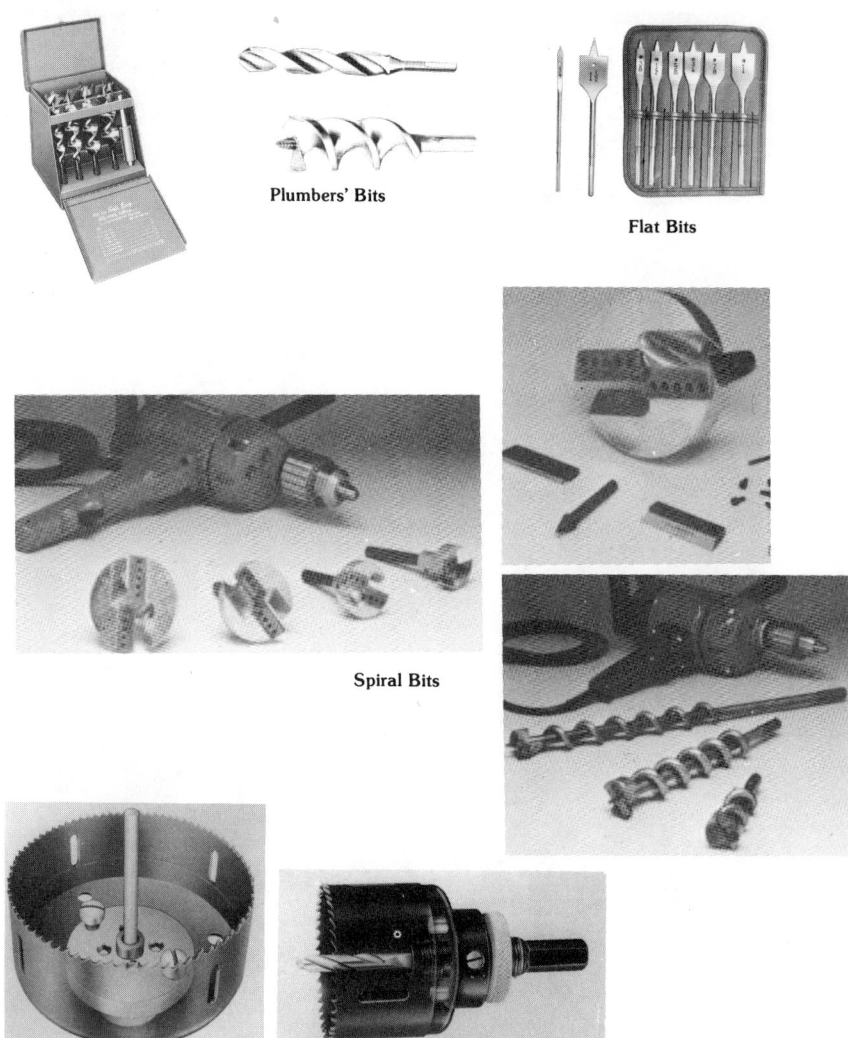

Plumbers' Bits

Flat Bits

Spiral Bits

Hole Saws

Hole Bits

2-8 Shovels and Ladders

Working outside and under buildings, the plumber (or plumber's helper) uses a variety of shovels, mattocks, and picks to dig trenches for storm and sanitary drains. The shovels shown in Fig. 2–26 are the "D" handle round-point shovel, long-handle round-point shovel, the same two types of square-point shovels, and a "D" handle drain spade. Shown on their handles are a standard cutter mattock and a pick mattock. Shown separate from the handle is a clay pick.

Outside, the plumber is often required to work on roofs. The extension ladder shown in Fig. 2–27 is of all-weather fiberglass to prevent deterioration. The hinged shoes at the bottom have rubber shoes to prevent slipping on hard surfaces. Inside, the plumber often uses step ladders such as that shown to work above the floor level. Hoists are used for lifting, dragging, and pulling into alignment.

Figure 2-26

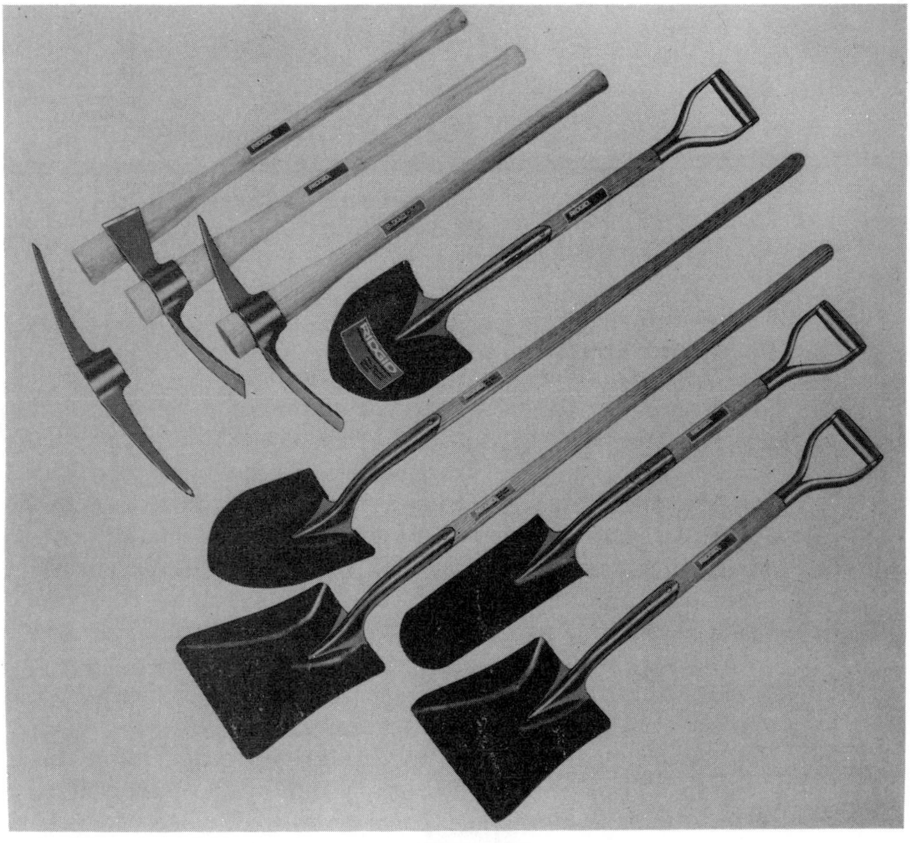

Shovels
(Ridge Tool Company)

Figure 2-27

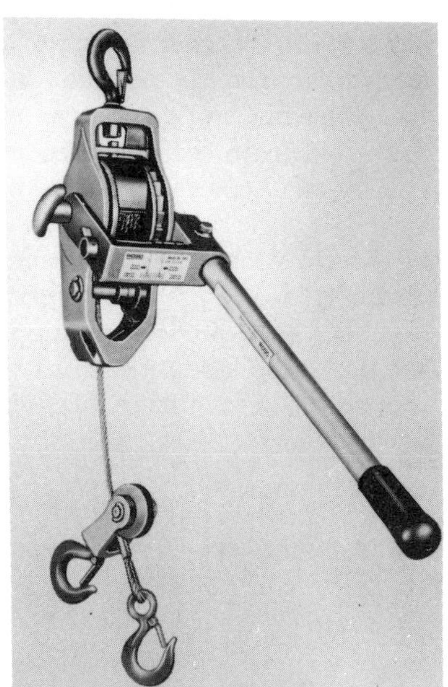

Hoist and Ladders (Ridge Tool Company)

2-9 Specialty Tools

There are many specialty tools for plumbing. A few of the more common are shown in Fig. 2–28 and are described here. The grappler hook is used to retrieve rocks, balls, and other objects that are blocking drains. Pulling the handle through the sleeve causes the sawtooth jaws to grasp the blockage. The closet auger is used to remove blockage from the closet trap. The obstruction is removed by turning the auger as the cable is fed through the closet trap. The trap spoon is used to guide cleaning tools into pipes when one is working back from a vented lawn or curb trap. The spoon is inserted in the line as shown on the left. As the spoon reaches the pipe to be cleaned, it drops to the lower pipe edge (center drawing) to direct the cleaning tool into the line. The spoon is then retrieved simply by removing it as shown. Flat sewer tapes for cleaning drains are standard in 25-ft to 100-ft (7.62 m–30.5 m) lengths. The grip handle helps force the tape back and forth. A number of interchangeable ends are available to clear the obstruction. Standard-length sewer tapes can be connected to clear lines to 300 ft (91.4m). Hand spinner drain-cleaning tools are used to navigate and clear

Figure 2-28

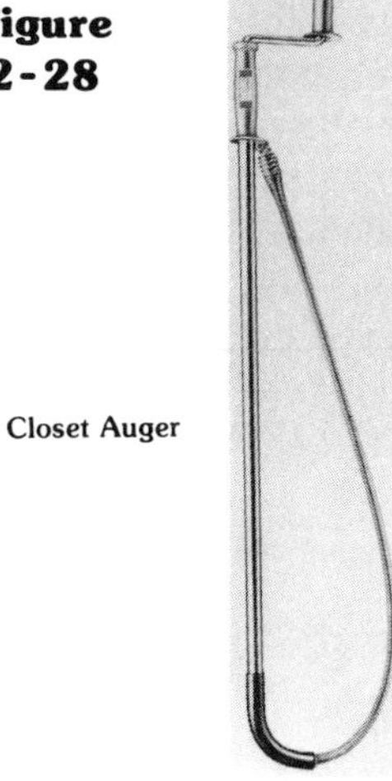

Closet Auger

Sewer Tape

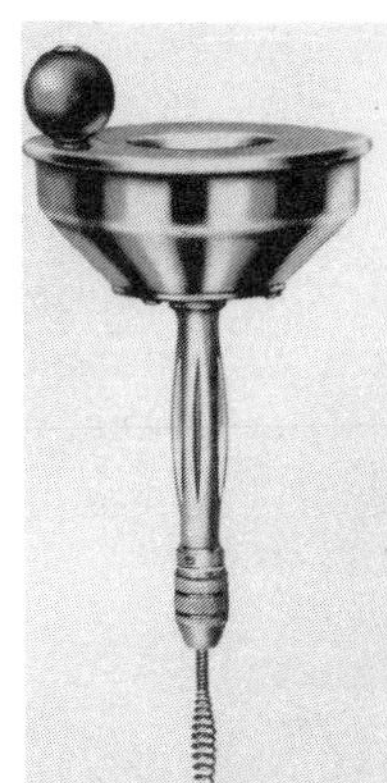

Hand Spinner

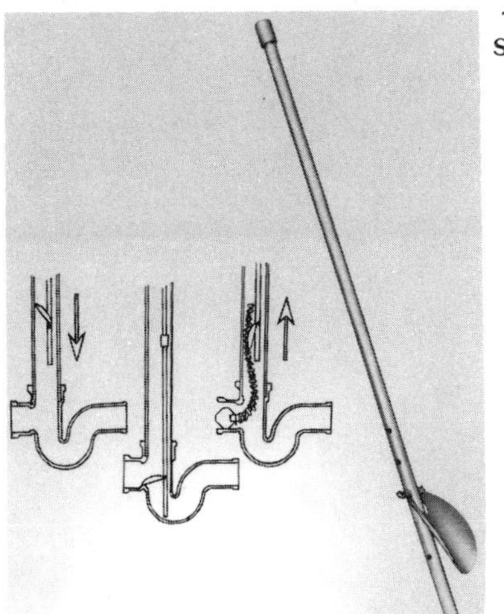

Trap Spoon

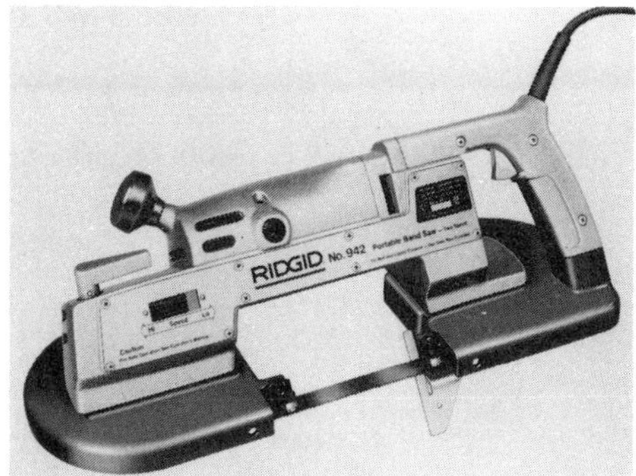

Portable Band Saw

Specialty Tools (Ridge Tool Company)

traps and bends in sinks, tubs, drinking fountains, water closets, urinals, and pipe vents.

2-10 Shop Tools

Although work at the job site usually requires portable tools, work at the shop is done with stationary tools and production equipment. This permits storage, layout, and fabrication areas, interspersed with work stations and equipment for cutting, grinding, drilling, and pipework.

The heavy duty bandsaw shown in Fig. 2–29 is used for cutting stock and pipe to 12 in. rectangular or 7 in. round. Depending upon the material, cutting speeds are adjusted from 65 ft/min for high-speed steels, to 280 ft/min for plastic, copper, soft brass, aluminum, and other light materials. The heavy-duty pipe and bolt threading machine shown uses a power drive coupled with swing-away cutter, reamer, and universal dies. A reversible vane pump recirculates thread-cutting oil. The lathe-type carriage advance

Figure 2-29

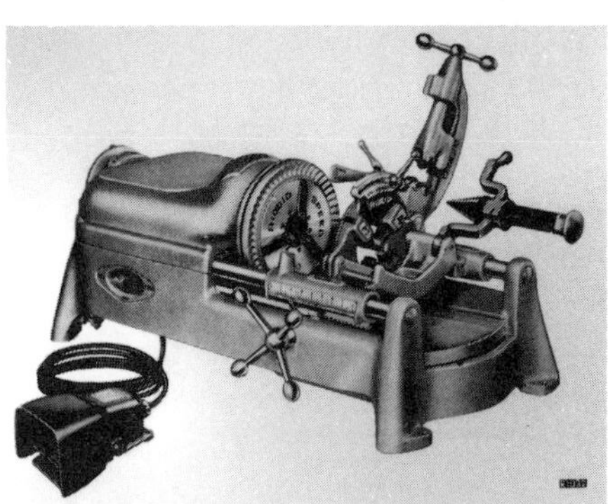

Threading Machine with Automatic Chuck

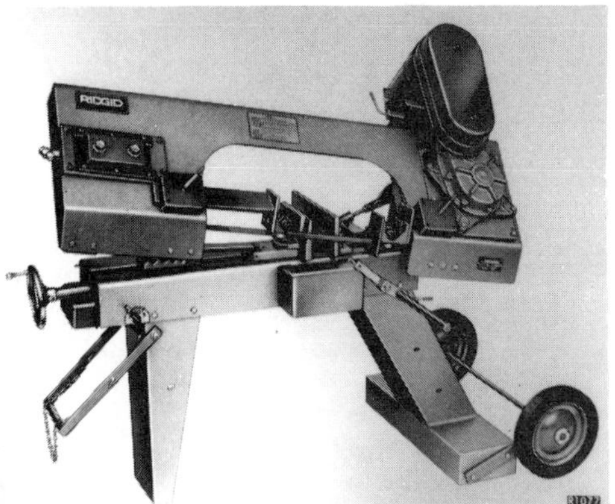

Metal Cutting Band Saw

Bench Grinder Pedestal Grinder

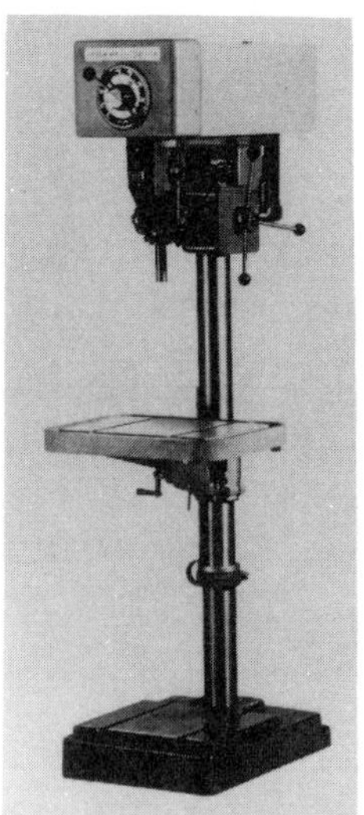

Drill Press

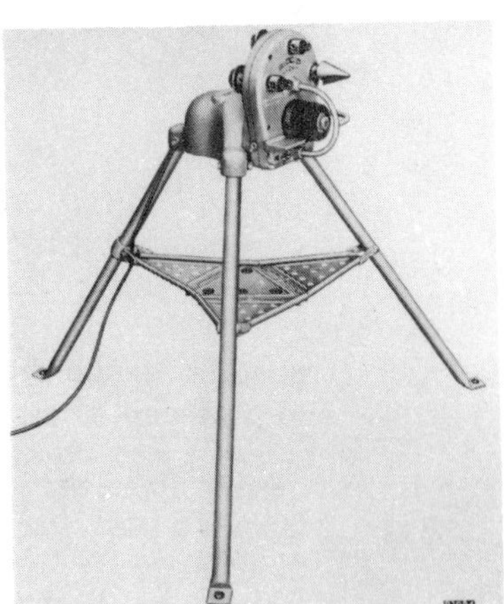

Copper Tube Cleaning Machine

Shop Tools

brings the tools along the scaled guides to cut and thread to measured lengths. Pipe to 2 in. can be threaded. The copper cleaning machine shown brushes, reams, faces, and deburrs pipe tube and fittings for assembly work. The large brush cleans the outside of the tube, whereas the smaller brushes are sized to accommodate fittings. The reamer is used to resize tube after cutting. The heavy-duty bench grinder is a general shop tool, as is the drill press. Bench grinders are available in $\frac{1}{4}$-hp through 1-hp models with wheel facing from $\frac{5}{8}$ in. to 1 in. The general-purpose model shown is a $\frac{1}{2}$-hp model with 7-in. x $\frac{3}{4}$-in. wheels in rough and smooth grit. The drill press shown in Fig. 2–29 is used for drilling flat steel, bar stock, and pipe. Although general shop work can be done with a $\frac{1}{2}$-in. drill chuck and 15-in. bed (with $7\frac{1}{2}$-in. clearance from the bit to the post), a $\frac{3}{4}$-in. drill chuck and longer reach table provide additional capacity to handle larger work.

2-11 Summary of Practice

The plumber is a craftsman who uses hand and power tools at the work site and in the shop. Tools used at the work site must be transported by truck along with the many items necessary to make the installation or repair. The personal hand tools of the plumber that are necessary are transported in one or more tool boxes and carried from the truck to the house. Power tools are usually supplied by the company with the inventory of the truck. Although it is more convenient to have everything necessary to complete the work on the truck, it must also be realized that tools and materials which are carried in large quantities or which are seldom used constitute a "dead inventory" on the truck. Fabrication in the shop can reduce transporting both equipment and the material to the work site. Making up subassemblies to standard specifications also has the advantage of increasing work efficiency, thereby increasing profits. The well-equipped shop has storage and production areas that can be used to fabricate subassemblies and work during inclement weather or periods when the work site is inaccessible.

REVIEW QUESTIONS

1. What benefit is there to assembling plumbing systems in the shop?
2. What is the purpose of a tool box?
3. What is meant by saw tooth pitch?
4. What is meant by thread pitch?
5. From experience, catalogs, and interviews, list the basic hand tools every plumber should own, together with their current prices.
6. Describe the purpose of five specialty tools used in plumbing.

7. From experience and interviews, list the 10 tools most commonly used on a residential repair call.
8. What is the most accurate and predictable method of cutting cast iron and malleable iron soil pipe?
9. When and why is threaded pipe reamed?
10. What is a drop-head threader?
11. What is the purpose of using oil when one is cutting threads?
12. Why can one set of pipe dies be used to thread more than one size of pipe?
13. How is tube cut with a tubing cutter?
14. What is the difference between a metal and plastic tube cutter?
15. Name five different tools that can be used to assemble chrome pipe, fittings, and fixtures.

3 Plumbing Mathematics

3-1 Introduction

Mathematics is the science of numbers. It has utility in making measurements with such tools as rulers and gauges, in computing the offset dimensions of pipe systems, and in making financial transactions.

Plumbing mathematics has two components; one is the numbers themselves, and the second is the dimensional units that give meaning to those numbers, such as feet, inches, meters, or centimeters. Usually the number is given first, followed by the dimensional units. For example, a pipe may be 2 ft (61 cm) long or may deliver water at 15 gal/min (68 1/min). Both the numbers and the dimensional units must be given attention when measurements and calculations are made, for if either one of the two components is incorrect, the answer is incorrect.

Numbers are used in much the same fashion everywhere. Dimensional units, however, are not. In America, the English system of dimensional units is predominant, although conversion to the International System of Units, abbreviated SI units, is in progress. Elsewhere in the world, SI units are used.

Conversion to the SI system may be "hard" or "soft." "Hard" metric conversion refers to the fabrication of products and components in round-numbered metric sizes; "soft" conversion implies continuing to employ customary sizes, but describing them in metric measurements, which usually are awkward numbers of units. At the present time, America is undergoing a "soft" conversion in the plumbing industry, although pipe and tube in metric sizes are now manufactured for export and for other industries here which are making the "hard" conversion.

Numbers themselves may be whole numbers, or parts of whole numbers. Parts of whole numbers may be given as fractions, for example, $\frac{1}{2}$ or $\frac{1}{4}$, read "one-half" or "one-quarter"; or given as decimals, for example 0.50 or 0.25, read "fifty one-hundredths" or "twenty-five one-hundredths."

Units are used in plumbing to describe length, or distance, width, volume, mass (amount of a substance), time, velocity, acceleration due to gravity, and weight.

Operations with numbers are indicated by signs: for example, addition with a + (plus), subtraction with a – (minus), multiplication with an

× or • (dot), division with a ÷ , percentage with a %, and square root with a $\sqrt{\ }$ (radical sign). Right-angle trigonometry expressions are used to compute offsets. These operations can be simplified to multiplication by using the trigonometric functions.

3-2 Fractions, Decimals, and Signed Numbers

A common fraction is a part of a whole number. It is represented by placing the part over the whole. The top portion is called the *numerator* and the bottom is called the *denominator*. Figure 3–1 illustrates several ways to describe $\frac{1}{4}$ (pronounced one-fourth). If the numerator and denominator can be divided by the same number, the fraction can be reduced, and the divisor is called a *common factor*. Any fraction in which the numerator and denominator are multiplied by the same number is an equivalent fraction. For example, the fraction

$$\frac{12}{16} = \frac{6}{8} = \frac{3}{4} = \frac{15}{20} = \frac{30}{40}$$

That is, the fraction $\frac{12}{16}$ was reduced by dividing the numerator and denominator by 2, the common factor, as was the fraction $\frac{6}{8}$. Both the

Figure 3-1

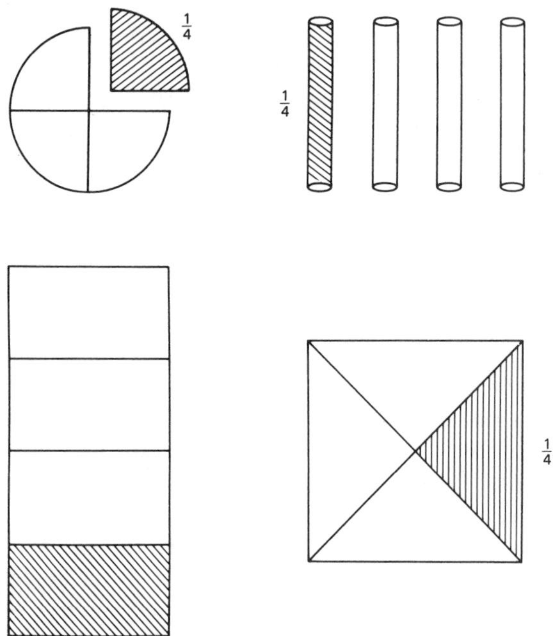

One-Fourth

numerator and denominator of the fraction $\frac{3}{4}$ were multiplied by 5, and then by 2 to produce the equivalent fractions $\frac{15}{20}$ and $\frac{30}{40}$. All are equivalent fractions.

The numerator and denominator of a fraction are divisible by 2 if the last number is a 2 or 0. They are divisible by three if the sum of the numbers is divisible by three. For example, the numbers in 12 add to three, which is divisible by three. The sum of the numbers in 16, which is 7, however, is not divisible by 3. If the last number in the numerator and denominator is either 5 or 0, the fraction is divisible by 5 or 10.

A fraction is *proper* if the numerator is smaller than the denominator. If the denominator is smaller than the numerator, the fraction is *improper.* The fractions, $\frac{2}{3}$, $\frac{3}{4}$, $\frac{5}{6}$, and $\frac{7}{8}$ are proper fractions, whereas the fractions $\frac{5}{3}$, $\frac{5}{2}$, and $\frac{15}{4}$ are improper fractions. If the numerator and denominator of the fraction are the same, the fraction is a 1, because they are both divisible by the same number. Improper fractions can be written as *mixed numbers.* A mixed number consists of a *whole number* and a *proper fraction.* Improper fractions are converted to mixed numbers by dividing the denominator into the numerator and adding that portion remaining to the quotient as a proper fraction. For example, $\frac{5}{3} = 1\frac{2}{3}$, $\frac{5}{2} = 2\frac{1}{2}$, and $\frac{15}{4} = 3\frac{3}{4}$. Mixed numbers are converted to improper fractions by multiplying the denominator by the whole number, adding the numerator of the proper fraction to that product, and placing that sum over the denominator. For example,

$$\frac{7}{8} = \frac{(8 \times 1) + 7}{8} = \frac{15}{8}$$

Fractions are added and subtracted by (1) converting them to proper or improper fractions, (2) finding the least common denominator into which each denominator will divide, and then (3) performing the operation indicated. For example,

$$\frac{3}{4} + \frac{7}{8} = \frac{6}{8} + \frac{7}{8} = \frac{6 + 7}{8} = \frac{13}{8} = 1\frac{5}{8}$$

As another example,

$$1\frac{3}{4} + 2\frac{7}{8} = \frac{7}{4} + \frac{23}{8} = \frac{14}{8} + \frac{23}{8} = \frac{14 + 23}{8} = \frac{37}{8} = 4\frac{5}{8}$$

Fractions are multiplied by multiplying the respective numerators and denominators, and then are reduced by using common factors. For example,

$$\frac{3}{4} \times \frac{2}{5} = \frac{6}{20} = \frac{3}{10}$$

As another example,

$$1\frac{3}{4} \times 2\frac{7}{8} = \frac{7}{4} \times \frac{23}{8} = \frac{161}{32} = 5\frac{1}{32}$$

Fractions are divided by inverting the denominator, and then performing the operation as multiplication. For example,

$$\frac{\frac{3}{4}}{\frac{2}{5}} = \frac{3}{4} \times \frac{5}{2} = \frac{15}{8} = 1\frac{7}{8}$$

As another example,

$$\frac{1\frac{3}{4}}{2\frac{7}{8}} = \frac{\frac{7}{4}}{\frac{23}{8}} = \frac{7}{4} \times \frac{8}{23} = \frac{56}{92} = \frac{28}{46} = \frac{14}{23}$$

A *decimal fraction* is one written without the denominator using a decimal point. The decimal point is signified by a period (.). Decimal fractions are possible because of the decimal system of numbers which permits ten *digits* in each designated place. The digits are 0, 1, 2, 3, 4, 5, 6, 7, 8, and 9. The *places* are shown in Fig. 3–2. The number shown in the figure is written 38 465.1785, and is pronounced thirty-eight thousand, four hundred sixty-five and seventeen hundred eighty-five ten-thousandths. Notice that *and* is inserted at the decimal to connect the whole number to the decimal. Notice also that when more than four digits are placed together that they are set off in groups of three digits separated by a space (in the English system they are set off by a comma). That is, 38465 is written as 38 465 because it contains five digits. The number 1785, however, may remain as it is written because it contains only four digits. Each place to the left of the decimal has a value 10 times greater than the preceding one, whereas each place to the right of the decimal has a value of one-tenth the preceding digit.

Decimals and fractions are convertible to each other. To convert a fraction to a decimal, the fraction is simply divided out. For example, $\frac{1}{4}$ equals

Figure 3-2

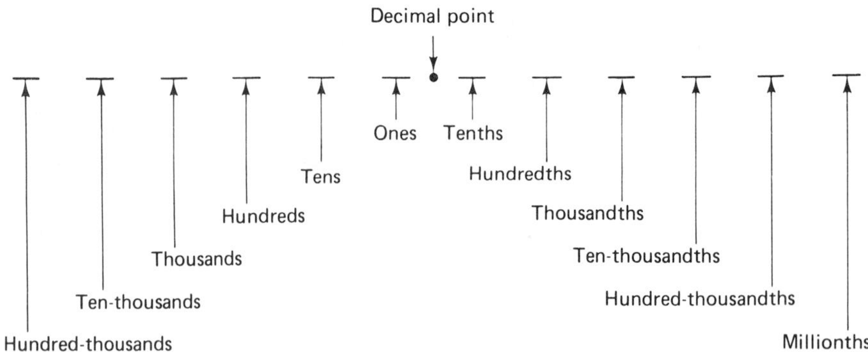

Places in the Decimal System of Numbers

0.25 or twenty-five hundredths. That is,

$$\frac{1}{4} = \begin{array}{r} 0.25 \\ 4\;\overline{)1.00} \\ -8 \\ \hline 20 \\ -20 \\ \hline \end{array}$$

As another example, $\frac{3}{4}$ equals 0.75, or seventy-five hundredths. That is,

$$\frac{3}{4} = \begin{array}{r} 0.75 \\ 4\;\overline{)3.00} \\ -28 \\ \hline 20 \\ -20 \\ \hline \end{array}$$

To convert from decimals to fractions, simply write the number as a proper fraction and reduce it, using common factors. For example, 0.125 equals one hundred twenty-five thousandths. That is,

$$\frac{125}{1000} = \frac{25}{200} = \frac{5}{40} = \frac{1}{8}$$

As another example, 0.375 equals three hundred seventy-five thousandths. That is,

$$\frac{375}{1000} = \frac{75}{200} = \frac{15}{40} = \frac{3}{8}$$

Notice that both decimals were written first as fractions and then were reduced by using 5 as the common factor.

Decimal numbers are rounded off to discard unnecessary places, but still retain the desired accuracy. It is important, however, that the accuracy of the number is not distorted or sacrificed in the process. For use in plumbing measurements, rounding to two places is usually appropriate. Machine work usually requires three places. If a number is to be rounded off to fewer places than the total number available, the following rules should be used:

1. If the first digit discarded is less than 5, the last digit retained should not be changed. For example, 9.012 343 rounded to six digits would be 9.01 234, to five digits 9.0123, and to four digits 9.012.
2. When the first digit discarded is greater than 5 (including any digits following with a value other than 0), the last digit retained should be increased by 1. For example, 8.012 678 rounded to six digits would be 8.01 268, to five digits 8.0127, and to four digits 8.013.
3. When the first digit discarded is exactly 5 followed only by zeros, the last digit retained should be increased by one unit if it is an odd

number, but not adjusted if it is an even number. For example, 5.312 545 rounded to six places would be 5.31 254.

Decimals are added and subtracted by placing them one above the other with the decimal points aligned, and then performing the operations in the usual manner. For example,

$$\begin{array}{r} 1.375 \\ +0.125 \\ \hline 1.500 \end{array}$$

As another example,

$$\begin{array}{r} 1.375 \\ -0.125 \\ \hline 1.250 \end{array}$$

In addition and subtraction, decimals are rounded to one more place than the least accurate of the numbers.

Decimals are multiplied by (1) placing them one above the other, (2) performing the operation indicated, (3) establishing the decimal place, and then (4) rounding the number off. The rounding rule for both multiplication and division is that the product or quotient shall contain no more digits than the fewest of any number used in the operation. For example,

$$\begin{array}{r} 0.375 \\ \times 0.15 \\ \hline 1875 \\ 375 \\ \hline 5625 \end{array}$$

Counting decimal places gives 0.05 625. This is because the numbers multiplied contain a total of five decimal places. But since the multiplier (0.15) contains only two digits, the answer must be rounded off to two digits to become 0.06.[1] As another example,

$$\begin{array}{r} 1.375 \\ \times 0.125 \\ \hline 6875 \\ 2750 \\ 1375 \\ \hline 171875 \end{array}$$

The answer should contain six decimal places and becomes 0.171 875. But since the multiplier (0.125) contains only three digits, the answer is rounded off to become 0.172.

Division is accomplished in much the same way as multiplication. The object is to find out how many times one number, the divisor, will fit into the other, the dividend. The first step is to set up the problem and move the decimal point in the divisor to the right the number of places that are necessary to make it a whole number. This also requires that the decimal point in the dividend be moved to the right the same number of places so that the sense of the problem will not be changed. Then the division operation is performed, the decimal point in the answer being kept even with the new one established in the dividend. Remember that the answer should have the same number of places before rounding as there are in the divisor. For example, setting up the following division problem and moving the decimal point in both the divisor and dividend, we have

$$0.15\overline{)0.375}$$

Now performing the division operation, keeping the decimal point in the answer even with the new one established in the dividend, we obtain

```
        2.5
0.15,)0.37,5
      -30
        75
      - 75
```

The answer indicates that 0.15 can be divided into 0.375, 2.5 times. This is checked by multiplying 2.5 × 0.15. Also notice that, since the divisor has two digits, the same as the answer, no rounding is necessary. As another example,

```
         11.
0.125,)1.375,
      -125
        125
      -125
```

Notice again that 0.125 divides evenly 11 times into 1.375 and that the answer may be checked by multiplying 0.125 × 11. Since the answer contains only two digits, which is less than either the divisor or the number being divided, no rounding is necessary.

Percentage is often used in computing discounts, profit, markup or markdown, and tax. It can also be useful when one is computing the thermal expansion of pipe, which is a characteristic that makes it elongate with a rise in temperature or shorten with a drop in temperature. Computing percentage is done by using multiplication, where one number, for example, dollars and cents, is multiplied by a percentage. But before this can be done, both numbers must be the same. Since dollars and cents are decimals, percentage, which means one hundredth, must be converted to a decimal. This is accomplished by moving the decimal point two places to the left. Now both numbers are decimals and they can be multiplied, the rules for establishing the decimal place and rounding the answer being used. Percent is changed to a decimal by moving the decimal point two places to the left as follows:

$$10 \text{ percent} = 10.0\% = 0.10$$

The following two problems show how typical tax and discount problems are figured by using 5 percent and 20 percent as multipliers.

$$\begin{array}{r} \$35.95 \\ 0.05 \\ \hline 17975 = \$1.80 \end{array} \qquad \begin{array}{r} \$89.75 \\ 0.20 \\ \hline 179500 = \$17.95 \end{array}$$

A decimal can be changed to a percentage simply by moving the decimal point two places to the right. That is, 0.15 = 15.0% and 0.105 = 10.5%. This conversion is useful in figuring percentage of elongation. For example, if a 20-ft. length of plastic tubing elongates 1.340 in. with a rise in temperature of 100°F (37.7°C), what percentage of its length does this represent? First, figuring the expansion for each inch of pipe, we have (20 ft × 12 in/ft) ÷ (1.340 in) = 0.0055 in/in. This means that the pipe elongates about 55 units for each 10,000. That is hard to understand. However, when this is converted to a percentage by moving the decimal point two places to the right, 0.0055 in/in = 0.55%, or about one-half of one percent for the temperature rise given. This may seem easier. Continuing, if the pipe were 1000 ft long, it would elongate by (1000 ft × 0.0055 in/in = 5.5 ft) more than 5 ft.

3-3 Square Root

Computing the square and square root of a number occurs frequently when one is solving for the long side (hypotenuse) of right triangles. This is necessary to find the travel length of an offset, or the distance between the diagonal (opposite) corners of a rectangular building when the length and width dimensions are known.

A number is squared when it is multiplied by itself. When a number is cubed, it is multiplied by itself twice, and so on. The number being multiplied is said to be raised to a power. If it is squared, it is raised to the power 2. If cubed, the power is 3. The power to which the number is raised is written to the upper right of the number. Following are numbers raised to powers, and the operations that occur.

$$2^2 = 2 \times 2 = 4$$
$$2^3 = 2 \times 2 \times 2 = 8$$
$$3^4 = 3 \times 3 \times 3 \times 3 = 81$$

The important concept to remember is that the number is multiplied the number of times given by the power. For convenience, Appendix H lists the powers and roots of numbers.

The square root of a number is the quantity of which the given quantity is the square. The square root of 4 is 2. It is indicated by a radical sign with the root to which the number is to be taken written in the depression at the left. That is,

$$\sqrt[2]{16} = \sqrt[2]{4 \times 4} = \sqrt[2]{4^2} = 4$$

Notice that 16 is the product of 4×4, which is the same as 4^2. And since the root to which the number is to be taken and the square are both 2, the radical is lifted and the answer is simply 4. Also notice that Appendix H can be used to solve for both the square roots and cube roots of numbers. When the root to which the number is to be taken is 2, the 2 placed in the depression is deleted, and it is understood that the square root of the number is to be taken.

Many squares and cubes of numbers, as well as their respective square and cube roots, are apparent by inspection. That is, just looking at some numbers gives rise to the answer. For example, the numbers 4, 9, 16, 25, and 36 are perfect squares for the numbers 2, 3, 4, 5, and 6. Often, however, the perfect square or square root are not so evident, and then it is important to be able to follow a simple set of rules so that any number can be solved for its square root. Here are the rules:

1. Starting from the decimal point, set the number for which the square root is to be determined off in groups of two, using a caret indicator. Add zeros if necessary to complete each group of numbers on the ends. For example, the number 1254.64 would be set off as 12 54.64 and would be written under the radical sign as

$$\sqrt{12_{\wedge}54.64}$$

2. Next, find the largest square that can be subtracted from the first two numbers. The square root of that number is the first digit in the answer. In the previous example given, the largest square that can be subtracted from 12 is 9, and the square root of that number is 3. This is the first digit in the answer. Notice that the next largest square would be $4 \times 4 = 16$, which is larger than 12.

3. Now, place the first digit of the answer over the first group of two numbers, square it, and subtract the square from the two numbers, bringing down with the remainder the next group of two numbers. Continuing with the example, we have

$$\begin{array}{l} \quad 3 \\ \sqrt{12_{\wedge}54.64} \\ \quad 9 \\ \hline \quad 3\ 54 \end{array}$$

4. Determine the remaining digits in the answer in the following manner:
   a. Double the present answer and place with that number a 0. This number becomes a trial divisor.
   b. Add to the trial divisor digits from 1 to 9, such that when the trial divisor is multiplied by the digit, the product can be just subtracted from the remainder.
   c. Now, place the trial digit over the next two numbers and subtract the product of the digit and the trial divisor from the remainder,

bringing down the new remainder with the next group of two numbers.

d. Continue doubling the answer, placing with that a 0, determining the trial digit, and subtracting the product of the trial digit and the divisor from the remainder until the desired accuracy is reached. Notice that the decimal point retains its position in the answer.

Completing the example, the 3 is doubled and with that is placed a 0 to make up the trial divisor.

```
                   3
                 √12‸54.64
trial              9
divisor        60 | 3 54
```

Adding digits to the trial divisor starting with 1 and multiplying them by that same digit indicates that 60 + 5 = 65 multiplied by 5 equals 325, which can just be subtracted from the remainder. Notice that 66 multiplied by 6 equals 396, and this number is larger than 354. The trial divisor then is 6, and the operation of multiplication and subtraction from the remainder is continued.

```
      3  5. 4
    √12‸54.64
      9
 65 |3 54
    |3 25
704 |29 64
    |28 16
      1 48
```

The final answer is checked by squaring it and comparing that number to the original number. That is, the answer is checked by squaring 35.4 and comparing that with 1254.64.

Another means of determining the square root of a number is with a calculator. Some are equipped with a square root function and make the computation automatically. Trial multiplication is another method. It is known, for example, that the square root of 1254.64 must be between 30 and 40, which square to 900 and 1600, respectively. A first trial might be to square 35 with the calculator and if that is less than the number for which the square root is to be determined, then a trial number slightly larger, for example, 35.2 could be tried. This method is typically faster than the long hand method.

Figure 3-3

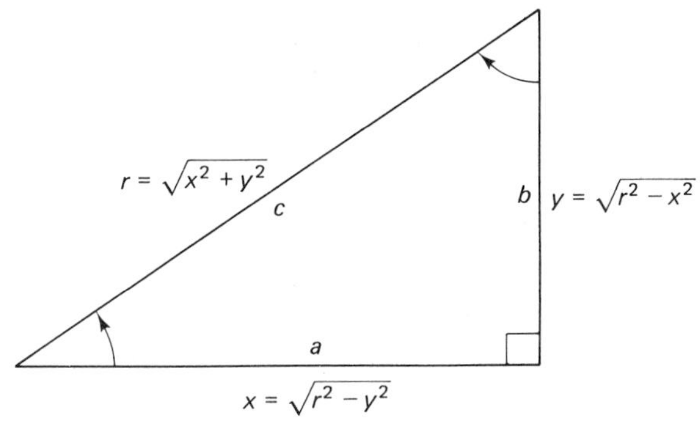

Pythagorean Theorem

3-4 Pythagorean Theorem

The Pythagorean theorem is a rule in mathematics that defines the relative lengths of the three sides of a right triangle. It is written as

$$c^2 = a^2 + b^2 \tag{3-1}$$

This means that if the two sides which make up the right angle are squared and added together, they will equal the square of the long third side, called the *hypotenuse*. Thus, the square root of this square will equal the length of the hypotenuse. Where the sides of the right triangle shown in Fig. 3-3 are labeled x, y, and r, the Pythagorean theorem can be rewritten as

$$r^2 = x^2 + y^2 \tag{3-2}$$

or

$$r = \sqrt{x^2 + y^2} \tag{3-3}$$

The formula can also be solved for x or y. That is,

$$x^2 = r^2 - y^2 \quad \text{and} \quad x = \sqrt{r^2 - y^2} \tag{3-4}$$

or

$$y^2 = r^2 - x^2 \quad \text{and} \quad y = \sqrt{r^2 - x^2} \tag{3-5}$$

An example will make this clear.

Example 3-1 The right triangle in Fig. 3-4 gives x as 4 and y as 3. Solve for the hypotenuse.

Solution

Substituting in Eq. (3-3), we have

Figure 3-4

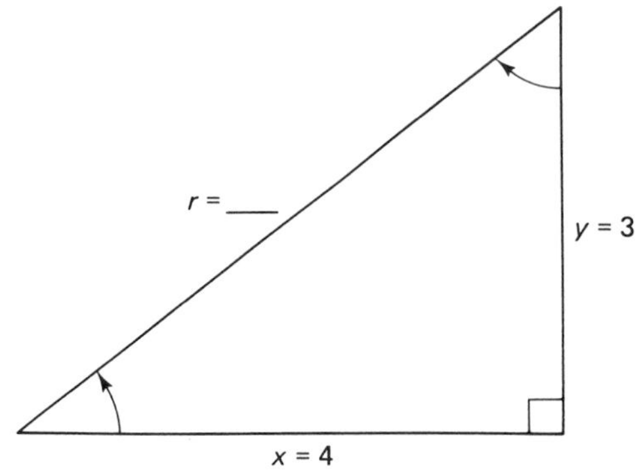

$$r = \sqrt{x^2 + y^2}$$
$$r = \sqrt{4^2 + 3^2}$$
$$r = \sqrt{16 + 9} = 5$$

and the triangle is a 3, 4, 5 right triangle.

Example 3-2 The triangle in Fig. 3-5 has three equal sides. If one side is given a length of 2, determine the length of the line that would divide the triangle into two right triangles.

Solution

If a line is drawn perpendicular from one of the sides which can be designated as the base, to the intersection of the other two sides, the

Figure 3-5

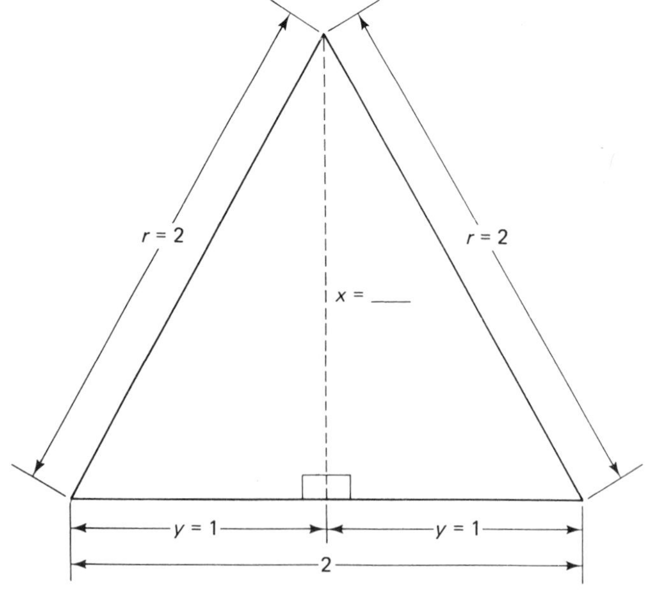

triangle with three equal sides is divided into two equal right triangles. Notice that if the sides are of length 2, the base is divided equally into two parts of length 1. The length of the third side, then, designated x, can be computed from Eq. (3–4).

$$x = \sqrt{r^2 - y^2}$$
$$x = \sqrt{2^2 - 1^2} = \sqrt{3}$$

where the square root of 3 is 1.732.

3-5 Right Angle Trigonometry

Right angle trigonometry defines the relationships between the sides and angles of right triangles. In plumbing, the most common angles encountered are 45, 90, and 180 degrees. Ninety degrees is a right angle. One hundred eighty degrees is a straight angle.

Figure 3-6

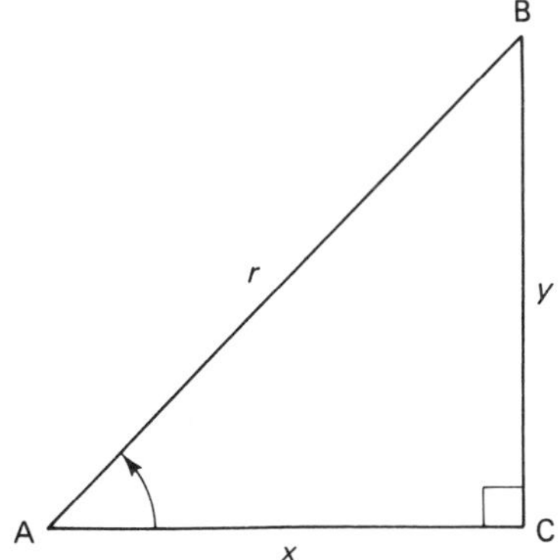

Right Angle Trigonometry

The sides and angles of a right triangle (Fig. 3–6) are labeled with respect to the position of the right angle, and the angle in question. The angles are typically designated A, B, and C. The sides are commonly designated x, y, and r. The long side, r, is the hypotenuse. If the angle of interest is A, the *opposite side* is y, and the *adjacent side* is x. If the angle of interest were B, then the triangle in Fig. 3–6 would be relabeled such that the opposite side from B would be y and the adjacent side would be x. The important relationship to remember is that the side *opposite* the angle of interest is always y, and the side *adjacent* to the angle of interest is always x. This has importance in defining the trigonometric functions.

The plumber most often uses two of the six trigonometric functions listed in Appendix I. These are

$$\text{Sine } A = \frac{y}{r} = \frac{\text{opposite side}}{\text{hypotenuse}} \tag{3-6}$$

$$\text{Cosine } A = \frac{x}{r} = \frac{\text{adjacent side}}{\text{hypotenuse}} \tag{3-7}$$

They are typically abbreviated sin A and cos A.

If any two of the three sides or angle are known, this relationship can be used to solve for the third. To make related calculations it is necessary to know the angle in question, to establish the sin or cos relationship, and then to solve for the side or angle of interest. It is also helpful if one is familiar with the relative lengths of the sides for the angles of common triangles. These can be established by inspection of the triangle and application of the Pythagorean theorem. For example, knowing that all triangles have internal angles that sum to 180 degrees, it can be seen in Fig. 3–7 (a) that a right triangle with a 45-degree angle must have two such angles, i.e., 45° + 45° + 90° = 180°. Thus, the adjacent and opposite sides would have the same length. If they are given a value of 1, then the hypotenuse would have a value of

$$r = \sqrt{x^2 + y^2} = \sqrt{1^2 + 1^2} = \sqrt{2} = 1.414$$

And so, before making computations, it is helpful to label the right angle, to label the angle of interest, and then to label the sides with the letter designations x and y, with their appropriate lengths. In this way, the relationships between the sides and the angle can be reduced to their simplest terms and the sense of the problem rather than the numbers can be given the most thought. Figure 3–7 reduces the more common angles and lengths of sides to their simplest form. Appendix J lists the values for other angles.

Typically, the plumber wishes to solve for the length of the offset r, when the value of the adjacent side x or opposite side y are given or can be measured. Notice that this requires solving the trigonometric function for r. This is easily accomplished by first writing the relationship for the angle and then locating r in the desired position by using a rule in algebra which permits inverting both sides of an equation and then cross-multiplying through the equal sign. For example,

$$\sin A = \frac{y}{r}$$

Inverting both sides of the equation yields

$$\frac{1}{\sin A} = \frac{r}{y}$$

Cross-multiplying through the equal sign solves for r:

$$r = \frac{y}{\sin A}$$

Figure 3-7

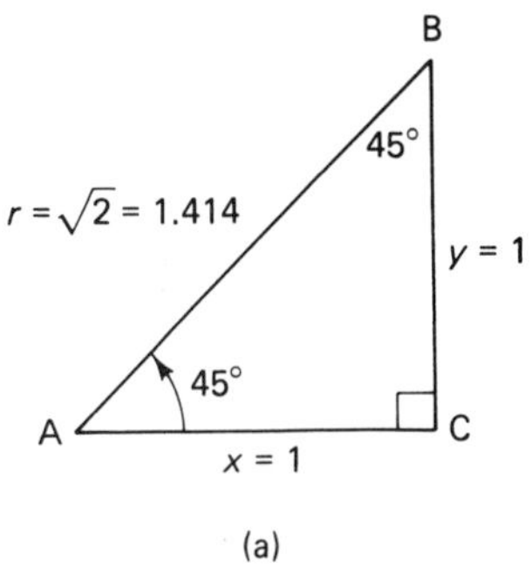

(a)

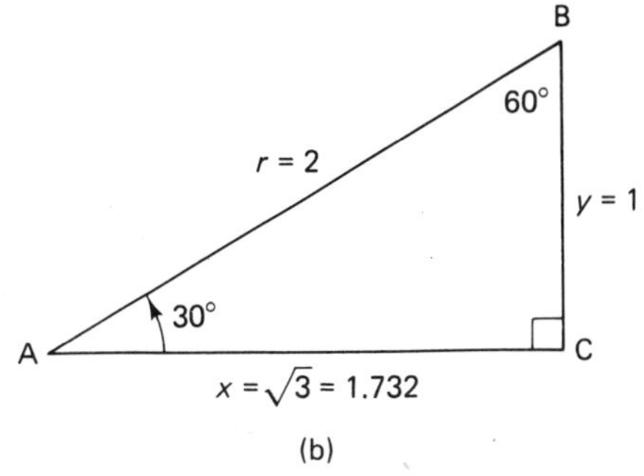

(b)

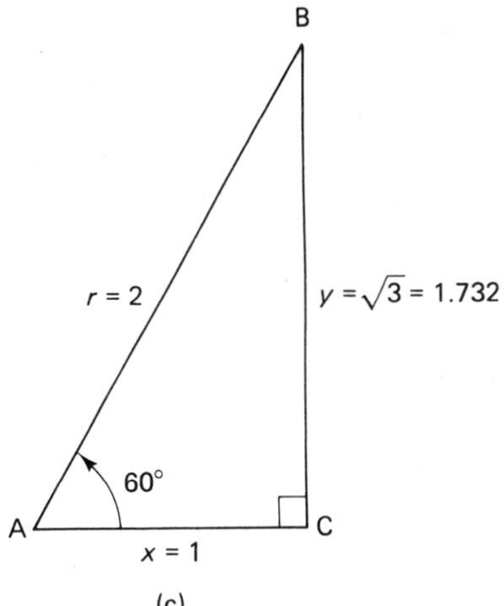

(c)

Lengths of Sides for Common Angles

Thus, for a 45-degree angle, r would have a value

$$r = \frac{y}{0.707} = 1.414 \times y \tag{3-8}$$

That is, multiplying the measured value of y by 1.414 would give the value of r.

Example 3-3 A right triangle with an angle of 60 degrees has an opposite side 48 in. long. Compute the length of the hypotenuse.

Solution

The triangle is shown in Fig. 3-7(c). Since the opposite side is given, the sine of the angle is used. That relationship is

$$\sin A = \frac{y}{r}$$

$$r = \frac{y}{\sin A} = \frac{y}{0.866} = y \times 1.155$$

and

$$r = 48 \text{ in.} \times 1.155 = 55 \text{ in.}$$

3-6 "Figuring" Offsets

An *offset* is constructed when a pipe must be routed around an obstacle along its run. The pipe is said to "break" *up, down,* or *over* to clear the obstruction. After routing around the obstacle, the pipe continues its run in the same direction but along another line. Figure 3–8 illustrates the most common 45-degree angle offset and the terms that are used to describe its components. The amount of offset depends upon the size of the obstacle to be cleared and the available space into which the pipe must be placed. This is usually measured with a plumber's rule, which reads the offset distance on one side and the travel length for a 45-degree angle on the other (Fig. 3-9). A 10-in. offset, for example, read from the inch side of the rule would result in a travel length of $14\frac{1}{8}$ in. on the 45-degree scale.

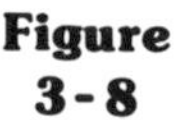

Figure 3-8

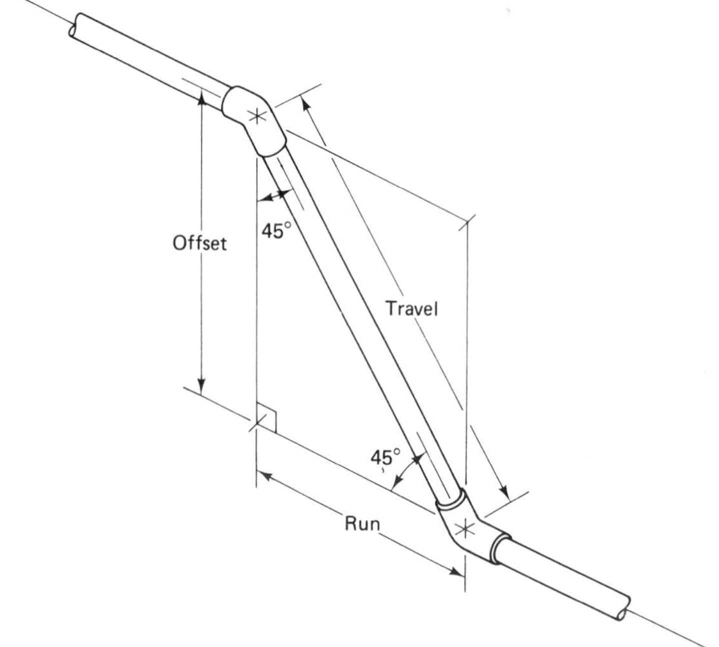

Forty-five-Degree Offset

Figure 3-9

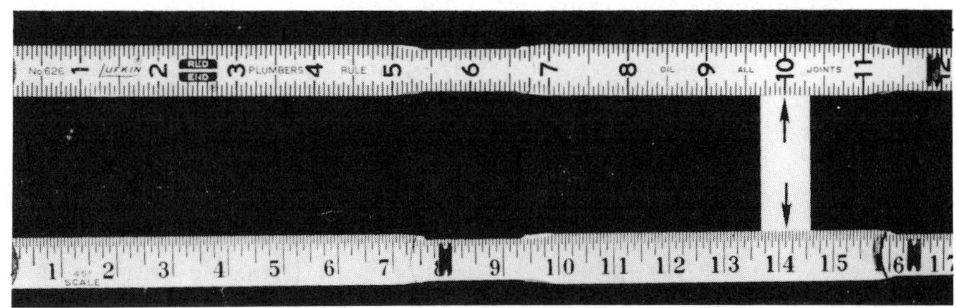

Measuring an Offset with a Plumber's Rule

Often, offsets are figured by cutting the pipe run, fitting the 45-degree elbow, and then using a piece of pipe inserted into the fitting as a measuring piece, making the necessary allowances at the end of the break for the second fitting and depth of threads. Another method is to use a string or measuring tape to determine the length. These are suitable when the installation is quicker and result in work that is neat in appearance. However, as pipe sizes become larger and other obstructions, such as walls, prevent using a rule, measuring tape, or string, simple computations will determine the length of pipe for the travel both quicker and more accurately than the trial cut and try method.

Computed offsets determine the travel distance between centers (C–C). That is, the distance is taken along the centerline of the pipe from the center point in the elbow where the pipe breaks to the point in the second elbow where it resumes the run. This means that the computed length of travel must be reduced by half the length of each fitting at the ends, and to that must be added the length necessary to "make up" the connections (inside the fittings) at each end. The dimensions of the fittings and the depth of thread to be added to each end are usually taken from tables and data that describe respective fittings of different sizes.

It is helpful when one is figuring an offset to sketch the angle on a piece of paper to visualize the relationships between the angle and the lengths of the sides. For angles such as 45 degrees, these relationships can be committed to memory, particularly if the length of travel is the unknown. For example, in this case C–C travel equals

$$\text{Travel} = 1.414 \times \text{offset} \qquad \text{(also 3–8)}$$

or

$$\text{Travel} = 1.414 \times \text{run} \qquad \text{(3–9)}$$

Example 3-4 Figure the length of the C–C travel for a 45-degree offset of 37 in.

Solution

The travel is computed from

$$\sin 45^\circ = \frac{y}{r} = \frac{\text{offset}}{\text{travel}} = 0.707$$

Solving for the travel, we have

$$\text{Travel} = 1.414 \times \text{offset} = 1.414 \times 37 \text{ in.} = 52.3\text{-in., C–C}$$

When a pipe is offset in two planes, it is called a *rolling offset.* That is, in addition to the offset, the pipe twists or rolls. This is illustrated in Fig. 3-10. Notice that the run will be the same but that both the offset and the travel will be longer. The angle of the fitting determines the angle of the break. Thus, the angle of the rolling offset will always be a known quantity. If the run or offset can be measured directly with a rule, determining the length of travel simply uses the computation described previously. If the set or roll are given or measured, then the solution to figure the run requires the use of squaring and square root to first compute the length of the offset. That is,

$$\text{Offset} = \sqrt{\text{set}^2 + \text{roll}^2} \qquad (3\text{–}10)$$

after which the length of travel (which is the third side of the triangle) is figured with right angle trigonometry in the usual manner. The key to the solution is a clear understanding of the problem. That is, in which direction does the pipe actually travel? An example will make this clear.

Figure 3-10

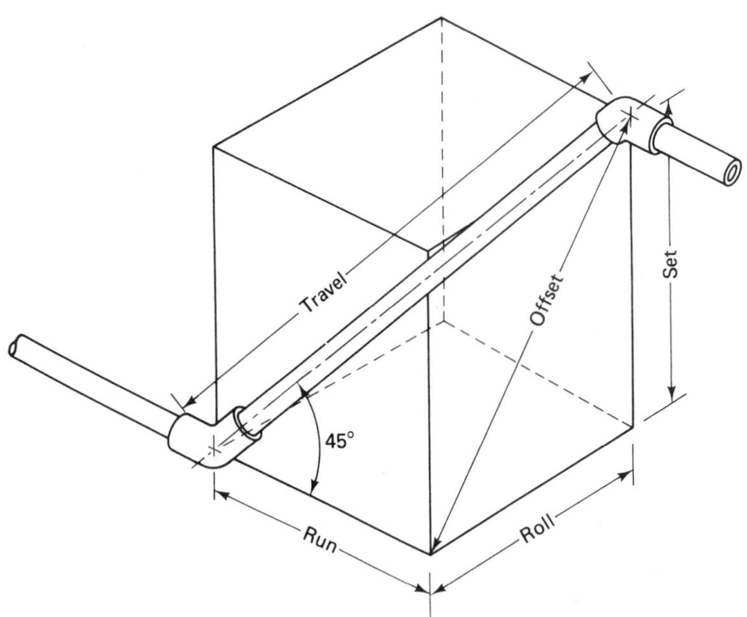

Forty-five-Degree Rolling Offset

Example 3-5 A 45-degree rolling offset is to have a roll of 36 in. and a set of 24 in. Compute the offset, run, and travel distance (center to center).

Solution

Notice there are three unknowns: the *offset,* which is computed from the set and roll, the *run,* which is computed from the angle (45 degrees) and the offset, and the *travel,* which is computed from the angle and either the run or the offset. Referring to Fig. 3–10, and solving for the length of the offset, we have

$$\text{Offset} = \sqrt{\text{set}^2 + \text{roll}^2}$$

$$\text{Offset} = \sqrt{(24)^2 + (36)^2} = 43.3 \text{ in.}$$

Next, figuring the length of the run using right angle trigonometry, we obtain

$$\tan 45^\circ = \frac{\text{offset } (y)}{\text{run } (x)} \tag{3–11}$$

$$\text{Run} = \frac{\text{offset } (y)}{\tan 45^\circ} = \frac{43.3 \text{ in.}}{1.000} = 43.3 \text{ in.}$$

Notice that although the right angle trigonometry was used to compute the run, knowing that the fitting makes an angle of 45 deg with the run, and knowing the properties of a 45-deg triangle ($x = 1, y = 1$, and $r = \sqrt{2}$) allows for the solution of the run by inspection. That is, the run = offset in a 45-deg triangle. This would not be the case, however, if the fitting were other than 45 deg.

Finally, solving for the travel

$$\text{Travel} = 1.414 \times \text{run} = 1.414 \times 43.3 \text{ in.} = 61.2 \text{ in.}$$

3-7 English and SI Metric System of Dimensional Units

The plumber is a craftsman in the construction industry, where the base units of measurement are length, mass, and time. A base unit is one that cannot be subdivided into other base units. Base units are used to compute derived units. For example, the base unit of length can be used to compute the derived units, area and volume. Other derived units used in the plumbing include weight, velocity, and acceleration.

In the English system of measurement, the units of length are the yard, the foot, the inch, and fractional parts of an inch. In the SI metric system, the base unit of length is the meter. The centimeter (one-hundredth part of a meter) and millimeter (one-thousandth part of a meter) are also used frequently to describe length. A meter is approximately 39.4 in. (Fig. 3–11).

The base unit used here for mass in the English system of measurement is the slug. In SI units, the base unit for mass is the kilogram. Mass is the

Figure 3-11

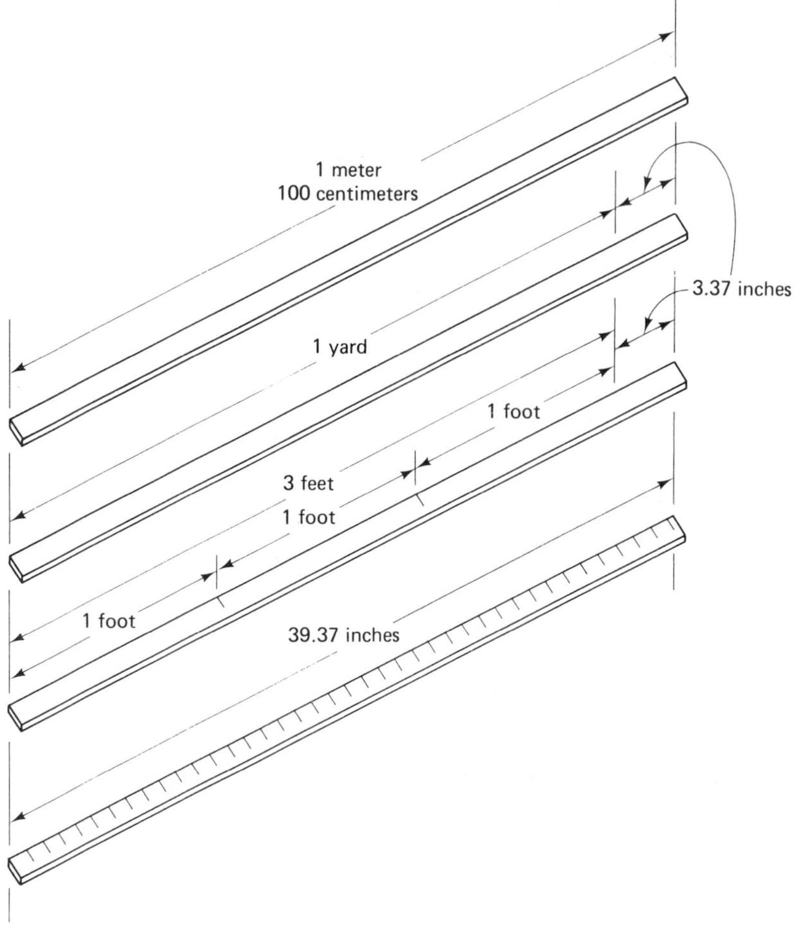

Units of Length

amount of a substance. It is not its weight. Rather, it refers to the molecular amount of a substance. Under ordinary circumstances, the mass of a substance never changes so long as its volume remains the same. Mass is defined by a standard prototype, the international kilogram maintained by the International Bureau of Weights and Measures (BIPM) in Paris, France, under the conditions specified in 1889. A duplicate in the National Bureau of Standards, U.S. prototype kilogram 20, serves as the mass standard for the United States. This is the only base unit still defined by an artifact (Fig. 3-12).

The unit of time in both the English and SI system of units is the second. Originally, the second was defined as a fraction of the solar day, but the definition was changed because measurements have shown that due to the irregularities in the rotation of the earth, the mean solar day

Figure 3-12

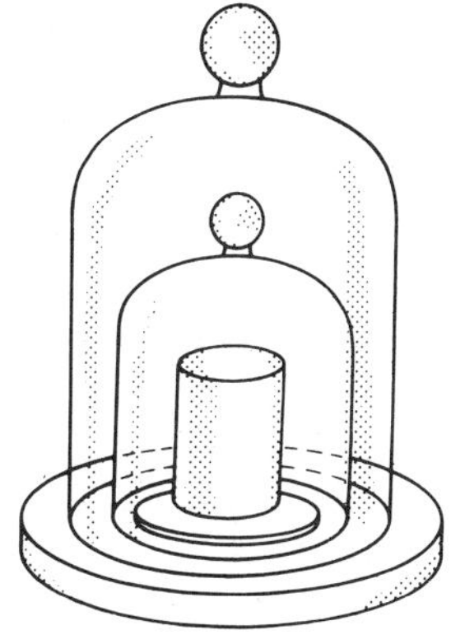

Unit of mass – kilogram

Unit of Mass—Kilogram

does not guarantee the desired accuracy for scientific work. Since 1967 the second has been defined in radiation units.

The velocity v of an object is equal to the distance traveled l divided by the time t. That is,

$$v = \frac{l}{t} \tag{3–12}$$

In the English system of units, velocity is measured in miles/hour, ft/s and in./s. In the SI metric system of units, velocity is measured in kilometers/hour (km/hr) and m/s. In the plumbing industry, the concern is typically one of how much time must be allocated to complete a certain task, such as getting to a job site, roughing in an installation, performing operations to assemble pipe, or setting a fixture. Where time t is the desired quantity, Eq. (3–12) becomes

$$t = \frac{l}{v}$$

That is, the time necessary to complete a task equals the distance divided by the velocity.

Example 3-6 Typically, a truck can average 20 mi/hr in light traffic. If two trips will be required to complete a job at the work site eight miles away, how many hours will be lost in travel?

Solution

Solving Eq. (3–12) for t, we obtain

$$t = \frac{}{v} = \frac{(2\text{ trips})\,(16\text{ mi})}{(20\text{ mi/hr})} = 1.6\text{ trip-hours}$$

Notice that the units are always included to ensure that the solution can be checked for correctness.

The weight of an object results from the gravitational pull of the earth acting on a given mass. Weight is derived in force units of pounds (lbf) in the English system of units, and in newtons (N) in the SI system of units. Whereas the force applied to an object may be in any direction, the force resulting from gravity is downward in the vertical plane. The weight of a body is affected by changes in gravity. As gravity increases, the weight of the body increases, and, conversely, as gravity decreases, the weight of the body decreases.

A body accelerates when it speeds up or slows down. When it maintains a steady speed, that is, a constant velocity, its acceleration is zero. By definition, force in pounds is that force which will cause an object with a mass of 1 slug to fall with an acceleration of 32.2 feet per second per second (ft/s^2). That is, the pound force is a gravitational unit, defined by the acceleration of a free-falling body. The newton force, a metric unit, is defined as that force which, when applied to a body having a mass of one kilogram, will cause it to accelerate (in any direction) at one meter per second per second (m/s^2). It is independent of the force of gravity acting on the body, that is, weight, even though both are measured in force units. This is a basic difference in the English and SI metric systems. That is, the English system is a gravitational system deriving mass from force and the acceleration due to gravity, whereas in the SI metric system, the prototype kilogram is used to define weight as well as acceleration. The acceleration constant due to gravity in the SI system is 9.806 m/s^2 (Fig. 3–13).

Weight (measured in force units), mass, and acceleration of a body are related by Newton's second law of motion. Simply stated, the law says that

Figure 3-13

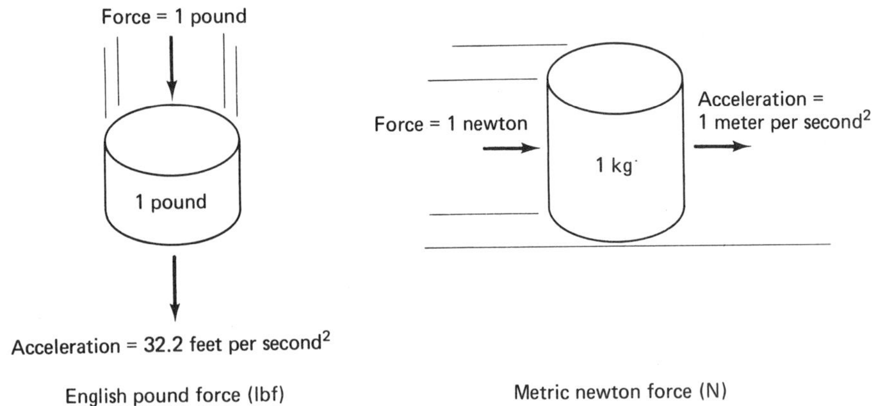

Units of Force

if a body of mass M is acted upon by a force F, the acceleration a is equal to the value of the force divided by the mass. In notation, where the force acting upon the body is due to gravity g, and the force is the weight w of the object,

$$\text{Gravitational constant } g = \frac{\text{weight } w}{\text{mass } M}$$

or simply

$$g = \frac{w}{M} \tag{3-13}$$

This basic formula defines the relationships among the weight, mass, and gravitational constant acting on an object. Because the gravity constant in either English or SI units will be known, it allows a convenient method of determining the weight or mass of an object. This becomes increasingly important when the mass of an object is given, for example, in kilograms, and the weight of the object is desired in pounds force. Table 3–1 gives the comparison of English and SI metric systems.

Table 3-1 **Comparison of English and SI Metric Systems**

System	*Weight*	*Mass*	*Acceleration*
English	pound	slug	feet per second squared
SI metric	newton	kilogram	meters per second squared

Example 3-7 A load of pipe with a mass of 10,000 kg must be transported by truck. What is the English unit weight equivalent?

Solution

Solving Eq. (3–13) for w, we have

$$w = Mg = (10{,}000 \text{ kg})\,(9.8 \text{ m/s}^2) = 98\ 000 \text{ N}$$

Knowing that 1 lbf = 4.445 N, we obtain

$$w = \frac{98\ 000 \text{ N}}{4.445 \text{ N/1bf}} = 22\ 047 \text{ lbf}$$

Table 3–2 summarizes the base and derived units in the English and SI metric system that are commonly encountered in plumbing.

Table 3-2 **Base Units and Derived Units with English and SI Metric Symbols**

BASE UNITS

Quantity	*Name*	English Symbol	*Name*	*SI Symbol*
length	foot	ft	meter	m
	inch	in.	centimeter	cm
mass	slug	slug	millimeter	mm
time	second	s	kilogram	kg
			second	s

DERIVED UNITS

Quantity	*Name*	English Symbol	*Name*	*SI Symbol*
area	square foot	ft^2	square meter	m^2
volume	cubic foot	ft^3	cubic meter	m^3
speed	miles per hour	mi/hr	kilometers per hour	km/hr
velocity	feet per second	ft/s	meters per second	m/s
acceleration	feet per second per second	ft/s^2	meters per second per second	m/s^2
force	pound	lbf	newton	N
weight	pound	lbf	newton	N

3-8 Summary of Practice

The plumber uses numbers and mathematics to measure, to figure dimensions at the work site, and to make financial transactions. Numbers are usually accompanied by the units they represent. In the English system of units, linear measurements and computations are made in inches and feet. In the SI metric system, the meter, centimeter, and millimeter are the units of linear measurement. Financial transactions are made in dollars or other local currency.

The numbers that the plumber uses may be described as whole numbers or fractions. Fractions are subdivided into common fractions and decimal fractions. Proper fractions are common fractions with a value less than one, whereas improper fractions are common fractions that have a value greater than one. Decimal fractions incorporate a decimal point and decimal arithmetic to describe parts of whole numbers.

Operations with numbers are indicated by the sign that accompanies them. The most difficult of these is square root, which is used to figure the length of one side of a right triangle when the other two are given. Right angle trigonometry describes the relationships of the angles and sides of right triangles and simplifies this process. By itself, right angle trigonometry can be used to figure the angles and side lengths of pipe offsets, center to center. In combination with the Pythagorean theorem, which uses square root in its solution, rolling offsets can be solved for the length of travel, run, and offset. Familiarity with the properties of right triangles is necessary to solve offset problems, that is, the trigonometric relationships, and for familiar angles, the relative lengths of the adjacent side, opposite side, and hypotenuse. The most common angle in offset pipe work is 45 degrees.

REVIEW QUESTIONS AND PROBLEMS

1. How does a plumber use mathematics?

2. What are "hard" and "soft" metric conversions?

3. Represent $\frac{7}{8}$ with four other equivalent fractions.

4. Perform the following operations:

a. $8\frac{3}{4} + 2\frac{5}{8}$
b. $8\frac{3}{4} - 2\frac{5}{8}$
c. $8\frac{3}{4} \times 2\frac{5}{8}$
d. $8\frac{3}{4} \div 2\frac{5}{8}$

5. Construct a table that converts the 16 fractional parts of an inch to decimals.

6. Construct a table that converts the 16 fractional parts of an inch to millimeters. (Note. 1 in. = 25.4 mm.)

7. Round the following numbers to five, four, and three digits.

a. 654 321
b. 176 542
c. 123 456
d. 213 782
e. 479 264

8. Convert the two improper fractions in Problem 4 to decimals and then multiply them together. Compare the answer with that which was computed in 4c.

9. Discount each of the following prices 40 percent and then add 5 percent tax to their total.

$ 8.63
9.75
5.43
87.56
156.78
27.43
18.57

10. Compute the square root of the following to two decimal places by hand calculation.

a. 291
b. 327
c. 575
d. 175.26
e. 1541

11. A simple 45-deg pipe offset breaks 12 in. (30.48 cm) to clear an obstruction. Figure the length of the run and travel, C–C.

12. A rolling offset that uses a 45-deg. elbow has a set of 18 in. (45.7 cm) and a roll of 24 in. (61 cm). Figure the lengths of the offset, run, and travel, C–C.

13. If a rolling offset breaks 36 in. (91 cm) at 45 deg and rolls 45 deg, figure the lengths of the run, set, roll, and travel.

14. At $33/hr, how much does it cost to have a truck make four round trips from the shop to a work site located 12 mi away, if the average speed is estimated at 22 mi/hr?

15. Describe the difference between the *mass* and *weight* of an object.

REFERENCES

[1]See also *ASTM Metric Practice Guide Handbook* 102, U.S. Department of Commerce, National Bureau of Standards; and ASTM Recommended Practice E 29 for Designating Significant Places in Specified Limiting Values, *1966 Book of ASTM Standards,* Part 30, p. 13.

4 Pipes and Fittings

4-1 Introduction

Much of the plumber's work involves assembling pipe and fittings. In addition to cutting, joining, and supporting pipe, the plumber must have information and experience to be familiar with the many materials used for pipes and fittings. These include the properties, sizes, appropriate uses, and restrictions imposed by plumbing codes.

Pipes and fittings are made from many materials, including clay, wood,[1] concrete, lead, iron, copper, brass, steel, plastic, bituminous fibers, asbestos, and borosilicate glass.

Plumbing systems are piped with both *pipe* and *tube*. Pipes are different from tubes in the way they are sized. The nominal or designated sizes used for pipe lines and piping systems correspond to the inside diameter of the pipes for sizes to 12 in., and to the outside diameter for sizes above 12 in. Below 12 in., the inside diameter *approximately equals* the nominal size. For example, regular thickness pipe designated as 1 in. has an inside diameter of 1.049 in. Above 12 in., the designated sizes *equal* the outside diameter. For example a 14-in. pipe has an outside diameter of 14 in. Tubing sizes, on the other hand, are always sized from the outside diameter. Copper water tube sizes, for example, have an outside diameter that is always $\frac{1}{8}$ in. larger than the designated size, regardless of the size of the tube or its wall thickness.

Pipe and tubing are classified by their composition and properties (for example, whether the product is made from cast iron, steel, copper, or plastic) and by its uses (for example, below or above ground and for drain, waste, vent, or potable water). Plumbing codes are consulted to determine where pipes and tubes of various materials and with different properties can be used.

Size and flow pressure determine flow capacity. The properties of the materials are used to determine wall thicknesses that will withstand expected pressure and deliver the necessary flow rate. The pressure ratings for pipe are either written on the pipe itself or taken from tables. Pressure ratings are given as the actual pressure in lbf/in.^{2} (kg/cm^2 or Pa), derived from a schedule of pressures from the wall thickness, for example, by

Figure 4-1

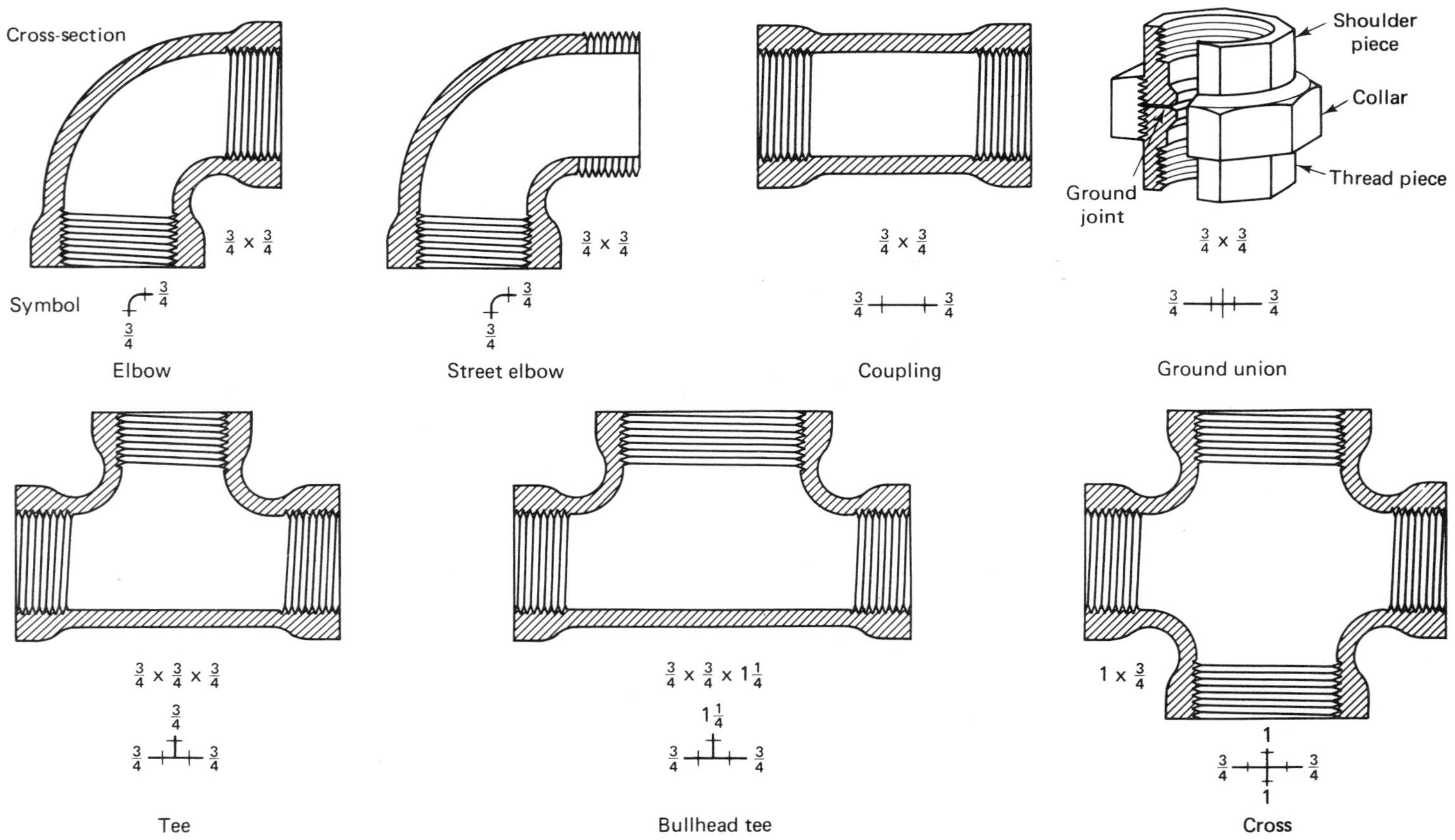

How Fittings Are Read

Schedule 40 (regular thickness) or Schedule 80 (extra strong). Or they are derived from the Standard Dimension Ratio (SDR), which, unlike Schedule piping, has a uniform pressure rating regardless of the diameter size.

Fittings of all types are read by the diameter size of the pipe connecting them. Through fittings, such as couplings and elbows, are read starting with the larger diameter, followed by the smaller diameter. In branch fittings, such as tees and Y's, the run is given first, followed by the branch connection. In a cross-tee, the run is read first, starting with the largest opening, followed by the cross connection. Y-connections are read starting with the largest opening and giving the run size first, followed by the size of the Y run. Outside threads are cited as *male,* whereas inside threads are cited as *female.* Figure 4–1 illustrates how several fittings are dimensioned and read.

4-2 Vitreous Clay Pipe and Fittings

Clay pipe is used for underground sanitary and storm sewers for municipalities, for drainage of streets, for culverts and conduits, for building sewers, for meter boxes, and on the farm for irrigation and a multitude of projects.

Clay pipe is one of the oldest materials used, tracing its history back 5000 years to the Babylonians, Cretans, and Romans. Although manufacturing methods have improved, the clay itself is essentially the same in structure. Clay is a chemically inert residue left by nature from pulverized rocks and soil. The raw material is mixed and shaped in molds under pressures of approximately 125 lbf/in.^{2} (862 kPa), and after drying is burned in kilns at temperatures of approximately 2000°F (1093°C). Firing fuses the material in a bond called vitrification, thus the term *vitrified.* In addition to being inert, vitrified clay is strong and nearly chemical-proof. This is particularly important for chemically active sewage, industrial wastes, and other liquids that form corroding acids. The only exception to this is hydrofluoric acid, which attacks clay pipe. Clay pipe is joined by using compression-type gaskets, which ensure leak- and root-proof sewer installations. Because it is rigid, no special side support is necessary in the trench. As with all pipe used for gravity drains and waste lines below ground, the bottom of the trench must be firm to keep the slope uniform after backfilling.

Clay pipe is available in sizes to 42 in. (1.07 m), and its use is restricted to gravity systems—as opposed to pressure systems. Plumbing codes restrict the use of clay pipe to sewers and drains below ground outside the building.

Extra heavy and standard strength vitrified clay pipe conforming to ASTM Specifications C–700[2] are available in 2-ft and 3-ft joint lengths in more than a dozen diameter sizes from 4 in. to 42 in. So too are a number of common and special fittings. Several of these are shown in Fig. 4–2.

Figure 4-2

Elbow

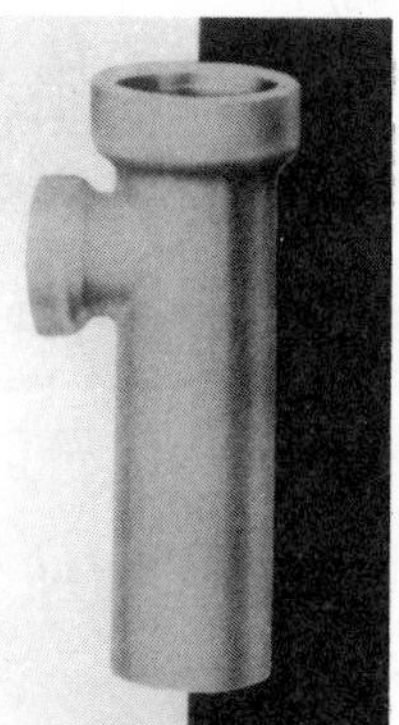

T-Branch

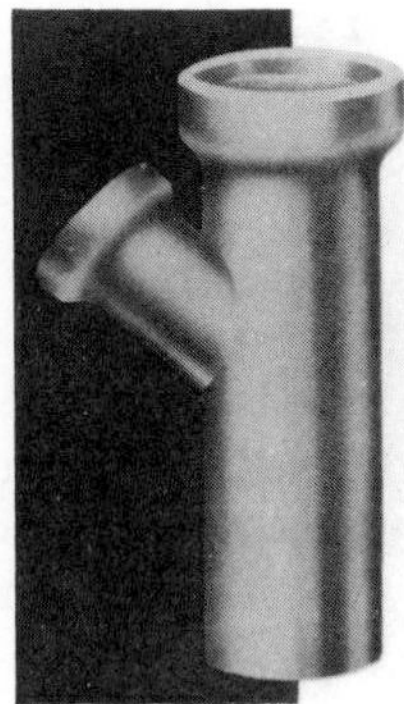

Y-Branch

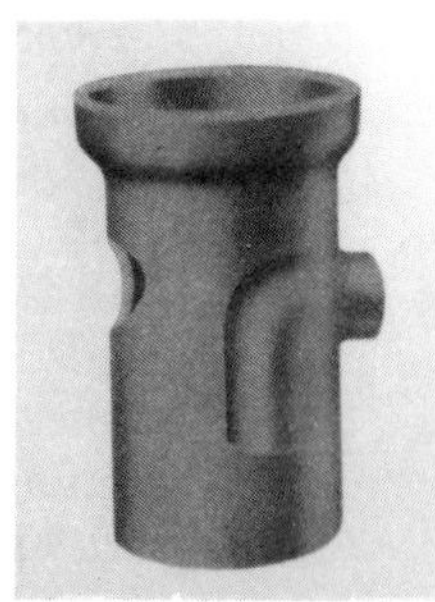

Grease Trap

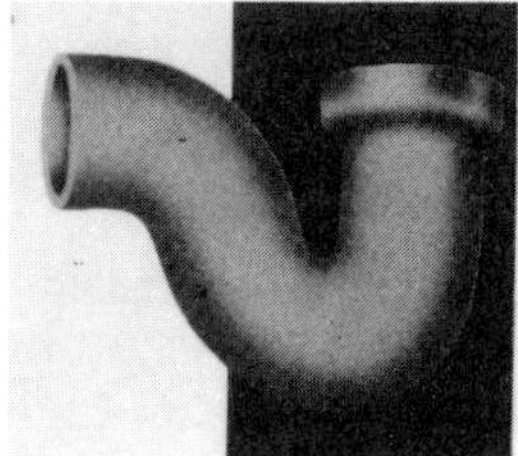

P-Trap

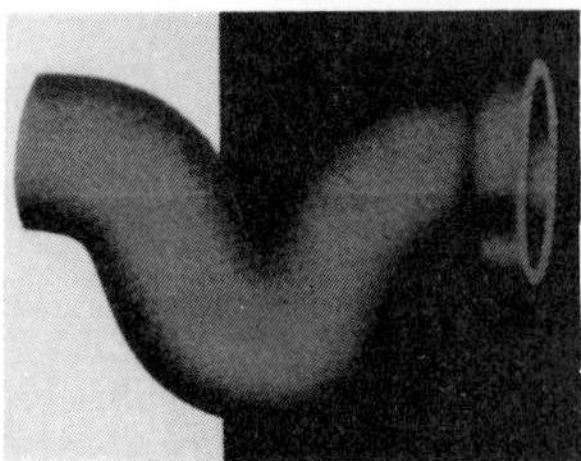

Running Trap

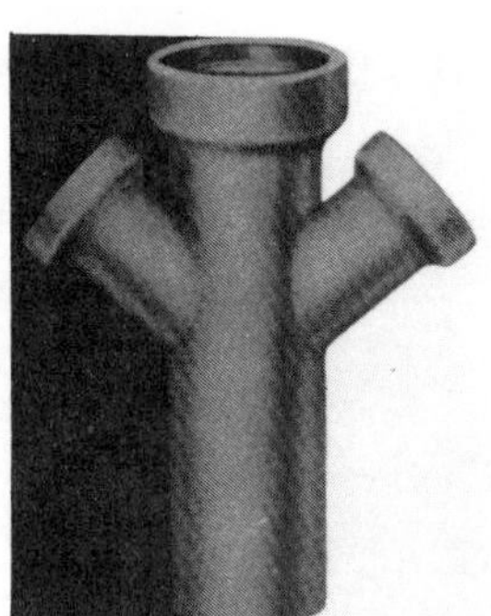

Double X-Branch

Vitreous Clay Pipe Fittings (National Clay Pipe Institute)

Strength and resistance are determined from ASTM Standard C–301[3]; the performance of compression joints for clay pipe that are made from ring sealing elements is determined from ASTM C 425.[4] Installation procedures and recommended practices for installing vitrified clay pipe lines are covered in ASTM Standard C 12.[5]

4-3 Cast Iron Soil Pipe Fittings

Cast iron pipe and fittings are available in bell and spigot design and in no-hub design to route drain, waste, and vent systems. Fittings are available for attachment to threaded pipe and standard plumbing fixtures.

Bell and spigot cast iron pipe[6] is joined by inserting the plain end of the pipe in the bell end of another pipe or fitting, and making the joint mechanically solid and leakproof, either by caulking or by using a compres-

Figure 4-3

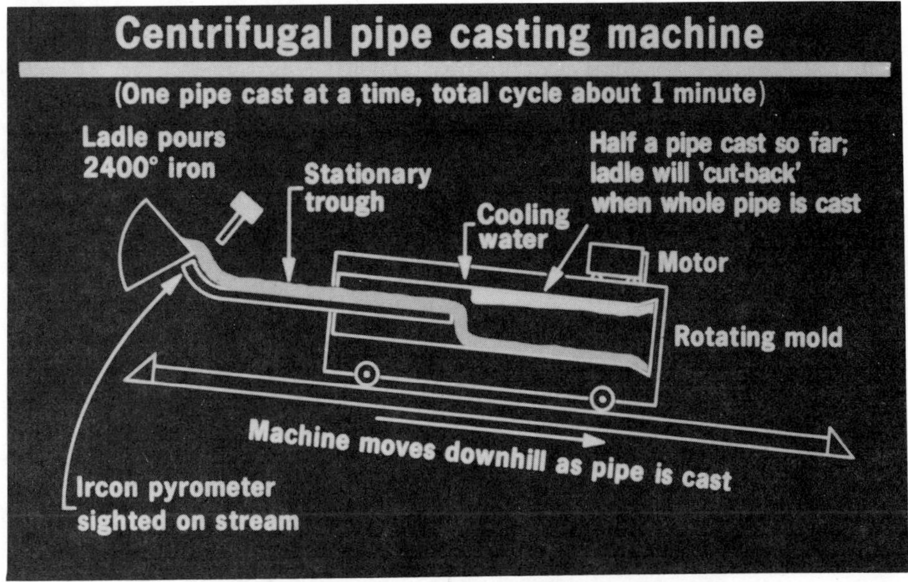

Centrifugal Cast Iron Pipe Casting Process
(United States Pipe and Foundry)

sion gasket. No-hub cast iron pipe[7] is assembled by inserting both ends of two pipes or fitting ends into a rubber sleeve, which is then held in place by a stainless steel band and clamps.

Cast iron for pipe is one of the oldest materials used both above and below ground for vents, drains, soil pipes, and sewers. It is strong, durable, and relatively inexpensive, and it provides the quietest system available inside buildings for drains and waste. For practical purposes cast iron pipe will outlast the building in which it is installed.

Modern manufacturing methods include centrifugal casting to provide a dense and uniform pipe with close tolerances. Centrifugal casting is a process whereby molten iron is poured from a stationary trough into a rotating mold. The process is illustrated in Fig. 4-3. The dimensional accuracy and improvement in the density of the pipe permit the wall thickness to be reduced, with a subsequent reduction in weight, which has been considered to be a drawback in cast iron pipe. Fittings are produced in sand cast and metal casting molds.

Hub and spigot cast iron soil pipe is available in standard barrel lengths of 5 ft and 10 ft, in diameters from 2 in. to 15 in. Double-hub pipe, which has a hub on each end to reduce waste when one is fabricating a system, is available in 5-ft and 30-in. lengths. Two weights of hub and spigot pipe are available, extra heavy (XH) and service weight (SV). *Extra* heavy and service weight pipe and fittings are *not* interchangeable, because

Figure 4-4

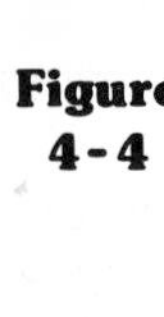
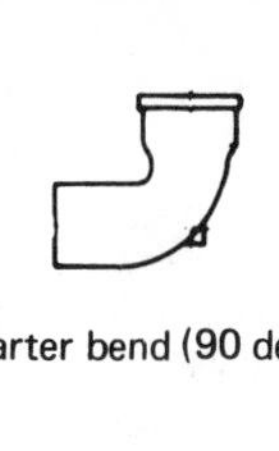

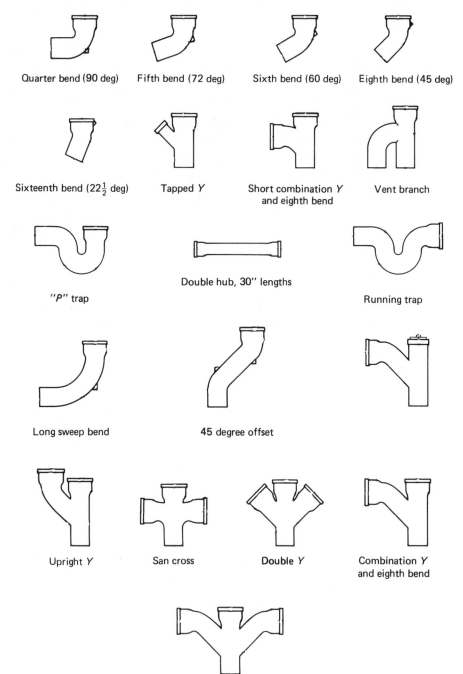

Quarter bend (90 deg) Fifth bend (72 deg) Sixth bend (60 deg) Eighth bend (45 deg)

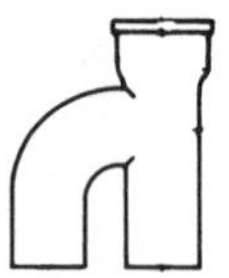

Sixteenth bend ($22\frac{1}{2}$ deg) Tapped *Y* Short combination *Y* and eighth bend Vent branch

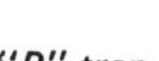
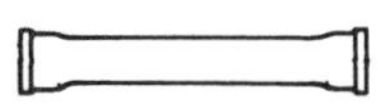

"*P*" trap Double hub, 30" lengths Running trap

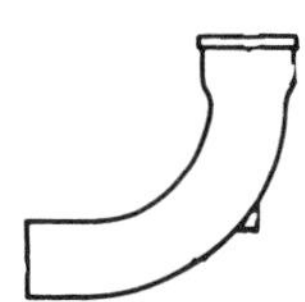
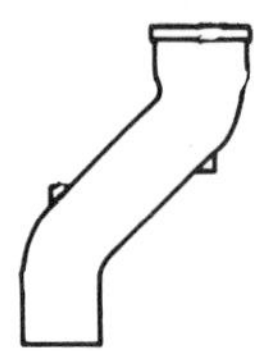
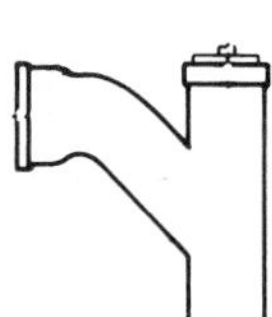

Long sweep bend 45 degree offset

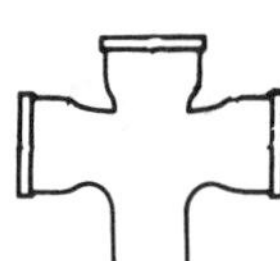

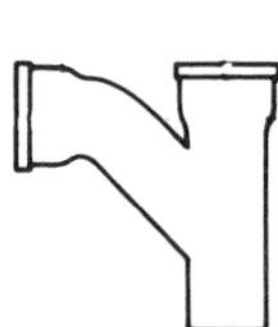

Upright *Y* San cross Double *Y* Combination *Y* and eighth bend

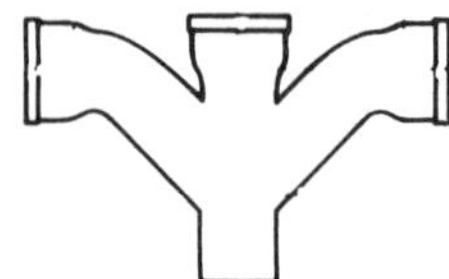

Double combination *Y* and eighth bend

Bell and Spigot Cast Iron Soil Pipe Fittings (Tyler Pipe Company)

Table 4-1

Dimensional Details for Service Weight Hub Cast Iron Soil Pipe (Tyler Pipe Company)

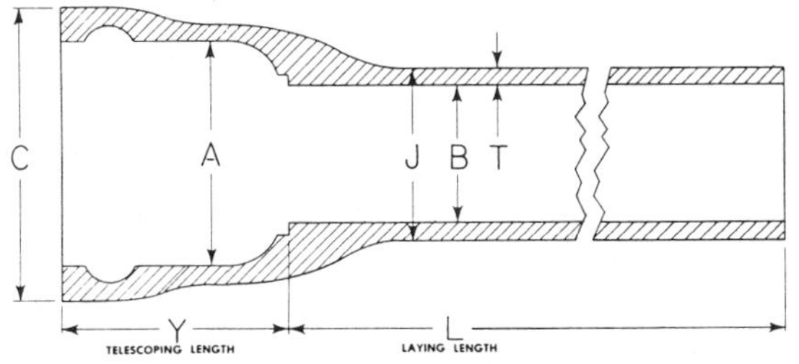

	Telescoping Length	*Hub I.D.*	*Hub O.D.*	*Barrel O.D.*	*Barrel I.D.*	*Nominal Thickness*	
Size	*Y*	*A*	*C*	*J*	*B*	*T*	
2″	2.50	2.94	3.62	2.30	1.96	0.17	
3″	2.75	3.94	4.68	3.30	2.96	0.17	
4″	3.00	4.94	5.68	4.30	3.94	0.18	
5″	3.00	5.94	6.68	5.30	4.94	0.18	+.06
6″	3.00	6.94	7.68	6.30	5.94	0.18	−.03
8″	3.50	9.25	10.13	8.38	7.94	0.23	
10″	3.50	11.38	12.44	10.50	9.94	0.28	
12″	4.25	13.50	14.56	12.50	11.94	0.28	+.09 −.06
15″	4.25	16.75	17.91	15.62	15.00	0.31	

5-FT. AND 10-FT. SINGLE AND DOUBLE HUB

Size	*Laying Length (L)* 5-Ft. Single Hub	*Laying Length (L)* 5-Ft. Double Hub	*Laying Length (L)* 10-Ft. Single Hub	*Shipping Weight* 5-Ft. Single Hub	*Shipping Weight* 5-Ft. Double Hub	*Shipping Weight* 10-Ft. Single Hub
2″	5′0″	4′9½″	10′0″	20	21	38
3″	5′0″	4′9¼″	10′0″	30	31	56
4″	5′0″	4′9″	10′0″	25	54	75
5″	5′0″	4′9″	10′0″	52	54	98
6″	5′0″	4′9″	10′0″	65	68	124
8″	5′0″	4′8½″	10′0″	100	105	185
10″	5′0″	4′8½″	10′0″	145	150	270
12″	5′0″	4′7¾″	10′0″	190	200	355
15″	5′0″	4′7¾″	10′0″	255	270	475

the outside diameter of extra heavy pipe is larger. Extra heavy cast iron soil weight pipe now accounts for less than five percent of the sales of cast iron soil pipe used in plumbing, and is no longer required by major plumbing codes. The dimensional details for service weight hub and spigot pipe are given in Table 4-1. The more common fittings for hub and spigot cast iron soil pipe are shown in Fig. 4-4.

The hand-caulked oakum fiber and molten lead joint is leakproof, strong, and rootproof. This is due to the waterproofing characteristic of the oakum and the mechanical strength of the lead, which is poured into the joint over the oakum and then rammed into the joint with a caulking tool and hammer to form a solid metal insert. The compression joint uses a neoprene rubber gasket instead of oakum and lead and can be deflected as much as 5 degrees after assembly without leakage or failure.

No-hub cast iron soil pipe and fittings are assembled by inserting both ends of the pipe and fittings into a rubber sleeve, which is then held in place by a stainless steel band and clamps tightened to 60 lb-in. of torque. The weights of service weight bell-hub and spigot cast iron soil pipe and no-hub cast iron soil pipe are approximately the same. No-hub pipe and fittings have a spigot bead and gasket positioning lug. The positioning lug is optional on pipe but is required on fittings. The rubber gasket is also available as a rubber reducing sleeve, which replaces an extra fitting when reducing from 2-in.-diameter to $1\frac{1}{2}$-in.-diameter pipe.

Ductile iron pipe[8,9], made of cast iron that has undergone a molecular change by a process introduced during the molten stage, has increased strengths, ductility, and impact resistance. Figure 4–5 illustrates the cast iron molecule and the ductile iron molecule, which exhibits superior strength because of the manner in which spheroidal graphite molecules are formed. Ductile iron pipe is used to transport raw and treated water, natural gas, domestic sewage, industrial chemicals and wastes, and many other liquid, gaseous, and even solid materials. To date, there are over 35,000 miles of ductile iron pipe installed in corrosive environments in sizes 3 in. through 54 in. in the United States with service histories of over 20 years. Thickness compliance is checked against ANSI Standard A 21.50.[10]

4-4 Steel Pipe and Iron Fittings

Steel pipe and wrought iron butt-welded and seamless pipe are used throughout the plumbing industry for water lines, drainage, soil, waste, and vent pipes, in sizes from $\frac{1}{8}$-in. to 6-in. (15.2-cm) diameter. It is available plain (black iron pipe) or zinc coated (galvanized).

Figure 4-5

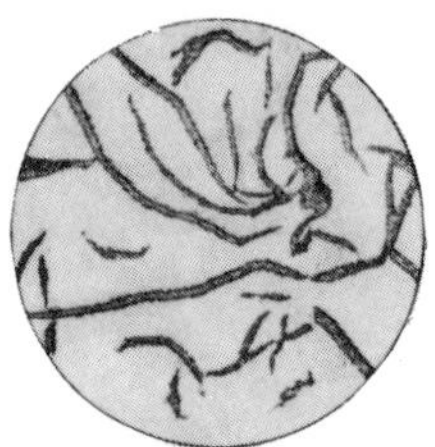

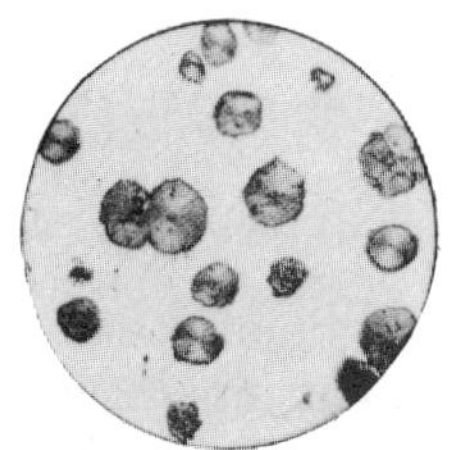

Cast Iron and Ductile Iron Molecular Structure (United States Pipe and Foundry)

Butt-welded steel pipe in standard 21-ft (6.4-m) lengths is made from a flat strip, which is made tubular as it is passed through forming rolls and is then resistance-welded at the seam. Seamless pipe is made by piercing a red-hot solid round billet of steel with a mandrel as the pipe is pulled through a rolling mill. Seamless pipe is available in lengths from 16 ft to 22 ft.

In sizes to 12-in. diameter, the nominal pipe sizes are used that conform approximately to the inside diameter of the pipe. Above 12-in. diameter, sizes conform to the actual outside diameter of the pipe. The outside diameter of all pipes of the same nominal size is the same, and variations in wall thickness between standard wall pipe (Schedule 40) and extra strong wall pipe (Schedule 80) are reflected in changes in the inside diameter of the pipe and wall thickness. The dimensions of nominal pipe sizes to 6-in. diameter are shown in Table 4–2 for Schedule 40 and Schedule 80 pipe. Most of the threaded pipe installed by plumbers is butt-welded Schedule 40 steel pipe.

Galvanized and black iron pipe in 21-ft lengths (joints) comes threaded on both ends with a coupling. This allows assembly of one joint with the next. Threaded pieces of pipe in short lengths are referred to as *nipples.*

Table 4-2 **Dimensions of Schedule 40 (Standard Weight) and Schedule 80 (Extra Heavy) Steel Pipe**

	Dimensions of Schedule 40 (Standard Weight) Steel and Wrought Iron Pipe			*Dimensions of Schedule 80 (Extra Strong) Steel and Wrought Iron Pipe*		
Nominal Diameter, Inches	*Actual Inside Diameter, Inches*	*Actual Outside Diameter, Inches*	*Weight per Foot, Pounds*	*Actual Inside Diameter, Inches*	*Actual Outside Diameter, Inches*	*Weight per Foot, Pounds*
$\frac{1}{8}$	0.269	0.405	0.25	0.215	0.405	0.31
$\frac{1}{4}$	0.364	0.540	0.43	0.302	0.540	0.54
$\frac{3}{8}$	0.493	0.675	0.57	0.423	0.675	0.74
$\frac{1}{2}$	0.622	0.840	0.86	0.546	0.840	1.09
$\frac{3}{4}$	0.824	1.050	1.14	0.742	1.050	1.47
1	1.049	1.315	1.68	0.957	1.315	2.17
$1\frac{1}{4}$	1.380	1.660	2.28	1.278	1.660	3.00
$1\frac{1}{2}$	1.610	1.900	2.72	1.500	1.900	3.63
2	2.067	2.375	3.66	1.939	2.375	5.02
$2\frac{1}{2}$	2.469	2.875	5.80	2.323	2.875	7.66
3	3.068	3.500	7.58	2.900	3.500	10.25
$3\frac{1}{2}$	3.548	4.000	9.11	3.364	4.000	12.50
4	4.026	4.500	10.80	3.826	4.500	14.98
5	5.047	5.563	14.70	4.813	5.563	20.78
6	6.065	6.625	19.00	5.761	6.625	28.57

Figure 4-6

90° Short Turn Elbow

90° Long Turn Double TYs

Straight Coupling

11 1/4° Elbow

45° Short Turn Elbow

60° Elbow

P-Trap with Cleanout

45° Double Branch

Straight Tee

90°Street Elbow

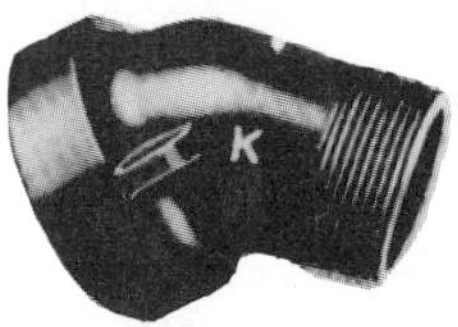

45° Street Elbow

22 1/2° Elbow

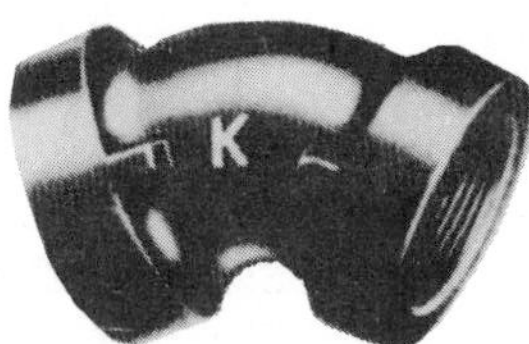

45° Long Turn Elbow

90° Long Turn Elbow

Cast Iron Recessed Fittings
(Kuhns Brothers Co., subsidiary of Nibco, Inc.)

They are available in close (all thread) or in $\frac{1}{2}$-in. increments to 6 in., and in 1-in. increments from 6-in. to 12-in. lengths. Cut and threaded lengths of pipe are available in 6-in. increments from 12 in. to 60 in. (1.5 m). Nipples and short lengths reduce assembly time and are particularly appealing to the home owner/builder who may not have access to cutting and threading tools.

Ferrous pipe fittings are cast iron, malleable iron, and ductile iron, and are available plain, painted black, or galvanized. Standard cast iron fittings[11] are made in sand molds and furnished either black or tar coated for use in vent piping. They are the least expensive of the cast fittings. Recessed cast iron plain, painted black, or tar coated fittings, also referred to as drainage fittings,[12] are constructed to provide a smooth waterway inside when the pipe is threaded into them. They are used for sanitary waste and storm drainage piping (Fig. 4–6). Short pattern ell's, tee's, and long sweep tee-Y's are also constructed at a slight pitch of $\frac{1}{4}$-in./ft to build in the correct drainage pitch for installed horizontal pipe. Cast iron recessed fittings are available in sizes from $1\frac{1}{4}$ in. to 12 in. Malleable iron fittings are used for water distribution. Malleable iron is cast iron that has undergone an annealing process (controlled cooling for 72 hours) after casting. This results in a change in the molecular structure and grain of the iron which makes it tougher and slightly elastic. Malleable galvanized iron fittings for plumbing are available in sizes to 6-in. diameter for piping water. Ductile iron fittings are made from a cast-ferrous material that solidifies with free carbon in a spheroidal graphite form. Ductile iron fittings are used in systems subject to shock loading, mechanical loading, or hydraulic loading caused by such conditions as a rapidly closing valve. Automatic fire protection sprinkler systems, for example, use ductile iron fittings.

Ductile iron differs from steel, which has no free graphite present. It also differs from cupola malleable iron, which has no free graphite present as cast—rather, graphite forms in agglomerates during the malleabilizing anneal. Ductile iron differs from gray iron, where free graphite is present in flake form. The presence of free graphite in a spheroidal form in ductile iron provides a material that has foundry and machining characteristics similar to gray iron, but with physical characteristics similar to steel. Table 4–3 lists the physical properties of four ferrous pipe fitting materials, indicating the superior strength and resiliance of ductile iron over malleable iron and cast iron.

Mechanical couplings hold grooved pipe sections and fittings together with a mechanical locking clamp. The seal is made by a coupling gasket, which is seated firmly around the circumference of the pipe by the clamp. Mechanical pipe coupling systems receive extensive use in commercial buildings and industrial plants, particularly in the transport of water, fire protection sprinkler systems, and industrial process systems.

In construction, the lock joint housing sections engage the complete circumference of the pipe grooves. Figure 4–7 illustrates the complete joint

Table 4-3

Physical Properties of Pipe Fitting Metals

	Tensile Strength lbf/in²	*Yield Strength lbf/in²*	*Elongation Percent*
Ductile Iron (A.S.T.M. A395)	60 000	40 000	18%
Cupola Malleable Iron (A.S.T.M. A197)	40 000	30 000	5%
Carbon Steel (A.S.T.M. A216 WCB)	70 000	36 000	22%
Cast Iron (A.S.T.M. A126)	21 000	21 000	0%

Figure 4-7

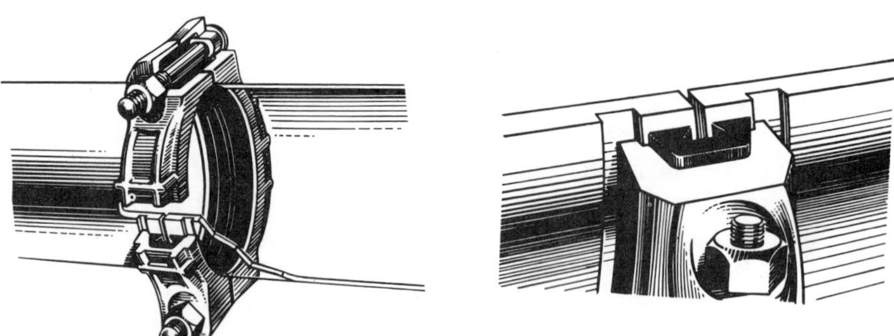

Grooved Mechanical Coupling Construction (Aerogroup Corporation)

and a cross section view of the clamp and sealing gasket position. Also notice that the joint allows for both expansion and flexing. Grooved couplings are available in sizes from $\frac{3}{4}$-in. diameter to 24-in. diameter, and for pressure rating from 300 lbf/in.^{2} to 2500 lbf/in.^{2} (2068 kPa to 17.24 MPa). Gasket compositions are available to be compatible with the variety of fluids transported, including neoprene, Buna-N, and Viton (Dupont) compounds.

4-5 Copper Tubing and Fittings

Copper tubing for plumbing is available in types K, L, M, and DWV (drain, waste, and vent) in drawn temper (hard) lengths to 20 ft, and annealed temper (soft) coils (except DWV) to 100 ft. Commercially available lengths are shown in Table 4-4.

Table 4-4

Dimensions and Weights For Copper Tubing

Nominal Size (in.)	Outside Diameter	Type K Inside Diameter	Type K Wall Thickness	Type K Weight/Foot	Type L Inside Diameter	Type L Wall Thickness	Type L Weight/Foot	Type M Inside Diameter	Type M Wall Thickness	Type M Weight/Foot	Type DWV Inside Diameter	Type DWV Wall Thickness	Type DWV Weight/Foot
$\frac{1}{4}$	0.375	0.305	0.035	0.145	0.315	0.030	0.126	—	—	—	—	—	—
$\frac{3}{8}$	0.500	0.402	0.049	0.269	0.430	0.035	0.198	0.450	0.025	0.145	—	—	—
$\frac{1}{2}$	0.625	0.527	0.049	0.344	0.545	0.040	0.285	0.569	0.028	0.204	—	—	—
$\frac{5}{8}$	0.750	0.652	0.049	0.418	0.666	0.042	0.362	—	—	—	—	—	—
$\frac{3}{4}$	0.875	0.745	0.065	0.641	0.785	0.045	0.455	0.811	0.032	0.328	—	—	—
1	1.125	0.995	0.065	0.839	1.025	0.050	0.655	1.055	0.035	0.465	—	—	—
$1\frac{1}{4}$	1.375	1.245	0.065	1.04	1.265	0.055	0.884	1.291	0.042	0.682	1.295	0.040	0.65
$1\frac{1}{2}$	1.625	1.481	0.072	1.36	1.505	0.060	1.14	1.527	0.049	0.940	1.541	0.042	0.81
2	2.125	1.959	0.083	2.06	1.985	0.070	1.75	2.009	0.058	1.46	2.041	0.042	1.07
$2\frac{1}{2}$	2.625	2.435	0.095	2.93	2.465	0.080	2.48	2.495	0.065	2.03	—	—	—
3	3.125	2.907	0.109	4.00	2.945	0.90	3.33	2.981	0.072	2.68	3.030	0.045	1.69
$3\frac{1}{2}$	3.625	3.385	0.120	5.12	3.425	0.100	4.29	3.459	0.083	3.58	—	—	—
4	4.125	3.857	0.134	6.51	3.905	0.110	5.38	3.935	0.095	4.66	4.009	0.058	2.87
5	5.125	4.805	0.160	9.67	4.875	0.125	7.61	4.907	0.109	6.66	4.981	0.072	4.43
6	6.125	5.741	0.192	13.9	5.845	0.140	10.2	5.881	0.122	8.92	5.959	0.083	6.10
8	8.125	7.583	0.271	25.9	7.725	0.200	19.3	7.785	0.170	16.5	7.907	0.109	10.6
10	10.125	9.449	0.338	40.3	9.625	0.250	30.1	9.701	0.212	25.6	—	—	—
12	12.125	11.315	0.405	57.8	11.565	0.280	40.4	11.617	0.254	36.7	—	—	—

Table 4-4 **Dimensions and Weights for (Metric) Copper Tubing (Cont'd)**

Iron Pipe In. Equivalent	*Copper MM*	*Outside Diameter*	*Inside Diameter*	*Wall Thickness*	*Kilograms/ Meter*
$\frac{1}{8}$	8	8.045	6.765	1.28	0.126
$\frac{1}{4}$	10	10.045	8.765	1.28	0.159
$\frac{3}{8}$	12	12.045	10.765	1.28	0.193
$\frac{1}{2}$	15	15.045	13.565	1.48	0.282
$\frac{3}{4}$	18	18.045	16.365	1.68	0.387
$\frac{3}{4}$	22	22.055	20.175	1.88	0.535
1	28	28.055	26.175	1.88	0.686
$1\frac{1}{4}$	35	35.070	32.590	2.48	1.142
$1\frac{1}{2}$	42	42.070	39.590	2.48	1.377
2	54	54.070	51.590	2.48	1.782
$2\frac{1}{2}$	66	66.730	63.580	3.15	2.748
3	79	79.430	76.280	3.15	3.283
4	104	104.833	101.580	3.25	4.353

Copper tube is made from 99.90 percent copper. Types K, L, and M are manufactured according to ASTM Standard B 88; and Type DWV is manufactured to ASTM B-306.[13] Hard temper copper tube is color marked as follows:

Green—Type K
Blue—Type L
Red—Type M
Yellow—Type DWV

Soft copper tube is not color marked, but cartons and shipping tags are so marked. Both hard tubing and soft tubing bear the names of the manufacturer and place of origin.

The actual size of the outside diameter of copper tube is always $\frac{1}{8}$ in. (0.125 in. or 3.175 mm) larger than the nominal size, which refers to the inside diameter. Put another way, the nominal size of copper tube is always $\frac{1}{8}$ in. smaller than the outside diameter. For example, $\frac{3}{8}$-in.-diameter tubing has an outside diameter of $\frac{1}{2}$ in.; $\frac{1}{2}$-in.-diameter tubing has an outside diameter of $\frac{5}{8}$ in., and so on. Types K, L, M, and DWV have different wall thicknesses, with type K being the thickest and type DWV being the thinnest. The dimensions for Types K, L, M, and DWV copper tube are shown in Table 4-4.

Copper tube is used for pressure piping in water distribution and heating systems, as well as for drain, waste, and vent systems (type DWV). It is joined above ground by soldering or brazing the tube to drawn or cast fittings and valves. Below ground, flared fittings are used to connect the water service. It is corrosion resistant except for some acids and fumes (for example, the ammonia fumes from urinal drain pipes), lightweight, readily bent, easy to join, and extremely dependable over a long service life.

The type of copper tube used depends upon the particular application. For underground water services, use type M for straight lengths and type L soft temper where coils are more convenient. Water distribution systems use type M for above and below ground. Drainage and vent systems use type DWV above and below ground for waste, soil, and vent lines, roof drainage, building drains, and building sewers. For radiant panel and hydronic heating, and for snow melting systems, use type L soft temper where the coils are formed in place or prefabricated. Type M should be used where straight lengths are joined with fittings. Hot water heating and low pressure steam systems use Type M for sizes up to $1\frac{1}{4}$ in., and type DWV for sizes of $1\frac{1}{4}$ in. and larger. Type L is commonly used for condensate return lines. In each case local and regional codes should be consulted for requirements.

Copper tube is most often joined above ground by soldering with capillary fittings, although brazing is preferred where high pressure and temperature conditions are encountered—for example, in the construction of solar collector panels. Mechanical joints that use flared fittings are used underground in places where heat is impractical or where the joint is to be disconnected (broken) from time to time.

Copper tube solder joint pressure fittings[14] are either cast bronze or wrought copper. Pressure fittings are illustrated in Fig. 4–8. Casting and sand marks, or forming tool marks, indicate the method by which the fittings have been manufactured. Cast bronze fittings are made from copper alloy red brass which consists of 81 percent copper, 7 percent lead, 3 percent tin, and 5 percent zinc per ASTM Specification B584. Wrought copper fittings are made from copper tube and red bronze mill products as per ASTM Specification B75 for alloys 120 and 122. Cast and wrought copper fittings are joined in the same manner, and costs are comparable. Preference and availability usually determine which type is used, although cast fittings are slightly less dependable because of pin holes that sometimes result from the casting process. Solder (and brazing) joint fittings are available in all standard tube sizes from $\frac{1}{8}$-in. diameter to 12-in. diameter. Wrought fittings are preferred for brazing.

Flared-tube fittings are shown in Fig. 4–9. They are available in sizes from $\frac{3}{8}$-in. diameter to 3-in. diameter. Flared joints provide a metal-to-metal contact similar to a ground union joint after they have been tightened once. They are used extensively with type K and L soft copper below ground to bring the water service into the building.

DWV fittings are used for drain, waste, and vent connections. Several typical fittings are shown in Fig. 4–10. They are characterized by a sweeping design which promotes the flow of water and waste, and are available in sizes from $1\frac{1}{4}$-in. diameter to 8-in. diameter. Assembly is by soldered capillary joint.

Figure 4-8

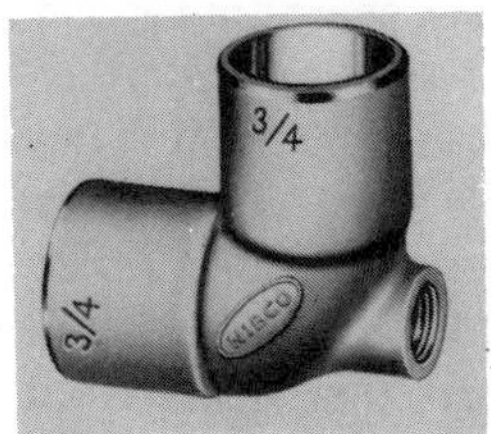

Baseboard Tee (C × F × C)

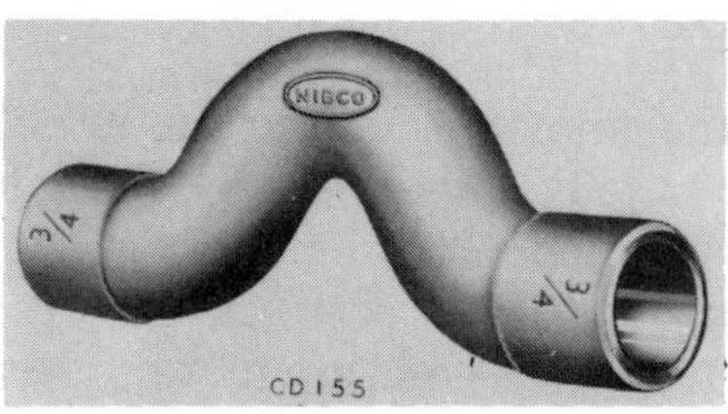

Cross-over

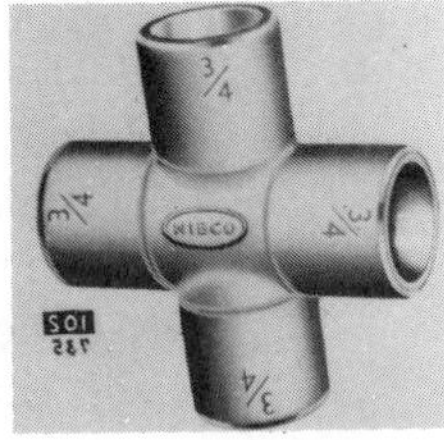

Cross

(Wrought) Tee

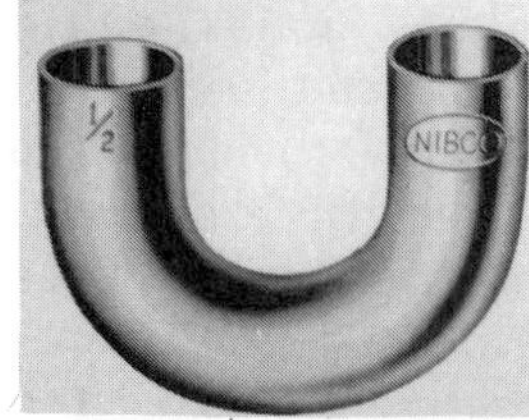

Return Bend

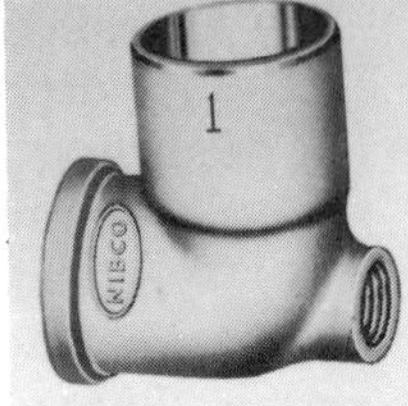

Baseboard Tee (F × F × C)

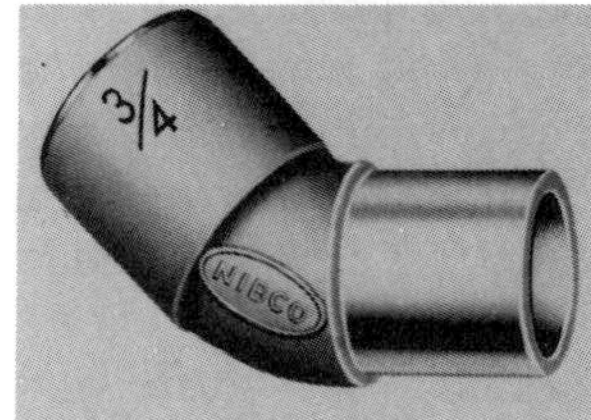

45° Fitting Ell

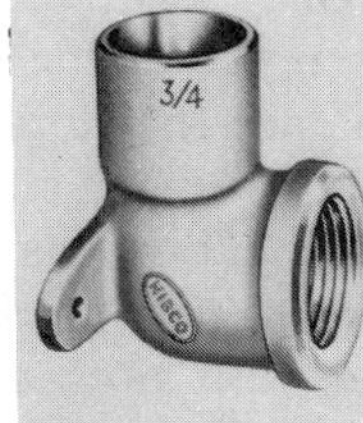

90° Drop Ell

Reducer

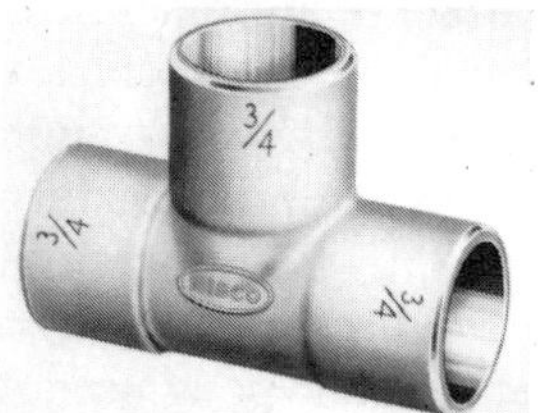

Tee

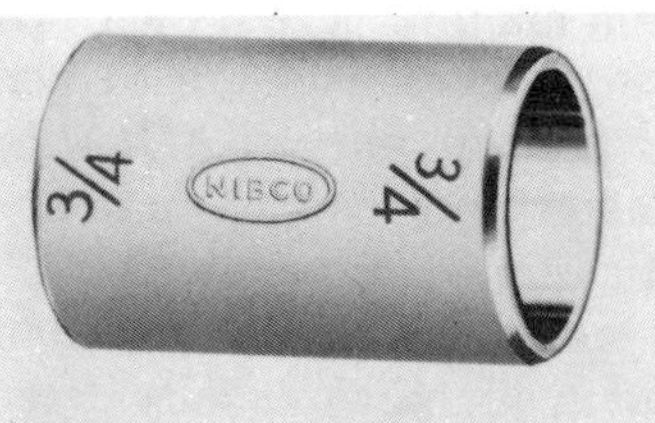

Coupling

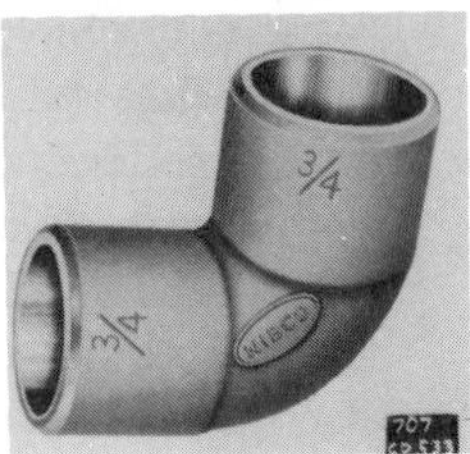

90° Ell

Tube Cap

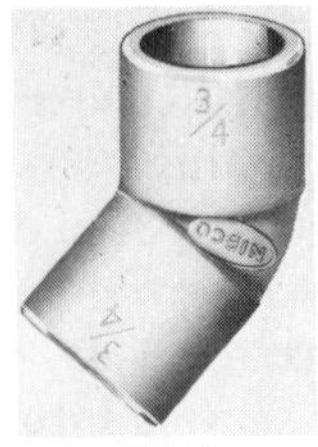

45° Ell

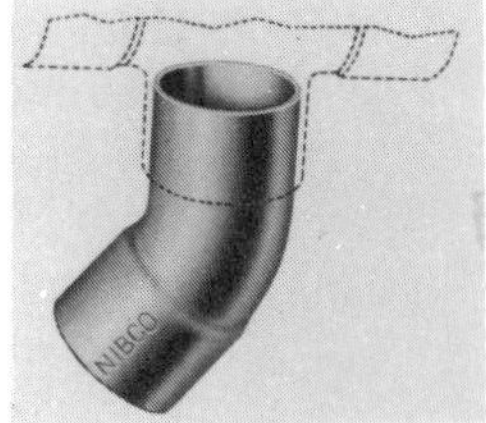

45° Fitting Ell

Copper Tube Pressure Fittings (Nibco)

Figure 4-9

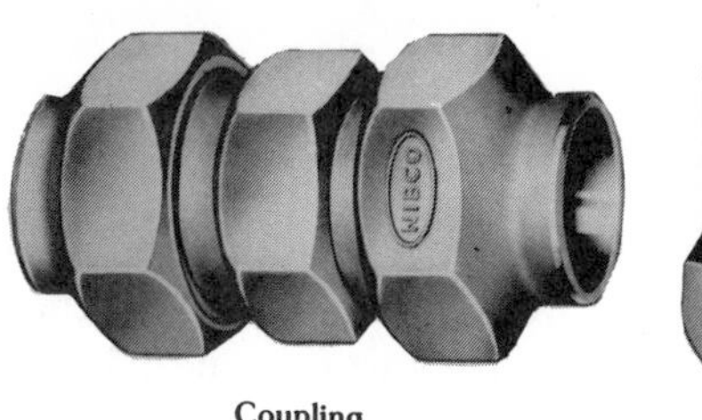

Coupling

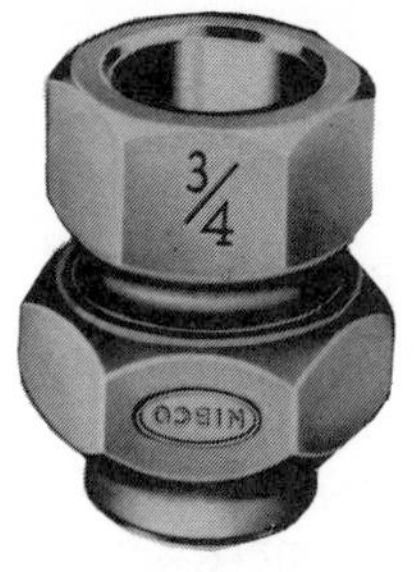

Two Part Coupling

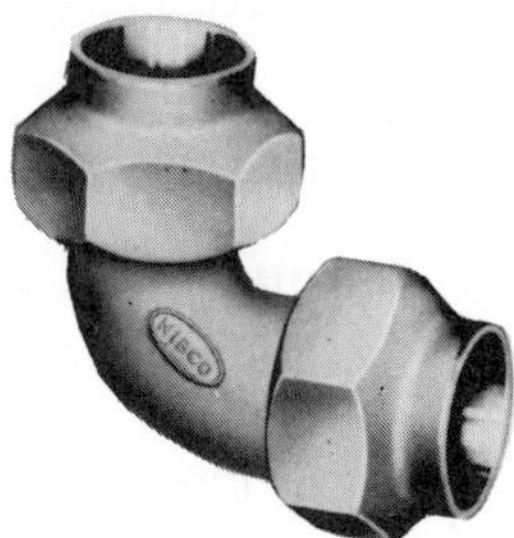

90° ell

Flare Tube Fittings (Nibco)

Figure 4-10

45° Elbow

90° Elbow

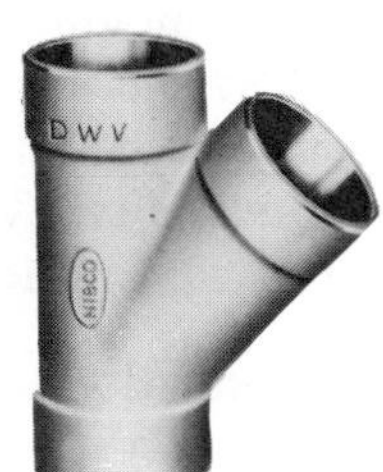

Wye

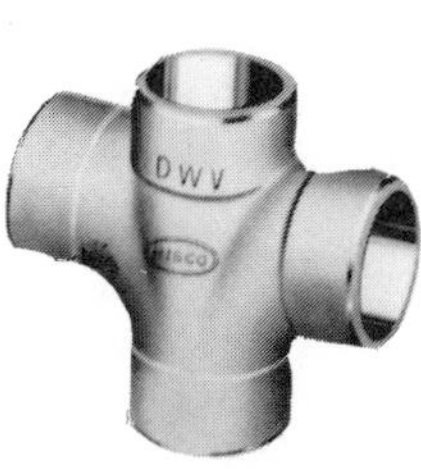

Double Tee

Test Tee

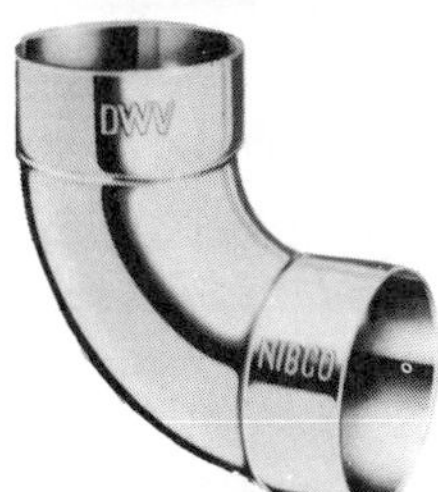

(Wrought) 90° Elbow

(Wrought) Wye

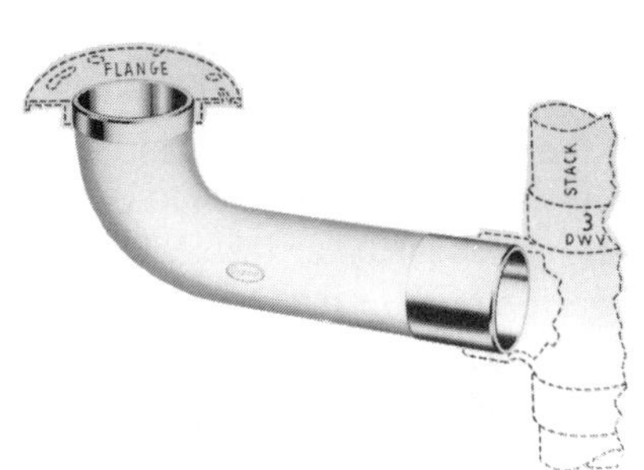

Closet Elbow

P-Trap

DWV Tee with 90° Left and Right Inlets

Wrought Copper and Bronze Drain, Waste and Vent Fittings (Nibco)

4-6 Plastic Pipe and Fittings

Plastic pipe and fittings have been available since the 1960's for plumbing drain, waste, and vent systems, as well as for pressure pipe and water distribution systems. They are also used for carrying industrial fluids such as gases, chemicals, and solutions that are corrosive to most metal pipes.

Advantages of plastic pipe include lower installed cost than most metal pipes, ease of installation, light weight, desirable flow characteristics, insulating qualities, and corrosion resistance. Modern plastics also have high impact strength, and good operating strength at temperatures up to 180°F (82.2°C) and 100 lbf/in.^{2} (690 kPa). High thermal expansion and reduction in strength at elevated temperatures are disadvantages.

Plastic pipes are extruded from thermoplastic plastic. The fittings are made by injection molding, a process that forces plastic liquid into cooled metal molds. Thermoplastic plastics soften and can be reshaped when heated, as opposed to thermosetting plastics, which cannot be remelted and reformed after initial molding.

Several plastic pipe and fitting compounds are available. These include:

ABS (Acrylonitrile-Butadiene-Styrene)—a rigid plastic used for drain waste and vent systems

PBC (Polyvinyl Chloride)—a rigid plastic used for drain, waste, and vent systems, as well as for pressure piping

PE (Polyethylene)—a flexible plastic and one of the oldest used for water and gas service as well as irrigation and chemical wastes

CPVC (Chlorinated Polyvinyl Chloride)—a rigid plastic related to PVC and formulated to withstand the higher temperatures encountered in domestic hot water and industrial applications

PB (Polybutylene)—a relative newcomer to the flexible plastics family; used for hot and cold water distribution, especially in retrofit (replumbing) applications.

Plastic in plumbing systems has been approved by model regional plumbing codes such as the Southern Building Code Congress (SBCC) Standard Plumbing Code. Many state codes now approve CPVC for plumbing domestic hot and cold water distribution systems. The FHA (Federal Housing Administration) approves CPVC in newly constructed single and double family units and rehabilitated structures of six stories or less. This fact notwithstanding, approvals in a specific area for plastic drain, waste, and vent systems, as well as for hot and cold water distribution systems, should be checked with local plumbing codes and ordinances before an installation is made.

ABS-DWV pipe and fittings for above and below ground have been available for over 25 years, and estimates to date place the number of drain, waste, and vent systems at 4 million units. Currently, ABS accounts for about 25 percent of the residential DWV market. This does not include the mobile home and recreational vehicle industry, which uses ABS-DWV systems almost exclusively. Figure 4-11 illustrates several common ABS-DWV fittings.

Figure 4-11

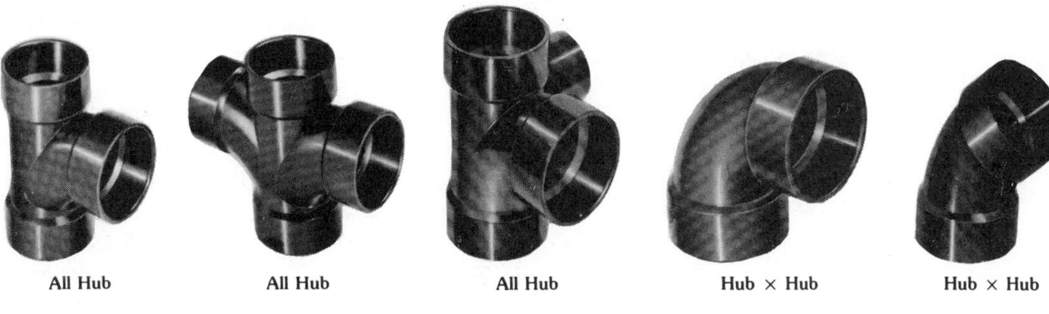

All Hub | All Hub | All Hub | Hub × Hub | Hub × Hub

Hub × Hub | Hub × Hub | All Hub | MPT | Hub

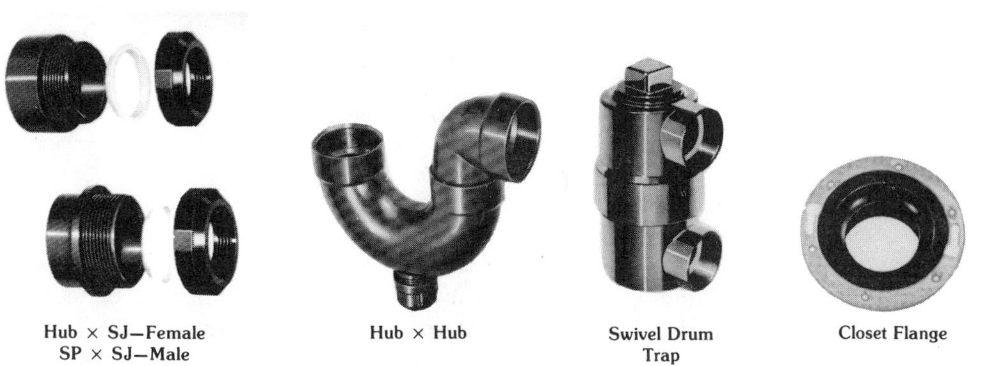

Hub × SJ—Female
SP × SJ—Male | Hub × Hub | Swivel Drum Trap | Closet Flange

ABS-DWV Fittings

ABS–DWV pipe and fittings are available in sizes $1\frac{1}{4}$ in. to 6 in. in 10- and 20-ft joints. Schedule 40 pipe and fittings are black and marked on both sides in indelible ink with the ASTM Standard 2661, company name and/or trademark, pipe size, and the symbol ABS–DWV. ABS pipe and fittings are solvent-welded with a solvent known as MEK (methyl-etyl-ketone). The solvent chemically etches the surface of both the pipe and fittings so that when they are joined the two surfaces are fused into each other. Properly made joints are stronger than either the pipe or the fittings. Depending upon weather and temperature conditions, setting time is two to five minutes, with three minutes being the average. Water tests may be applied after joints are fully set.

Table 4–5 gives the thermal expansion properties for ABS plastic pipe for *changes* in temperature of 40° to 100°F (approximately 20° to 50°C). For example, if the highest temperature expected is 110°F, and the lowest temperature is 60°F, the temperature *change* will be 50°F. For a length of run of 60 ft the chart indicates that the installation should provide 2.410 in. (about $2\frac{3}{8}$ in.) for linear expansion. For this reason, horizontal and vertical runs exceeding 30 ft (or at every floor in wet stacks) should have two restraining fittings (both sides of the wall, or above and below the floor) installed. Figure 4–12 illustrates the assembly of a restraining fitting. After the solvent is applied to the clean surface, the flange assembly is pressed together, down against the floor, and the clamp is tightened snugly. Another means to compensate for expansion and contraction is to use an expansion joint that consists of a premade piston and sleeve assembly installed in the stack or run.

Table 4-5 **Thermal Expansion Properties of ABS Plastic Pipe**

LENGTH (Feet)	*40°*	*50°*	*60°*	*70°*	*80°*	*90°*	*100°*
20	.536	.670	.804	.938	1.072	1.206	1.340
40	1.070	1.340	1.610	1.880	2.050	2.420	2.690
60	1.609	2.010	2.410	2.820	3.220	3.620	4.020
80	2.143	2.680	3.220	3.760	4.290	4.830	5.360
100	2.680	3.350	4.020	4.700	5.360	6.030	6.700

Chart shows length changes in inches per degrees temperature change.

PVC pipe is used above and below ground for water distribution, as well as for DWV installations. It is white. The fittings are white, gray, or black. Maximum service temperature is 140°F (60°C). PVC pipe and fittings are available in schedule 40 from $\frac{1}{2}$-in. to 8-in. diameters. Schedule 80 pipe and fittings are available in diameter sizes from $\frac{1}{4}$ in. to 12 in. Both are available in 10-ft and 20-ft lengths and can be solvent-welded or joined by heat fusion. The solvent-welded joint is the most widely used for pressure

Figure 4-12

Plastic Pipe Restraining Fitting (Tyler)

service and involves an initial etching and softening of the pipe with a purple primer followed by the application of the solvent cement. Schedule 80 PVC pipe can be joined by threaded connections. PVC–DWV and ABS–DWV pipe are both used for drain, waste, and vent systems, and both have pitched fittings. The one used depends upon local availability and price. It must be remembered that the two materials and solvents are *not* interchangeable, and one type or the other should be used throughout a system to avoid confusion.

Figure 4–13 illustrates several common PVC pressure fittings. It must be remembered that PVC pressure fittings, PVC–DWV fittings, and PVC–S/D thin wall sewer and drain fittings are not interchangeable, and where permitted and appropriate, adapters must be used to join the two.

Table 4–6 gives the thermal expansion properties for PVC plastic pipes for *changes* in temperature of 40° to 100°F (approximately 20° to 50°C). For example, if the highest temperature expected is 100°F, and the lowest is 50°F, the temperature difference will be 50°F, and each 20-ft run of pipe can be expected to change length by 0.348 in. (about $\frac{3}{8}$ in.).

PE pressure plastic pipe is black and is available in coils or rolls from 100 ft to several hundred feet, and in diameter sizes from $\frac{3}{4}$ in. to 2 in.

Figure 4-13

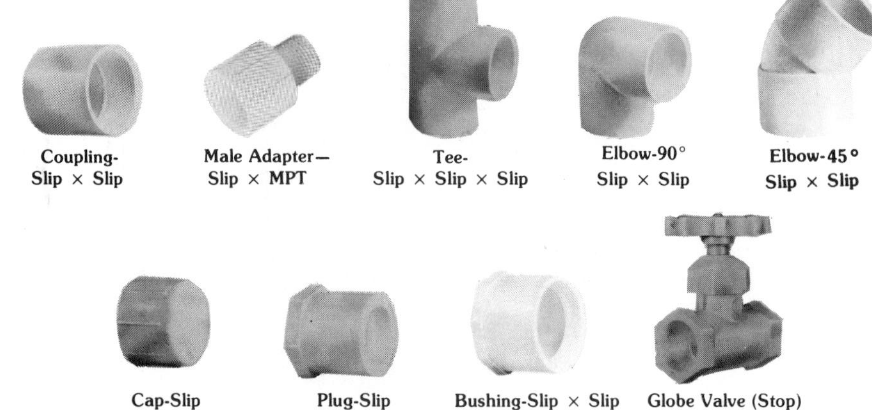

PVC (Polyvinyl Chloride) Pressure Fittings

Table 4-6

Thermal Expansion Properties of PVC Pipe

LENGTH (Feet)	*40°F*	*50°F*	*60°F*	*70°F*	*80°F*	*90°F*	*100°F*
20	.278	.348	.418	.487	.557	.626	.696
40	.557	.696	.835	.974	1.114	1.253	1.392
60	.835	1.044	1.253	1.462	1.670	1.879	2.088
80	1.134	1.392	1.670	1.949	2.227	2.506	2.784
100	1.392	1.740	2.088	2.436	2.784	3.132	3.480

Chart shows length changes in inches per degrees temperature change.

It is used both below and above ground for water distribution, including water service, piping, sprinkler systems, irrigation systems, and mobile home hookup. PE pipe is joined with insert fittings made of plastic, nylon, brass, and galvanized steel adapters. The fittings are inserted in the end of the square-cut pipe and clamped with a stainless steel hose clamp. PE pipe is available in several grades and densities ranging from low density to ultra-high molecular weight. This is an important consideration when PE pipe for a specific application is selected, in that so-called "bargain" PE pipe can be expected to be of low density, and care must be exercised to assure that the grade selected has both the pressure range and life expectancy required by the application.

Four-inch-diameter PE rigid drain pipe and fittings are available in 10 ft lengths, with or without drain holes, for use below ground as downspout drains, foundation footing drains, sewer to house connections, storm drains, house to septic tank connections, and land drainage. Polyethylene sewer and drain pipe is manufactured in compliance with ASTM D3350

Figure 4-14

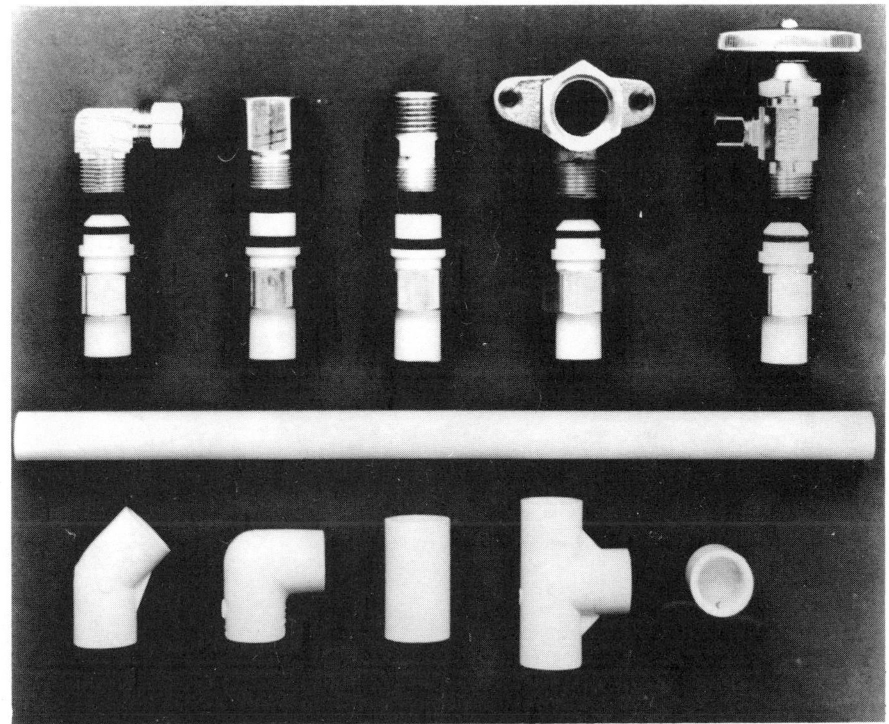

CPVC (Chlorinated Polyvinyl Chloride) Hot and Cold Pressure Fittings (B. F. Goodrich Chemical Company)

standards for materials, dimensions, workmanship, chemical resistance, and markings.

CPVC pipe was introduced in the United States in 1960 for use above ground for hot and cold water distribution. It is available in $\frac{1}{2}$-in. diameter and $\frac{3}{4}$-in. diameter copper tube sizes for use as supply lines for branches and risers to fixtures. It is not recommended for use under concrete slabs. Figure 4-14 illustrates several CPVC fittings.

CPVC will withstand water temperatures in excess of 180°F (82.2°C) at ordinary supply pressures and up to 100 lbf/in.^{2} without harmful effects. Its low thermal conductivity, which is only a fraction of that of copper tube, serves to reduce heat loss, thereby ensuring quick delivery of hot water in the system. Allowance, however, must be made for longitudinal expansion of the pipe itself, particularly in hot water pipes. For example, a 10-ft length of CPVC pipe will expand $\frac{1}{2}$ in. in length when the internal temperature is increased from 73°F to 180°F (22°C to 82.2°C), and allowance or expansion joints for this linear movement must be made when one is installing long horizontal or vertical runs. Joints should not butt against walls or obstructions that would limit movement, and strap

Figure 4-15

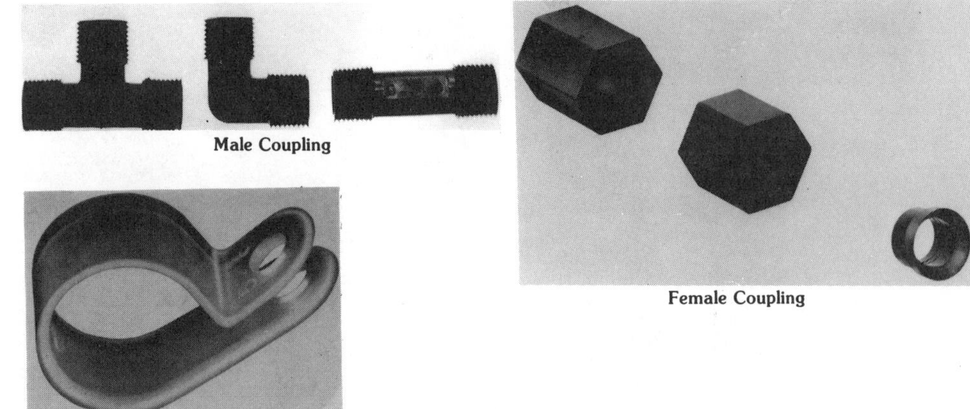
Male Coupling

Female Coupling

Tube Hanger

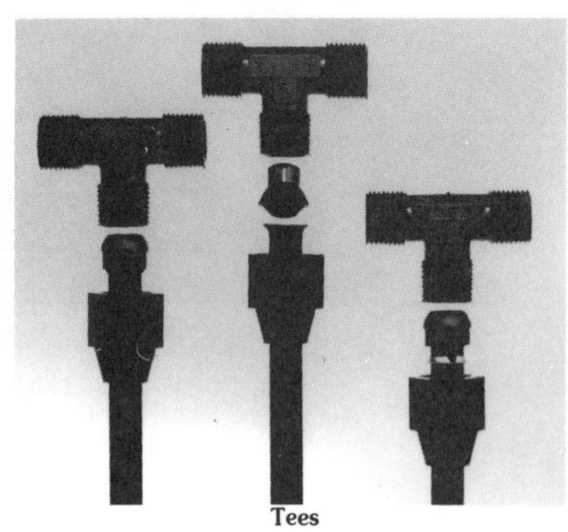
Drop Ells

Tees

PB (Polybutylene) Pressure Fittings

hangers that permit linear movement should be used to secure the pipe against studs and partitions. To prevent sagging, the pipe should be supported every 32 to 36 in. with strap hangers at least $\frac{3}{4}$ in. wide. The local plumbing code should also be consulted for specific requirements.

CPVC is connected by using a primer followed by a solvent cement. When CPVC is connected to fixtures, transition compression fittings are recommended. In the case of connecting the hot water heater, galvanized nipples of 6- to 12- in. length also should be used to separate the CPVC pipe from the water heater.

Polybutylene (PB) flexible water service tubing is available in diameter sizes from $\frac{3}{4}$ in. to 2 in. in coil lengths to 500 ft. It is used above and below ground for both hot and cold water service. It is manufactured in copper tubing sizes and joined by flare, crimp, and compression fittings, or by fusing. Other applications include well service tubing, well pipe, irrigation pipe, sprinkler systems, and snow melting pipe. A number of PB fittings are shown in Fig. 4–15.

As with other plastic pipes, the linear thermal expansion is high, approximately 0.85 in. (2.16 cm) for each 10°F (5.5°C) change in temperature for each 100 ft of pipe. For example, if a 40-ft run of polybutylene flexible pipe has an increase in temperature of 40°F, it could be expected to lenghten by about (0.85 in. × 40°/10° × 40 ft/100 ft = 1.36 in.). Because PB tubing is flexible, "snaking" during installation rather than expansion joints is used to account for changes in length due to thermal expansion. Normal installation would ordinarily leave enough flexibility to account for this. The working pressure for hot and cold water PB tubing under ASTM Standard D 3309 is 160 lbf/in.^{2} at 73.4°F (23°C) and 100 lbf/in.^{2} at 180°F (82.2°C). An additional advantage that PB tubing has over other plastic pipe is its ability to expand when frozen full of water and then to contract when thawed, leaving the pipe undamaged.

There are many standards which govern the manufacture, dimensioning, and testing of plastic pipe. Several of those applicable to plumbing are listed in Table 4–7. The standards are useful when one is specifying plastic pipe for a particular system and indicate what to be aware of when inspecting a piece of pipe or fitting with an identifying standard: what the material is, where it is appropriately used, and which joining techniques or solvents are applicable.

4-7 Summary

The most common piping materials used by plumbers for drain, waste, and vent systems above ground are cast iron, copper, and plastic compounds. Clay pipe is used extensively underground for sewers and storm drains. For piping hot- and cold-water distribution systems copper and plastic materials are the most common. Steel and iron pipe are used extensively for fire protection systems, and are connected by using threaded fittings, grooved couplings, or welded fittings made of cast, malleable, or ductile iron.

Pipe specifications cover both materials and dimensional specifications. Plumbing codes refer to these specifications often when listing sizes, approved uses, and installation practices. Standards organizations, such as the American Society for Testing and Materials, designate approval numbers and codes for specifications, and these are stenciled or stamped on the pipe or containers to inform suppliers and plumbers about the standards to which the materials and dimensions conform. Several of these specifications have been listed in this chapter and in other places throughout the text.

Table 4-7

Plastic Pipe Standards (Plastic Pipe Institute)

Material	*Type*	**Standards*	*Applications*	***Operating Temperature*	*Joining Methods*
ABS Acrylonitrile-Butadiene Styrene	Rigid	ASTM D1527-* (Sch. 40, 80 & 120 ASTM D2282-* (SDR–PR) ASTM D2661-* (DWV) ASTM D2680-* (Composite) ASTM D2750-* (Conduit) ASTM D2751-* (Sewer) ASTM D2465-* (Sch. 80 Thread. Fittgs.) ASTM D2468-* (Sch. 40 Socket Fittgs.) ASTM D2469-* Sch. 80 Socket Fittgs.)	Drain, Waste & Vent (DWV) Building Sewers & Sewer Mains Electrical & Communications Conduit Water Piping	100°F (Pressure) 180°F (Non-Pressure)	Solvent Cement Threading (Schedule 80 & 120) Transition Fittings
PE Polyethylene	Flexible	ASTM D2104-* (Sch. 40) ASTM D2239-* (SDR–PR) ASTM D2447-* (Sch. 40 & 80 O.D.) ASTM D2737-* (Tubing) ASTM D2609-* (Insert Fittgs.) ASTM D2610-* (Butt Fuse Fittg. Sch. 40) ASTM D2611-* (Butt Fuse Fittg. Sch. 80) ASTM D2683-* (Socket-Type Fusion Fittg.) ASTM D3035-* (SDR–PR O.D.) ASTM D3197-* (Insert-Type Fusion Fittg.)	Water Service & Water Mains Gas Service & Gas Mains Chemical Waste Irrigation Systems	100°F (Pressure) 180°F (Non-Pressure)	Insert Fittings Socket Fusion Butt Fusion Transition Fittings
PE Polyethylene	Rigid	ASTM D3350 (Drain Pipe & Fittings	Foundation Storm and Drains	(Non-Pressure)	Solvent Cement
PB Polyethylene		ASTM D2663-* (SDR–PR) ASTM D2666-@ (Tubing) ASTM D 3000-* (SDR–PR O.D.) ASTM D3000-* (System)	Water Service & Water Mains Gas Service & Gas Mains Irrigation Systems	180°F (Pressure) 200°F (Non-Pressure)	Insert Fittings Socket Fusion Butt Fusion Transition Fittings

PVC Polyvinyl Chloride	Rigid	ASTM D2666-* (Sch. 40, 80 & 120 ASTM D 2241-* (SDR–PR) ASTM D2665-* (DWV) ASTM D2672-* (Bell End) ASTM D2949-* (3-inch Thin Wall DWV) ASTM D3033-* (PSP Sewer) (PSP Sewer) ASTM D3034-* (PSM Sewer) ASTM D2464-* (Thread. Fittgs. Sch. 80) ASTM D2466-* (Socket Fittgs. Sch. 40) ASTM D2467-* (Socket Fittgs. Sch. 80) ASTM D2740-* (Tubing) ASTM D3036-* (Line Couplings) ASTM D2729-* (Sewer Pipe & Fittgs.)	Water Service & Water Mains Gas Service & Gas Mains Building Sewers & Sewer Mains Drain, Waste & Vent (DWV) Electrical & Communications Conduit Industrial Process Piping Irrigation Systems	100°F (Pressure) 180°F (Non-Pressure)	Solvent Cement Elastomeric Seal Threading (Schedule 80 & 120) Mechanical Couplings Transition Fittings
CPVC Clorinated Polyvinyl Chloride	Rigid	ASTM D2846-* (System)	Hot & Cold Water Distribution Chemical Process Piping	180°F @ 100 PSIG (SDR–11)	Solvent Cement Threading (Schedule 80 & 120) Mechanical Couplings Transition Fittings
PP Polypropylene	Rigid		Chemical Waste Chemical Process Piping	100°F (Pressure) 180°F (Non-Pressure)	Mechanical Coupling Socket Fusion Butt Fusion Threading (Schedule 80 & 120)
SR Styrene Rubber Plastic	Rigid	ASTM D2852-* (Drain Pipe & Fittgs.) ASTM D3298-* (Perforated)	Septic Tank Absorption Fields Sub-soil Dewatering Systems Storm Drains	150°F (Non-Pressure)	Solvent Cement Transition Fittings Elastomeric Seal

* ASTM Standards *only* are shown . . . The year of latest revision has been omitted. Refer to the current list of ASTM Standards as published by ASTM to determine the latest year of revision. For standards promulgated by ANSI, NBS and others, refer to PPI Technical Report TR–5. Plastics raw material standards are not shown here. Refer to TR–5.

** Pressure ratings for plastics piping is based on an operating temperature of 73°F. Higher temperatures require a de-rating of the pipe. Limitations for pressure pipe shown here are recommended maximums. Non-pressure piping limits may be exceeded if time of exposure is short.

Pipe to 12 in. in diameter is sized from the inside diameter, which is referred to as the *nominal size.* Twice the wall thickness is added to that dimension to determine the outside diameter. The thickness of the pipe wall determines its mechanical strength and pressure rating. Cast iron soil pipe designations refer to weight, such as SV (service weight) when describing wall thickness. Iron pipe sizes are given by a schedule number, with Schedule 40 being the most common. Copper tube is sized to have the outside diameter always $\frac{1}{8}$ in. greater than the nominal size. Variations in wall thickness are given by the letter designations K, L, M, and DWV; as the wall thickness increases, the inside diameter of the pipe decreases for any given nominal size, since the outside diameter remains the same. Plastic pressure tubing is made in copper tube sizes; that is, the outside diameter is the same. This permits fittings standardization within the industry and a relatively easy transition between the two. Plastic tubing wall thickness and pressure ratings are commonly given by wall thickness schedules, for example, Schedule 40 and Schedule 80. But there is a trend to designate wall thicknesses from a standard dimension ratio (SDR) such that the pressure rating is constant regardless of pipe size. The formula for the standard dimension ratio is

$$\text{SDR} = \frac{D}{t}$$

where the standard dimension ratio equals the average outside diameter D divided by the minimum wall thickness t. An SDR of 11, for example, is commonly designated for PB (polybutylene) flexible water service tubing.

The number and complexity of piping materials introduced to the marketplace have made it necessary for the tradesman to become familiar with all their designations, uses, methods of joining, and code allowances.

REVIEW QUESTIONS AND PROBLEMS

1. Technically, what is the difference between *pipes* and *tubes?*
2. What is meant by *nominal* pipe and tube size?
3. What materials are used for pipe and tube?
4. What two means are commonly used to establish the pressure rating for pipe?
5. How are fittings read?
6. Name common uses for clay pipe. How is it joined?
7. Name two types of case iron soil pipe and describe how each is joined.
8. What are the differences among: (a) cast iron, (b) ductile iron, (c) cupola malleable iron, and (d) carbon steel?

9. What advantages does "spin casting" have over the traditional method of making cast pipe?

10. How does a grooved coupling system connect and seal?

11. How are copper tube types designated, and what does each designation mean?

12. Name five types of plastic pipe and tube and indicate where each is used.

13. What are advantages and disadvantages of plastic pipe and tube?

14. What is meant by the *standard dimension ratio?*

15. What would be the SDR of a pipe with an outside diameter of 6.625 in (16.83 cm) and a wall thickness of 0.560 in (14.22 mm)?

REFERENCES

[1] Those not familiar with applications of wood pipe will be interested in the paper, "Redwood Pipe," Sec III, File 3D4, Sheet 2, available from California Redwood Association, One Lumbard Street, San Francisco, CA 94111. Another source of information is Simpson Timber Company, Tank and Pipe Division, 2301 N. Columbia Boulevard, Portland, OR 97217.

[2] ASTM C 700 Vitrified Clay Pipe, Extra Strength, Standard Strength and Perforated.

[3] ASTM C 301 Testing Vitrified Clay Pipe.

[4] ASTM C 425 Compression Joints for Vitrified Clay Pipe and Fittings.

[5] ASTM C 12 Installing Vitrified Clay Pipe and Lines.

[6] ANSI A 40.1 Cast Iron Soil Pipe and Fittings.

[7] CISPI 301 Specification Data for Hubless Cast Iron Sanitary System With No-Hub Pipe and Fittings; and CISPI 310 Cast Iron Soil Pipe Institute's Patented Joint For Use In Connection With Hubless Cast Iron Sanitary System.

[8] W. Harry Smity, "The Maturation of Ductile Iron Pipe," *Cast Iron Pipe News,* 1975.

[9] "Ductile Iron Pipe Design," United States Pipe and Foundry, (Birmingham, Alabama), 1977.

[10] ASTM A 21.50 Thickness Design of Ductile Iron Pipe.

[11] ASTM A 126 Gray Iron Castings for Valves, Flanges and Pipe Fittings.

[12] ANSI B 16.12 Cast-Iron Screwed Drainage Fittings.

[13] ASTM B 88 Seamless Copper Water Tube—Types K, L and M. ASTM B 306 Streamline Copper Tube, Type DWV.

[14] Fittings for copper water tube used in plumbing and heating comply with the following standards: Cast bronze threaded fittings (ANSI B 16.15), cast bronze solder joint pressure fittings (ANSI B 16.18), wrought copper and bronze solder joint pressure fittings (ANSI B 16.22), cast copper alloy solder joint drainage fittings DWV (ANSI B 16.23), cast copper alloy fittings for flared copper tubes (ANSI B 16.26), wrought copper and wrought copper-alloy solder joint drainage fittings (ANSI B 16.29), cast bronze solder joint fittings for SOVENT drainage fittings (ANSI B 16.32).

5 Cutting, Joining, and Supporting Pipe

5-1 Introduction

A great deal of the plumber's time is used to measure, cut, fit, install, and support pipe. Although the method used to connect the pipe is determined largely by the material and code restrictions, the actual installation and appearance result from the plumber's skill and pride in workmanship. That is, how the joint is made and how the pipe runs within the building structure and system reflect the plumber's ability to take proper measurements, to select the necessary fittings, to assemble the pipe and fittings correctly, and finally, to anchor them securely in place so that the installation will withstand the expected conditions of service for the life of the structure.

Plumbing codes require that plumbing installations shall be gastight and watertight for the pressures expected by test, with the exception of those portions which may be perforated or may have open joints for the purpose of collecting or conveying ground or seepage water to underground storm drains. This means that before any portion of the system is covered, either within or outside the structure, it must pass an inspection pressure test that conforms to the code, to ensure the integrity of the joints, as well as of the pipe and fittings material.

Piping connections fabricated from straight sections of rigid pipe and fittings require that measurements must be made which allow not only for the required location of the pipe, for example, to connect to the hot water tank in a specific position, but must also make allowances for the fittings and connections where the piping goes into these fittings. The offset shown in Fig. 5-1, for example, indicates that to fit a pipe, the measurement is made to determine the distance between the centers of the two pipes, followed by determining the length of pipe that is necessary to join the two tee and elbow fittings. The distance between the centers of the two pipes is called the *center-to-center* (C–C) distance, whereas the distance between the two fittings is called the *face-to-face* (F–F) distance. To the face-to-face measurement must be added the lengths at each end that are necessary to join the pipe inside the fittings. The length necessary to join the pipe to the fittings is called the *make-up length* (M.U.L.). The steps used to determine the length of pipe necessary to join an offset, then, are

Figure 5-1

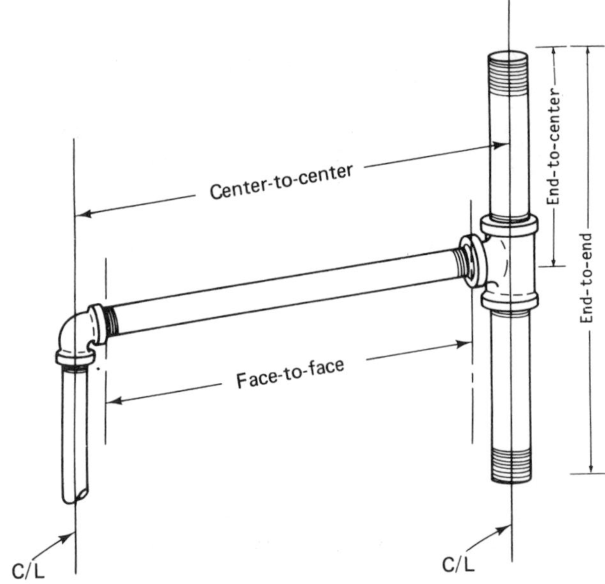

Measuring Pipe Lengths

1. Measure the distance between the pipes center to center (C–C).
2. Measure the fittings from the center line of the pipe to the face (C–F).
3. Measure the make-up length (M.U.L.) for each fitting.
4. Finally, determine the pipe length from:

$$\text{Pipe length} = (\text{C–C}) - 2(\text{C–F}) + 2(\text{M.U.L.}) \qquad (5\text{–}1)$$

Example 5-1 The distance between the centerlines of two water pressure tubes to be connected is 21 in. (53 cm). The measurement from the center of the fittings to the face is found to be $1\frac{1}{4}$ in. (3 cm), and the tube will recess into the fittings at each end $\frac{5}{8}$ in. (1.6 cm). How long must the connecting length of pipe be?

Solution

Refer to Fig. 5–2. Substituting the values in Eq. (5–1), we have

$$\text{Pipe Length} = (21.0 \text{ in.}) - 2(1.25 \text{ in.}) + 2(0.625 \text{ in.})$$

and

$$\text{Pipe length} = 19.75 \text{ in.} = 19\tfrac{3}{4} \text{ in. } (50 \text{ cm})$$

It will also be noticed from Fig. 5–1 that, depending upon the circumstance, it may be necessary for pipe work which is partially installed to

Figure 5-2

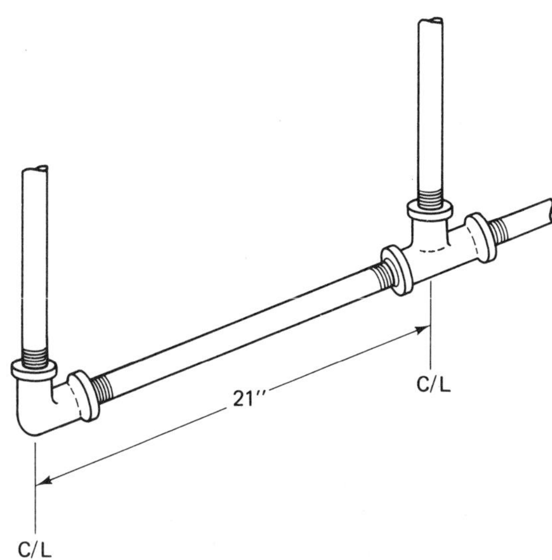

make measurements from end to end, end to center, or face to end. Notice, for example, that end-to-end measurement must still make allowance for the length of the tee fitting as well as the depth of thread that must be made up. For this reason, it is usually more efficient and accurate to rough-sketch the detail, take measurements, apply mathematics to determine the desired lengths, and then cut the pipe to the desired length, rather than use the cut and try method.

5-2 Joining Cast Iron Soil Pipe

Cast iron soil pipe is used below and above ground to transport drain water, soil, and waste, as well as to vent the system. Below ground it is important that the joints are made and remain mechanically sound and leakproof because of the possibility that root fibers will enter and block the pipe. Inside the structure, leaks can damage the building as well as human health, should waste pollute the water supply, or gases (which are sometimes explosive) permeate the building.

Cast iron hub and spigot soil pipe is joined with a lead caulk joint or compression gasket whereas no-hub soil pipe is joined by a neoprene rubber gasket and band clamps. The lead caulk joint is constructed of oakum which is yarned in the joint, followed by pouring of the molten lead to fill the joint. When the lead has cooled and hardened, the joint is caulked with a hammer by pounding inside and outside caulking irons to set the joint.

Keep in mind that the flow through hub and spigot pipe is from the hub end to the spigot end. Where short pieces will be used from 5-ft lengths, double-hub pipe insures that both pieces will be usable, thus reducing the amount

Figure 5-3

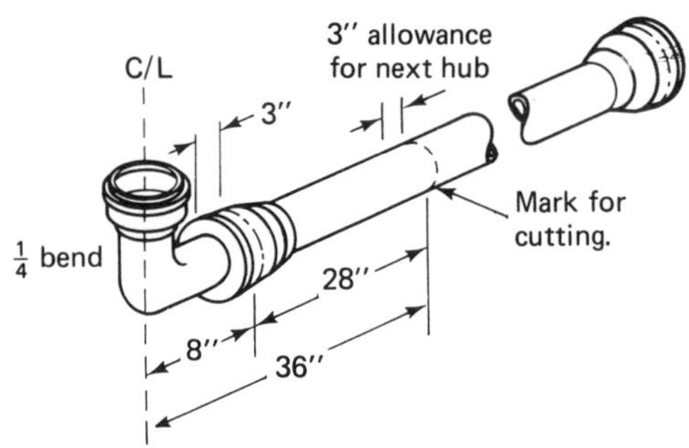

Measuring Cast Iron Hub and Spigot Pipe

of waste. When hub and spigot pipe is measured, allowance must be made for the spigot end to seat in the hub of the next piece. This prevents the oakum from entering the pipe. The make-up length will be $2\frac{1}{2}$ in. for 2-in. pipe, $2\frac{3}{4}$ in. for 3-in. pipe, and 3 in. for 4-in. pipe.

Figure 5-3 illustrates making the measurement for a 36-in. length from a piece of double-hub pipe. Notice that the 36-in. measurement is made from the center of the $\frac{1}{4}$ bend to the end (allowing for the next hub), and from this distance is subtracted the 8-in. distance from the centerline of the fitting to the face of the fitting. Then added to the length is the 3-in. make-up length for the hub. Thus the length of the piece should be 31 in.

After the pipe is marked, it can be cut by hand with a hack saw or hammer and chisel, a process which is painfully slow and costly because of the labor involved, with a power saw, or with a cast-iron soil pipe cutter, as shown in Fig. 5-4. Care must be exercised to assure that the piece being cut off does not fly and hit someone. This occurs because the roller cutters are squeezed into the surface and separate the pipe lengthwise. Regardless of the method of cutting, the outside and inside edges should be rounded with a rasp or peened with a hammer to remove the sharp corners and rough edges, inside and out.

The steps illustrated in Fig. 5-5 to make a vertical lead caulk joint are

1. Seat the spigot end of the pipe squarely in the hub.
2. Yarn and pack the oakum into the pipe joint.
3. Pour the molten lead into the joint in one continuous pour.
4. Caulk the joint with a hammer and caulking irons.

Seating and spacing of the spigot end into the hub are important to keep the joint in alignment. This prevents obstructions from being formed inside the pipe, and oakum from entering the pipe through the space between the spigot and hub. The pipe should be supported to hold it steady for making the joint.

The oakum is yarned into the joint evenly with a yarning iron and packed to a depth of about 1 in. from the top of the hub. The yarning iron

Figure 5-4

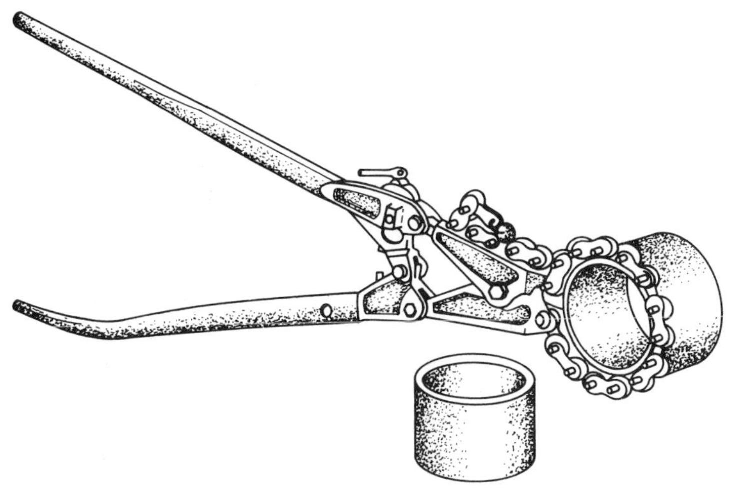

Cutting Cast Iron Soil Pipe

and 16-oz hammer are used to pack the oakum evenly and tightly in the joint. No loose fibers should stick up where they would become part of the lead joint when it is poured. This would weaken the joint. When the oakum is correctly packed, it will hold the pipe upright without supports, although this is not recommended practice.

The lead is dipped with the ladle from the plumber's furnace where it is heated and poured in one pass. The lead should be hot enough past the melting point to retain its liquid state throughout the pour without the formation of bubbles or lumps. Heating the lead red hot, however, will cause it to form excessive slag on the surface and even burn the oakum when it is poured. To be safe, the joint, including the pipe and oakum, must be dry, and the ladle should be preheated before dipping the lead. Moisture and dampness can cause an explosion as the water vaporizes. The same is true of wet or frozen lead when it is added to the furnace. To be safe, the lead should be placed where it can warm and dry. *For safety, leather gloves and a face shield or goggles are required protection.* After the joint is poured, it will begin to harden almost immediately. In fact, if the lead is not hot enough, it will set while being poured and will have to be redone. After several minutes the hardened joint can be caulked with a hammer and caulking irons.

Lead caulk joints require about $1\frac{1}{2}$ lbf (3.3 kg mass) of lead for a 2-in. soil pipe joint, $2\frac{1}{2}$ lbf (5.5 kg mass) of lead for a 3-in. soil pipe joint, and $3\frac{1}{2}$ lbf (7.7 kg mass) of lead for a 4-in. soil pipe joint.

The first step in caulking the joint is to compact the lead joint by hammering it a few times in three or four places. This is done to wedge the joint initially and keep it stable when the caulking irons are used. Begin caulking the joint with the outside iron first, followed by the inside iron. Caulking is

Figure 5-5

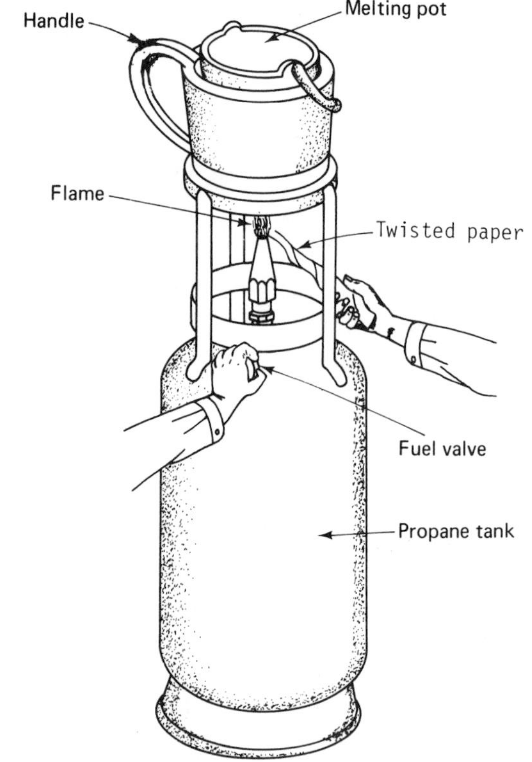

Lighting a melting furnace

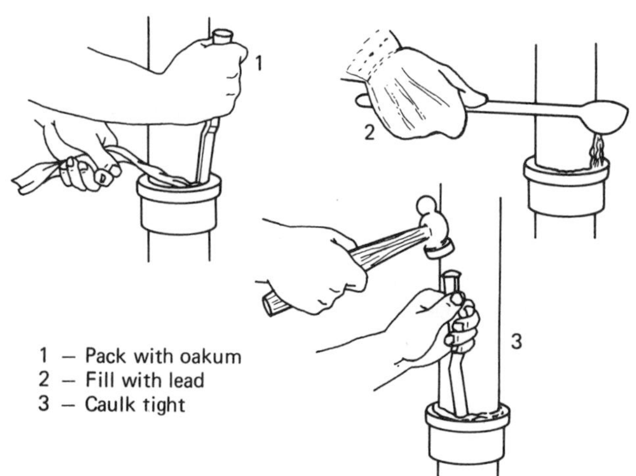

Steps to Make a Lead Caulk Joint

done by placing the tool against the lead at the hub and pipe and driving it down slightly with the 16-oz. ball peen hammer. Proceed slowly around the pipe, first using the outside iron around the inside of the hub and then using the inside iron against the pipe. The caulking iron should overlap the previous position each time it is moved. Compressing the lead with the outside and inside irons spreads it tightly between the pipe and hub, making it both gas and watertight. Some caution should be exercised not to overcaulk the joint, which can compress the lead to the extent that it will break the hub, or work it to the place that it will crack and lose its sealing qualities.

Pouring a horizontal joint requires clamping a runner around the pipe and snugging it against the hub by tapping it with a hammer. This keeps the molten lead in the joint while it cools. A small piece of oakum is used to fill the triangular space underneath where the runner is clamped. The clamp is then fixed to hold the runner ends out of the way so that the molten lead can be poured into the joint through the gate. The excess lead that remains at the gate after the pour hardens is removed with a hammer and cold chisel before caulking. On horizontal joints caulk the inside first.

Cast iron hub and spigot, service weight (SV) cast iron soil pipe, and less frequently, extra-heavy (XH) cast iron soil pipe are also assembled by using compression gaskets. The gaskets are inserted into the hub end of the pipe or fitting and lubricated, and then the spigot end of the pipe or fitting is inserted and shoved into the joint until it bottoms. The steps to make the compression joint are as follows. Figure 5-6 illustrates the compression gasket assembly in cross section.

Figure 5-6

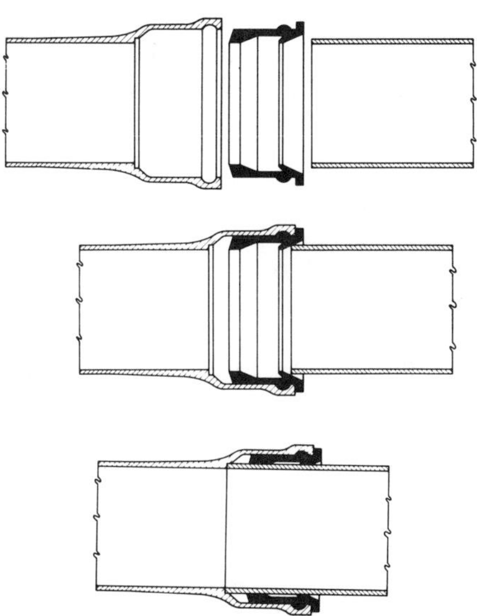

Compression Gasket Assembly in Cross-Section (Tyler Pipe Company)

1. The hub end and spigot end must be clean for the gasket to seal. They should be wiped or brushed clean. If a pipe has been cut to length, the end must be filed or peened to remove the sharp edge, which could cut the gasket during assembly.

2. Insert and lubricate the inside of the gasket. Small gaskets will have to be folded inward (using the thumbs) to reduce their size sufficiently to allow insertion in the hub end. Push the gasket into place either by hand, or with a wooden paddle. Apply a thin coat of approved lubricant to the inside of the gasket, but not to the spigot end of the pipe.

3. Using one of the approved methods which will be described, shove the spigot end of the fitting into the gasket until it bottoms.

On smaller pipes to 4 in., the pipe can be driven into the joint with a lead maul applied against the driving lug (on fittings) or a sledge hammer against a block laid across the hub end of the pipe being inserted. Depending upon conditions, this may be the best option to assemble the joint, although it leaves some doubt about what actually occurs as the spigot end of the pipe or fitting moves through the gasket. That is, the gasket could be twisted, torn, or cut, or the pipe or fitting could be broken by excessive misalignment or force applied during the process. The second and more desirable method is to use an assembly tool, which allows better observation and control during the assembly process. Figure 5-7 illustrates this method.

No-hub cast iron soil pipe is cut to the assembled lengths (without a make-up length for the joint), since the joint is made by using a sleeve gasket and clamp assembly which fits over the ends of both pipes and/or fittings being joined together. A slight gap at the ends is necessary to accommodate the stop at the center of the sleeve.

The steps to assemble no-hub pipe and fittings are as follows:

1. Pipe and fittings are measured so that they will butt squarely together, by a small space (approximately $\frac{1}{4}$ in.) when they are assembled. They must be clean and free from burrs, which would impair assembly, damage the sleeve gasket, or cause leaks.

2. The neoprene sleeve coupling and clamp-shield assembly are slipped over one end; then the remaining pipe or fitting is inserted into the other until the two come together against the inside shoulder sleeve. No lubricant is used.

3. The clamps are then tightened to a torque[1] of 60 in.-lb (6.78 m-N) with the preset hand or assembly torque wrench. This allows the joint and sleeve to set and still retain a tightness of 48 in.-lb (5.42 m-N).

The assembly procedure is shown in Fig. 5-8.

Figure 5-7

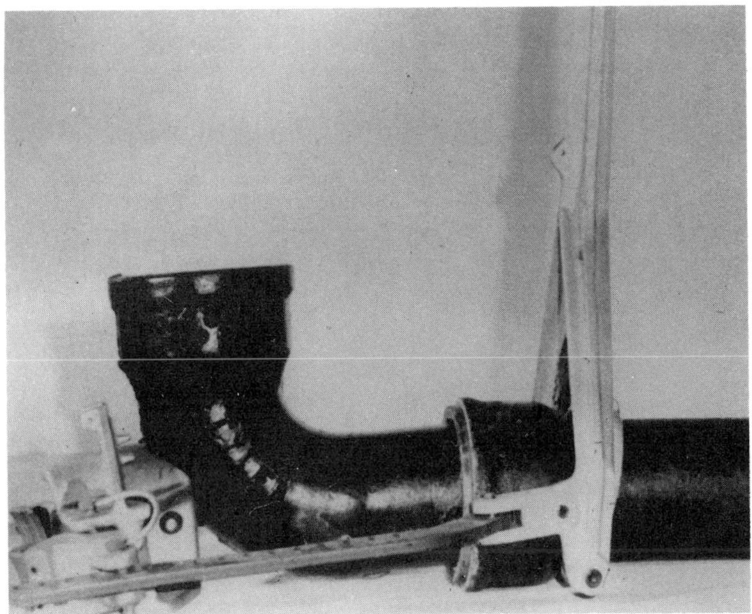

Using a Compression Gasket Assembly Tool (Tyler Pipe Company)

5-3 Galvanized Steel Threaded Joints

American National Standard taper pipe threads (NPT) are tapered and have a pitch of 60 deg when measured from the axial plane. The taper is one thread in sixteen and is $\frac{3}{4}$-in. per ft increase in diameter, which is also an angle of taper with the centerline of 1 deg 47 min.

Tapered pipe threads produce a pressure-tight joint when assembled with joint compound or sealing tape. When they are tight, they also pro-

Figure 5-8

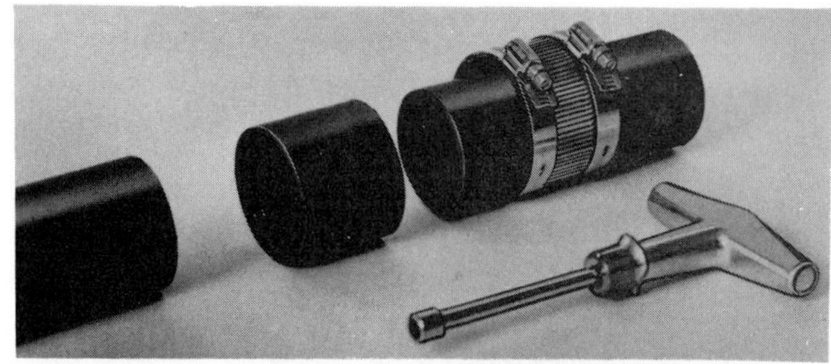

Assembly Procedure for No-Hub Cast Iron Soil Pipe (Tyler Pipe Company)

duce a mechanically rigid piping system, which is an important consideration when one is supporting and hanging metal piping that has a substantial mass. A profile of an NPT threaded connection is shown in Fig. 5-9, where it is noticed that because the thread is tapered, not all of it is usable.

Figure 5-9

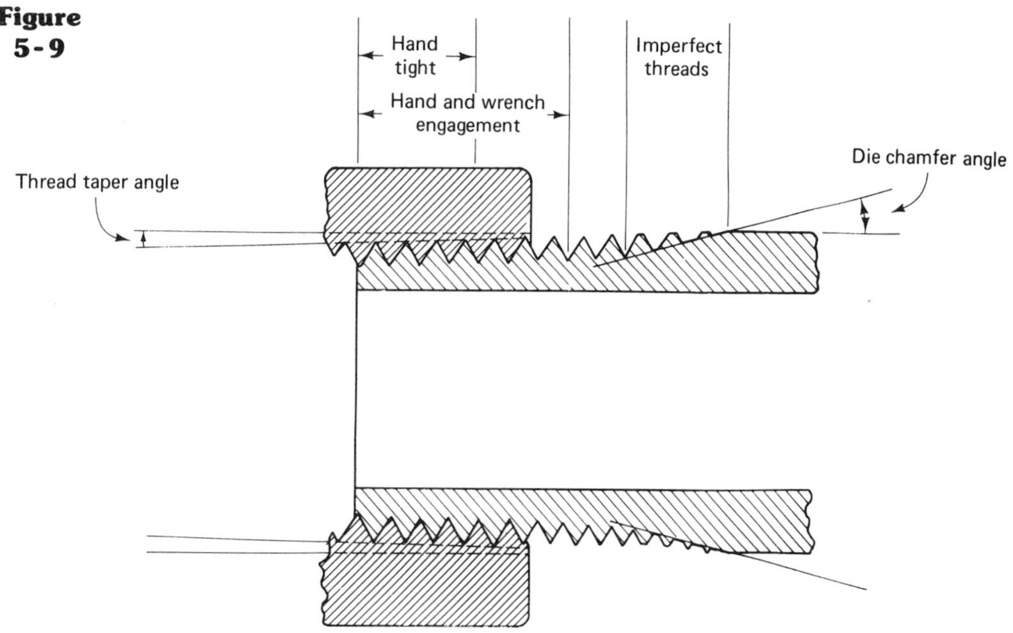

American National Standard Taper Pipe Thread (NPT)

The first few threads (two or three turns) are engaged by hand, and then the joint is wrench-tightened, still leaving the imperfect threads showing past the joint.

The imperfect threads have a lip angle or rake (pitch) of 15–25 deg, depending upon the type and condition of the die. Threading dimensions are given in Table 5-1. When one is threading the pipe for insertion into a fitting, either the total number of threads can be counted to establish the proper length of the thread, or the total length of the thread dimension can be measured directly and marked on the pipe.

The imperfect threads result from the pitch of the die, which is necessary for it to engage the pipe when the thread is started. In plumbing work these will show. In pipe railing joints, the fittings have a machined recess that covers the last scratch or imperfect thread on the pipe.

The procedure for cutting, threading, and assembling threaded metal pipe is as follows (Fig. 5-10):

1. Mount the pipe in a pipe vise or threading machine and mark the total length of thread from information in Table 5-1.
2. Cut the pipe to the desired length with a wheeled pipe cutter.
3. Use the reamer to resize the pipe to its original diameter and remove the metal, which otherwise would act as an obstruction to flow.

Figure 5-10

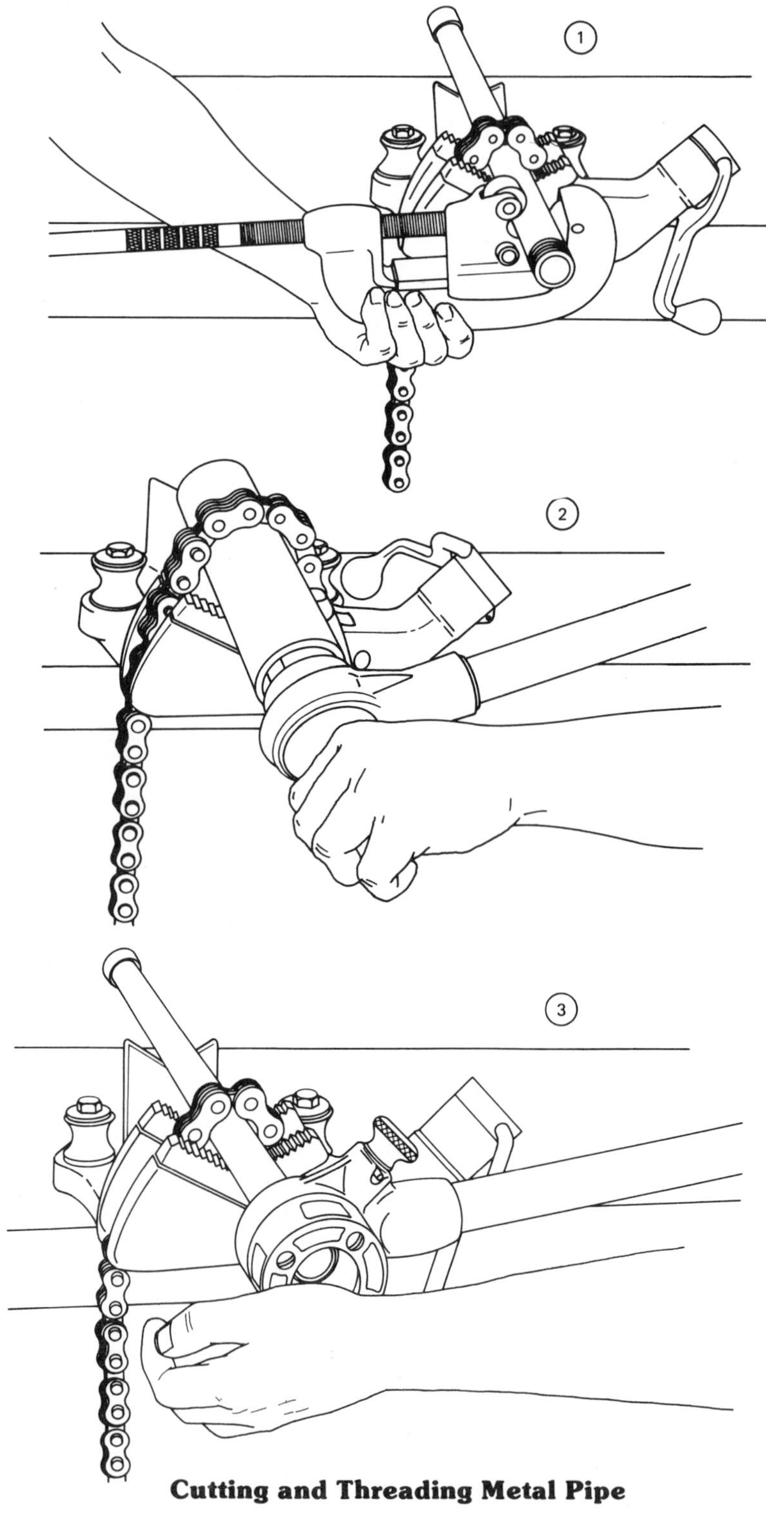

Cutting and Threading Metal Pipe

Table 5-1 **American National Standard Taper Pipe Threads (NPT) Dimensions**

Nominal Pipe Size	*Threads Per Inch*	*Number of Usable Threads*[1]	*Hand Tight Engagement*[2] *(inches)*	*Thread Makeup-Wrench and Hand Engagement*[2] *(inches)*	*Total Length of External Thread (inches)*
$\frac{1}{8}$	27	7	$\frac{3}{16}$	$\frac{1}{4}$	$\frac{3}{8}$
$\frac{1}{4}$	18	7	$\frac{1}{4}$	$\frac{3}{8}$	$\frac{9}{16}$
$\frac{3}{8}$	18	7	$\frac{1}{4}$	$\frac{3}{8}$	$\frac{5}{8}$
$\frac{1}{2}$	14	7	$\frac{5}{16}$	$\frac{1}{2}$	$\frac{3}{4}$
$\frac{3}{4}$	14	8	$\frac{5}{16}$	$\frac{9}{16}$	$\frac{13}{16}$
1	$11\frac{1}{2}$	8	$\frac{3}{8}$	$\frac{11}{16}$	1
$1\frac{1}{4}$	$11\frac{1}{2}$	8	$\frac{7}{16}$	$\frac{11}{16}$	1
$1\frac{1}{2}$	$11\frac{1}{2}$	8	$\frac{7}{16}$	$\frac{3}{4}$	1
2	$11\frac{1}{2}$	9	$\frac{7}{16}$	$\frac{3}{4}$	$1\frac{1}{16}$
$2\frac{1}{2}$	8	9	$\frac{11}{16}$	$1\frac{1}{8}$	$1\frac{9}{16}$
3	8	10	$\frac{3}{4}$	$1\frac{3}{16}$	$1\frac{5}{8}$
$3\frac{1}{2}$	8	10	$\frac{13}{16}$	$1\frac{1}{4}$	$1\frac{11}{16}$
4	8	10	$\frac{13}{16}$	$1\frac{5}{16}$	$1\frac{3}{4}$
5	8	11	$\frac{15}{16}$	$1\frac{3}{8}$	$1\frac{13}{16}$
6	8	12	$\frac{15}{16}$	$1\frac{1}{2}$	$1\frac{15}{16}$
8	8	14	$1\frac{1}{16}$	$1\frac{11}{16}$	$2\frac{1}{8}$
10	8	15	$1\frac{3}{16}$	$1\frac{15}{16}$	$2\frac{3}{8}$
12	8	17	$1\frac{3}{8}$	$2\frac{1}{8}$	$2\frac{9}{16}$

[1] Rounded to the nearest thread.
[2] Rounded to the nearest $\frac{1}{16}$ inch.

4. Start the die squarely on the end of the pipe and apply hand pressure to the end to ensure that it starts promptly.
5. While the pipe is being threaded, apply a liberal supply of thread cutting oil through the die to flush the chips and cool the die.
6. If a fitting is to be connected, apply thread sealing tape or approved pipe thread compound to the outside threads, two or three threads back from the end to be sure that no compound enters the pipe.
7. Hand-tighten the fitting two or three turns, and then finish tightening the fitting with a pipe wrench.

While the pipe-threading operation is the same if either hand or powered equipment is used, caution must be exercised when one is using machine methods to ensure that the operator is not injured. The directions for operating the power vise or threading machine must be followed. For safe operation these should be posted conspicuously on or near the machine.

5-4 Copper Tube Joints

Copper tube is joined by soldered, brazed, and flared joints. The joints provide a pressure-tight system as well as the mechanical strength necessary for support and sustaining water shocks without failure.

Because of its excellent formability, copper pressure tube is often bent at the job site to clear obstructions and to make connections to fixtures and equipment. Copper tube can be bent by hand, with forming blocks, with a spring bender, or with a lever type bender, such as that shown in Fig. 5-11. The procedure for using the lever bender is as follows:

1. With the handles positioned at 180 deg as shown and the clip raised, insert the tube in the forming wheel groove.
2. Raise the forming shoe handle approximately 90 deg so that the tube holding clip can be latched over the tube, and then bring it down until the zero mark on the forming wheel is even with the front edge of the forming shoe.

Figure 5-11

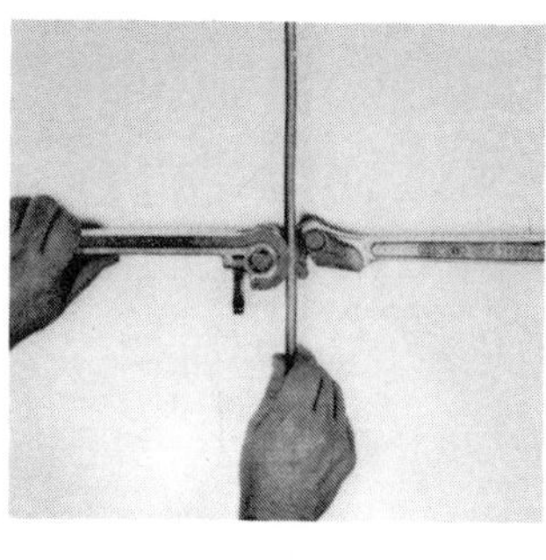

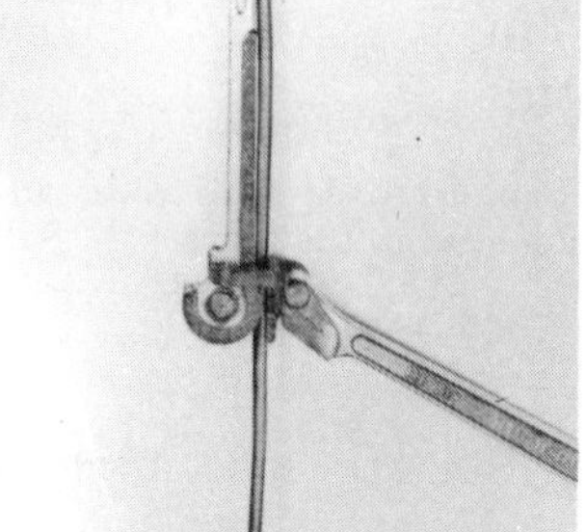

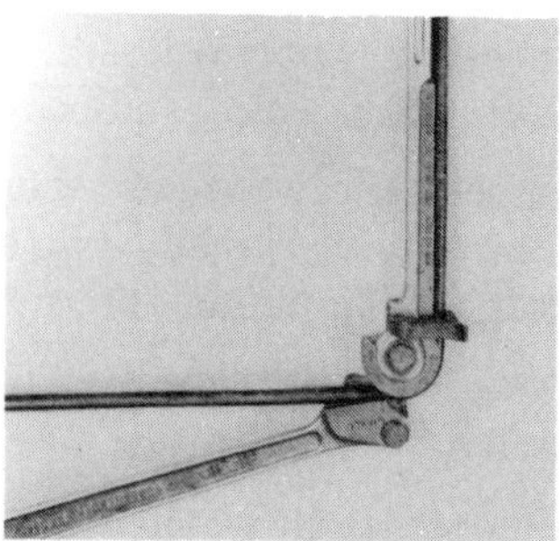

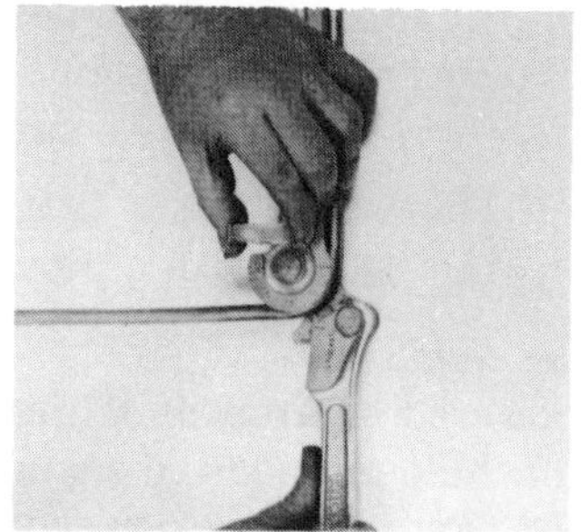

Bending Copper Tubing with a Lever Bender (Copper Development Association Inc.)

3. Bending is accomplished by pulling the forming shoe handle down smoothly until the degree of bend is indicated.
4. Remove the tube by returning the forming shoe handle to the 90-deg position, disengage the forming shoe, and lift the holding clip.

Soldered joints use capillary action to draw molten solder into the space between the fitting and the tube. Mechanical cleaning of the fitting and tube, as well as the flux, serve to clean the surfaces. The flux also wets the surfaces and promotes uniform spreading of the solder in the 0.002- to 0.005-in. (0.051 to 0.127 mm) clearance space between the tube and fitting.

Which solder is used depends upon the temperature and pressure conditions under which the joint must operate. The rated internal working pressure for copper tube joints is given in Table 5–2. Solder made of 50–50 (percent) tin-lead is suitable for most water distribution systems. Higher pressures and joint strength can be attained by using 95–5 tin-antimony solder. For continuous operation at temperatures above 250°F (121°C), for example in solar collector panels, brazing filler metals should be used. As the strength of the solder or brazing filler metal increases, the melting-solidification temperature range becomes more narrow, making it more difficult to work.

The flux is applied to the tube and fitting soon after cleaning to prevent further oxidation. Fluxes used with 50–50 and 95–5 solder are mildly corrosive liquid or petroleum-base pastes that contain zinc chloride and ammonium chloride. Although there are some fluxes advertised as "self-cleaning," caution should be exercised in their use, particularly near the face and eyes. There is also some uncertainty about their continued chemical action after the joint has been soldered.

The twelve accepted steps for soldering copper tube are listed and explained here and are shown in Fig. 5–12.

1. Measure the tube to length.
2. Cut the tube square.
3. Ream the cut end.
4. Clean the tube end.
5. Clean the fitting socket.
6. Apply flux to the tube end.
7. Apply flux to the fitting socket.
8. Assemble the joint.
9. Remove excess flux.
10. Apply heat to the joint.
11. Apply solder to the joint.
12. Allow the joint to cool.

Table 5-2

Rated Internal Working Pressure for Copper Solder Joints (Copper Development Association Inc.)

Rated Internal Working Pressures of Joints Made with Copper Water Tube and Solder Type Fittings (Pounds per Square Inch)

Solder or Brazing Alloy Used in Joints	*Service Temperature Degrees F.*	*Copper Water Tube K, L and M. Nominal Sizes (Inches)*					
		Water (a)					*Saturated Steam*
		$\frac{1}{4}$ to 1"	*1-$\frac{1}{4}$ to 2"*	*2-$\frac{1}{2}$ to 4"*	*5 to 8"*	*10 to 12"*	*All*
50–50 Tin-Lead (b)	100	200	175	150	130	100	
	150	150	125	100	90	70	
	200	100	90	75	70	50	
	250	85	75	50	50	40	15 (c)
95–5 Tin-Antimony	100	500	400	300	150	150	
	150	400	350	275	150	150	
	200	300	250	200	150	140	
	250	200	175	150	140	110	15 (c)
Brazing Alloys (Melting at or above 1000°F.)	100–150–200	*	*	*	*	*	
	250 (e)	300	210	170	150	150	
	350	270	190	150	150	150	120 (d)

The values in the above table are based on data in the National Bureau of Standards publications, *Building Materials and Structures Reports BMS 58* and *BMS 83.*

(a) Including other noncorrosive liquids and gases.
(b) ASTM B32, Alloy Grade 50A.
(c) This pressure is determined by the temperature of saturated steam at 15 lb pressure or 250°F.
(d) This pressure is determined by the temperature of saturated steam at 120 lb pressure or 350°F.
(e) For service temperatures lower than 250°F. the solders as above may be used.
(*) Rated internal pressure is that of the tube being joined.

Approximate Metric Values of Rated Internal Working Pressures of Joints Made with Copper Water Tube and Solder Type Fittings (kPa)

Solder or Brazing Alloy Used in Joints	Service Temperature Degrees C.	Copper Metric Water Tube Nominal Sizes Water (a) 8 to 28 mm	35 to 54 mm	66 to 104 mm	Saturated Steam All
50–50	40	94.4	82.8	71.0	
Tin-Lead (b)	65	71.0	59.3	46.9	
	90	47.6	42.8	35.2	
	120	39.9	35.2	23.4	6.9 (c)
95–5	40	237.2	189.7	142.0	
Tin-Antimony	65	189.7	166.2	130.3	
	90	142.1	118.6	94.5	
	120	94.5	82.8	71.0	6.9 (c)
Brazing Alloy	to 90	*	*	*	
(Melting at or above 540°C.)	120 (e)	142.1	99.3	80.7	
	176	128.3	90.3	73.1	56.6 (d)

The values in the above table are based on data in the National Bureau of Standards publications, *Building Materials and Structures Reports BMS 58* and *BMS 83.*

(a) Including other noncorrosive liquids and gases.
(b) ASTM B32, Alloy Grade 50A.
(c) This pressure is determined by the temperature of saturated steam at 6.9 kPa pressure or 120°C.
(d) This pressure is determined by the temperature of saturated steam at 56.55 kPa pressure or 180°C.
(e) For service temperatures lower than 120°C. the solders as above may be used.
(*) Rated internal pressure is that of the tube being joined.

Figure 5-12

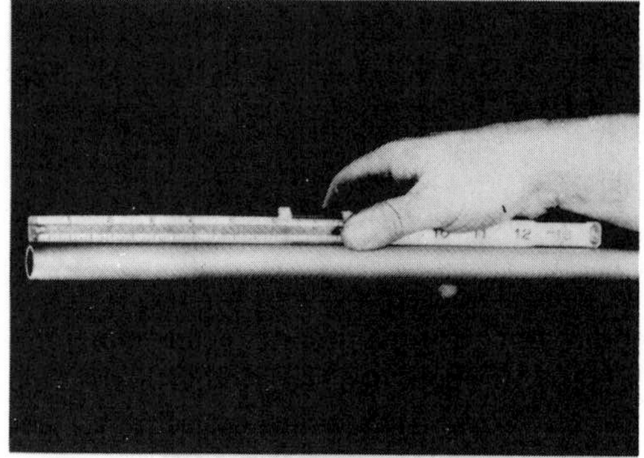

1. Measuring

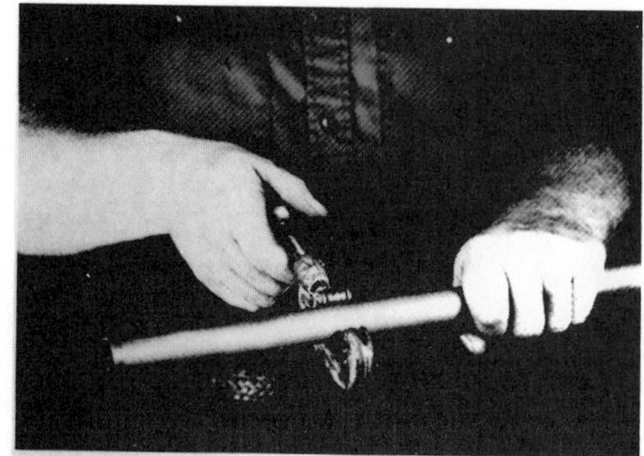

2. Cutting

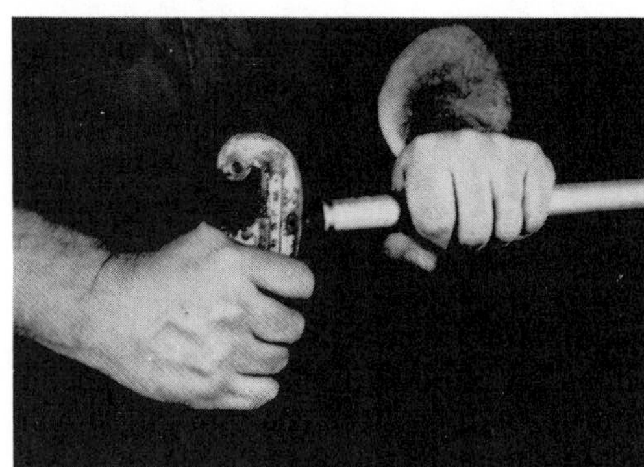

3. Reaming

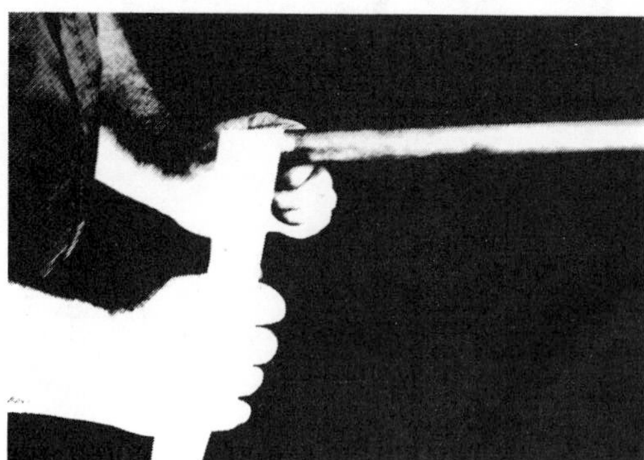

4. Cleaning tube end

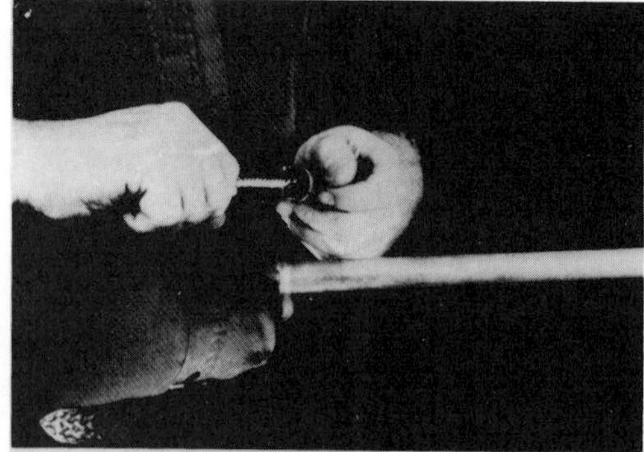

5. Cleaning fitting socket

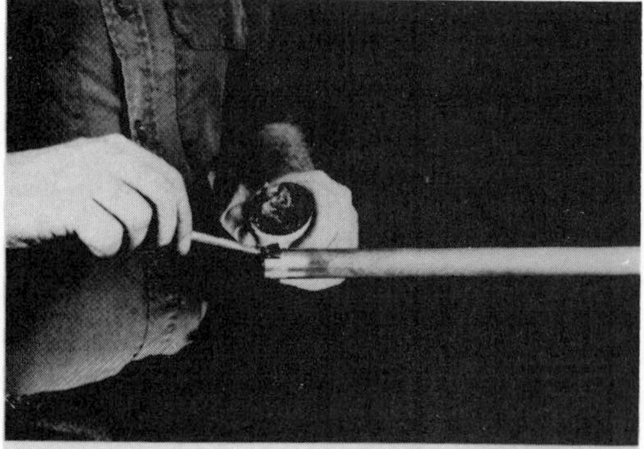

6. Fluxing tube end

Procedure for Making Copper Solder Joints (Copper Development Association Inc.)

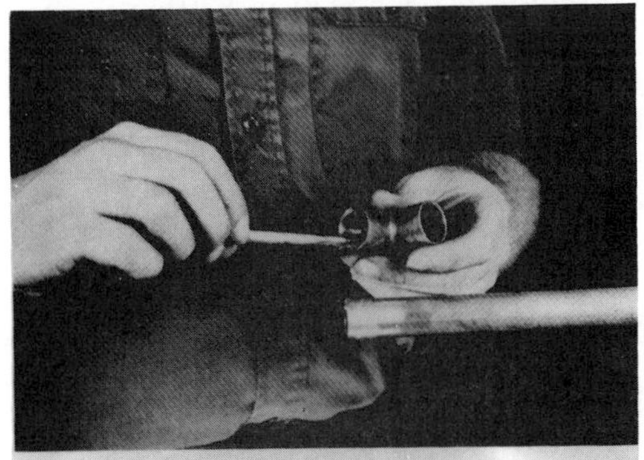
7. Fluxing fitting socket

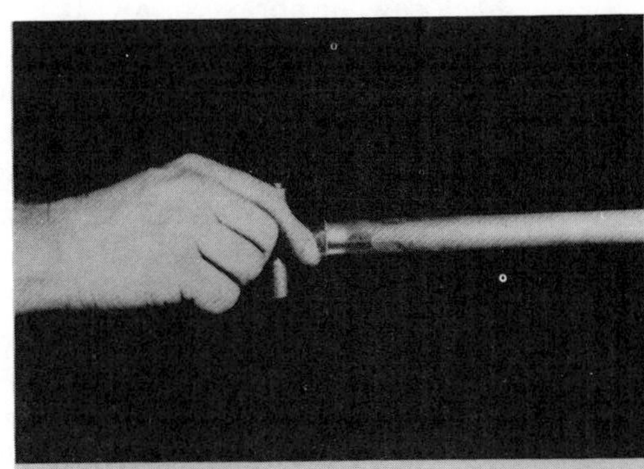
8. Assembling fitting and tube

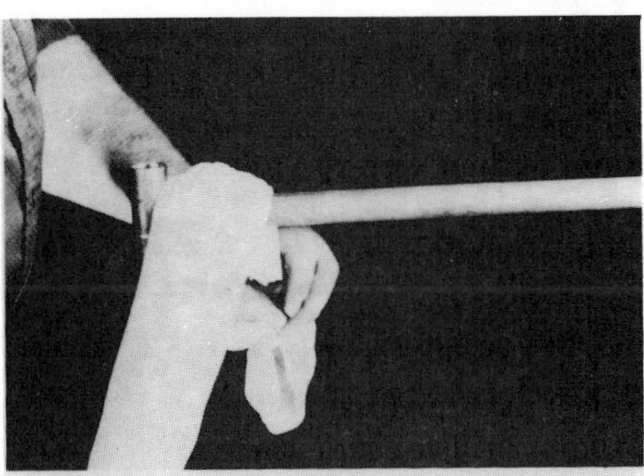
9. Removing excess flux

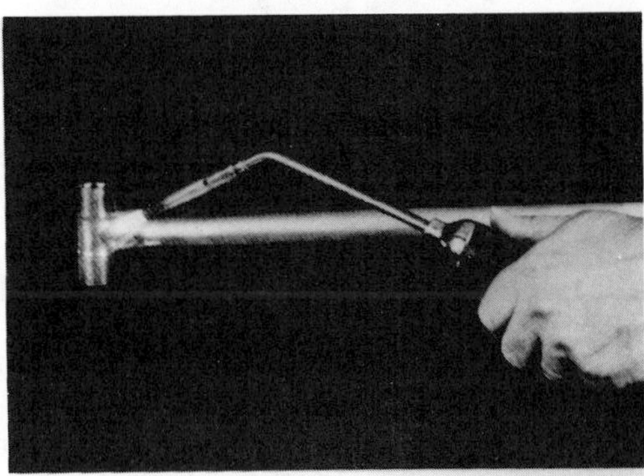
10. Heating the assembly

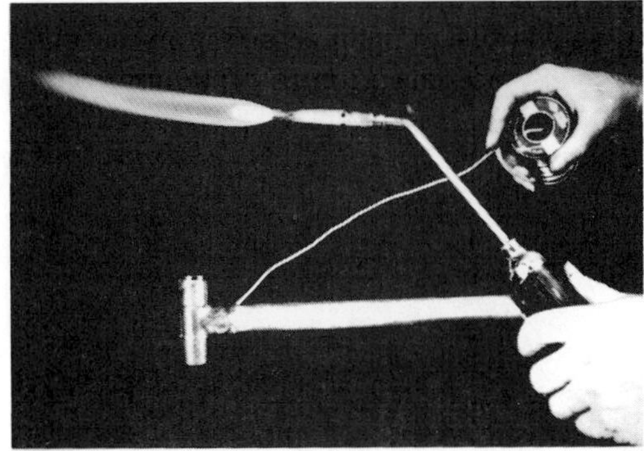
11. Applying solder

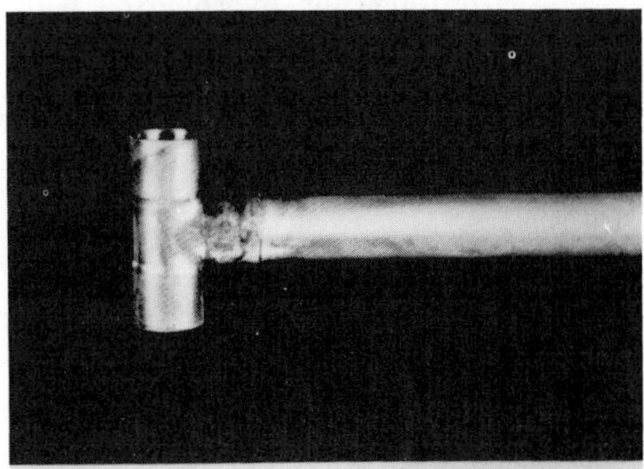
12. The finished joint

The tube is measured and cut to the exact length with a tubing cutter, hack saw, or abrasive saw. It is important that the length be correct to ensure that the tube will bottom in the fitting. This permits the solder joint to be the proper length. The tube must be cut square to ensure that it will bottom evenly. Tubing cutters are available in sizes to 8 in.

Reaming the cut end removes the burr of material, which can both reduce flow as well as damage fixtures should loose material migrate through the piping system. If the tube is out of round, this will adversely affect the solder joint, and a resizing tool consisting of a plug and sizing ring should be used to bring the tube to the true dimension and roundness (Fig. 5-13).

The mating surfaces must be clean of dirt and oil to make a good joint. The tube end can be cleaned with fine sand cloth or with an outside tubing cleaner. The fitting socket is cleaned similarly with either sand cloth or a brush cleaner.

Flux is applied to both the fitting and tubing with a small brush, swab, or rag. Using the fingers to apply and spread the flux is discouraged because of the possibility of later contact with the face and eyes. The flux should not be spread so thick that it deposits an excess amount inside the joint. When this occurs, there is the increased possibility that solder will also accumulate inside the joint, as well as flux, which can be carried through the piping system with the water supply. After the joint is fluxed and assembled, the excess is removed with a rag. It may be necessary to hold the fitting in position on the tube during this operation.

Heat is applied with a propane or butane torch, or with an air-acetylene torch. The flame is played on the fitting and moved about to spread the heat through the joint area. The joint should be heated evenly, but avoid directing the flame into the fitting itself. At frequent intervals, touch the solder to the tube at the joint to start the soldering operation as soon as the correct temperature is reached. While underheating will prevent the solder from melting, overheating will burn and destroy the flux, preventing the solder from being drawn into the joint as expected. If this does occur, disassemble the joint and reflux the tube and fitting. Care should also be exercised not to overheat cast fittings, which can sometimes crack.

Figure 5-13

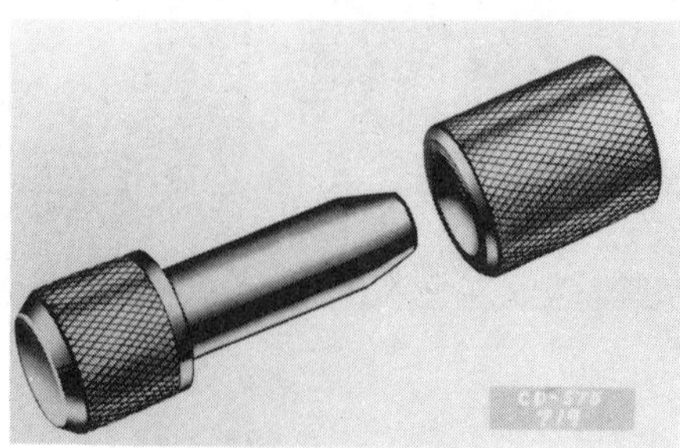

Copper Tube Resizing Tool (Nibco, Inc.)

When the joint has reached the correct temperature, the solder will melt when it comes in contact with the tube. If the cleaning and fluxing operations have been completed properly, additional heat and capillary action will draw the solder into the joint until it completely fills the space between the tube and fitting. If a fillet is specified at the edge of the fitting, this will require some additional solder, although the value of such a fillet is sometimes questioned. The amounts of solder and flux for typical joints of 95–5 solder per hundred joints is given in Table 5–3. The amounts shown include an allowance for 100 percent to cover wastage and loss for tube sizes up to 2 in., and 25 percent for tube sizes of $2\frac{1}{2}$ in. and above.

Table 5-3

Quantity of Solder per 100 Joints (Copper Development Association Inc.)

Nominal Size Joint (in.)	*Solder Required General Use (lb)*	*Solder Required Drainage Use (lb)*	*Nominal Size Joint (in.)*	*Solder Required General Use (lb)*	*Solder Required Drainage Use (lb)*
$\frac{1}{4}$	$\frac{1}{2}$	. . .	2	$2\frac{1}{2}$	$1\frac{1}{2}$
$\frac{3}{8}$	$\frac{1}{2}$	. . .	$2\frac{1}{2}$	$3\frac{1}{2}$	. . .
$\frac{1}{2}$	$\frac{3}{4}$	. . .	3	$4\frac{1}{2}$	$2\frac{3}{4}$
$\frac{5}{8}$	$\frac{7}{8}$	. . .	$3\frac{1}{2}$	5	. . .
$\frac{3}{4}$	1	. . .	4	$6\frac{1}{2}$	$4\frac{1}{4}$
1	$1\frac{1}{2}$	. . .	5	9	. . .
$1\frac{1}{4}$	$1\frac{3}{4}$	$1\frac{1}{4}$	6	17	. . .
$1\frac{1}{2}$	2	$1\frac{3}{8}$	8	35	. . .

Solder requirements in this table are based on estimate of weight of solder used to prepare 100 solder joints of sizes shown.
Two oz. of solder flux will be required for each pound of solder.

Tube Size (millimeters)	*Solder Required General Use (kilograms)*	*Flux (grams)*
8	.2	25
10	.3	36
12	.3	36
15	.4	50
18	.4	50
22	.5	63
28	.7	88
35	.9	113
42	1.0	125
54	1.2	150
66	1.6	200
79	2.0	250
104	3.0	375

125 grams of solder flux will be required for each kilogram of solder.

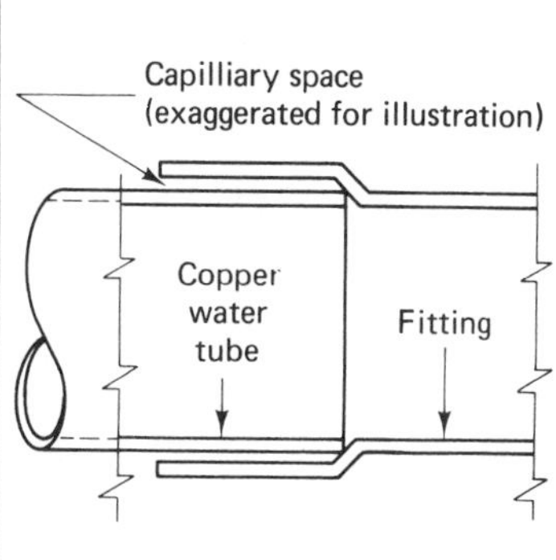

Capilliary Joint

Proper cooling of a soldered joint is important. Cooling too fast, for example by quenching, can crack cast fittings. Pressure-testing the assembled system with water will reveal if cracks do exist, and if so, the system must be drained to allow heating and resoldering, or replacement of faulty fittings. Placing a bleed or "stop and waste" valve at the low point in the system will greatly improve this operation if it is necessary to drain the system after the pressure test or at some time in the future when the system is repaired or extended.

Brazed joints are made with filler metals with melting temperatures in the 1100–1500°F (593–916°C) range. In this temperature range the metal changes from a solid to a liquid and is an indication of the time within which the metal must be worked. Metals with a narrow temperature range must be worked more quickly before they solidify. Those with a wider range give the plumber more time and a greater margin of error in bringing and holding the joint at the exact best temperature. Figure 5–14 illustrates the temperatures at which different brazing materials and flux react during the brazing cycle. Notice, for example, that the flux used for brazing begins to melt several hundred degrees above the melting temperature of 50–50 solder. Also notice that the brazing temperature more nearly approaches the melting temperature of the material used for the tube and fitting, and the strength of the joint approaches the strength of the tube and fitting. Because more heat is necessary and the flame must have a higher temperature for brazing, an oxyacetylene torch is recommended, although air-acetylene is sometimes used on smaller sizes. In either case, the flame is played on the tube and fitting to bring them evenly to the melting temperature of the filler metal. Although the procedure is nearly the same as for soldering with 50–50 solder, it is important that special attention be given to the application of heat, and feeding the filler metal to the capillary space at the proper temperature. It is at this point in the procedure that most errors are made in making the joint. If the filler metal fails to flow or has a tendency to ball up, it indicates oxidation on the metal surface or insufficient heat on the tube and fitting. If the work starts to oxidize during heating, there is too little flux. If the filler metal fails to enter the joint but rather tends to flow over the outside of the tube or fitting, it indicates that one of the two is overheated and/or the other is underheated.

Flared joints are used extensively below ground and where the joint must be broken from time to time. They are also used where it is difficult or inadvisable to use an open flame or heat necessary to construct a solder joint.

Flared joints make the seal with metal-to-metal contact between the machined cone of the fitting and the flared end of the tube. Two common methods of making the flare are shown in Figs. 5–15 and 5–16. The impact method uses the following procedure:

1. Cut the tubing square and ream it, being careful to remove all burrs that would interfere with the metal-to-metal seal when the tube is flared.

Figure 5-14

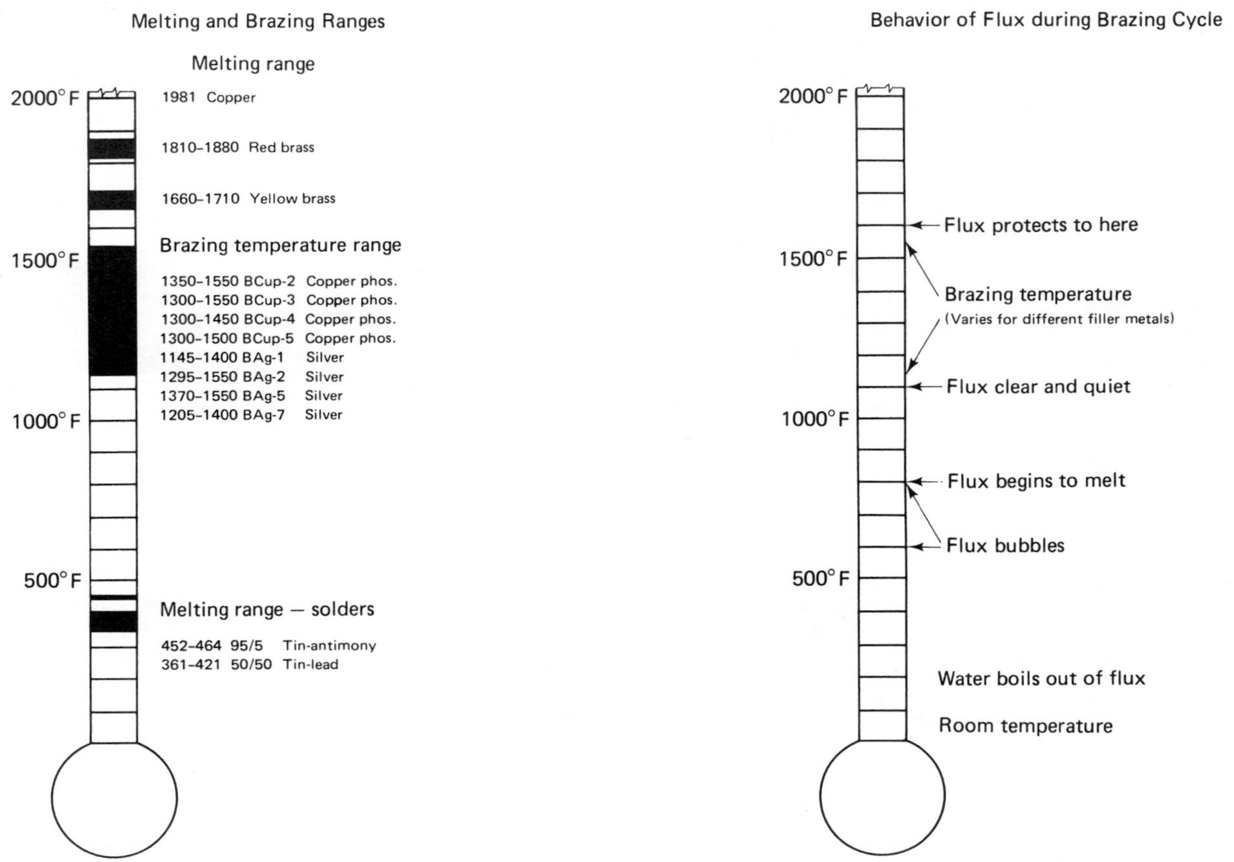

Melting and Brazing Ranges

Melting range

2000° F

1981 Copper

1810–1880 Red brass

1660–1710 Yellow brass

Brazing temperature range

1500° F

1350–1550 BCup-2 Copper phos.
1300–1550 BCup-3 Copper phos.
1300–1450 BCup-4 Copper phos.
1300–1500 BCup-5 Copper phos.
1145–1400 BAg-1 Silver
1295–1550 BAg-2 Silver
1370–1550 BAg-5 Silver
1205–1400 BAg-7 Silver

1000° F

500° F

Melting range — solders

452–464 95/5 Tin-antimony
361–421 50/50 Tin-lead

Behavior of Flux during Brazing Cycle

2000° F

Flux protects to here

1500° F

Brazing temperature
(Varies for different filler metals)

Flux clear and quiet

1000° F

Flux begins to melt

Flux bubbles

500° F

Water boils out of flux

Room temperature

Reaction of Brazing Materials with Temperature
(Copper Development Association Inc.)

Figure 5-15

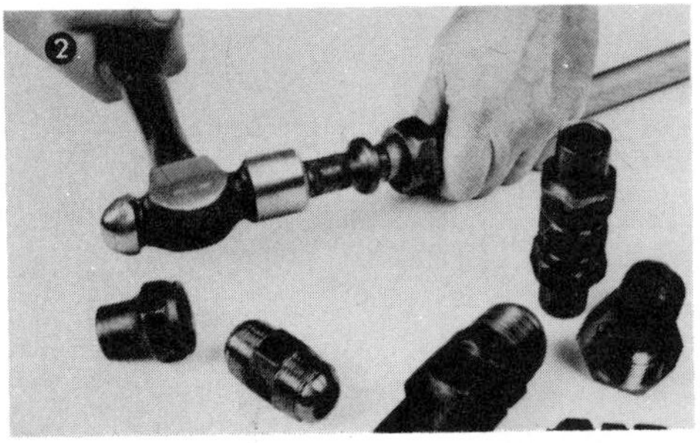

Impact Method of Making Copper Flare Joints (Copper Development Association Inc.)

Figure 5-16

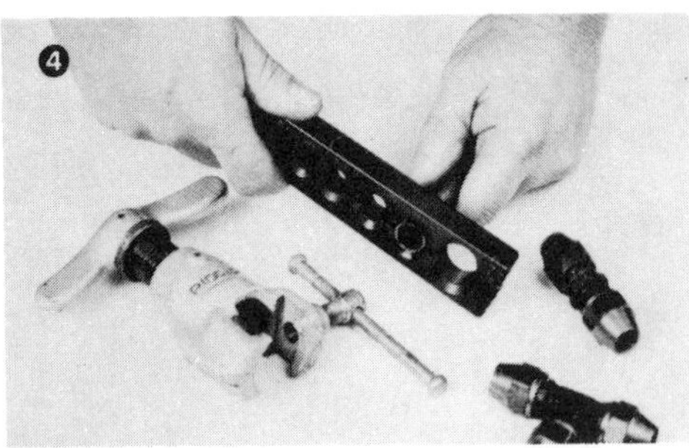

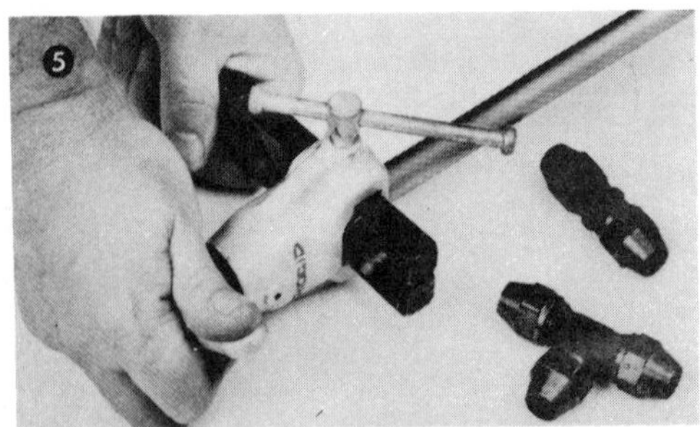

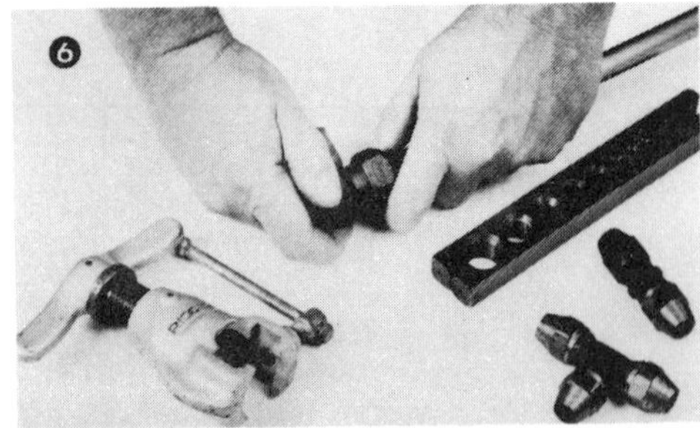

Screw-Type Method of Making Copper Flare (Copper Development Association Inc.)

2. Slip the coupling nut over the end of the tube, threads facing toward the end.
3. Insert the impact flaring tool in the end, strike it with a ball peen hammer moderately until the flare opens to the stop-ridge.
4. Remove the flaring tool and assemble the fitting by hand several threads. Then tighten with two wrenches, being careful not to twist the tubing as the flare is seated. No lubricant should be used, as this may introduce a safety hazard in the piping system.

The screw-type flaring tool makes the flare by forcing the flaring cone attached to the end of the compressor screw against the beveled chamber of the flaring block. The procedure is illustrated in Fig. 5-16 as follows:

1. Cut the tube square and ream it; then slip the coupling over the end, threads facing toward the end.
2. Clamp the tube in the flaring block so that the end of the tube is slightly above the face of the block. (After several practice flares have been made by the beginner, the skill of setting this dimension will be acquired easily.)
3. Place the yoke of the flaring tool over the block so that the beveled end of the compressor cone is over the end of the tube. (Also be sure that the cone turns freely on the end of the screw. If not, it will damage the flare by the swiping action as the screw is turned, resulting in galling of the flare).
4. Turn the compressor screw down firmly, forming the flare between the beveled chamber in the block and the compressor cone.
5. Remove the flaring tool and inspect the flare to see that it is even, smooth, and not split. This is very important, since copper work hardens, sometimes to the extent that improper flaring will split the end of the tube.
6. Assemble the fitting by hand and tighten, using two wrenches, or a vise and a wrench, to seat the flare against the cone.

Sovent copper DWV cast fittings and tube are assembled as subassemblies in shops and at job sites. The tube is cut with a tubing cutter or abrasive saw and cleaned with sandcloth. Fluxing, heating, and soldering sovent systems with 50-50 solder is accomplished by using techniques similar to those used for other soft soldering, except that additional heat is required to bring the large mass of copper adjacent to the joint to the required temperature. Figure 5-17 illustrates a sovent aerator fitting being soldered to a side branch line leading to the single stack system.

Figure 5-17

Sovent Aerator Assembly (Copper Development Association Inc.)

5-5 Plastic Pipe Joints

Of all the materials used in plumbing, plastic piping is probably the easiest and fastest to install. Correct techniques, however, are extremely important.

Plastics used for pressure piping and drain, waste, and vent systems are joined by heat fusion and compression fittings, as well as by solvent welding, with the solvent method being predominant in plumbing installations. Heat fusion is used primarily in production shops and on long underground lines where the cost of the specialized heating equipment and exacting assembly standards can be justified by reduced material costs and large job volume. The use of plastic compression fittings in plumbing is also increasing.

Solvent welding is done as a one- or two-step operation. If only one step is required, this will be applying the solvent directly to the pipe and fittings, and then they are joined. Where the operation requires two steps, first a solvent is used to clean and etch the surfaces of the tube and fittings, and then solvent is applied to both surfaces to be joined. For example, ABS drain, waste, and vent pipe is assembled by using the one-step procedure, whereas PVC and CPVC are joined by using the two-step procedure.

Rigid pipe and tube are measured to relatively close tolerances to join fittings, valves, and fixtures, whereas flexible tubing such as polybutylene gives more allowance because it can be pushed, pulled, or offset slightly to make required connections. Compression fittings, insert fittings, and flared fittings are used extensively with flexible tubing, because by its chemical composition it is often impervious to solvents.

Since plastic piping systems often connect to other pipe, tube, fixtures, and appliances which are not plastic, many transition fittings are used. This is particularly true in pressure piping systems.

Plastic pipe and tube are cut by using chop saws, tubing cutters, and hand saws. Tubing cutters for plastic pipe are of lighter construction than those for metal tubing, and the cutter wheel blade has a thinner and narrower profile to prevent collapsing the wall. Using a hand saw to cut plastic pipe eliminates the potential for this problem, but if the cut is not made square, preventing full extension into the fitting, or if the burr material from the saw cut is not removed from the pipe, allowing it to migrate with the water in pressure piping, both the integrity of the joint and system are questionable.

CPVC tubing for hot- and cold-water distribution is joined with the following procedure, which is illustrated in Fig. 5–18:

1. Cut the tubing square with a miter box or tubing cutter.
2. Remove the burrs or ridge from the outside and inside with a pocket knife or reamer.
3. Check the fit of the tube in the fitting.
4. Remove the surface gloss from the tube and fitting with primer or fine sandpaper. Wipe. (If sandpaper is used, wipe clean with a dry cloth.)
5. Brush cement on the tube and fitting.
6. Push the tubing into the fitting and give it a quarter turn to distribute the cement evenly, adjust for direction, and hold firmly in place until the cement sets. Wipe the excess off the fitting lip with a rag.

Rigid and flexible polybutylene tubing for hot- and cold-water distribution is joined by a number of methods, including insert fittings, fuse-welded mechanical compression cone fitting, and flared-tube compression fittings. The tools and layout for assembling PB tubing using mechanical compression cone fittings are shown in Fig. 5–19. The following procedure is used:

1. Cut the PB tubing to the desired length with a tubing cutter.
2. Remove burrs from the inside.

Figure 5-18

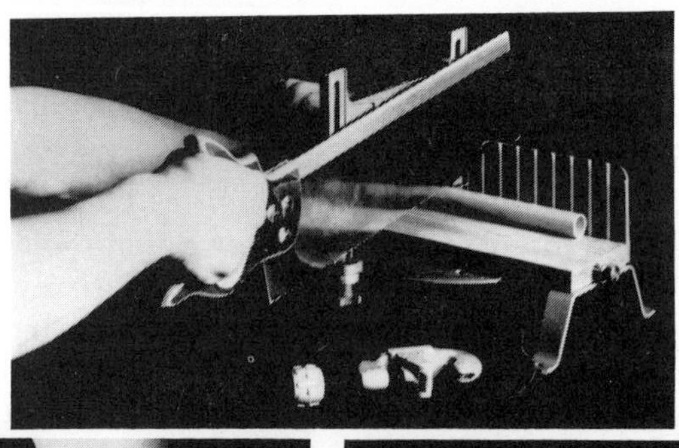

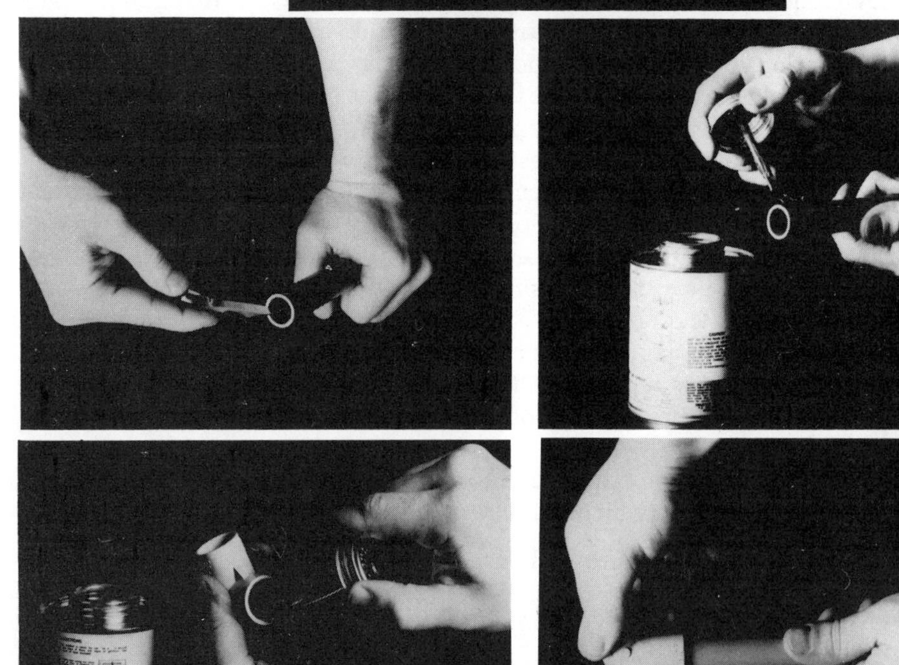

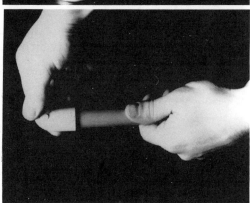

Solvent Welding CPVC Tubing (B. F. Goodrich Chemical Company)

3. Assemble the seal nut, compression nut, and seal cone over the tube, making sure that the tube end is flush with the end of the seal cone.
4. Assemble the fitting and tighten it by hand; then wrench-tighten it one to two full turns.

Insert fittings are used extensively to join polybutylene and polyethylene tubing. Both are assembled dry over metal and plastic fittings that have serrations. Metal clamps are used to secure the tubing over the serrated portion of the fitting. Rigid polyethylene tubing can be flared

Figure 5-19

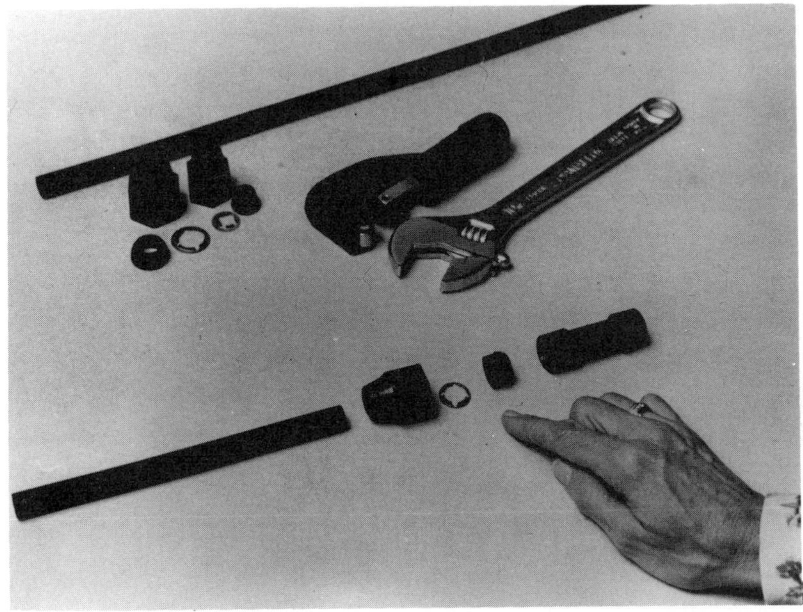

Polybutylene Tubing Assembly (QEST Products, Inc.)

and assembled in a similar fashion to copper tubing by using a special flaring tool.

5-6 Supporting and Hanging Pipe

Pipe must be supported both above and below ground to keep it on *line* and *grade* without leaking or breaking.

Below ground, clay tile, cast iron soil pipe, and plastic pipe are laid in trenches at a pitch of approximately $\frac{1}{4}$ in. per ft. Although small rocks themselves offer no particular hazard to clay tile and cast iron soil pipe, they can exert a high unit pressure if they are in the bottom of the trench, and so should be removed. The bottom of the trench should be smooth and have a uniform grade, and should be composed of undisturbed ground or a compacted backfill to ensure that there will be no settlement, which could cause sagging and leakage. The pipe must bear along its entire length, and this will require that at couplings and fittings the bottom of the trench must be scooped out so that they do not touch. Otherwise, the pipe would be supported primarily at the couplings, putting stress on the pipe barrel from the fill above. Figure 5-20 illustrates how the bottom of the trench is notched to receive the couplings.

Plastic soil pipe laid below ground is more flexible and has a tendency to conform to the bottom of the trench. It is also affected by rocks, which

Figure 5-20

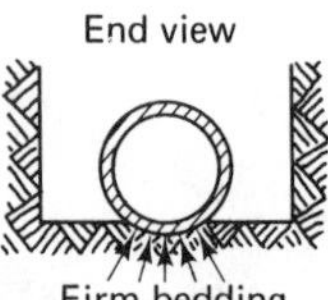

Proper bedding is imperative to develop the full load bearing capacity of the pipe.

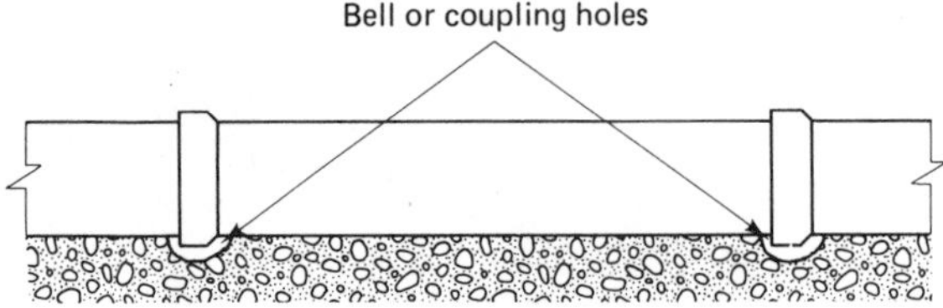

Provide uniform and continuous support of pipe barrel between bell or coupling holes for all classes of bedding.

Side view

Notched Trench to Receive Couplings

can puncture or crack it when the trench is backfilled. Thus, it is very important that, in addition to keeping the grade of the trench smooth and uniform, rocks and other hard objects be removed before the pipe is installed in the trench. The same is true for the first stage of backfill, which covers the pipe.

Backfilling a trench (Fig. 5–21) is done in stages to ensure that the pipe is adequately supported and not disturbed when the trench is filled. Stage one spreads fine dirt, sand, and gravel (no gravel around plastic) around and under the pipe up to the centerline to provide a bedding, and then it is

Figure 5-21

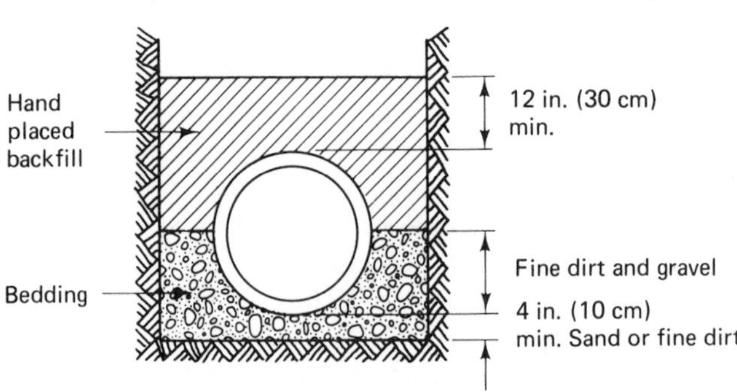

Backfilling a Trench

tamped lightly by hand; one must be careful not to disturb the pipe. This allows the bottom half of the pipe to support the load uniformly. The second stage of backfill covers the pipe by hand with approximately one foot of fine dirt, sand, and other fine fill material, and this is also tamped by hand in some cases. The third stage covers the trench, if it is not too deep, with other fill material, including small rocks, hard lumps of dirt, and other fill material. This stage is done either by hand or with machinery. Be careful not to disturb the pipe and avoid putting the machine tracks and pressure directly on the soft dirt in the trench. If several feet of dirt are required to fill the trench, tamping between layers is recommended to prevent voids or soft places in the trench fill.

Above ground, cast iron soil pipe, steel and iron pipe, plastic pipe and tubing, and copper tubing are supported both vertically and horizontally. Model plumbing codes such as the National Standard Plumbing Code specify that hangers and anchors shall be securely attached to the building construction at sufficiently close intervals to support the piping and its contents. Hangers and anchors are to be metal or other material with sufficient strength to support the pipe and its contents.

Vertical piping made of the following materials is supported as listed:

- Cast iron soil pipe—at the base and at each story height. The bases of cast iron stacks are to be supported on concrete, brick laid in cement mortar, metal brackets attached to the building construction, or by other methods approved by the local administrative authority. An alternate method is to support the stack at each floor level. That portion of the stack below the lowest floor is carried by floor clamps. This can be important as a cost saving procedure. Other piping materials are to be anchored so as to relieve the load from the stack at the base. Figure 5-22 illustrates two methods of supporting the base of a stack.

Figure 5-22

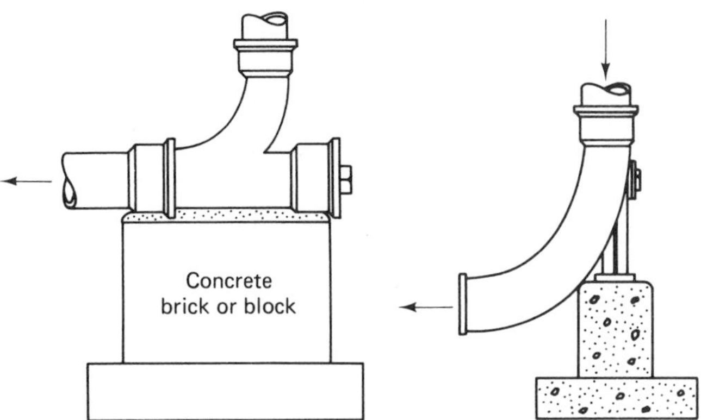

Stack Base Supports
(Cast Iron Soil Pipe Institute)

- Threaded pipe—at every other story height
- Copper tube—at each story height, but not more than 10-ft intervals
- Plastic pipe—at 3-ft intervals, at the end of branches, and such further support to keep vertical piping in alignment without restricting free movement of the pipe. Restraining joints and expansion joints should be installed as necessary in accordance with the manufacturer's recommendations.

Figure 5-23 illustrates several brackets for supporting vertical pipes.

Figure 5-23

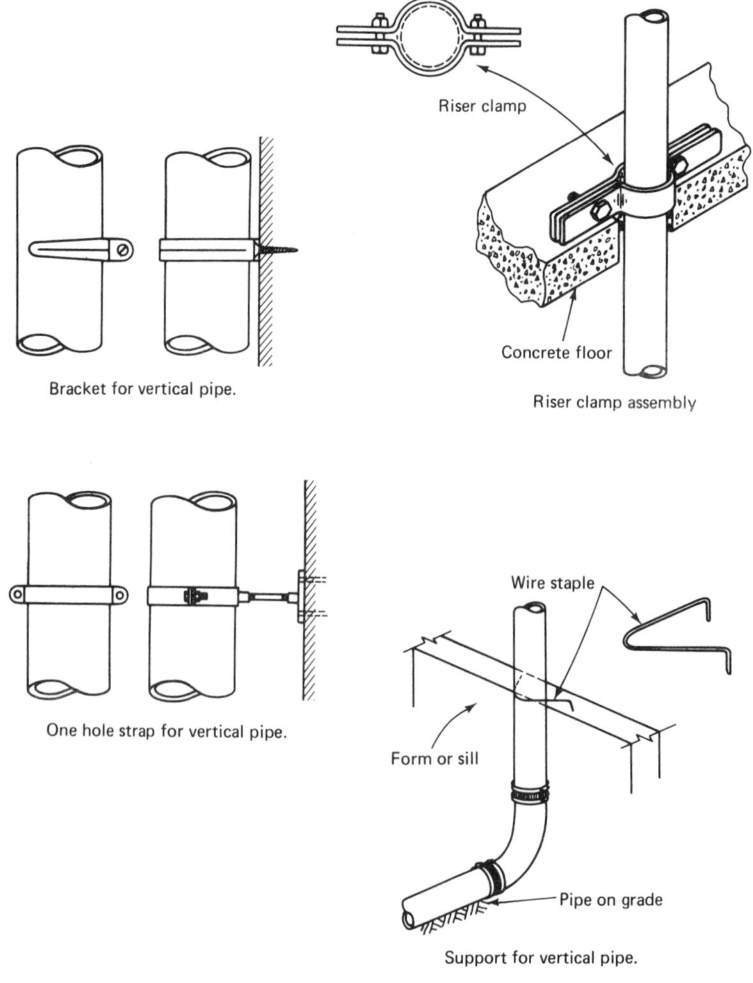

Vertical Pipe Supports (Cast Iron Soil Pipe Institute)

Horizontal piping made of the following materials is supported as listed:

- Cast iron pipe—five-foot intervals, except where 10-ft lengths of cast iron pipe are used, in which case 10-ft intervals are acceptable.
- Steel threaded pipe—at 8-ft intervals for diameters to one inch, and at 10-ft intervals for larger diameters.
- Copper tube—at 6-ft intervals for diameters to $1\frac{1}{4}$ inch, and at 8-ft intervals for larger diameters.
- Plastic pipe—at intervals close enough to prevent sag and misalignment. Support trap arms in excess of 3 ft as close as possible to the trap. Also consult standards organizations and manufacturer's instructions for the material used.

Figure 5–24 illustrates several hangers for supporting horizontal pipe. Figure 5–25 shows sway bracing for horizontal pipe.

Figure 5-24

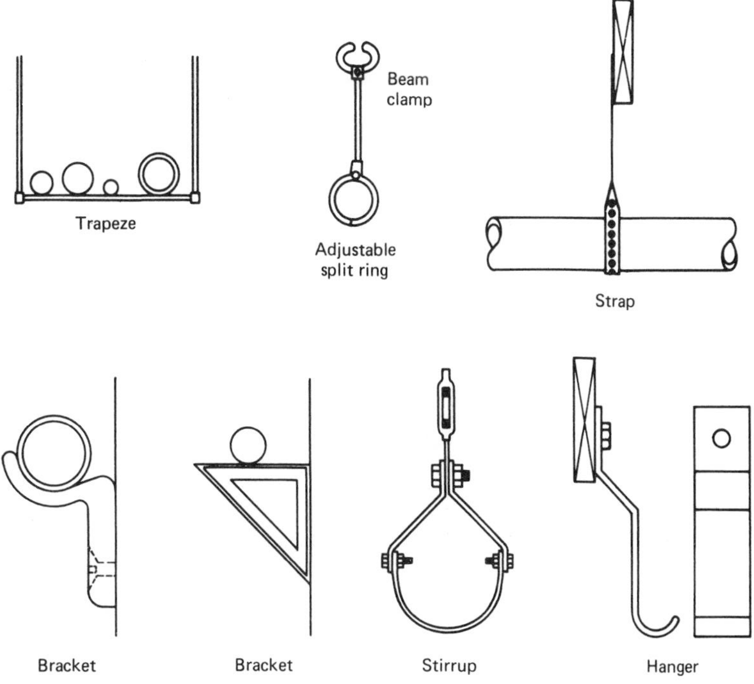

(Continued on following page)

Horizontal Pipe Hangers

Figure 5-24 (cont.)

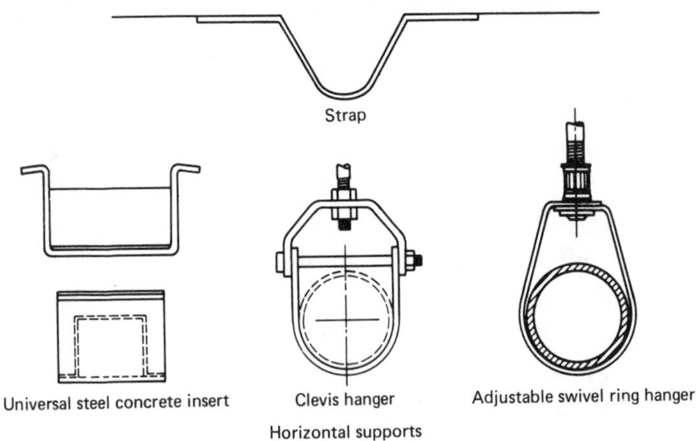

Horizontal supports

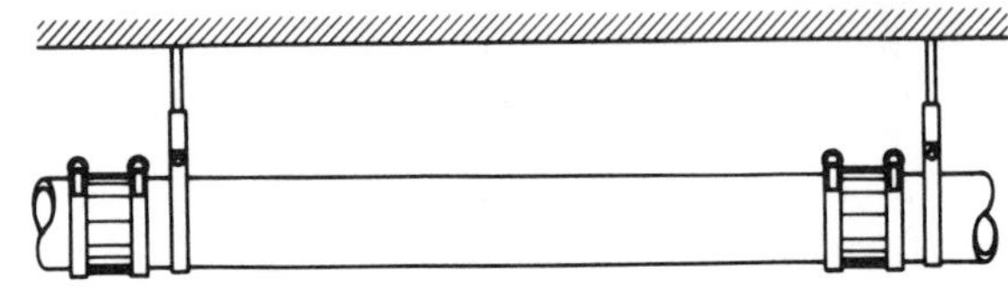
No-hub support

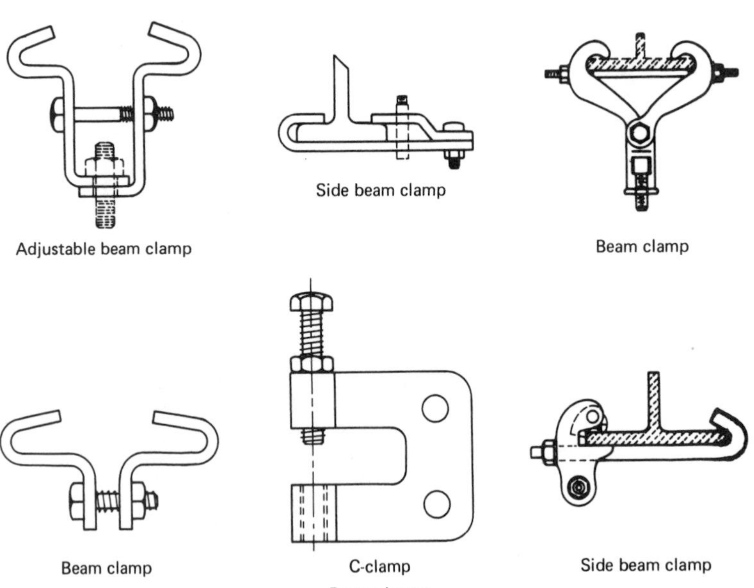

Beam clamps

Figure 5-25

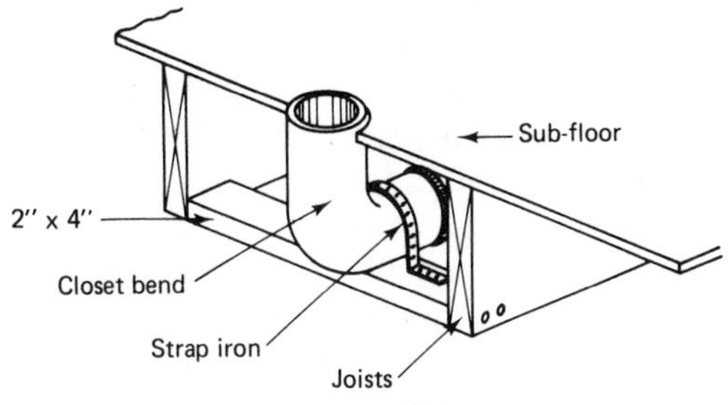

Bracing for closet bend

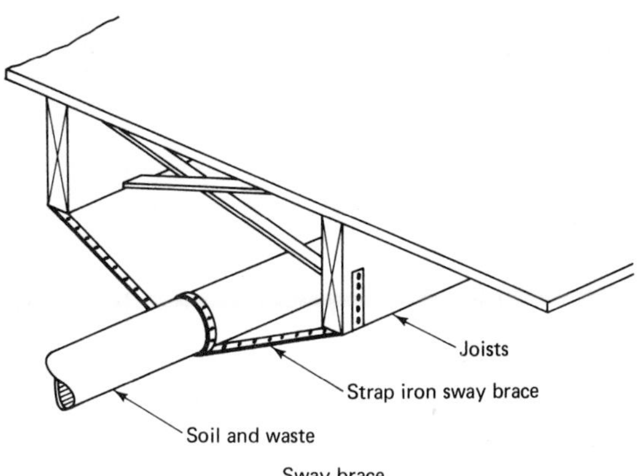

Sway brace

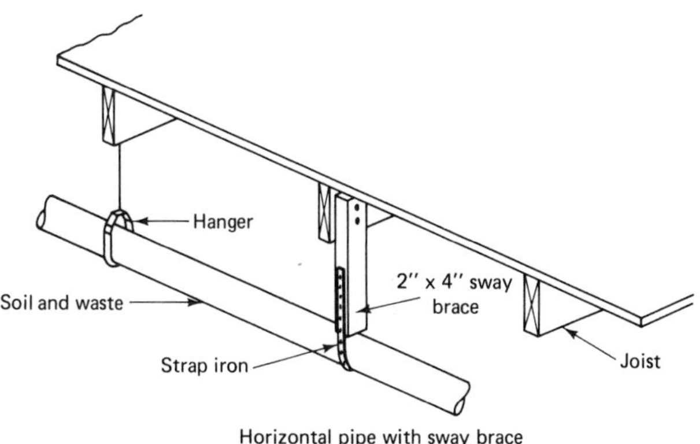

Horizontal pipe with sway brace

Sway Bracing

Plastic hangers for both DWV systems as well as pressure pipe have grown in acceptance because they permit movement from thermal expansion while still supporting the pipe. The metal straps used to support cast iron and other metal pipes are stronger than necessary to support plastic pipe because its weight is much less. The number of hangers and clamps is also increased for plastic pipe because of the tendency to sag; sometimes a hanger is used every two or three feet, or more often where joints and fittings should be supported. Figure 5-26 illustrates several hangers which are first attached to the pipe and then nailed to joists or plates.

Plastic clamps and insulators for pressure pipe are a recent development to replace metal brackets and many other materials that have been used to contain pressure pipes, especially behind walls where they are hidden. Metal strapping, mechanic wire, nails, and wedges have been used to secure pipes during the rough-in phase and sometimes result in rattling pipes, electrolysis, noise transmission, crimping, and abrasion. Figure 5-27 illustrates a family of clamps and their installation. Using a consistent set of clamps such as these, while a little more expensive initially, results in

Figure 5-26

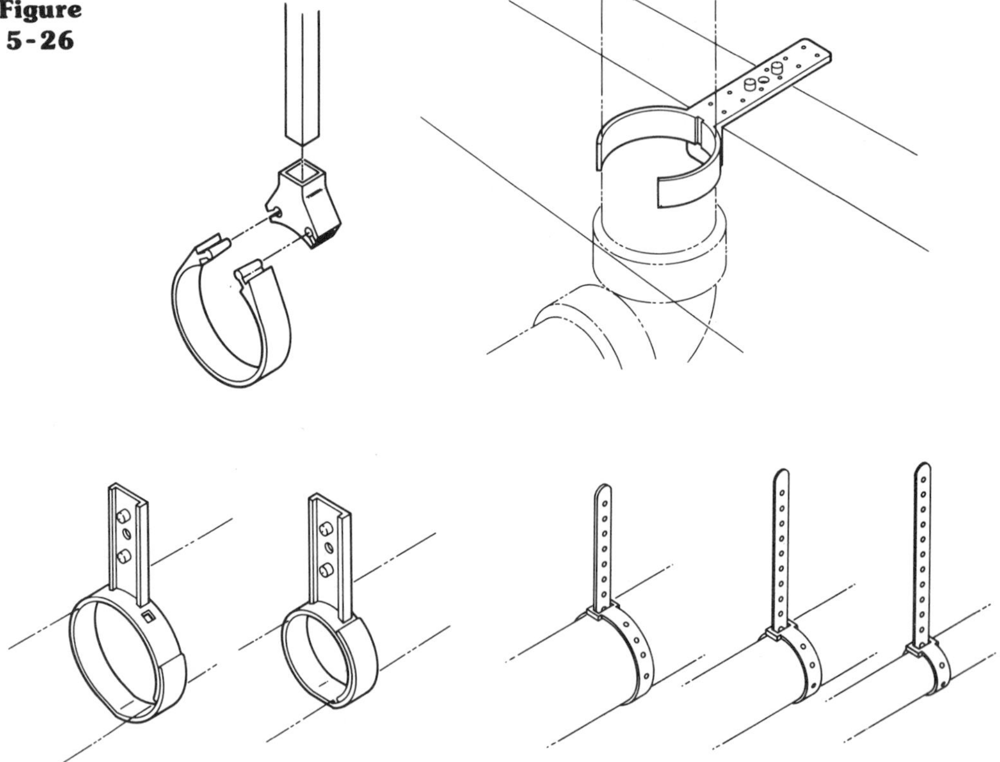

Plastic Soil Pipe Hangers (Specialty Products Company)

both a neater appearance and higher-quality installation. They also reduce installation time because they are ready made, thus resulting in some labor savings.

Figure 5-27

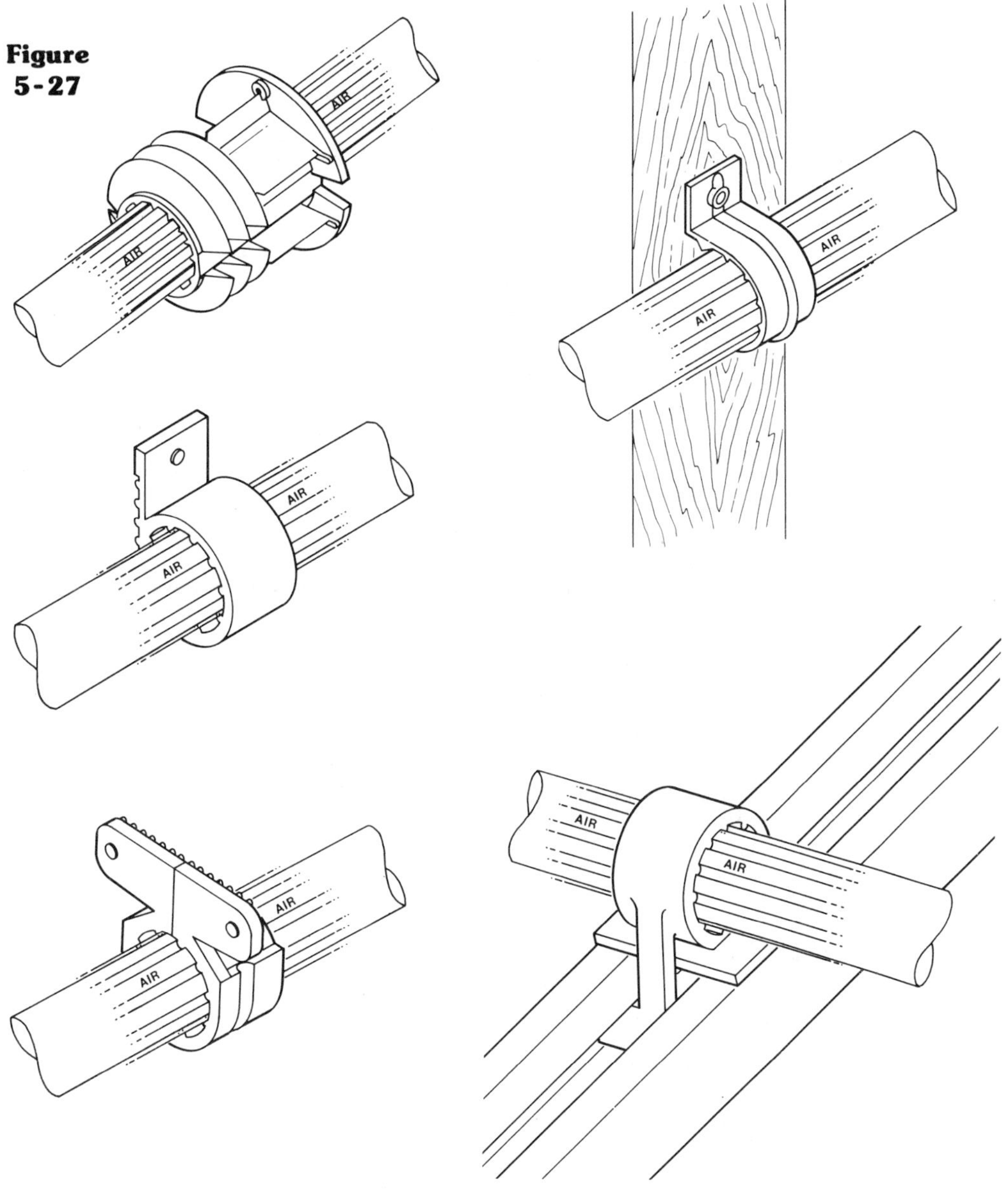

Plastic Pressure Pipe Clamps and Insulators (Specialty Products Company)

5-7 Summary of Practice

Piping systems consist of connected lengths of pipe and tube. The craftsman plumber assembles and then supports portions of the system above and below ground to transport water, waste, and soil to sewers and drains. The piping system must be gas- and watertight, as well as mechanically strong enough to support itself, the fluid it transports, and any backfill load for pipe underground, usually for the life of the building.

The selection of materials is influenced by the local plumbing code, costs of material and labor for assembly, and the knowledge and skill of the plumber. Thus, in addition to the craftsmanship necessary to fabricate a system that meets high standards, the plumber must have a knowledge of the materials that are available and specific techniques associated with their assembly. Typically, this knowledge is gained by reading trade journals and literature available, and from local jobbers and wholesale houses which supply the materials.

How well the plumber can perform the tasks of measuring, cutting, fabricating, installing, and then supporting pipe will affect both the appearance and the quality of the system. Usually the customer notices only the appearance of the plumber and the pipe if the system works, and it is this impression that often establishes his or her reputation as a craftsman. The point to be made is that not only must the system work, but it is to the benefit of the business that the plumber present a neat appearance, demonstrate skill in completing the tasks, and install a system that reflects pride in workmanship.

With traditional materials, length expansion of the pipe has never been a difficult problem to solve, but, with the newer plastics which expand more, on the order of five times as much for some plastics over copper, it is. This makes it very important that consideration be given to the design for expansion and installation of the system in advance, particularly where rigid plastic is used for tall stacks or horizontal runs. Typically, these will require both restraining fittings as well as expansion joints.

REVIEW QUESTIONS AND PROBLEMS

1. What are the steps to determine the length of pipe to join an offset?

2. In your own words, define the following measurements:
a. center-to-center dimension
b. center-to-face dimension
c. face-to-face dimension
d. make-up length

3. The distance between the centerlines of two water pipes to be connected is 31 in. (79 cm). The measurement between each centerline and the face of the fittings is found to be $1\frac{1}{2}$ in. (25 cm), and the recess into the fittings at each end is $\frac{3}{4}$ in. (19 cm). How long must the connecting length of pipe be?

4. Describe the three methods used to join cast iron soil pipe.

5. What purpose does double-hub cast iron pipe have?

6. What safety precautions must be observed when making a lead caulk joint?

7. What does caulking the joint with a caulking iron and hammer do to a lead caulk joint? What does "over caulking" do?

8. What advantages does a gasket joint have over a lead caulk joint?

9. Why is a torque wrench used to tighten no-hub clamps?

10. In your own words, describe how to cut, thread, and assemble threaded pipe, including a list of the necessary tools.

11. How is a capillary solder joint made?

12. What are the differences between 50–50 solder and 95–5 solder?

13. Describe how to make a flare joint, including a list of the necessary tools.

14. In your own words, describe the assembly procedure for each of the following plastic pipes:
    a. ABS rigid
    b. PVC rigid
    c. CPVC rigid
    d. PE flexible
    e. PB rigid
    f. PB flexible

15. What is the purpose of supporting and hanging pipes?

6 Science of Fluids

6-1 Introduction

The plumber who connects pumps, fixtures, and pipes to convey potable water, waste, soil, storm drainage, gas, compressed air, and other fluids applies many principles that govern the behavior of fluids. Waste and drainage flow are usually by gravity and plumbing vents are open to the atmosphere. Some of these are formulas; others are simple rules of thumb that have been derived from experience and more complex formulas applied to specific cases. Although the plumber does not typically engineer new systems and components that are installed, a knowledge of the scientific principles that are used to properly size and install them is helpful. This knowledge, coupled with the practical on-site experience, makes this craftsman a highly qualified and valuable person who can evaluate the compatibility of the system's various components and the probability that once installed together the entire system will function properly as a whole. Remember, it is the plumber who finally "makes it work." Replumbing and retrofitting do require engineering, that is, sizing and selecting such components as pumps, pipes, appliances, and fixtures, such that the remodeled system will operate both satisfactorily and safely as an entire unit. This is a large portion of the plumbing market. For example, many of the hot water heater sales are for replacement. Again this requires not only the application of the several rules of thumb, but an understanding of both the behavior of fluids as well as the operation of the components so that detailed calculations can be made for proper sizing.

6-2 Fluids: Liquids and Gases

Fluids are divided into two large categories; *liquids and gases.* Liquids such as water have definite mass and volume, but not definite shape. Gases such as air, when put in a closed container, will occupy the complete volume. If a liquid is poured into an irregular container, it will assume the shape of the inside. If it is poured into a shallow pan or on the floor, it will assume the shape of the surface and spread in all directions, with the free surface arranged in a plane perpendicular to the force exerted by the gravitational pull of the earth. This is considered to be *level* and, for practical purposes, is even with the horizon.

The volume occupied by a liquid is independent of its shape. The pressure and temperature, however, do affect its volume. Positive pressures applied to water, for example, reduce its volume about one percent for each 2000 lbf/in.^{2} (13.79 MPa) gauge pressure. Negative atmospheric pressures, or vacuum suction, however, significantly reduce the surface tension of water and other liquids and drastically reduce their ability to be pumped. Temperature has the effect of increasing the volume of liquids when it is raised, and decreasing the volume when it is lowered. The volume will change about two percent for each degree Fahrenheit of temperature change. Raising the temperature of liquids also reduces their surface tension and ability to be pumped. Relating the effects of pressure and temperature, water, for example, can be made to boil in a vacuum. Essentially this is what occurs when a pump cavitates; gas bubbles form at the intake and damage to the unit is imminent.

Gases have definite mass, but no definite shape. They expand or contract to fill the container holding them, and their volume is highly dependent on pressure and temperature. If a container such as an air chamber at a fixture holding one unit of a gas such as air under pressure is enlarged to $2V$ which doubles the available space, the contained gas will expand to occupy the entire volume, and the absolute pressure $p/2$ will be lowered proportionally. If the space again assumes its original volume, with no change in temperature, the gas will again occupy the total volume V and maintain its original pressure p. If the space further reduces the volume occupied by the gas by one-half $\frac{V}{2}$ with no change in temperature, the gas will still occupy all the space available and the absolute pressure will be doubled. This is an application of Boyle's Gas Law (Fig. 6–1).

6-3 Properties of Fluids: Density, Specific Weight, Specific Gravity, and Viscosity

The density of a substance is defined as the mass M of a body per unit volume V. Density is signified by the Greek letter rho (ϱ). In notation

$$\text{Density} = \varrho = \frac{\text{mass of body}}{\text{volume of body}} = \frac{M}{}$$

or simply

$$\varrho = \frac{M}{V} \tag{6-1}$$

In the English system of units, density is measured in slugs/ft^3, and in SI units, density is measured in kg/m^3. At freezing, 32°F (0°C), water has density of 1.94 slugs/ft^3 (1000 kg/m^3). The density of liquids such as water, oil, and nonfreezing solutions used in some solar systems can be determined by

Figure 6-1

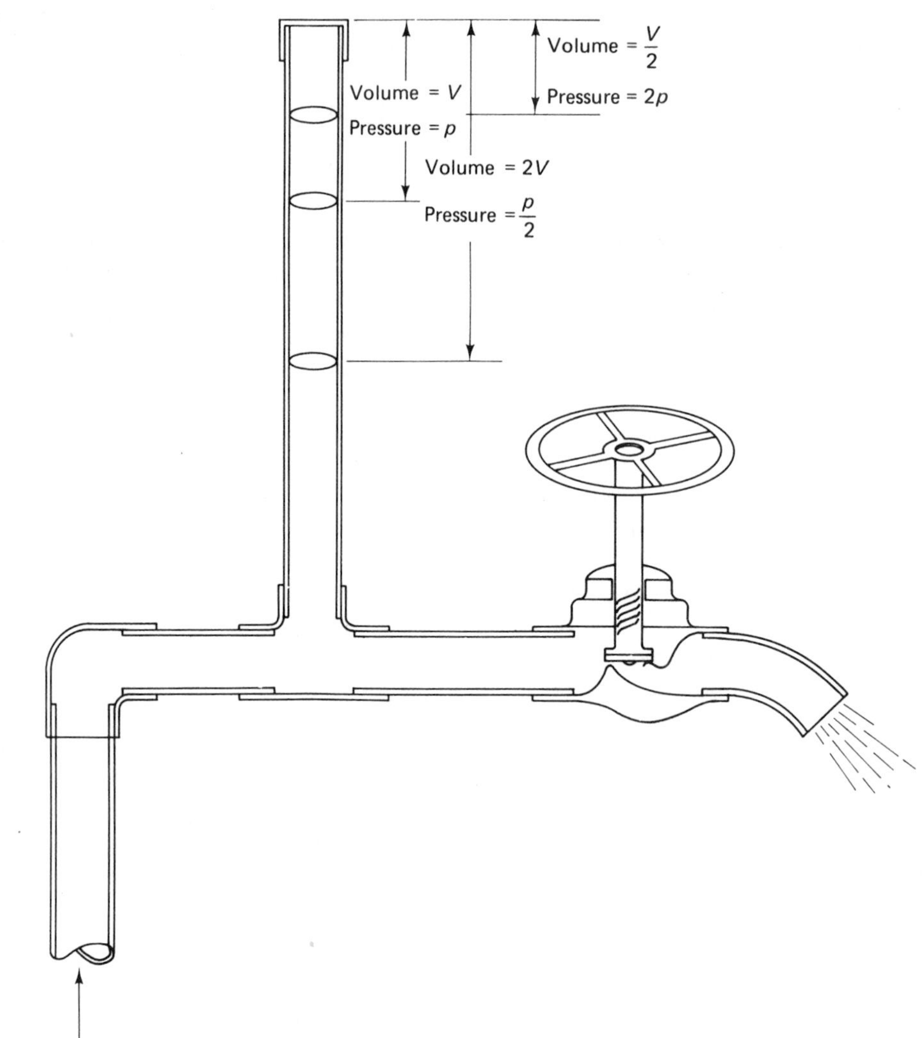

Pressure-Volume Relationships

several means. Among these are ASTM Test Method D 941, in which the density is calculated from the weight of a specified volume of the fluid, and ASTM Test Method D 1298, in which the density is read directly from a hydrometer lowered into the sample. Correction factors are applied to the results of these test methods to account for variance in temperature and the buoyancy of air. Since the volume of water and most other fluids varies with temperature, the results are corrected to some standard temperature, usually 60°F (15°C). This is because as the temperature increases, the volume also increases, and subsequently the density decreases. Corrected density values will thus be higher than actual readings for liquids where the density was determined at a temperature above 60°F (15°C), and lower for liquids where the density was determined at a temperature below 60°F (15°C). Appendix F gives the density of water at various temperatures in both English and SI units.

The specific weight of a substance is defined as the weight of a substance per unit volume. Specific weight is signified by the Greek letter gamma (γ). In notation,

$$\text{Specific weight} = \gamma = \frac{\text{weight of body}}{\text{volume of body}} = \frac{Mg}{V}$$

or simply

$$\gamma = \frac{Mg}{V} \tag{6-2}$$

In the English system of units, specific weight is measured in lbf/ft^3, and in SI units, specific weight is measured in N/m^3. At freezing, 32°F (0°C), water has a specific weight of 62.4 lbf/ft^3 (9802 N/m). The specific weight of water is used often in plumbing calculations.

Specific gravity (Sg) is defined as the ratio of the specific weight of a given substance to the specific weight of a standard substance. That is,

$$\text{Specific gravity} = \text{Sg} = \frac{\gamma \text{ of substance}}{\gamma \text{ of standard}}$$

or simply

$$\text{Sg} = \frac{\gamma}{\gamma_{std}} \tag{6-3}$$

The standard for computing the specific gravity of solids and liquids is water, and for gases is air. Both standards have an assigned value of 1 at standard conditions of 40°F (4°C), and 29.92 in. of water (76 cm of mercury, designated by Hg). For convenience, standard conditions can be approximated at room temperature and atmospheric conditions without introducing appreciable error.

Density and specific weight can be related by substituting the value for density ϱ into the formula for specific weight.
That is,

$$\gamma = \frac{Mg}{V}$$

$$\varrho = \frac{M}{V}$$

Substituting ϱ for M/V yields

$$\gamma = \varrho g \tag{6-4}$$

This relationship indicates that the specific weight of a substance equals its density multiplied by the acceleration due to gravity. It is the equivalent of saying that the force (weight) equals the mass multiplied by the acceleration ($F = Ma$) on a unit basis, which is essentially a restatement of Newton's second law of motion.

The relationship between density and specific gravity is defined from

$$\varrho = \frac{\gamma}{g}$$

and since $\text{Sg} = \frac{\gamma}{\gamma_{std}}$

it follows that

$$\varrho = \frac{(\gamma_{std})\,(\text{Sg})}{g} \tag{6-5}$$

Absolute viscosity, signified by the Greek letter mu (μ), is a measure of the internal resistance of a fluid to shear and indicates its relative resistance to flow. Viscosity is related to the internal friction of the fluid itself. Thick fluids flow much more slowly than thin fluids, indicating their internal friction is higher. Viscosity numbers are assigned to fluids to describe the relative differences in the ability of the fluid to flow in comparison with other fluids. Higher numbers are assigned to thicker fluids, lower numbers to thinner fluids. Viscosity is important in applications where the necessary flow rate is used to size pipes for a system. This would include the pressure pipes as well as drains, including storm drains.

In the English system, the unit of absolute viscosity is the lbf·s/ft^2 (read pound force second per foot squared). In the SI system, absolute viscosity units will be given in N·s/m^2 (read newton second per meter squared). Sometimes, the traditional cgs (centimeter-gram-second) metric system will be used where the units are in dyne·s/cm^2, which is called *poise* and given the symbol upper case P. Because the poise is a large unit of measure, the centipoise (10^{-2} poise) is used to simplify calculations.

Temperature affects the viscosity of a fluid inversely. That is, as the temperature increases, the viscosity decreases. Because of this, measures of viscosity must be reported with the temperature at which they were determined. Some common standard temperatures at which viscosities of water, oils, and air are determined are 100°F (37.7°C) and 210°F (98.88°C). The viscosity of water for different temperatures is given in Appendix F.

6-4 Pressure, Area, and Force Relationships

Pressure p is defined as force F per area A, where the units must be specified. In the English system, the units are

$$p = \frac{F}{A}\ \text{lbf/in.}^2 \tag{6-6}$$

sometimes designated as psi for convenience. In the SI system, the designated unit of pressure is the pascal (Pa) with units of N/m^2. Other units commonly used are the bar (14.5 lbf/in.^{2}) and the kgf/cm^2. Notice that the bar is approximately equal to one standard atmosphere, which is also sometimes used to designate pressure. Pressure unit designations are used to suit specific applications. Pressure reading gauges indicate the units on the dial. Many incorporate dual scales to display the pressure reading in more than one set of units, particularly now that the United States is undergoing the conversion to the metric system of units, although this is illegal in some countries that have converted totally to SI metric units. Much controversy has been generated over what will be the final designation for the unit of pressure worldwide in the future.

Pascal's law states that pressure is transmitted undiminished in all directions throughout a fluid, and that it acts normal (at right angles) to the surfaces of the walls of the vessel which holds it. Figure 6-2 illustrates this principle and related pressure, area, and force relationships, where the pressure in a confined system is

$$p = \frac{F_1}{A_1} = \frac{F_2}{A_2} \qquad (6\text{-}7)$$

Figure 6-2

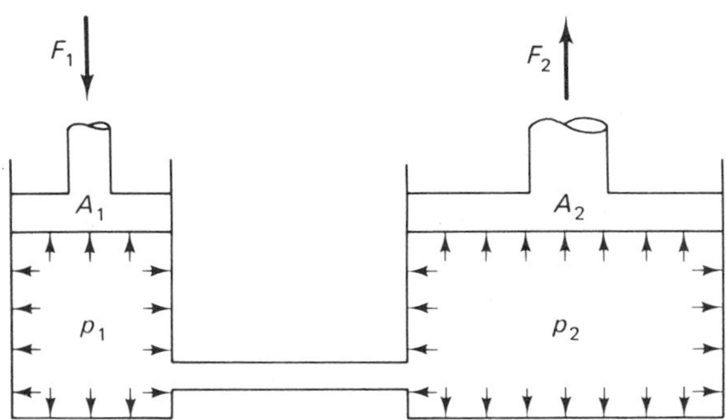

Pressure, Area, and Force Relationships

Example 6-1 A force of 60 lbf (266.7 N) is transmitted from a pump piston with an area of 1 in.^{2} (6.45 cm^2) to a cylinder piston with an area of 12 in.^{2} (77 cm^2). Compute the pressure in the system in lbf/in.^{2} and in Pa, and the force on the larger piston in lbf and N.

Solution:

Referring to Fig. 6-2, the pressure is computed from Eq. (6-6).

$$p = \frac{F_1}{A_1} = \frac{60\ \text{lbf}}{1\ \text{in.}^2} = 60\ \text{lbf/in.}^2$$

Converting lbf/in.^{2} to Pa (N/m^2) gives

$$\frac{60\ \text{lbf}}{\text{in.}^2} \times \frac{1\ \text{in.}^2}{6.45\ \text{cm}^2} \times \frac{10\,000\ \text{cm}^2}{1\ \text{m}^2} \times \frac{4.445\ \text{N}}{1\ \text{lbf}} = 413\,488\ \text{Pa (N/m}^2\text{)}$$

Typically, this would be written as 413 kPa (kilopascals). The force on the larger piston, F_2, is computed by solving Eq. (6-7) for F_2:

$$F_2 = \frac{F_1 A_2}{A_1} = \frac{(60\ \text{lbf/in.}^2)\ (12\ \text{in.}^2)}{(1\ \text{in.}^2)} = 720\ \text{lbf (3200 N)}$$

indicating that the multiplication of force is proportional to the ratio of the areas of the pistons.

6-5 Pressure, Specific Weight, and Height Relationships

The force exerted by a body at rest is attributed to the effect of gravity acting on its mass. That is,

$$F = w = Mg \text{ pounds force (N)} \tag{6-8}$$

where the acceleration due to gravity is 32.2 ft/s^2 (9.806 m/s^2).

When the volume of the body is specified, dividing the weight by the volume yields the specific weight [Eq. (6–2)]. That is,

$$\gamma = \frac{Mg}{V} = \frac{w}{V} \text{ lbf/ft}^3 \text{ (N/m}^3\text{)}$$

Substituting γV for F in Eq. (6–8), we derive

$$p = \frac{\gamma V}{A} \tag{6-9}$$

in lbf/in.^{2} (N/m^2) or other consistent pressure units.

If the volume component of Eq. (6–8) consists of a cube of base area A and height h, pressure becomes

$$p = \frac{\gamma A h}{A}$$

and the A's cancel to give

$$p = \gamma h \tag{6-10}$$

where p and h are in consistent units. This is the basic formula that relates pressure to the specific weight and height of the fluid, where γ is in lbf/ft^3 (N/m^3) and h is in ft (m).

If the Sg of the fluid is given rather the γ, which is often the case with liquids other than water, such as oil, the formula is modified to become

$$p = \gamma_{std}\text{Sg}h \tag{6-11}$$

These two formulas, then, are used to develop several other convenient relationships. In the English system of units, where water has a γ_{std} of 62.4 lbf/ft^3 and a Sg of 1,

$$p = (62.4 \text{ lbf/ft}^3)(1/144 \text{ in.}^2/\text{ft}^2)(h)$$

and

$$p = 0.433\, h \text{ in lbf/in.}^2 \tag{6-12}$$

Or, if h is given and pressure is to be solved for

$$h = (1/0.433)(p)$$

Figure 6-3

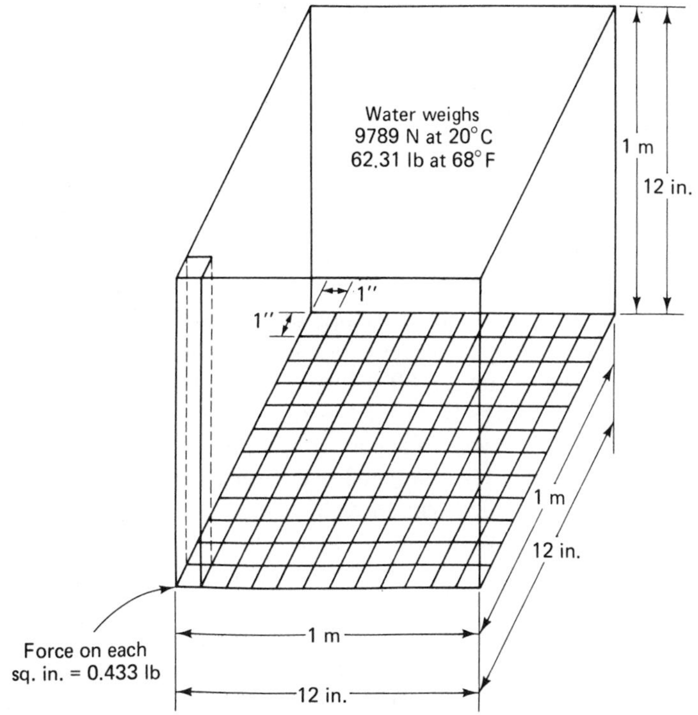

Pressure, Specific Weight, and Height Relationships

and

$$h = 2.31\,p \text{ in ft} \qquad (6\text{-}13)$$

called *pressure head* or *head.* It is illustrated in Fig 6-3.

Remember, however, for liquids other than water, the Sg must be included and Eq. (6-11) used, since its value will not be 1.

Example 6-2 Water at the third floor of an apartment building is 30 ft above ground level. Compute the static pressure on a gauge inserted in the pipe at the base of the building.

Solution:
From Eq. (6-12), the pressure is computed as

$$p = 0.433h = (0.433 \text{ lbf/in.}^2)(1/\text{ft})(30 \text{ ft}) = 13 \text{ lbf/in.}^2$$

Notice that the units from Eq. (6-12) were included so that the answer could be checked for consistency. Where possible, units should always be included in the calculation. This will avoid many errors.

6-6 Static Pressure in a Plumbing System

Water provided within municipalities has a supply pressure within the range of 25–80 lbf/in.^{2} (172–552 kPa). Above 80 lbf/in.^{2}, pipes, fixtures, and components of the system are not engineered for safety and operation, and a pressure-reducing valve is required. Most codes limit supply pressure to any fixture to 80 lbf/in.^{2} under no-flow conditions. The lower limit of operation for fixtures is about 15 lbf/in.^{2}, although flushometer water closets and urinals require 25 lbf/in.^{2} for blowout action and 15 lbf/in.^{2} for jet action. Private water systems using pumps are adjustable to establish high and low limits to the pressure such that the pump turns off at 50–60 lbf/in.^{2}, and turns on at 25–30 lbf/in.^{2}. The average static pressure is within this range. Buildings in which the pipes rise 40 feet or more reduce the pressure at the upper levels in accordance with Eq. (6–12), and where the pressure at the water main is near the lower limit, pumps may be necessary in the basement and at successive levels to maintain the required pressure necessary to operate fixtures in the system at the upper levels. For example, in a building 500 ft high, the total difference in pressure due only to elevation would be $h = 0.433 \times 500 = 217$ lbf/in.^{2}. This would be divided among four or more stages during the lift.

Although under ordinary circumstances the static pressure within the vent stack is considered to be atmospheric, that is, zero gauge pressure and 14.7 lbf/in.^{2} absolute pressure, there is the probability that if the drain plugs at its lower extremity, for example in the basement, the pressure caused by waste elevation in the stack could also pressurize the pipe, at least up to the first fixture. This would impose static pressures above 5 lbf/in.^{2} for each floor of the dwelling, and thus makes it imperative that joints in the waste pipe and vent system be pressure-tested after assembly to assure leak-free operation.

6-7 Pressure Measuring Devices

Both positive and negative (vacuum suction) pressures are encountered in plumbing systems. They are measured from a base line called a datum at one of the standard conditions, or at prevailing atmospheric conditions. Standard atmospheric pressure conditions at sea level are taken as 29.92 in. of mercury (symbol Hg) (760 mm Hg), which has a Sg of 13.6. Using Eq. (6–11) and $\gamma_{std} = 62.4$ lbf/ft^3, we compute a standard atmosphere from

$$p = \gamma_{std} \mathrm{Sg} h$$

$$p = (62.4 \text{ lbf/ft}^3\left(\frac{1}{144 \text{ in.}^2/\text{ft}^2}\right)(13.6)(29.92 \text{ in. Hg})\left(\frac{1}{12 \text{ in./ft}}\right)$$

and

$$p = 14.7 \text{ lbf/in.}^2$$

Figure 6-4

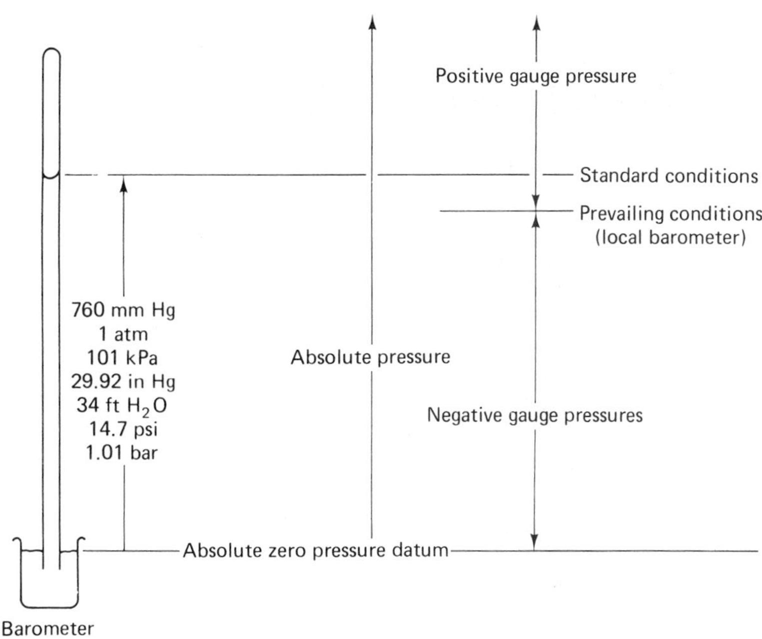

Standard Atmospheric Pressure

Negative pressures below atmospheric are read in inches of water (symbol H_2O) or inches of Hg from vacuum gauges placed in pump inlet lines or across orifices. Sometimes dual scales are provided on the same gauge for personnel convenience. Pressures below one inch of Hg are measured in microns μ (given the greek letter symbol mu) where

$$1 = \frac{1}{1\ 000\ 000} \text{ meter}$$

Figure 6–4 lists common values and units for standard atmospheric conditions. Gauge pressures above atmospheric conditions are read as positive gauge pressures, whereas pressures below atmospheric conditions are read with a negative sign. By convention, it is understood that positive gauge pressures do not include local atmospheric conditions, and where absolute pressure is used, it is signified as "p abs."

Manometers are simple gauges that measure heights or height differences in tubes filled with liquid. Figure 6–5 illustrates the application of a differential manometer measuring the pressure drop across an orifice in a pipe. They differ from bourdon gauges in that they measure pressure differences across two points rather than the gauge pressure at one point in the pipe using atmospheric conditions as the datum. Valving and fittings permit isolation and removal of the gauge from the system. Standard orifice plates supplied with the differential manometer conveniently scale differential pressures to the calibrated scale on the face of the instrument.

Figure 6-5

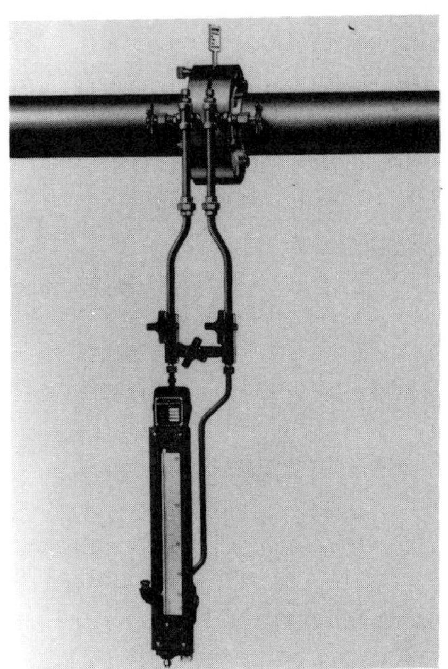

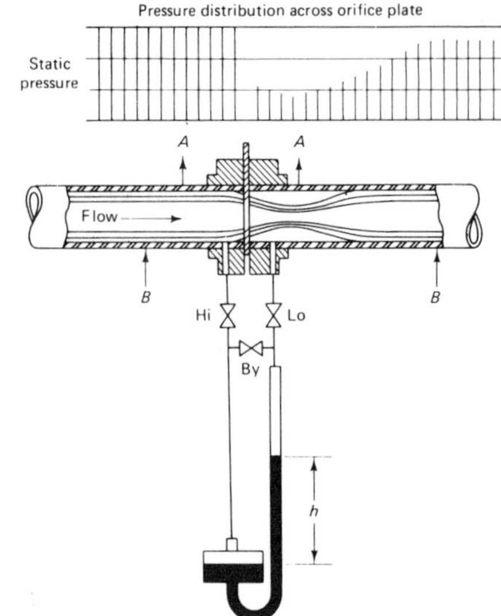

Differential Manometer

Bourdon gauges such as that shown in Fig. 6-6 measure both positive and negative pressures from local atmospheric conditions. They are constructed from a flattened bent tube attached to a movable pointer. Pressures greater than local atmospheric conditions cause the tube to straighten slightly, with the end moving linearly, and the pointer to indicate positive pressures. Vacuum suction pressures less than local atmospheric conditions cause the tube to adopt a closer bend and the pointer to indicate negative pressures. Positive pressures are scaled in lbf/in.^{2}, kPa, bars, ft of head, meters of head, kg/cm^2, and others, whereas negative pressures are usually scaled in inches or Hg or H_2O. To insure accuracy, bourdon gauges are sized to systems such that they register normally expected pressures at half scale; that is, the pointer reads half the highest value on the gauge dial at normal pressure.

6-8 Maximum Static Lift

Maximum static lift refers to the height that a column of water can be lifted with vacuum suction. At sea level and standard conditions of 60°F (15.6°C), this is approximately 34 ft (10 m).

Figure 6-6

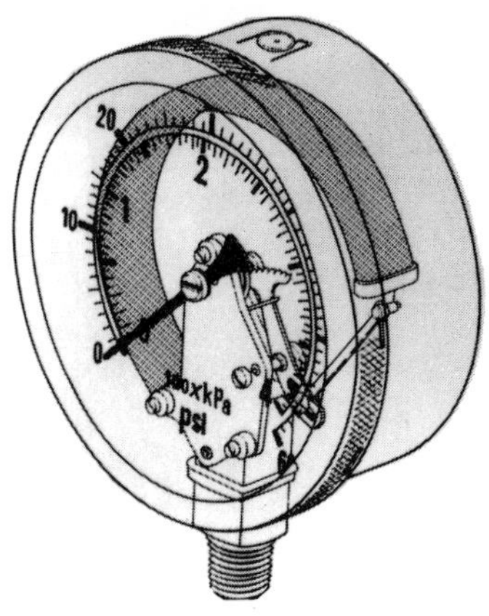

Bourdon Tube Pressure Gauge

If a tube filled with mercury is inverted and the end is submerged in a reservoir, a vacuum is formed at the top closed end. This is a simple barometer gauging atmospheric pressure. The weight of the mercury column has a tendency to fall, while atmospheric pressure acting on the surface of the reservoir tends to support it (Fig. 6-7). At sea level and standard conditions of 60°F (15.6°C) the column will stand at approximately 29.92 in. Converting this to the length of a column of water, using a Sg of 13.6 for Hg, we obtain

$$h = (29.92 \text{ in. Hg})(13.6 \text{ Hg Sg})(\tfrac{1}{12} \text{ ft/in.}) = 33.9 \text{ ft of } H_2O$$

The atmospheric pressure p_a acting on the reservoir supporting the column can be computed from Eq. (6-11). That is,

$$p_a = \gamma_{std}\text{Sg}h = (62.4 \text{ lbf/ft}^3)(13.6)(29.92 \text{ in. Hg})(1/1728 \text{ ft}^3\text{/in.}^3)$$

and

$$p_a = 14.7 \text{ lbf/in.}^2$$

This is commonly referred to as one standard atmosphere.

At elevations above sea level, the atmospheric pressure decreases. Table 6-1 lists the decreasing pressures in ft (m) and lbf/in.^{2} (kPa) to 10,000 ft (3048 m) at a temperature of 60°F (15.6°C). At 2500 ft above sea level at 60°F, the atmosphere would support a column of water of only 31 ft.

As the temperature of water (and other fluids) increases, its specific weight, which is about 62.4 lbf/ft^3 at 60°F, decreases. This is because the molecules within the liquid become increasingly active, spread out, and occupy more space with increases in temperature. Molecules near the surface break through and evaporate, exerting a partial pressure on the surface. This is referred to as the vapor pressure or saturation pressure. At sea level and 60°F, the vapor pressure p_v of water is 0.591 ft (0.180 m) or 0.256 lbf/in.^{2}

Figure 6-7

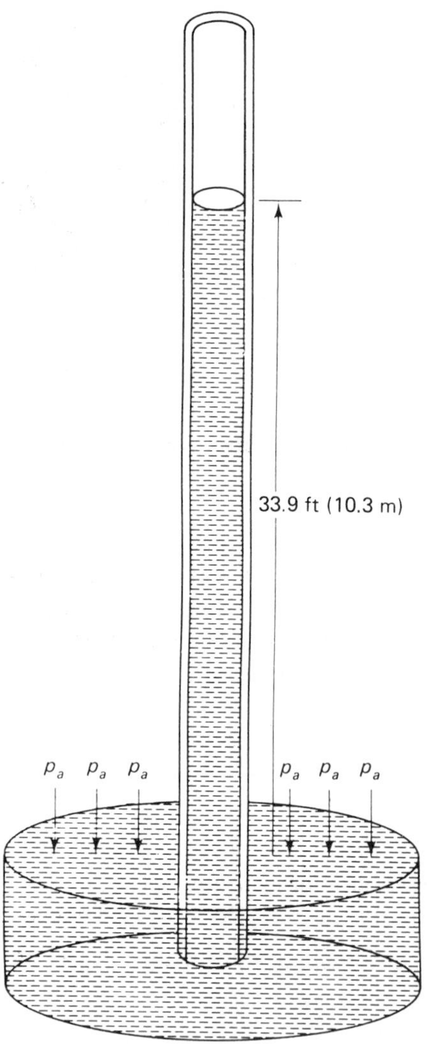

Maximum Static Lift

(1.766 kPa). Since the vapor pressure and degree of evaporation increase with temperature, water can be boiled either by heating or by reducing the vapor pressure with a vacuum. Appendix F lists the vapor pressure with the physical properties of water.

The relationship among the elevation (atmospheric pressure), vapor pressure, and height to which a column of water can be lifted is given by

$$h = p_a - p_v \qquad (6\text{-}14)$$

That is, first the atmospheric pressure due to elevation above sea level is

Table 6-1 **Atmospheric Pressure at Various Altitudes at 60°F**

Elevation, (ft)	(m)	h (ft, H_2O)	h (m, H_2O)	P_a (lbf/in^2)	P_a (kPa)
Sea Level	Sea Level	34.0	10.36	14.7	101
500	152	33.3	10.15	14.4	99
1000	305	32.7	9.97	14.2	98
1500	457	32.1	9.78	13.9	96
2000	610	31.5	9.60	13.6	94
2500	762	31.0	9.45	13.4	92
3000	914	30.4	9.27	13.2	91
3500	1067	29.9	9.11	13.0	90
4000	1219	29.4	8.96	12.7	88
5000	1524	28.4	8.66	12.3	85
6000	1829	27.4	8.35	11.9	82
7000	2134	26.4	8.05	11.4	79
8000	2438	25.5	7.77	11.0	76
10000	3048	23.7	7.22	10.3	71

computed, and then the vapor pressure resulting from increases in temperature is subtracted from it. An example will make this clear.

Example 6-3 At 2500 ft (762 m) and 160°F (71.1°C), to what height can a static column of water be lifted?

Solution:
With reference to Table 6–1, an elevation of 2500 ft will reduce the column height to 31.0 ft H_2O. With reference to Appendix F, a temperature of 160°F will raise the vapor pressure to 11.0 ft approximately. Substituting in Eq. (6–14), we have

$$h = p_a - p_v = 31.0 \text{ ft} - 11.0 \text{ ft} = 20.0 \text{ ft } H_20$$

Thus, the effects of elevation and temperature have reduced the maximum suction lift by 14 ft approximately.

6-9 Displacement and Buoyancy

Displacement V is the computed measure of the volume of an object. It refers to how much space an object occupies. In dry measure, it has the units of ft^3, m^3, in.^{3}, and cm^3. As a liquid measure it has the units of gallons, quarts, liters, and milliliters. However, the two may be used interchangeably, so long as the units are consistent.

A number of familiar objects in plumbing must have the volume computed. That is, it is important to know how much liquid a tank will hold,

Figure 6-8

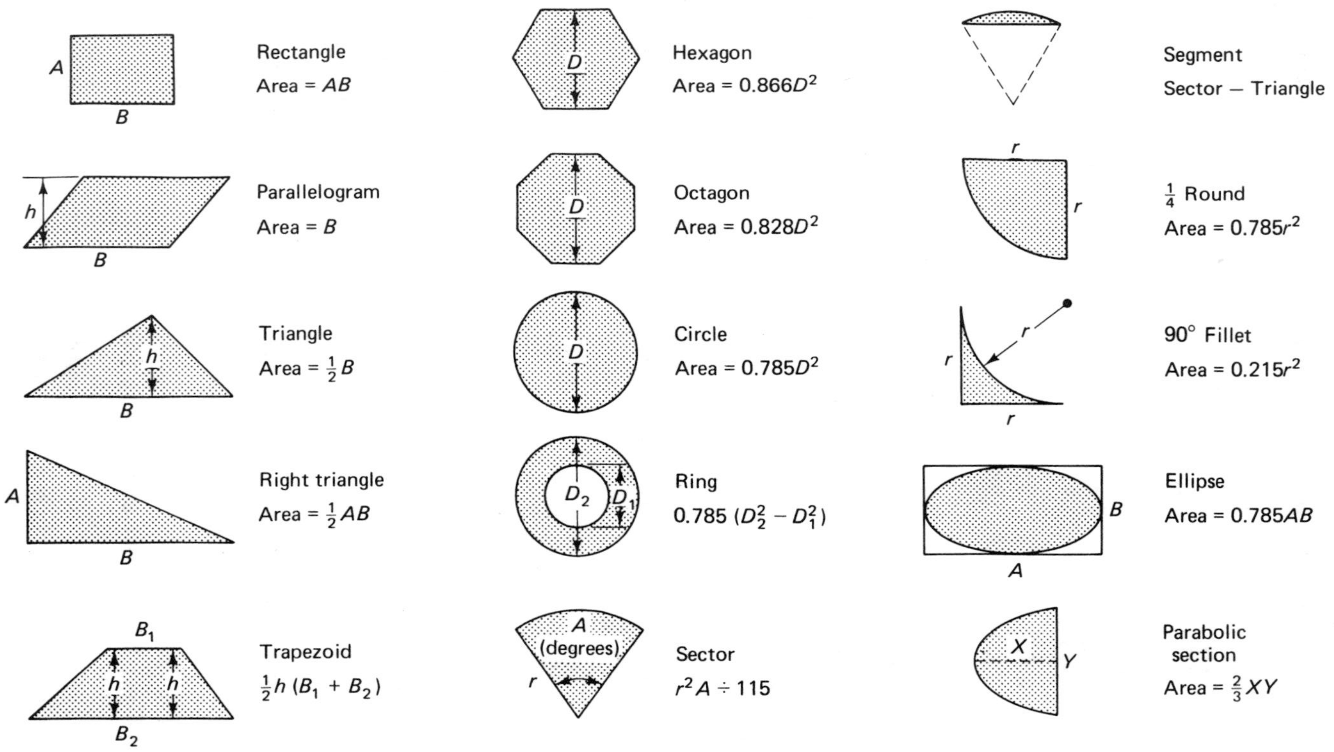
A
B
Rectangle
Area = AB
h
B
Parallelogram
Area = B
h
B
Triangle
Area = ½B
A
B
Right triangle
Area = ½AB
B₁
h
h
B₂
Trapezoid
½h (B₁ + B₂)
D
Hexagon
Area = 0.866D²
D
Octagon
Area = 0.828D²
D
Circle
Area = 0.785D²
D₂
D₁
Ring
0.785 (D₂² − D₁²)
A
(degrees)
r
Sector
r²A ÷ 115
Segment
Sector — Triangle
r
r
¼ Round
Area = 0.785r²
r
r
r
90° Fillet
Area = 0.215r²
B
A
Ellipse
Area = 0.785AB
X
Y
Parabolic
section
Area = ⅔XY

Common Areas and Volumes

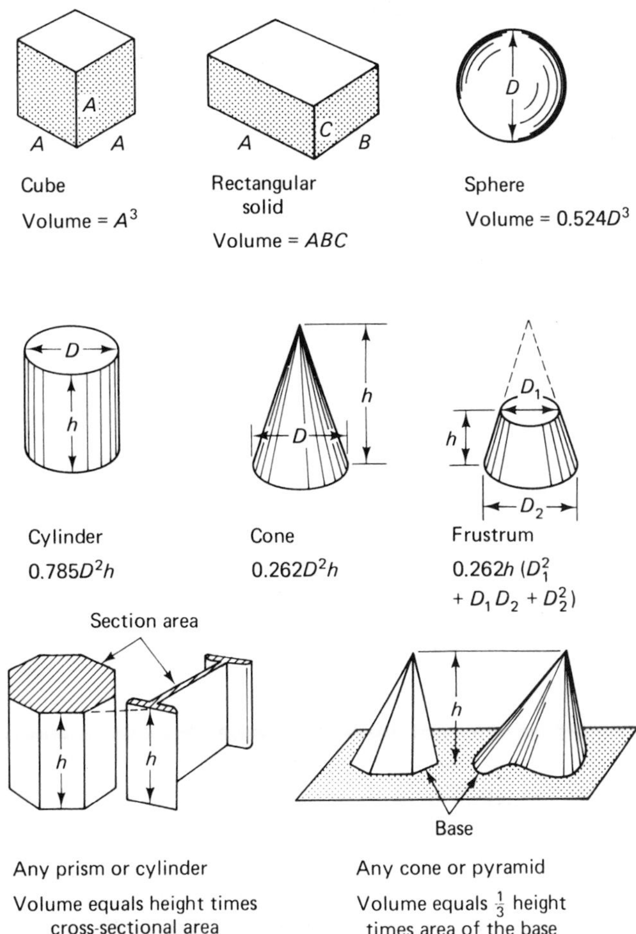

Figure 6–8 (Cont'd)

how much space it will occupy, and what surface area must be covered by insulation or other covering. Figure 6–8 illustrates several common areas and volumes of geometric shapes, together with the formulas for their computation. Notice that taking a number such as the diameter to a power, means to multiply that number by itself the number of times indicated. An example will make this clear.

Example 6-4 A cylindrical tank must go into a space with a maximum height of 6 ft (1.83 m) and hold 100 gal. Compute the diameter of the tank.

Solution:
From Fig. 6–8, the formula for computing the volume of the tank is

$$V = 0.785\ D^2h$$

Solving for D, we obtain

$$D^2 = \frac{V}{0.785h}$$

and

$$D = \sqrt{\frac{V}{0.785h}}$$

It must be remembered that to be correct, each of the factors D, V, and h must be in the same units. Since the diameter of the tank is sought, the volume must be converted to ft^3. Thus,

$$D = \sqrt{\frac{(300 \text{ gal})(0.134 \text{ ft}^3/\text{gal})}{(0.785)(6 \text{ ft})}}$$

$$D = \sqrt{8.54 \text{ ft}^2} = 2.92 \text{ ft}$$

or approximately 2 ft, 11 in. in diameter

Buoyancy determines whether an object floats or sinks in a liquid. When an object floats either above or below the surface in a static fluid, the resultant buoyant force equals the weight of the fluid displaced and acts vertically in the direction of the free surface. The resultant buoyant force is nearly independent of the relative pressure that the fluid exerts on the exterior of the object because of its depth in the fluid, and equals the difference between the force acting above the object and that acting below the object. Thus, if the buoyant force that a liquid such as water exerts is greater than or equal to the downward weight of the object, it will float at or in equilibrium below the surface, and if it is not, the object will sink toward the bottom. The force due to buoyancy is computed by using

$$F = \gamma V \qquad (6\text{–}15)$$

Figure 6–9 illustrates an application of buoyancy in plumbing.

Example 6-5 A round 4-in water closet float ball of negligible weight lifts the end of the float rod to depress the plunger in the ball cock valve. If the ball were submersed, how much upward force would it exert?

Solution:
From Fig. 6–8 the volume of a sphere is

$$V = 0.524D^3$$

Since the ball is submersed, the upward force equals the volume of the ball multiplied by the γ of the water. At standard conditions, this would be 62.4 lbf/ft^3. Substituting in Eq. (6–15) using consistent units, we obtain

Figure 6-9

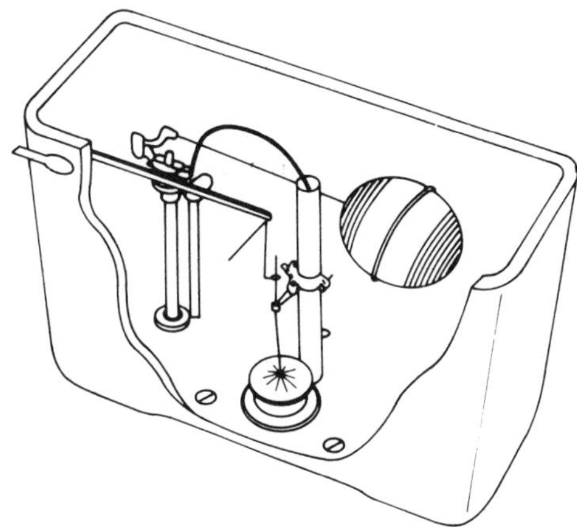

An Application of Buoyancy (Ball Cock Float Ball)

$$F = \gamma V$$

$$F = (62.4\ \text{lbf/ft}^3)(0.524)(0.333\ \text{ft})^3 = 1.21\ \text{lbf}$$

This is the upward force, if it is assumed that the weight of the ball is negligible, that the water would exert on the ball.

6-10 Summary and Applications

The operation of a plumbing system is determined by the behavior of the fluid. Usually, the fluid is a liquid, water, and for practical purposes its γ can be assumed to be 62.4 lbf/ft^3 at room temperature without introducing appreciable error. At higher temperatures, for example at 160°F, the density is lowered and vapor pressure is increased, and allowances for these changes must be made. Such a change would reduce the ability of a pump to draw a static head from about 34 ft at standard conditions to about 20 ft at an elevation of 2500 ft above sea level.

The static pressure exerted by water is in direct proportion to its height. For every foot of height, the pressure will rise about 0.433 lbf/in.^{2} on a pressure gauge. If the pressure in lbf/in.^{2} is known, the height can be computed by multiplying it by 2.31, which is the quotient of 1/0.433.

Displacement is a measure of size in cubic units and determines the space that an object will occupy. Buoyancy is the product of the density and volume of water that an object displaces. Floats, such as that used in water closets, rely on buoyancy. If the object is lighter than the volume displaced, it will float; if it is heavier, it will sink. Wood floats because it is lighter than an equal volume of water. A brick sinks because it is heavier than the water it displaces.

REVIEW QUESTIONS AND PROBLEMS

1. What are the differences between liquids and gases?
2. What is the difference between the density and specific weight of a substance?
3. A 2-in. schedule 40 steel pipe carries water at a pressure of 50 lbf/in.^{2} (344.8 kPa). Compute the force against a cap plug threaded over the end.
4. Neglecting friction, at what pressure would a pump have to operate at the outlet side to maintain a street pressure of 50 lbf/in.^{2} on the sixth floor 70 ft above?
5. Compute a table giving the theoretical maximum static lift possible in feet for the elevations given in Table 6–1 (down) and temperatures (across) from freezing (32°F) to 180°F in 20° increments, i.e., 32°F, 60°F, 80°F, 100°F, 120°F, 140°F, and 160°F.
6. Compute the cross-sectional area of the material in a 2-in. schedule 40 steel water pipe.
7. Compute the volume of a cylindrical tank 30 in. in diameter and 6 ft long.
8. What is the capacity of the tank in Problem 7 in gallons and liters?
9. Assuming a round 5-in.-diameter water closet float of negligible weight leaks, how much water could it retain and still exert a buoyant force of 1 lbf?
10. List two additional applications of buoyancy in plumbing.

7 Flow in Pipes

7-1 Introduction

Much of the plumber's work involves constructing piping systems that convey fluids, including water, air, gas, and waste. Some notion of how these fluids negotiate the confines of a pipe is necessary to insure not only that the fluid reaches its destination at fixtures and sewers without leaking, but that it is available with sufficient flow capacity to render the system operational.

The size of the pipe determines its average velocity for a given flow rate. Increasing the size reduces the velocity. Reducing the size increases the velocity. Reducing the size also increases internal friction. Friction, in turn, causes the pressure to drop as the fluid negotiates the piping system. Velocities in pressure pipes should not exceed 6–8 ft/s (1.8–2.4 m/s). In waste pipes, although the velocity must be sufficient to promote scouring of the inner surfaces, if the pipe is too small, it has a tendency to clog with the material being carried by the water stream. If the pipe is too large, there is the tendency for the solid matter to settle out and thus cause an obstruction. Larger than necessary pipes for pressure and waste increase material costs unnecessarily.

The flow rate through a pipe running full is computed from the area of cross section and the average velocity of the flow. As the pressure against the fluid is increased upstream, the average velocity will increase when the pipe is opened to the atmosphere downstream. This increase in velocity promotes friction between the molecules that make up the fluid stream and, in turn, generates friction and losses as the fluid travels through the length of the pipe. This has the effect of lowering the pressure near the outlet.

Sizing pipes, then, requires an understanding of how the available pressure, pipe size, friction losses, and desired flow rate vary with each other. The object is to be able to size pipes so that they provide the necessary flow with the available pressure, at the least necessary cost, with the velocity maintained within the recommended limits. Local plumbing codes also must be considered and followed when there is a variance between computed and required standards.

7-2 Cross-section Area of a Pipe

The cross-section area of a pipe equals

$$A = \pi r^2 = \frac{\pi D^2}{4} \qquad (7\text{-}1)$$

where $\pi = 3.14$

r = radius of the pipe

D = diameter of the pipe

Notice in Fig. 7-1 that $4r^2 = D^2$ or $r^2 = \frac{D^2}{4}$. That is, the diameter squared is four times larger than the radius. Computing the area of the passage within a pipe is almost always simplified by using the diameter rather than the radius. This is true because pipe and tube sizes most often refer to the nominal size not to the actual internal diameter. Internal diameters are taken from tables that give the actual sizes for the pipe specified.

7-3 Velocity of Fluids in Pipes

When fluid flows in a pipe, its volume V is displaced from one place along the pipe to another. If successive particles passing a fixed point in the stream have the same velocity v, then the flow is said to be steady. The

Figure 7-1

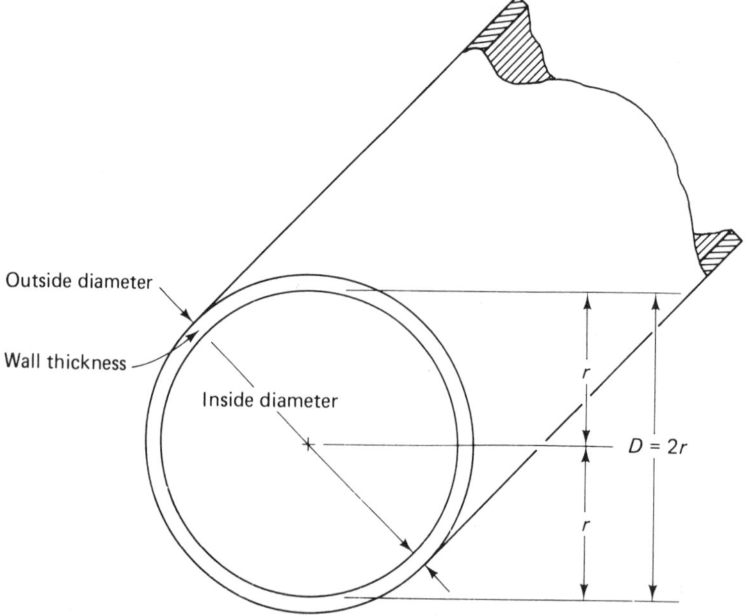

Pipe Wall Diameters

average velocity of the stream equals the distance it travels in the pipe l, divided by the time t. The velocity of a smooth flowing stream at any one place along the pipe is greatest near the center of the pipe cross section. At the wall of the pipe the velocity is zero. Near the wall there is flow, but at the wall there is no flow (Fig. 7–2). Thus, the velocity of flow through the cross section of the pipe is not uniform and an average velocity for the total flow must be used rather than the velocity at any one place in the pipe. This is derived by separating the volume flow V into its components of area and length. In notation

$$\frac{V}{t} = \frac{A \times l}{t} \tag{7-2}$$

Notice that since $l/t = v$, by substitution

$$\frac{V}{t} = A \times v$$

Solving for the average velocity of the stream within the pipe, we obtain

$$v = \frac{V}{A \times t} \tag{7-3}$$

where the velocity v is computed in ft/s (m/s), the volume V is observed in gallons (liters), the time t is measured in seconds and the area A is computed from the actual cross section of the pipe. An example will make the computation clear.

Figure 7-2

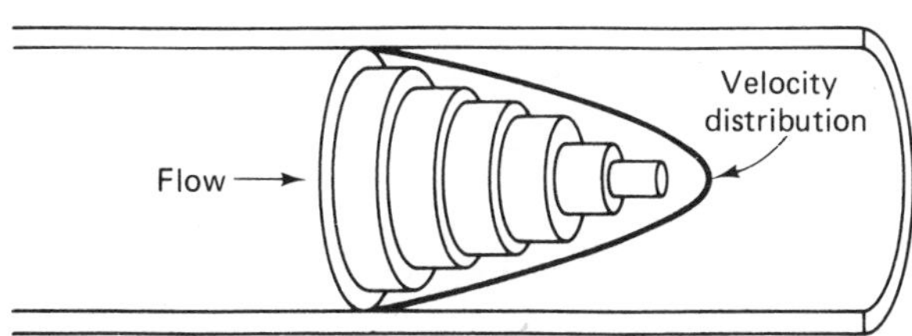

Velocity Distribution across a Pipe

Example 7-1 A 1-in. schedule 40 threaded steel pipe is observed to fill a 55 gal drum in 5 minutes. Compute the average velocity of the fluid through the pipe (refer to Fig. 7–3).

Solution:

The inside diameter of the schedule 40 pipe is 1.049 in. and the area is computed as

$$A = \frac{\pi D^2}{4} = \frac{(3.14)\ (1.049)^2}{4} = 0.864 \text{ in.}^2 \ (5.57 \text{ cm}^2)$$

The volume of the drum (55 gal) and the time (min) are converted to consistent units.

Figure 7-3

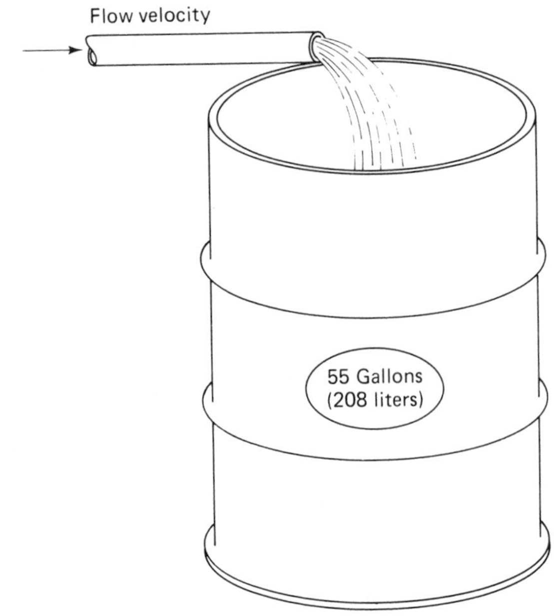

Flow Velocity

$$V = (55 \text{ gal})(231 \text{ in.}^3/\text{gal}) = 12\,705 \text{ in.}^3$$

and

$$t = (5 \text{ min})(60 \text{ sec/min}) = 300 \text{ s}$$

Finally, solving for the average velocity using Eq. 7–3, we obtain

$$v = \frac{V}{A \times t} = \frac{(12\ 705 \text{ in.}^3)}{(0.864 \text{ in.}^2)(300 \text{ s})} = 49 \text{ in./s} = 4.08 \text{ ft/s } (1.24 \text{ m/s})$$

7-4 Flow Rate

The flow rate q in a branch circuit describes the volume flow V per unit time t.
That is,

$$q = \frac{V}{t} \tag{7-4}$$

It is measured in gallons per minute (gpm) and liters per minute l/min). Flow rate is used to measure the capacity of pipes of different sizes and materials when the velocity and pressure losses due to friction are given. This is common in plumbing applications where pressures of approximately 40 psi can be assumed or controlled by pressure regulation.

It is also noticed that the flow rate q can be stated in terms of the velocity and the area, since from Eq. (7–3) the volume V equals

$$V = A \times v \times t \tag{7-5}$$

That is,

$$q = \frac{A \times v \times t}{t} = Av \tag{7-6}$$

7-5 Limitation of Velocity

The limitation of water velocity to 10 fps (3.05 m/s) through supply pipes has its basis in engineering practice and plumbing codes.[1] Excessive velocity and turbulence results in objectional whistling noises in the system, generation of water shocks that damage piping and equipment, and accelerated deterioration from erosion-corrosion. To insure a margin of safety, the recommended practice is to limit the velocity in supply mains to 8 fps, and in piping to headers, risers, and water outlets controlled by quick-acting shut-off valves to 4 fps. This latter limitation is to insure that damaging water pressures will not be generated by the sudden closing of the valve, which can initiate the reverbrating shock wave in the system.

Sizing Table 7–1 is based on velocity limitations of 4 fps and 8 fps. It lists the pipe sizes associated with delivering the necessary flow rate supply in gpm, and the equivalent load demand in water supply fixture units. Tabled values are for pipe and tube having a fairly smooth finish. Column A is used for systems that do not have flush valves. Column B is used for those systems that do. The load values in Columns A and B correspond to values in Table 8–4, which are used to estimate the total demand and water service size for a system in wsfu and gpm.

It is seen from Table 7–1 that 1-in. standard size galvanized iron and steel pipe will deliver 10.8 gpm (37.85 1/min) at a velocity of 4 fps (1.22 m/s), and 21.6 gpm (81.8 1/min) at a velocity of 8 fps (2.44 fps). These velocities would supply a demand for 6.1 wsfu and 25.3 wsfu, respectively, for systems that do not have flush valves.

7-6 Continuity Equation

The continuity equation governing the flow of fluid in a pipe is derived from the principle that during steady flow conditions, the mass or amount of fluid flowing past any section in the pipe is the same. This means that if the diameter size of the pipe is reduced, and the amount of fluid flowing is the same, the velocity will be increased. If the pipe is enlarged, the velocity will decrease. In the case of water and hydraulic fluids, which are nearly incompressible, the density of the fluid does not change much with changes in pressure. In low-pressure gas systems, the density can also be assumed to remain approximately constant without introducing appreciable error. As

Table 7-1

Sizing Table Based On Velocity Limitation (Reprinted from National Standard Plumbing Code)

		Velocity = 4 feet per second				*Velocity = 8 feet per second*			
Nominal Size (in.)	*Actual I.D. (in.)*	*Flow (gpm) q*	*Load (WSFU) 1 * Col. A*	*Load (WSFU) 2 * Col. B*	*Friction (psi/100′) p^3 **	*Flow (gpm) q*	*Load (WSFU) 1 * Col. A*	*Load (WSFU) 2 * Col. B*	*Friction (psi/100′) p^3 **
Copper Water Tube, Type L									
$\frac{1}{2}$	.545	2.9	1.0	—	8.2	5.8	2.5	—	29.0
$\frac{3}{4}$	.785	6.0	2.5	—	5.2	12.0	7.3	—	18.7
1	1.025	10.3	5.5	—	3.9	20.6	22.5	7.0	13.7
$1\frac{1}{4}$	1.265	15.7	11.5	5.0	3.0	31.4	58.0	15.5	10.7
$1\frac{1}{2}$	1.505	22.8	28.5	8.0	2.5	45.6	109.0	38.0	8.7
2	1.935	38.6	82.0	26.0	1.8	77.2	261.0	138.0	6.3
$2\frac{1}{2}$	2.465	59.5	172.0	75.0	1.4	119.0	474.0	356.0	4.9
3	2.945	85.0	300.0	178.0	1.1	170.0	750.0	692.0	4.0
4	3.905	149.0	636.0	544.0	0.8	298.0	1759.0	1759.0	2.8
Copper Water Tube, Type K									
$\frac{1}{2}$	.527	2.7	.75	—	8.5	5.4	2.3	—	31.0
$\frac{3}{4}$	.745	5.5	2.3	—	5.6	11.0	6.3	—	20.2
1	.995	9.7	5.3	—	4.1	19.4	19.5	5.8	14.4
$1\frac{1}{4}$	1.245	15.2	10.8	5.0	3.1	30.4	54.0	14.0	11.1
$1\frac{1}{2}$	1.481	21.5	25.0	7.8	2.6	43.0	98.0	34.0	9.2
2	1.959	37.6	78.0	24.0	1.8	75.2	251.0	130.0	6.5
$2\frac{1}{2}$	2.435	58.2	166.0	69.0	1.4	116.4	460.0	340.0	5.2
3	2.907	82.8	289.0	161.0	1.2	165.6	725.0	663.0	4.2
4	3.857	146.0	609.0	528.0	0.8	292.0	1705.0	1705.0	3.0

				Galvanized Iron and Steel Pipe, Standard Pipe Size					
$\frac{1}{2}$	.622	3.8	1.5	—	8.2	7.6	3.7	—	31.0
$\frac{3}{4}$	.824	6.7	3.0	—	6.0	13.4	8.4	—	22.5
1	1.049	10.8	6.1	—	4.6	21.6	25.3	7.7	17.2
$1\frac{1}{4}$	1.380	18.6	17.5	6.0	3.4	37.2	77.3	23.7	12.8
$1\frac{1}{2}$	1.610	25.4	37.0	9.3	2.9	50.8	132.3	52.0	10.8
2	2.067	41.8	93.0	29.8	2.2	83.6	293.0	171.6	8.4
$2\frac{1}{2}$	2.469	59.8	174.0	75.6	1.8	119.6	477.0	361.0	6.8
3	3.068	92.0	335.0	209.0	1.4	184.0	842.0	806.0	5.4
4	4.026	158.6	688.0	615.0	1.1	317.2	1980.0	1930.0	4.1
				Schedule 40 Plastic Pipe, (PE, PVC & ABS)					
$\frac{1}{2}$	.622	3.8	1.5	—	6.8	7.6	3.7	—	24.2
$\frac{3}{4}$	.824	6.7	3.0	—	5.1	13.4	8.4	—	18.0
1	1.049	10.8	6.1	—	3.7	21.6	25.3	7.7	13.2
$1\frac{1}{4}$	1.380	18.6	17.5	6.0	2.8	37.2	77.3	23.7	9.6
$1\frac{1}{2}$	1.610	25.4	37.0	9.3	2.3	50.8	132.3	52.0	8.2
2	2.067	41.8	93.0	29.8	1.7	83.6	293.0	171.6	6.1
$2\frac{1}{2}$	2.469	59.8	174.0	75.6	1.4	119.6	477.0	361.0	4.8
3	3.068	92.0	335.0	209.0	1.1	184.0	842.0	806.0	3.8
4	4.026	158.6	688.0	615.0	0.8	317.2	1930.0	1930.0	2.8

[1] * Col. A applies to piping which does not supply flush valves.
[2] * Col. B applies to piping which supplies flush valves.
[3] * Friction loss p corresponding to flow rate q for piping having fairly smooth surface condition after extended service, applying the formula: $q = 4.57\,(p)^{0.546}(D)^{2.64}$

stated, the continuity equation can be written for a plumbing system as

$$Q = A_1 v_1 = A_2 v_2 = \text{constant} \tag{7-7}$$

The continuity equation as illustrated in Fig. 7-4 is useful in computing the equalization of flow through pipes of different sizes. For example, it may be necessary to know how many one-inch pipes would be required to deliver the flow from one three-inch pipe. As another example, when one pipe is reduced to another, it is necessary to know how much the velocity is increased in the smaller pipe to ascertain if that exceeds the velocity limitation imposed on the system or network. These two examples will make the computation clear.

Example 7-2 How many 1-in. galvanized standard pipes would be required to handle the delivery from a 3-in. galvanized standard pipe without increasing the velocity of the fluid?

Solution:
Refer to Fig. 7-5. Notice that the sense of the problem asks that the area of the 3-in. pipe must be equated to the areas of the number of 1-in. pipes necessary to equal that area. Since the flow and velocity remain constant, Eq. (7-7) is written

$$Q = A_1 v_1 = A_2 v_2 = \text{constant}$$

and A_1 must be set equal to A_2. This means that dividing the area of the 3-in. pipe by the area of the 1-in. pipe will give the number of 1-in. pipes necessary. Performing that computation, using 3.068 and 1.049 as the diameters of the two pipes, we obtain

$$A_1 = \frac{(3.14)(3.068\text{ in.})^2}{(4)} = 0.86\text{ in.}^2$$

Dividing A_1 by A_2, we have

$$\frac{A_1}{A_2} = \frac{7.38\text{ in.}^2}{0.86\text{ in.}^2} = 8.58$$

Figure 7-4

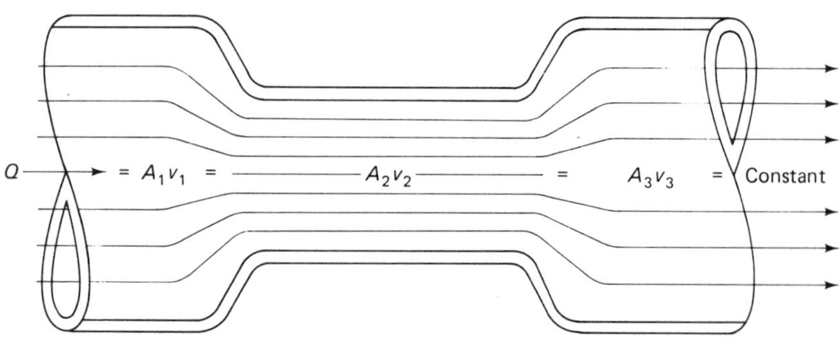

Continuity Equation

Thus, to carry the delivery from the 3-in. pipe, 9 one-inch pipes would be required.

Example 7-3 A $2\frac{1}{2}$-in. galvanized standard pipe delivering water at 5 ft/s (1.5 m/s) is connected to a $1\frac{1}{4}$-in. pipe. What will be the velocity in the smaller pipe? (Refer to Table 7–1 for the pipe internal diameters)

Figure 7-5

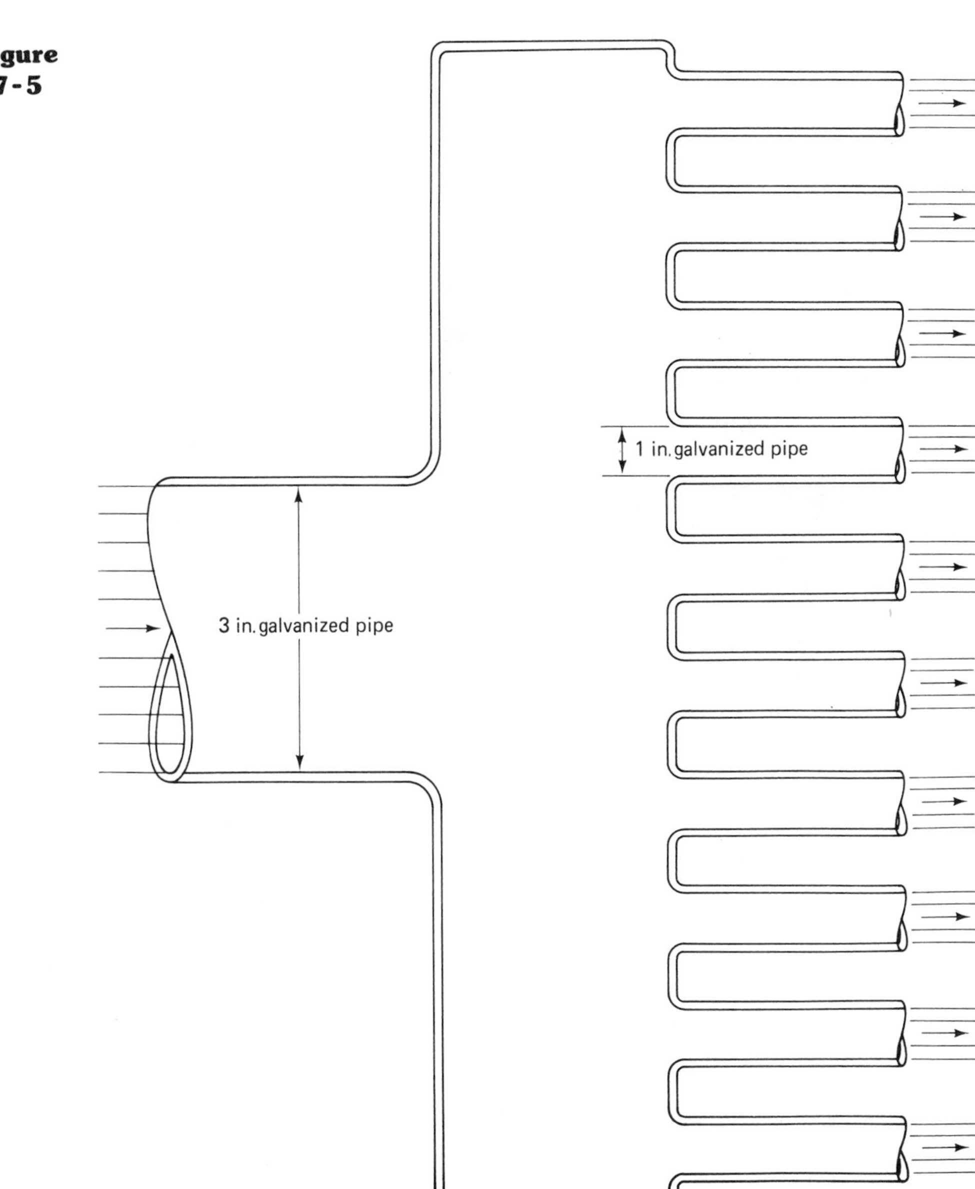

Solution:

In this example, the problem asks to solve the continuity equation for v_2, the velocity of the fluid in the smaller pipe. That is,

$$A_1v_1 = A_2v_2$$

and

$$v_2 = \frac{A_1v_1}{A_2} = \frac{(4)(3.14)(2.469 \text{ in.})^2(5 \text{ ft/s})}{(4)(3.14)(1.380 \text{ in.})^2} = 16 \text{ ft/s } (4.88 \text{ m/s})$$

Notice that this is greater than the velocity limit imposed by common plumbing practice and codes, and the size of the second pipe would have to be increased.

7-7 Flow Losses through Pipe, Meters, Fittings, and Valves

Losses occur in plumbing systems by three main causes:

1. Raising the fluid to a higher elevation.
2. Friction losses through the developed length of pipe.
3. Friction losses through the water meter fittings and valves.

Raising fluid to a higher level results in a pressure loss of approximately 0.433 lbf/in.^{2} for each foot lifted (9797 Pa for each meter lifted). Lifting water 40 ft to the fourth floor of an apartment building, for example, would cause the pressure to drop approximately 17 psi (117 kPa). Thus, a water pressure of 40 lbf/in.^{2} at ground level would be only 23 lbf/in.^{2} at the fourth floor. This would result in less available flow from the same size pipe. The static pressure drop is not a flow loss in and of itself, but instead causes a loss because it reduces the pressure available to propel the water when the outlet is opened.

Uniform friction loss, which occurs as the fluid flows through the pipes, is computed as a pressure drop per 100 or 1000 feet of pipe. The plumber figures the pressure loss to the last fixture and then compares the value of the pressure remaining to that necessary for its proper operation. If the pressure is too low to operate the fixture, the pipe size (or pressure) can be increased.

The friction from flow in pipes brings to mind the notion that the fluid rubs the boundary surface as if a close-fitting solid plug were being pushed through a tube, generating heat losses at the boundary. This is not the case. What occurs is a rubbing action between the fluid particles themselves. Because the fluid flows in a turbulent manner, the particles are intermingling across the stream as the total mass of fluid moves through the length of pipe. The velocity is still greatest near the center of the stream and zero at the wall, but near the wall, because of imperfections in the surface, there is a greater increase in the friction within the fluid as it shears internally. Thus, pipe with a rougher internal surface generates more friction than one that is smooth, and increases in the velocity of the fluid also increase the value of the friction.

Flow rate charts for uniform pipe friction have been developed for the several types of pipe encountered, and take into account the actual internal diameters and acceptable range of velocity limitation. The representative flow charts shown in Figs. 7-6, 7-7, and 7-8 are used for copper water tube, galvanized iron and steel standard weight pipe, and schedule 40 plastic water pipe. The charts are read by entering at the left margin with the required delivery in gal/min to the farthest run. Proceeding to the right across the chart to the velocity lines, this limitation is incorporated and used to determine the pipe size necessary to deliver the required flow rate. Finally, reading across the bottom of the chart directly below this intersection gives the pressure loss due to friction per 100 ft of pipe. This is the pressure lost uniformly along the developed length of the pipe. An example will clarify the use of the pipe friction charts.

Figure 7-6

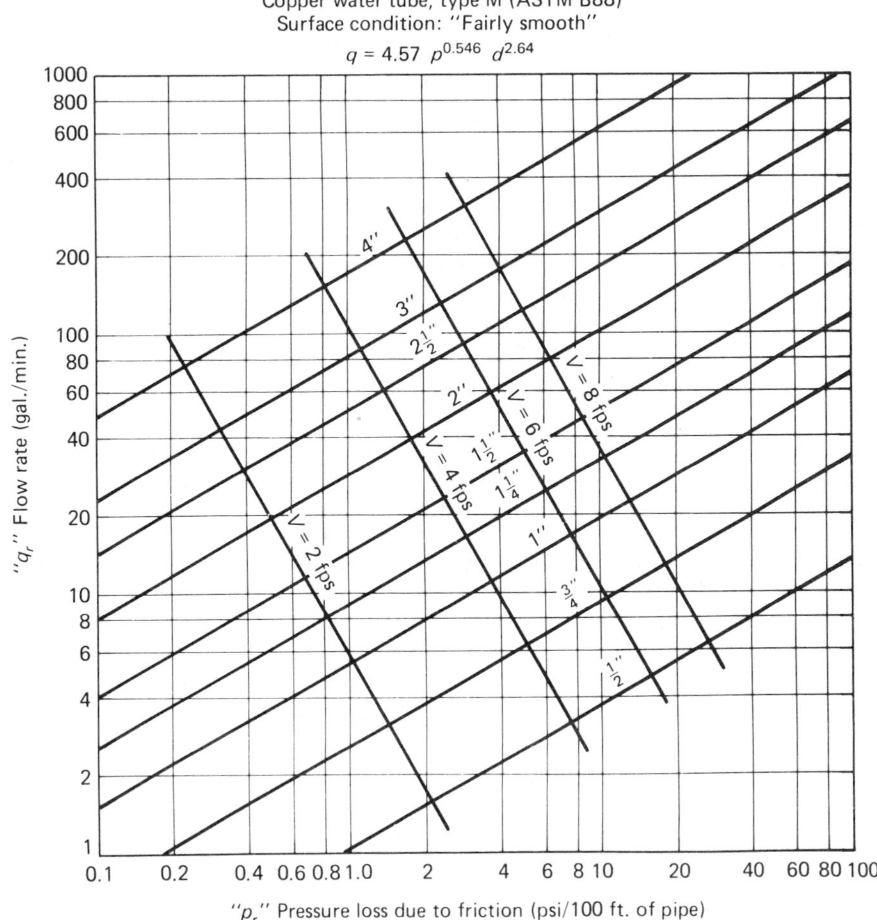

Pressure Loss due to Friction
(Reprinted from *Plumbing-Heating-Cooling Business*)

Figure 7-7

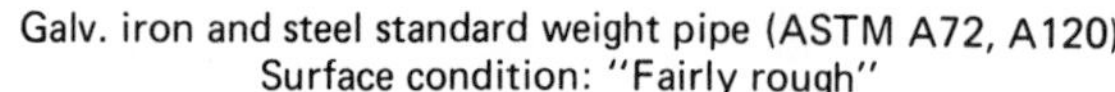

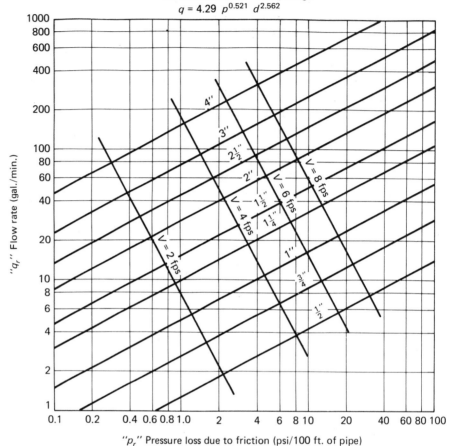

Pressure Loss due to Friction
(Reprinted from *Plumbing-Heating-Cooling Business*)

Example 7-4 The farthest fixtures in a cold water distribution system are 110 ft distant from the meter. If the demand is 30 gal/min (94.6 l/min), and the velocity limitation is 6 ft/sec, what would be the necessary uniform size and friction loss through a type M fairly smooth copper tube? (Notice that the meter and fittings are not included in this calculation.)

Solution:

Entering the chart in Fig. 7-6 at the left margin with 30 gal/min and reading across to the 6-ft/sec line, the necessary pipe size is approximately $1\frac{1}{2}$ in. Reading across the bottom of the chart directly below

Figure 7-8

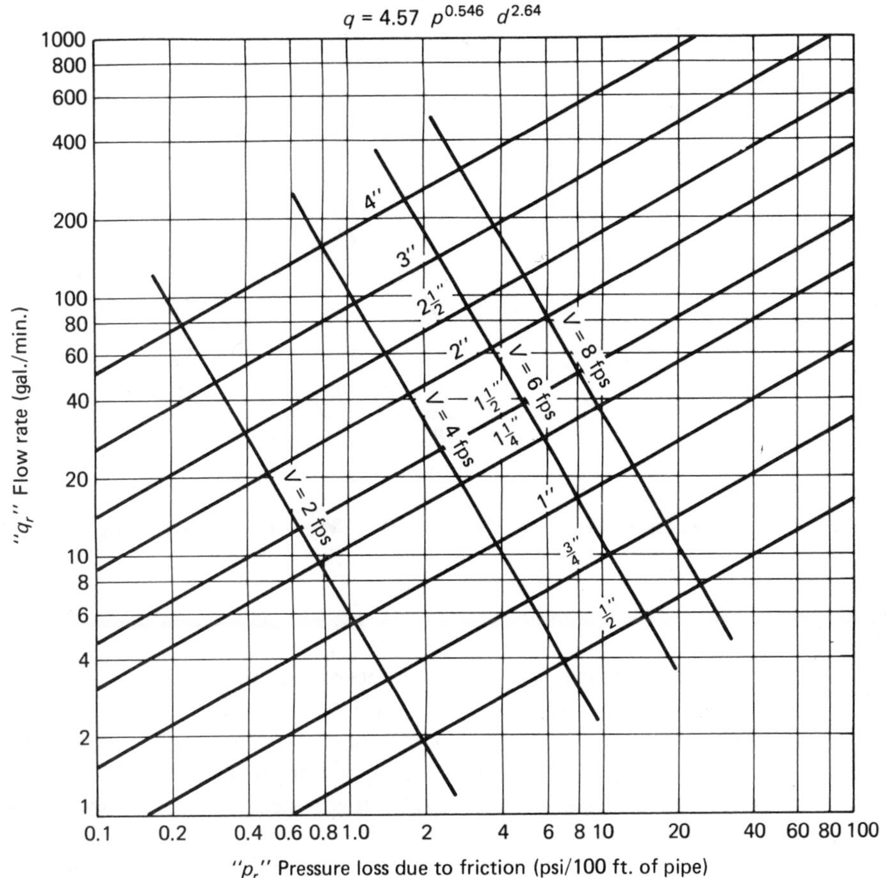

Pressure Loss due to Friction
(Reprinted from *Plumbing-Heating-Cooling Business*)

this intersection, the pressure loss per 100 ft of straight pipe would be approximately 5 lbf/in.². Since the run is 110 ft long, the pressure loss would be about 5.5 lbf/in.². That is, 5 lbf/in.² × 1.1 = 5.5 lbf/in.².

Friction loss through the water meter reduces the pressure from the street main or other source of supply to the building plumbing system. How much the pressure is reduced depends upon the size of the meter and the delivery. Figure 7–9 is a friction chart for meters of the displacement

Figure 7-9

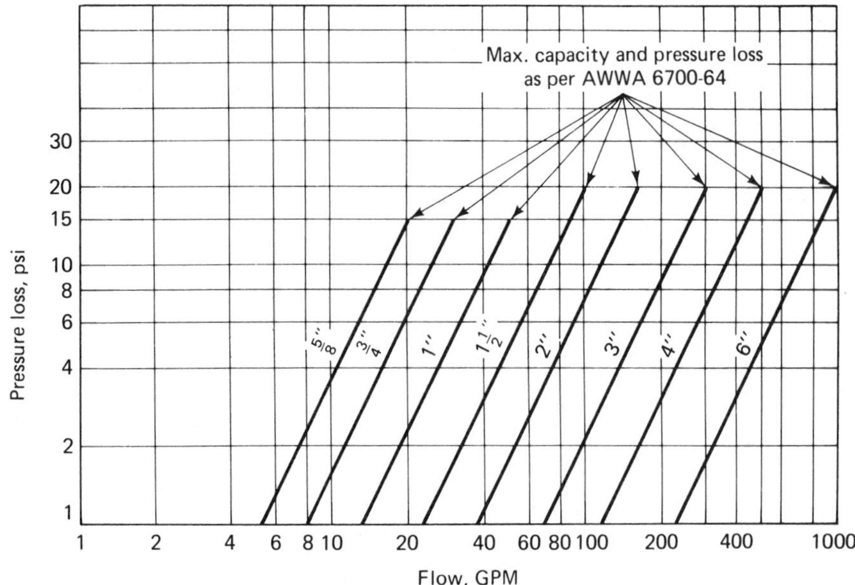

Pressure Loss in Cold-Water Meters of the Displacement Type

type. It is useful to size the meter, given the supply demand in gal/min necessary to meet the load requirements of the system, and acceptable pressure losses during peak periods. This is usually in the 4-to-10-lbf/in.^{2} range. Excessive pressure losses through a water meter are avoided by selecting a meter of the next larger size. The pressure available to the plumbing system in the building for a given flow rate equals the pressure in the water service line minus the pressure drop through the meter. Friction losses for the plumbing system within the building are subtracted from this value, rather than the pressure in the building service line.

Friction losses through the fittings and valves of the plumbing system are computed as equivalent lengths of pieces of pipe of the same diameter size. Thus, the *total length* of pipe for a system consists of the *developed length* of all pipe sections, plus the *equivalent length* for the fittings and valves connecting the pipe. In copper tube installations, friction losses through the fittings and valves typically equal about half the value of the frictional loss computed for the developed length. In threaded pipe systems, this value often increases to 75 percent of the frictional loss in the developed length. Table 7–2 gives the equivalent length of straight pipe in feet for each fitting and valve in the various sizes. The values for threaded fittings are taken as twice those for soldered fittings. The total equivalent length is the sum of the values for all fittings and valves. This value is

Table 7-2

Allowance for Friction Loss in Valves and Fittings Expressed as Equivalent Length Tube (Copper Development Association Inc.)

Fitting Size, inches	*Standard Ells*		*Equivalent Length of Tube, feet* *90° Tee*				
	90°	*45°*	*Side Branch*	*Straight Run*	*Coupling*	*Gate Valve*	*Globe Valve*
$\frac{3}{8}$	0.5	0.3	0.75	0.15	0.15	0.1	4
$\frac{1}{2}$	1	0.6	1.5	0.3	0.3	0.2	7.5
$\frac{3}{4}$	1.25	0.75	2	0.4	0.4	0.25	10
1	1.5	1.0	2.5	0.45	0.45	0.3	12.5
$1\frac{1}{4}$	2	1.2	3	0.6	0.6	0.4	18
$1\frac{1}{2}$	2.5	1.5	3.5	0.8	0.8	0.5	23
2	3.5	2	5	1	1	0.7	28
$2\frac{1}{2}$	4	2.5	6	1.3	1.3	0.8	33
3	5	3	7.5	1.5	1.5	1	40
$3\frac{1}{2}$	6	3.5	9	1.8	1.8	1.2	50
4	7	4	10.5	2	2	1.4	63
5	9	5	13	2.5	2.5	1.7	70
6	10	6	15	3	3	2	84

NOTE: Allowances are for streamlined soldered fittings and recessed threaded fittings. For threaded fittings, double the allowances shown in the table.

added to the developed length to make up the total length. Friction loss is then computed for the total length, as in Example 7-4. This would have the effect of increasing the length of the run, thereby adding to the pressure loss which was read from Fig. 7-6 across the bottom in lbf/in.^{2}/100 ft of pipe. For example, if the equivalent lengths of all elbows, tees, couplings, and valves in the 110-ft run were added together, the increase in length could be expected to be approximately 50 ft, and the pressure loss for the entire length and the fittings would be 5 lbf/in.^{2} $\times$ 1.6 = 8 lbf/in.^{2}.

7-8 Summary of Practice

Pipes that convey fluids are sized to carry the supply or waste to or from fixtures and appurtenances. Friction losses reduce the flow by lowering the available pressure. The total pressure losses in the water supply that are accounted for by the lift, frictional losses in the developed length, and equivalent length of fittings and valves, are subtracted from the pressure on the building side of the meter to determine the available pressure to the fixture farthest from the meter. These and the velocity are used to determine the size of the pipes required to plumb the building main and branch circuits.

REVIEW QUESTIONS AND PROBLEMS

1. What effect does doubling the diameter of a pipe have on the area?
2. Compute the relative cross-section areas of 1-, 2-, 4-, and 8-in. schedule 40 steel pipe, schedule 40 PVC pipe and type M copper tube. Complete the following table.

Relative Cross Section Areas where Diameters Double

Nominal Size	*Schedule 40 Steel Pipe*	*Schedule 40 Plastic Pipe*	*Type M Copper Tube*
1 in.			
2 in.			
4 in.			
8 in.			

3. What volume will a $1\frac{1}{2}$- in. (38.8-mm) Type M copper tube discharge in 5 min if the velocity is 4 ft/s (1.22 m/s)?
4. How long would it take to fill a 300-gal (1136-l) tank using a $\frac{3}{4}$-in. (20.6-mm) Type M copper tube, if the velocity is limited to 8 ft/s (2.44 m/s)?
5. What is the flow rate from a $\frac{1}{2}$-in. (14.6-mm) Type M copper tube if the flow rate is 5 ft/s (1.52 m/s)?
6. Describe the conditions that limit the velocity of water in supply pipes.
7. How many $\frac{1}{2}$-in. (14.6-mm) Type M copper tubes would be necessary to handle the delivery from a $\frac{3}{4}$-in. (20.6-mm) Type copper tube without increasing the velocity of the fluid?
8. If a 1-in. (28.8-mm) Type M copper tube delivers water at 3 ft/s to a $\frac{3}{4}$-in. (20.6-mm) Type M copper tube, what will be the velocity of the water in the smaller tube?
9. What is the necessary pipe size and total friction loss in a developed length of Type M copper tube carrying 20 gal/min (76 l/min) a distance of 80 ft (24.4 m)? Assume that the equivalent length of all fittings and valves is 50 percent of the developed length.
10. What would be the flow rate and pressure loss through all the 1-in. (threaded) fittings in Table 7–2, given a velocity limitation of 6 ft/s (1.8 m/s)? (*Note:* Use Fig. 7–7.)
11. In a high rise building, and where the water pressure in the main at street level is 60 lbf/in² (414 kPa), how far up the building will the water rise?

REFERENCES

[1] National Association of Plumbing-Heating-Cooling Contractors, *National Standard Plumbing Code,* 1978, Sec. 10.14.1 and Appendix B.6.

8 Cold and Hot Water Distribution

8-1 Introduction

The cold and hot water distribution system transports clean, safe water to plumbing fixtures for drinking, bathing, cooking, processing of food, and medical purposes. Clean and safe water is called *potable* water in the plumbing trade, and it is the plumber's responsibility under all codes to protect the potable water supply by preventing its contamination by other materials or fluids that could cause poisoning or disease.

The amount of water needed for a dwelling is determined by its occupancy and the number of plumbing fixtures that are to be installed. The water used by each fixture can be represented by a measure called *fixture units,* and these are used to size the pressure pipes. A fixture unit is approximately 7.5 gal/min (34.1 l/min) or 1 ft^3/min (0.028 m^3/min). The water supply pressure needed to operate the plumbing fixtures, sometimes called flow pressure, and the elevation to which the water supply must be lifted within the dwelling determine the supply pressure necessary at the ground level. If the pressure is too high, it must be reduced to protect the pipes and fixtures. If it is too low, some method must be used to safely boost the pressure. A pressure of 40 lbf/in.^{2} (276 kPa) is adequate to supply water to a dwelling if the pipes are sized correctly.

Water hammer is a knocking or banging sound in the pipes. It is caused by high pressure waves that transmit hydraulic shocks in the lines. Water hammer starts when water in the pipes is shut off abruptly by a valve. Depending upon the pressure, flow velocity, and how quickly the water is shut off, the severity of water hammer shocks can rupture the system.

8-2 Potable Water Supply and Its Protection

Water is composed of hydrogen and oxygen, that is, H_2O. However, it carries with it in solution many other substances and microscopic organisms. It is potable if it is safe for human consumption, and this safety standard is determined by the Public Health Service Drinking Water Standards,[1] or regulations of the public health authority having jurisdiction in the area.

Table 8-1 **Standards for Potable Water***

Physical Characteristic or Chemical Substance	*Concentration Scale or Parts per Million by Weight*
Color	20 (platinum cobalt scale)
Turbidity	10 (silica scale)
Arsenic	0.05 ppm
Copper	3.0 ppm
Chloride	2.5 ppm
Chromium	0.05 ppm
Fluoride	1.5 ppm
Iron and Manganese	0.3 ppm (combined)
Lead	0.1 ppm
Magnesium	12.5 ppm
Selenium	0.05 ppm
Sulfate	2.5 ppm
Zinc	15.0 ppm

* Standards of the United States Public Health Service for Potable Water in Interstate Traffic.

The physical characteristics of water include its turbidity (clearness from particles in suspension), color, odor, and taste. The chemical analysis of water determines the amount of soluble mineral substances. Table 8-1 lists standards for both physical and chemical characteristics for the potable water supply.

Water is supplied from a municipal system, private wells, cisterns, and lakes; and, depending upon its quality, is usually desirable in that order. This is to say that the quality and desirability of municipal water is usually the highest, whereas the water from a lake to a private residence has the lowest quality and desirability. Although the quality of water from a municipal system is almost always assured, water from wells, cisterns, and lakes usually requires further treatment to improve its purity, remove or neutralize the effects of minerals, and ensure its softness. Water filters and softeners are installed in the dwelling by the plumber to accomplish this. Periodic service is also required by the owner or a service man.

Because a water supply is clean and safe for human consumption when it is first connected is no assurance that it will remain that way, and several precautions must be taken to keep it clean and free from disease. Most of these are specified in the codes having jurisdiction in the area. Water-borne diseases that can contaminate a water supply include cholera, typhoid fever, paratyphoid fever, amoebic dysentery, bacillary dysentery, hookworm, flukes, and diarrhea.[2] One of the first disease outbreaks attributed to the drinking water was the Asiatic cholera epidemic of 1854 in London. Water was also blamed for the epidemic of typhoid fever in Plymouth, Pennsylvania in 1885. In the early 1930's effluent from the sanitary waste lines was drawn into the potable water supply and became responsible for a major outbreak of amoebic dysentery in Chicago among hotel guests which left nearly 100 dead and over one thousand more in a prolonged state of debilitation.[3] In the late 1960's disease from human waste drawn into the potable water supply through sprinkler heads at the Holy Cross College (Massachusetts) football field left all but one member of the team with infectious hepatitis. The point to be made is that clean water in a system can become contaminated during use. Usually this is caused by what is known as a *cross connection:* a piping arrangement that allows the potable water supply to be connected to a line or other source having contaminants. The cross connection usually occurs when pressure drops in the supply water line, and siphoning reverses the flow, causing the contaminants to enter the water supply.

It has been recognized for some time that there is danger of contamination outside the building under ground where broken water and sewer lines can become cross-connected. Less obvious but just as likely is the chance that cross connections can exist within a building as well. For example, a rubber hose connected to a fresh water supply and left in a laundry sink, or connected to sprinkler heads below the surface of the lawn, can siphon contaminants back into the potable water supply if the supply pressure drops. Swimming pools are another potential source for a cross connection with the fresh water supply. Most codes outlaw such connections unless backflow preventers are installed which stop a reverse flow. Among other applications covered by backflow prevention are flush valves, hose bibbs, fire protection systems, commercial ice makers, hand-held personal showers, kitchen sink spray hoses, solar energy panels, water-connected coffee vending machines, and domestic refrigerators. One major manufacturer of backflow preventers estimates that as many as 100 thousand potential cross connections are made each day in the United States alone.[4]

Cross connections are of two types: *inlet connections* and *direct pressure connections.* Inlet connections are used for filling sinks, spraying gardens or watering lawns. They are open to the atmosphere. *Backflow* results from siphonage when the pressure drops in the system and line connected to the inlet due to a vacuum or partial vacuum. This is referred to as *back-siphonage.* It can be caused by gravity, undersized piping, or a vacuum. Back-siphonage from gravity will occur if the supply is lower than the fixture. Undersized piping increases the velocity and can draw water out of branch pipes. A venturi tee operates on this principle. Back-siphonage and reverse flow from a vacuum occur when the pressure in the supply line is reduced below atmospheric conditions. A negative pressure can be caused by a fire truck pumping from a pressure line, overusage at a lower level in the system, or a break in a main at a lower level which draws a vacuum when it empties. A typical cross connection of the back-siphonage type is shown in Fig. 8–1 where a hose bibb can siphon washer waste into the potable water system. *Direct pressure connections* cause backflow by raising the pressure above that of the system supply, thus forcing contaminants into the fresh water system under pressure. Boiler

Figure 8-1

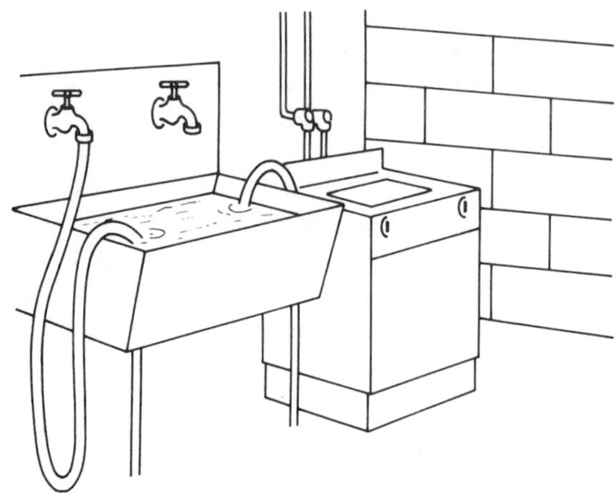

The indoor service sink with an open hose left hanging in the tub could allow washer waste to be siphoned back into the potable water system.

Back-Siphonage Cross Connection

feed pumps connected to the fresh water system can cause a backflow, as can a pump primer line. Both of these devices can raise the system pressure above the supply pressure, causing the flow to reverse (Fig. 8–2).

Backflow of a foreign liquid or liquefied substance into the potable water supply constitutes *pollution* of the supply. Pollution does not necessarily constitute a health hazard, although the quality of the water is impaired. If the substance also causes a health hazard, for example if it is toxic (poison) or carried disease, it constitutes a *contamination* of the supply. Back-siphoned sewage or boiler water driven by a feed pump into the water supply is considered to be contamination of the supply. Plumbing codes specify that supply faucets be placed above the flood rim of plumbing fixtures such as sinks, bathtubs, and lavatories and be separated by a prescribed air gap, to prevent back-siphonage. Codes also specify other precautions. In addition, there are four other types of backflow preventers used in systems to protect the potable water supply. Their placement is also prescribed by plumbing codes. They are

1. An air gap
2. Atmospheric vacuum breaker
3. Pressure-type vacuum breaker
4. Double check valve assembly
5. Reduced pressure zone devices

These devices are shown in Fig. 8–3.

Figure 8-2

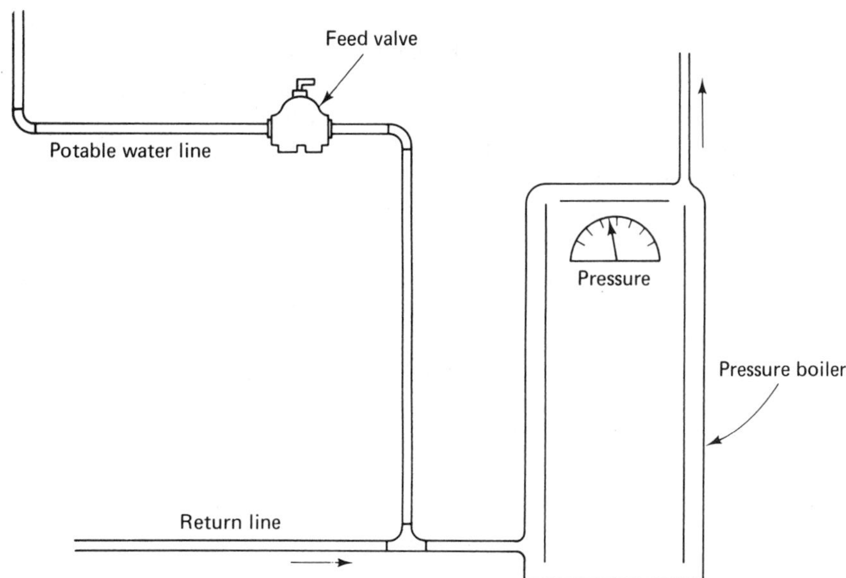

Direct connection to potable water supply allows back flow when the supply pressure is reduced below the boiler line pressure, permitting boiler water to flow into the potable water supply.

Direct Pressure Cross Connection

Figure 8-3

Atmospheric Vacuum Breaker

Atmospheric Vacuum Breaker

Pressure-type Vacuum Breaker with Vent

Pressure-type Vacuum Breaker

Reduced Pressure Zone Backflow Preventer

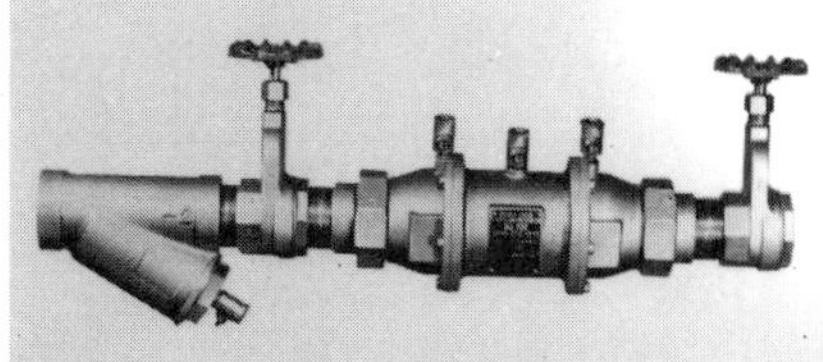

Double Check Valve Backflow Preventer

Backflow Preventers (Watts Regulator Co.)

An air gap is the physical separation of the potable and nonpotable system by an air space. The vertical distance between the supply pipe and the flood level rim should be two diameters of the supply pipe, but never less than 1 inch. The air gap can be used on a direct or inlet connection and for all toxic substances.

Atmospheric vacuum breakers may be used only on connections to a nonpotable system where the vacuum breaker is never subjected to back-pressure and is installed on the discharge side of the last control valve. It must be installed above the usage point. It cannot be used under continuous pressure. A typical application is a portable appliance (Fig. 8-4). Hose connection vacuum breakers also may be used on sill cocks and service sinks. The action of a hose connection vacuum breaker under operating conditions is illustrated in Fig. 8-5.

Figure 8-4

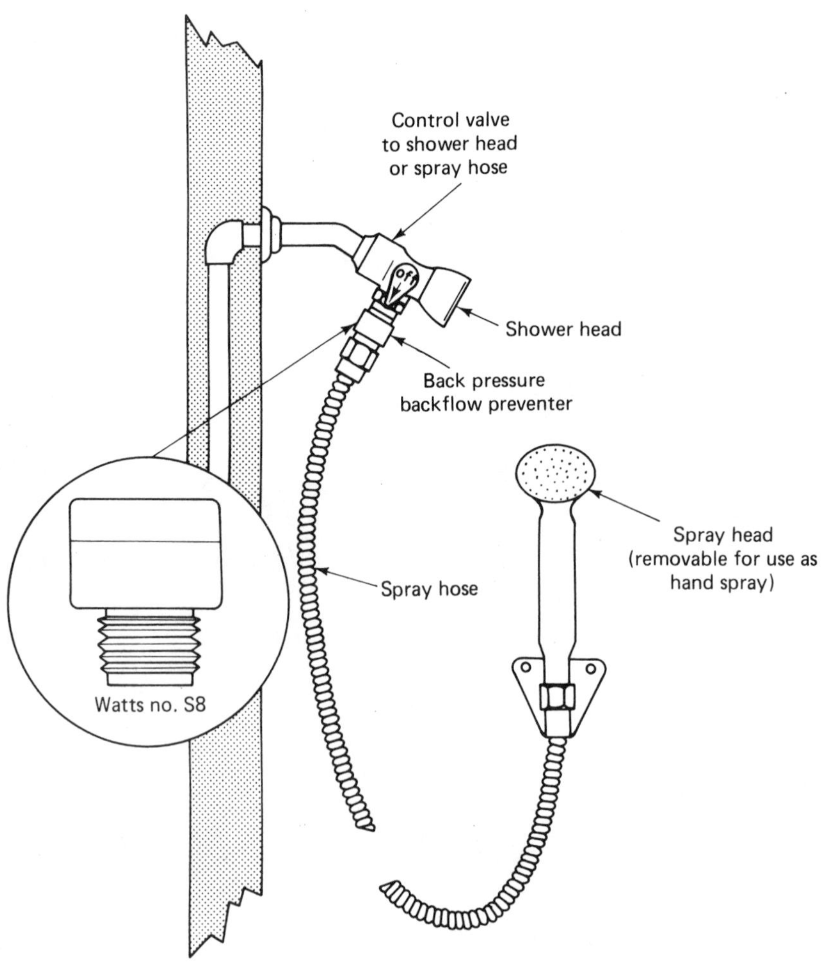

Hand-held Shower Vacuum Breaker

Figure 8-5

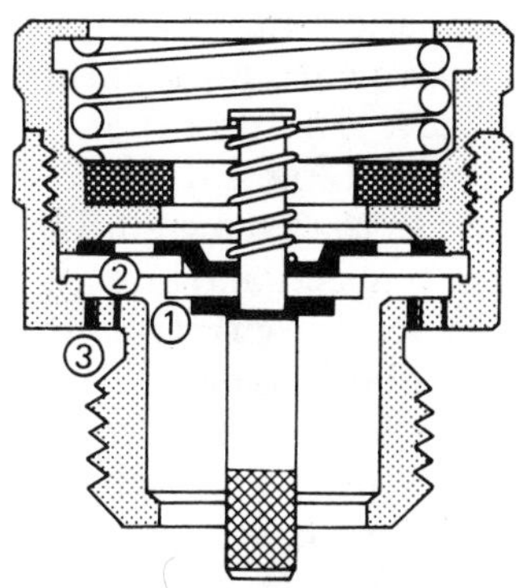

In closed position with supply valve shut off, disc (1) seats against diaphragm (2). Atmospheric ports are open (3) during no flow, preventing back syphonage or back flow of water.

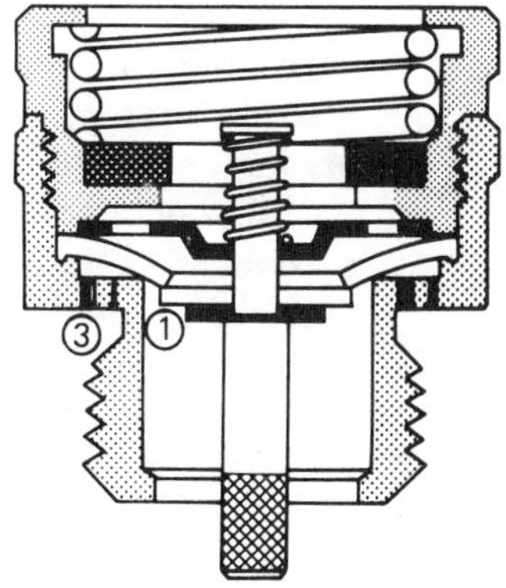

Before flow begins, atmospheric ports (3) are sealed off before lower disc (1) opens to permit flow.

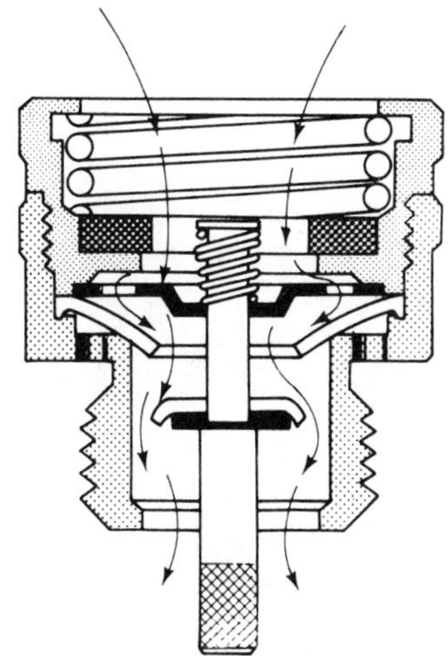

Fully opened valve illustrates poppet action to provide high capacity with minimum pressure drop through the valve.

Action of a Hose Connection Vacuum Breaker

Pressure-type vacuum breakers may be used as protection for connections to all types of nonpotable systems where the vacuum breakers are not subject to back-pressure. These units may be subjected to continuous supply pressure. They must be installed above the usage point. Backflow preventers with an intermediate atmospheric vent may be used as an alternate equal for $\frac{1}{2}$-in. and $\frac{3}{4}$-in. pressure-type vacuum breakers and in addition, provide protection against back-pressure. The operation of a backflow preventer with intermediate vent is shown in Fig. 8-6.

Double check valve assemblies such as that shown in Fig. 8-7 may be used as protection for all direct pressure connections through which foreign material might enter the potable system in concentrations that would constitute a nuisance or be aesthetically objectionable, such as air, steam, food, or other material that does not constitute a health hazard.

Reduced pressure zone devices may be used on direct connections that may be subject to back-pressure or back-siphonage, and where there is the possibility of contamination by material that does constitute a potential

Figure 8-6

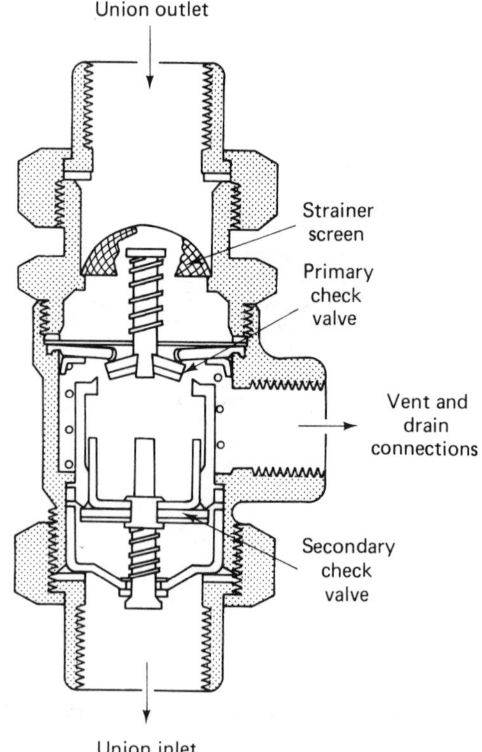

Pressure-type Vacuum Breaker

health hazard. The construction of a typical reduced pressure principle backflow preventer is shown in Fig. 8-8. The operation of the valve is illustrated by the four cross-section views and text in Fig. 8-9.

8-3 Sizing the Potable Water Supply

Plumbing codes specify that dwelling plumbed for potable water will have ample water supplied with sufficient pressure to operate in a safe and satisfactory manner. Their purpose is to ensure that water will be provided to each plumbing fixture and outlet according to its demand and pressure requirements. Thus, the potable water system is designed from the demand of the fixtures, and considers such factors as the pressure available in the

Figure 8-7

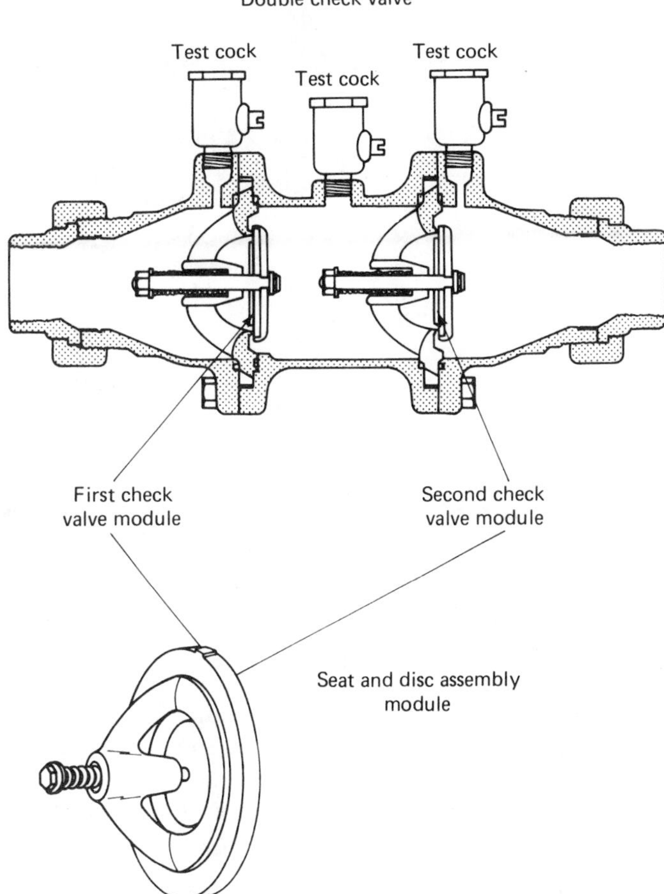

Double Check Valve Backflow Preventer

Figure 8-8

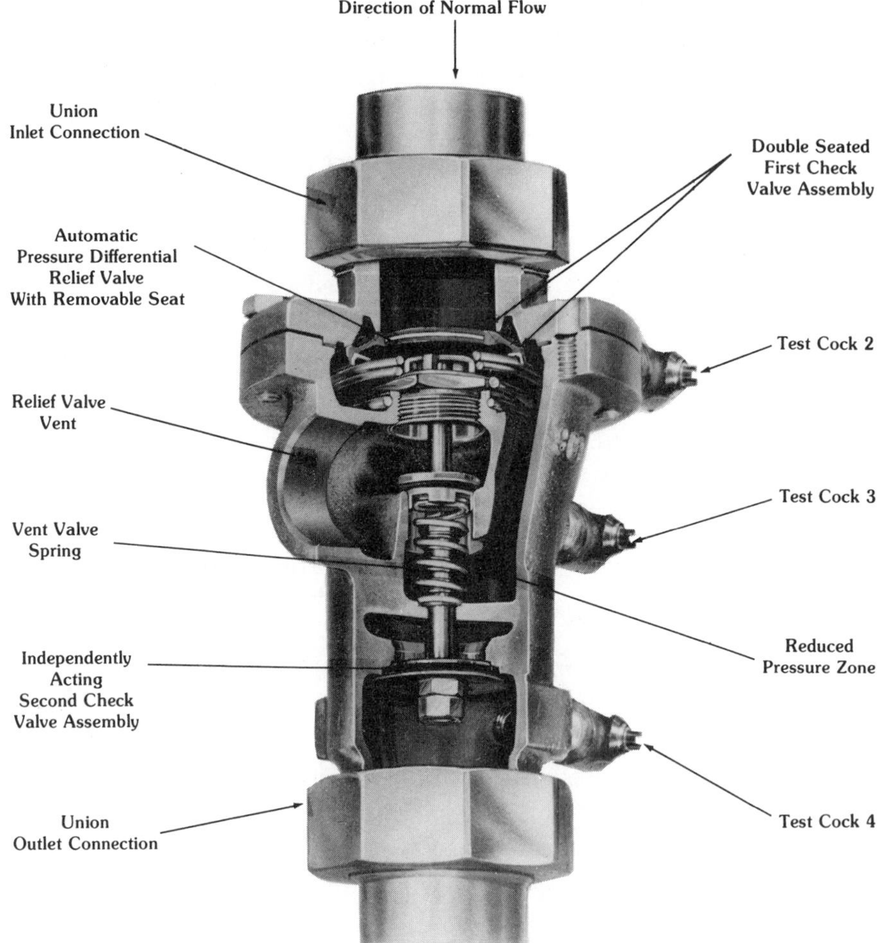

Construction of a Reduced Pressure Principle Backflow Preventer (Watts Regulator Company)

corporation water main, the building height and friction losses, and a velocity limitation that prevents damage to the pipes.

The potable water supply is sized first for the demands of the total system, after which the principal branch lines (both hot and cold water) are sized. At each step, charts are used to convert pressure losses and system demand into pipe sizes. The charts in Chapter 7 are used for this purpose and keep the velocity of the water within the desired limitation.

The steps used to size the potable water supply are

1. Determine the pressure available at the street water main.
2. Calculate the total system demand in water supply fixture units (wsfu's).

Figure 8-9

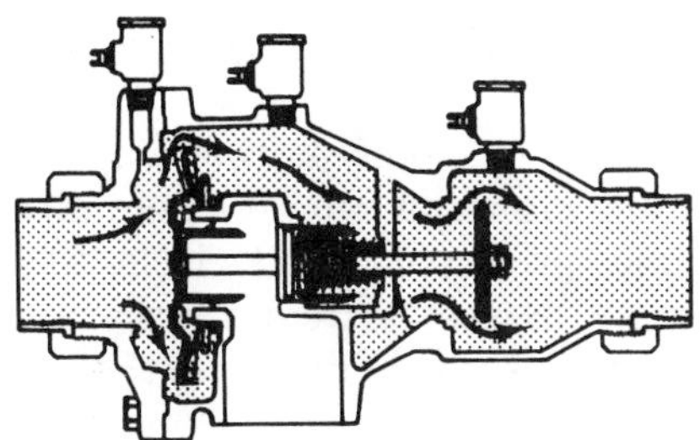

1 Normal Flow
Both check valve assemblies are open supplying water to downstream equipment. The relief valve is held closed by supply pressure on the first check valve assembly. Water pressure in the intermediate zone is 5–11 psi lower than the supply pressure.

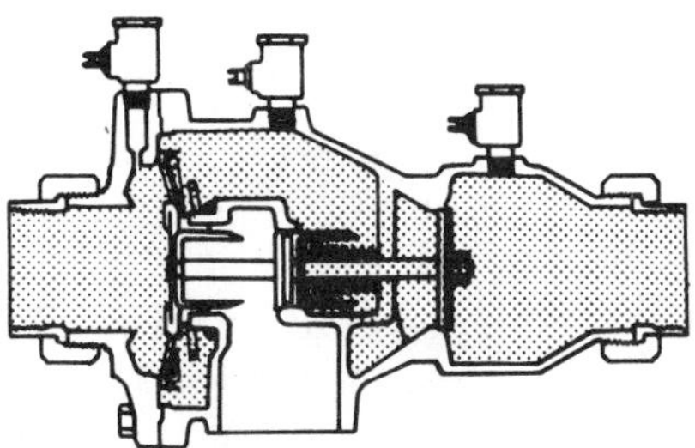

2 Static Pressure
Both check valves are closed as no water is flowing to downstream equipment. Supply pressure in the inlet side of the unit is approximately 5–8 psi higher than the intermediate zone pressure; thus the relief valve is held closed by the supply pressure.

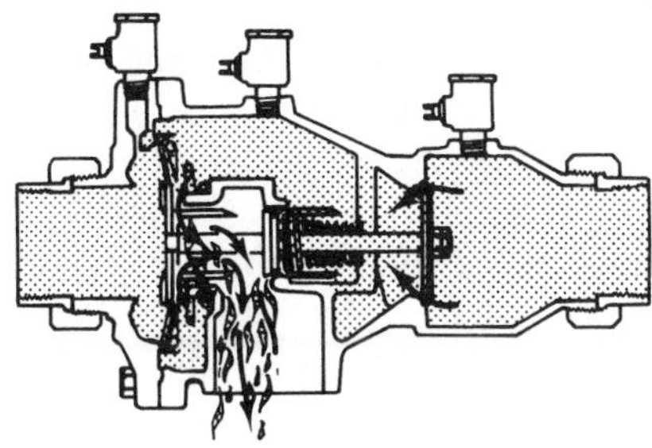

3 Back Pressure
with a "fouled" second check valve—Assume that a pump downstream has caused a sudden back pressure and a foreign object has lodged itself on the second check valve preventing it from seating tightly. The first check valve assembly will close immediately protecting the system. However, when leakage through the second check valve increases the pressure in the intermediate zone to within 2–6 psi of the supply pressure, the relief valve opens, discharging water out of the vent. This discharge would continue as long as the condition remained providing a visual trouble signal.

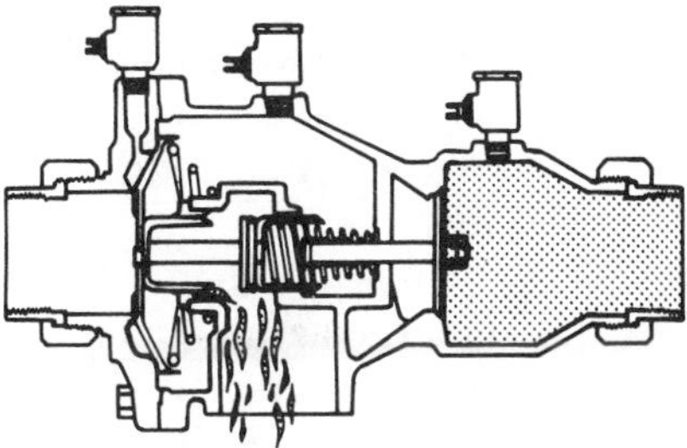

4 Back Siphonage
With a negative supply pressure, both the first and second check valves are closed tightly. The relief valve is now open allowing pressure in the zone to drop to atmospheric and discharging water trapped in the intermediate zone through the vent port. The back-up check valve is now in a closed position.

Operation of a Reduced Backflow Preventer

3. Size the meter from the total demand.
4. Compute the pressure losses through the highest fixture, including the meter, static head to the fixture, and fixture pressure requirements.
5. Size the building supply pipe (coming out of the meter) from the total demand, available pressure, and pressure loss in the longest developed length (to the farthest fixture).

6. Size the principal branches for the hot and cold water, using the respective demands and minimum branch and supply pipe/tube sizes.

The pressure at the street main is usually in the range of 45–60 $lbf/in.^2$ (310–414 kPa). If the pressure is higher, it must be regulated so as not to exceed 80 $lbf/in.^2$ (552 kPa), and preferably it should be lower to provide pipes and fixtures some margin of safety from transient pressures and pressure shocks. The pressure regulator is installed on the house side of the water meter at the owner's expense if it is for the protection of the plumbing in the dwelling. A typical pressure regulator is shown in Fig. 11–19. The pressure in the street main can be determined by consulting the water department. Product literature is used typically to determine the pressure range of well pumps used for private systems.

The total system is sized from fixture demand. Table 8–2 lists the various fixtures and the number of fixture units that must be allowed for each. The total number of fixtures for the dwelling are added together to determine the total wsfu demand. Notice that Table 8–2 lists values for both hot and cold water. If the value of either hot or cold is needed separately to size a principal branch, a value of 75 percent of the total value is used. Also notice that values are given for private dwelling use and commercial use.

The number of fixture units that a specific plumbing fixture requires is not the actual amount of water that any one fixture will need at any one time, or that all fixtures within a dwelling will require at any one time. Rather, it is a number that incorporates the fixture's actual demand, the average time water will be flowing through the fixture, and how often the fixture is used. The water supply fixture unit (wsfu) translates these factors into an average use value, and, summing them together for all fixtures within a dwelling, gives a measure of the total demand on the system. The same is true for each principal branch, both hot and cold water, which places a demand on that part of the system.

Total fixture demand is used to size the water meter. Table 8–3 gives the range of flow for displacement meters of the various sizes. Table 8–4 gives the demand in gal/min associated with the number of fixture units. A meter size is chosen that is less than the maximum to prevent excessive friction losses through the meter. When one is summing wsfu's for a dwelling, it is also important to figure separately and to add in those fixtures that are considered to be used continuously. These include sill cocks, air conditioning circulators, and lawn sprinklers.

The pressure loss to the highest fixture and the flow pressure demand of that fixture (which is also technically a loss when the fixture is flowing) are used to size the building supply pipe. This is the first piece of pipe coming out of the water meter. All other pipe within the building will be this size or smaller. The pressure loss to the highest fixture will be the total of the following:

Table 8-2

Fixture Demand in Fixture Units (National Standard Plumbing Code)

Fixture	*Occupancy*	*Type of Supply Control*	*Load Values, in Water Supply Fixture Units*		
			Cold	*Hot*	*Total*
Water closet	Public	Flush valve	10.		10.
Water closet	Public	Flush tank	5.		5.
Urinal	Public	1″ flush valve	10.		10.
Urinal	Public	$\frac{3}{4}$″ flush valve	5.		5.
Urinal	Public	Flush tank	3.		3.
Lavatory	Public	Faucet	1.5	1.5	2.
Bathtub	Public	Faucet	3.	3.	4.
Shower head	Public	Mixing valve	3.	3.	4.
Service sink	Offices, etc.	Faucet	2.25	2.25	3.
Kitchen sink	Hotel, restaurant	Faucet	3.	3.	4.
Drinking fountain	Offices, etc.	$\frac{3}{8}$″ valve	0.25		0.25
Water closet	Private	Flush valve	6.		6.
Water closet	Private	Flush tank	3.		3.0
Lavatory	Private	Faucet	0.75	0.75	1.
Bathtub	Private	Faucet	1.5	1.5	2.
Shower stall	Private	Mixing valve	1.5	1.5	2.
Kitchen sink	Private	Faucet	1.5	1.5	2.
Laundry trays (1 to 3)	Private	Faucet	2.25	2.25	3.
Combination fixture	Private	Faucet	2.25	2.25	3.
Dishwashing machine	Private	Automatic		1.	1.
Laundry machine (8 lbs.)	Private	Automatic	1.5	1.5	2.
Laundry machine (8 lbs.)	Public or General	Automatic	2.25	2.25	3.
Laundry machine (16 lbs.)	Public or General	Automatic	3.	3.	4.

Note: For fixtures not listed, loads should be assumed by comparing the fixture to one listed using water in similar quantities and at similar rates. The assigned loads for fixtures with both hot and cold water supplies are given for separate hot and cold water loads and for total load, the separate hot and cold water loads being three-fourths of the total load for the fixture in each case.

Table 8-3

Water Flow for Meters of the Fixed Displacement Type

Meter Size (in.)	*Capacity* (gal/min)*
$\frac{5}{8}$	20
$\frac{3}{4}$	30
1	50
$1\frac{1}{2}$	100
2	160
3	300
4	500
6	1000

* American Water Works Association Standard C700

1. Pressure loss through the meter (flow loss).
2. Pressure loss due to elevation to the highest fixture (static loss).
3. Pressure required to operate the fixture (flow loss across the fixture when it is operating).

Pressure losses through cold water meters of various sizes for a given demand (total wsfu's) are given in Fig. 7–9. To this must be added the static pressure drop resulting from changes in elevation. This amounts to about 0.433 lbf/in.^{2} (2.99 kPa) for each foot of elevation. At the fourth floor of an apartment building, for example, an elevation of approximately 50 ft above the main would reduce the pressure by 21.7 lbf/in.^{2} (150 kPa). If the pressure at the main were 50 lbf/in.^{2} (345 kPa), it would be reduced to 28.3 lbf/in.^{2} or 65 ft of elevation. The flow pressure required to operate the various fixtures, including the highest fixture, are listed in Table 8–5. An ordinary basin faucet, for example, requires a flow pressure of 8 lbf/in.^{2} (55 kPa). Thus, the pressure available to supply water to the highest fixture (and that pressure used to size the first pipe out of the meter) is the pressure available at the main, minus the losses due to flow friction through the meter, static losses from elevation, and the flow pressure required to operate the uppermost fixture. An example will make this clear.

Example 8-1

The maximum water demand for a building is calculated to be 30 wsfu's. If the pressure at the street main is 60 lbf/in.^{2} (414 kPa), and the highest fixture 35 ft (10.7 m) above the main is a faucet requiring 8 wsfu's, compute (a) the size of the meter, and (b) the pressure available to supply water to the highest fixture.

Table 8-4 **Water Supply Demand Associated with Fixture Units (National Standard Plumbing Code)**

Supply Systems Predominantly for Flush Tanks		*Supply Systems Predominantly for Flushometers*	
Load (Water Supply Fixture Units)	*Demand (Gallons per Minute)*	*Load (Water Supply Fixture Units)*	*Demand (Gallons per Minute)*
6	5		
8	6.5		
10	8	10	27
12	9.2	12	28.6
14	10.4	14	30.2
16	11.6	16	31.8
18	12.8	18	33.4
20	14	20	35
25	17	25	38
30	20	30	41
35	22.5	35	43.8
40	24.8	40	46.5
45	27	45	49
50	29	50	51.5
60	32	60	55
70	35	70	58.5
80	38	80	62
90	41	90	64.8
100	43.5	100	67.5
120	48	120	72.5
140	52.5	140	77.5
160	57	160	82.5
180	61	180	87
200	65	200	91.5
225	70	225	97
250	75	250	101
275	80	275	105.5
300	85	300	110
400	105	400	126
500	125	500	142
750	170	750	178
1000	208	1000	208

Table 8-5 **Pressures and Flow Rates to Operate Various Fixtures (National Standard Plumbing Code)**

	Flow Pressure lbf/in²	*Flow Rate gal/min*
Lavatory faucet	8	2
Sink faucet	8	4
Bathtub faucet	8	6
Laundry tub	8	5
Shower	8	5
Tank closet	8	3
Flushometer, water cl.	15	15–35
Flushometer, urinal	15	15
Sill cock or hydrant	10	5
Drinking fountain	8	$\frac{3}{4}$

Figure 8-10

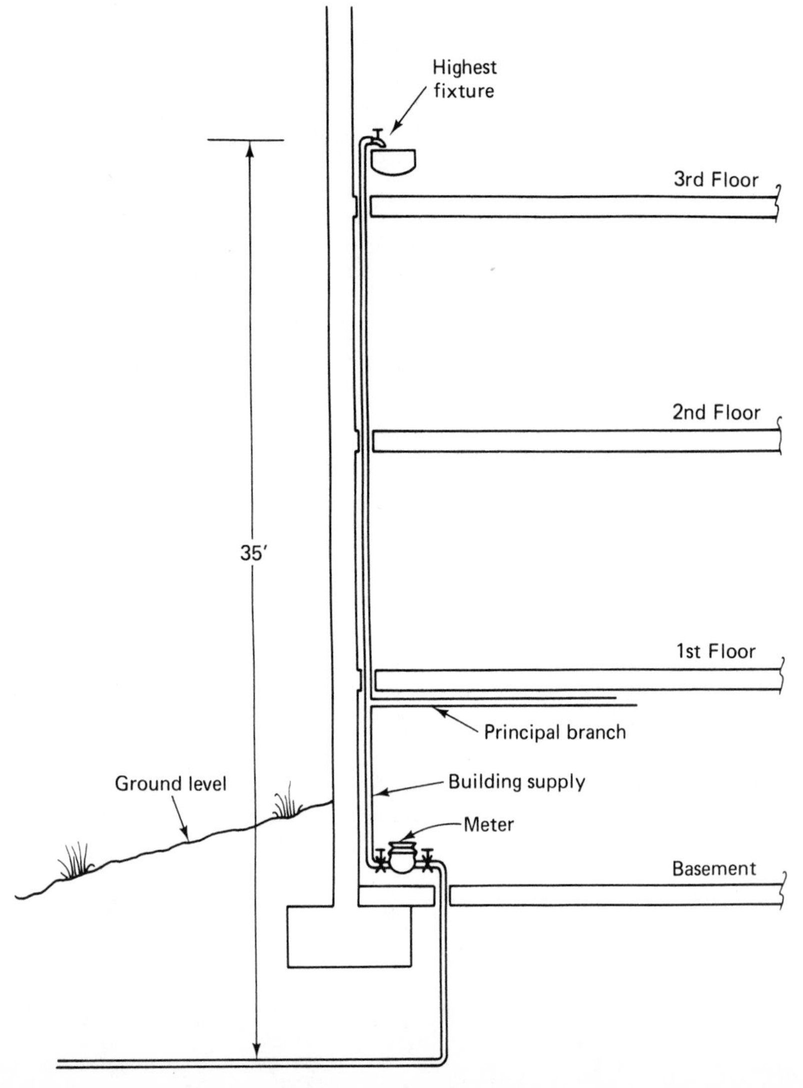

Solution:

(a) With reference to Table 8–4, the demand associated with 30 wsfu's is approximately 20 gal/min (76 l/min), and Table 8–3 lists a capacity of 30 gal/min (114 l/min) for a $\frac{3}{4}$-in. water meter that would have a capacity of 20 gal/min in its mid-range without excessive friction.

(b) The pressure loss through the meter is given by Fig. 7–9 and at the demand flow would be approximately 7 lbf/in.^{2} (48 kPa).

The pressure loss due to elevation is 0.433 lbf/in.^{2} (2.99 kPa) for each foot of elevation or 15.16 lbf/in.^{2} (105 kPa) for 35 ft.

The flow pressure necessary to operate the faucet is given in Table 8–5 as 8 lbf/in.^{2}. Summing the pressure losses to the highest fixture, we have

Meter loss	7 lbf/in.^{2}
Elevation loss	15 lbf/in.^{2}
Fixture loss	8 lbf/in.^{2}
Total	30 lbf/in.^{2}

Thus, the available pressure to supply water to the highest fixture is $60.00 - 30 = 30$ lbf/in.^{2} (206 kPa).

The size of the building supply pipe is computed from

1. Total building demand.
2. Pressure available to overcome friction in the pipe.
3. The friction loss per 100 ft of pipe in the longest run.
4. The type of pipe or tube used to plumb the system.
5. The velocity limitation, which is held to less than 8 ft/sec (2.44 m/s).

The flow friction loss per 100 ft in the longest run is computed from the formula

$$\text{Flow friction loss per 100 ft} = \frac{100 \times \text{flow pressure loss}}{\text{developed length} + \text{fittings equivalent length}} \qquad (8\text{–}1)$$

where the pressure loss is the pressure available to supply water to the system after losses to the highest fixture are subtracted from the pressure in the main. The developed length consists of the actual pipe length taken along its center line, and the equivalent length is the comparable length of all the fittings and valves. (Friction loss through an equivalent length of pipe is also explained in Section 7–7.) The equivalent lengths of fittings and valves are given in Table 7–2. A rule of thumb commonly used is to multiply the developed length of pipe by 50 percent for tubing and 75 percent for threaded pipe to arrive at an estimate of the equivalent length for the fittings. For example, if the developed length of tube to the farthest fixture (longest run) were 220 ft from the meter, all valves, elbows, tees, and couplings added together could be estimated to equal the equivalent of another

110 ft of tube. Substituting these values in Eq. (8–1), and using the pressure remaining from the previous example to overcome friction losses, we have

$$\text{Flow friction loss per 100 ft} = \frac{100 \text{ ft} \times 30 \text{ lbf/in.}^2}{220 \text{ ft} + 110 \text{ ft}} = 9 \text{ lbf/in.}^2$$

Given the demand of the system and the pressure loss due to friction per 100 ft of pipe, the size of the building supply is computed from values given in Figs. 7–6, 7–7, and 7–8 for the various types of pipe and tube. If, for the present example, the building were plumbed with copper tube, Fig. 7–6 would be used. If the demand of 20 gal/min from the previous example were used with a flow friction loss of 9 lbf/in.^{2} per 100 ft of tube, the size of the building supply pipe would be read from the graph as approximately 1 in. to maintain a flow velocity of less than 8 ft/sec. Thus, a supply pipe size of 1 in. would be selected.

Now that the building supply pipe size has been selected, no pipe within the building need be any larger.

8-4 Sizing the Cold and Hot Water Supply

Just as the building supply pipe was sized from the total demand in wsfu's, so too are the cold and hot water supplies sized from the demand of the principal branches in cwsfu's (cold water supply fixture units) and hwsfu's (hot water supply fixture units). The method used to size the principal branches is to make the computations starting from the use at the fixtures and work backward through the principal branches to the building supply pipe.

There are also some general rules that govern the size of the pipes. In general, the following apply:

- The minimum size *service main* and *building supply* to a dwelling will be $\frac{3}{4}$ in.
- The minimum size *supply* to any fixture is $\frac{1}{2}$ inch, although $\frac{3}{8}$-inch and individual $\frac{1}{4}$-inch supply lines are used in the open (not covered by the wall).

The minimum size for fixture supply pipes is given in Table 8–6.

To size the individual branches and supply pipes for both hot and cold water, use the following procedure:

1. Draw a schematic of the system and identify each fixture and supply pipe.
2. Divide the system into *principal branches* that can be sized, and list the fixtures for each.

Table 8-6

Minimum Size for Fixture Supply Pipes (National Standard Plumbing Code)

Type of Fixture or Device	*Pipe Size (inch)*
Bathtubs	$\frac{1}{2}$
Combination sink and tray	$\frac{1}{2}$
Drinking fountain	$\frac{3}{8}$
Dishwasher (domestic)	$\frac{1}{2}$
Kitchen sink, residential	$\frac{1}{2}$
Kitchen sink, commercial	$\frac{3}{4}$
Lavatory	$\frac{3}{8}$
Laundry tray, 1, 2, or 3 compartments	$\frac{1}{2}$
Shower (single head)	$\frac{1}{2}$
Sinks (service, slop)	$\frac{1}{2}$
Sinks, flushing rim	$\frac{3}{4}$
Urinal (flush tank)	$\frac{1}{2}$
Urinal (direct flush valve)	$\frac{3}{4}$
Water closet (tank type)	$\frac{3}{8}$
Water closet (flush valve type)	1
Hose bibbs	$\frac{1}{2}$
Wall hydrant	$\frac{1}{2}$

3. Assign load values to each of the fixtures from Table 8–2.
4. Size the hot and cold water pipes *on* the schematic for each branch, starting from the fixtures, and work backward through the principal branch lines to the building supply pipe. Fixtures that have a continuous demand (such as sill cocks) should be added to the demand for the branches.

Example 8-2

Figure 8–11 is a schematic for a two-story building copper tubing water supply. The dwelling has two baths, a kitchen and wet bar, a laundry room, water heater, and three sill cocks. Water pressure in the main is 45 lbf/in.^{2} (310 kPa), the highest fixture is 26 ft above the main, and the farthest fixture is 133 ft from the main. For this system, compute the following:

a. Total system demand
b. Size of the meter
c. Size of the building main
d. Size of the principal branches

Solution

a. With reference to the schematic and Table 8–2, the demand from each fixture and branch is as follows:

Figure 8-11

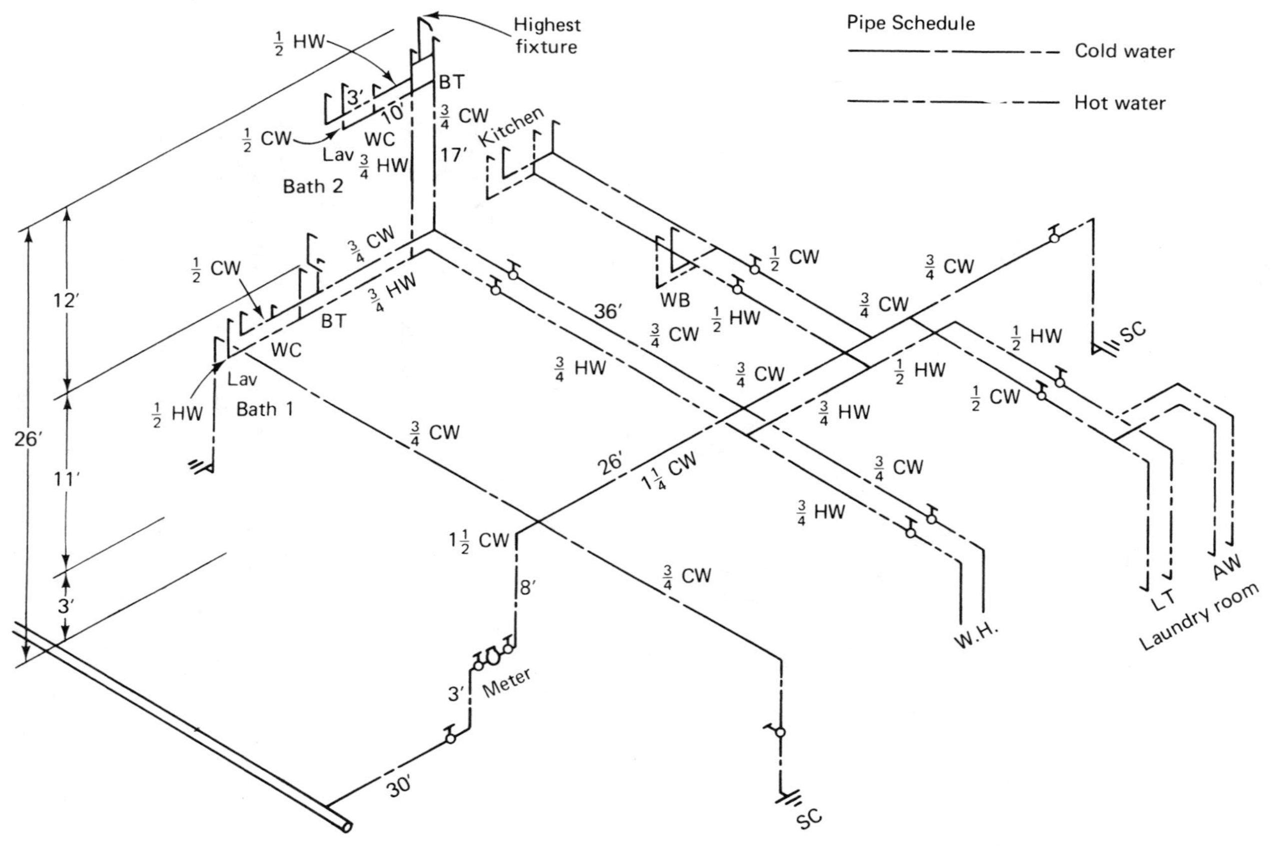

Building Water Supply Schematic

	wsfu	*cwsfu*	*hwsfu*
Upstairs Bath			
Tub and shower	2	1.5	1.5
Water closet	3	3	
Lavatory	1	0.75	0.75
Subtotal	6	5.25	2.25
Downstairs Bath			
Tub and shower	2	1.5	1.5
Water Closet	3	3	
Lavatory	1	0.75	0.75
Subtotal	6	5.25	2.25
Kitchen & Wet Bar			
Kitchen sink	2	1.5	1.5
Dishwasher	1		1
Wet bar	1	0.75	0.75
Subtotal	4	2.25	3.25
Laundry Room			
Automatic Washer	2	1.5	1.5
Laundry Tray	3	2.25	2.25
Subtotal	5	3.75	3.75
Total Fixture Demand	21	16.5	11.5
Total fixture demand in gal/min estimated from Table 8–4	15 gal/min	12 gal/min	9 gal/min
Sill Cocks			
Three @ 5 gal/min	15 gal/min	15 gal/min	—
Total system demand in gal/min	30 gal/min	27 gal/min	9 gal/min

b. The size of the meter is taken from Table 8–3, which gives the flow for displacement meters of the various sizes. Entering the demand of 30 gal/min, a 1-in. meter would be sufficient to handle the flow to 50 gal/min, without excessive friction loss and with capacity for future expansion of the system.

c. The size of the building main is determined from the pressure loss through the meter at full demand, the pressure loss due to elevation to the highest fixture, the pressure required to operate the highest fixture, and the friction loss in the farthest run. These are as follows:

- From Fig. 7–9, the pressure loss through the 1-inch meter at a flow rate of 30 gpm is approximately — 5 lbf/in.^{2}
- The static pressure loss due to elevation is computed from (26 × 0.433 lbf/in.^{2} per ft) — 11.3 lbf/in.^{2}
- The shower fixture loss in the upstairs bathroom is taken from Table 8–5 to be — 8 lbf/in.^{2}

Loss to the highest fixture — 24.3 lbf/in.^{2}

- The pressure remaining to overcome friction losses is (45 lbf/in.^{2} − 24.3 lbf/in.^{2}) = 20.7 lbf/in.^{2} — 19.7 lbf/in.^{2}
- The longest run consists of the developed length shown on the schematic to be 133 ft from the main, plus an equivalent length for the fittings of 66.5 ft, which for copper tubing is an estimate of 50 percent of the developed length.

The flow friction loss in the farthest run is computed from Eq. (8–1):

$$\text{Flow friction loss per 100 ft} = \frac{(100\ \text{ft})(20.7\ \text{lbf/in.}^2)}{(133\ \text{ft} + 66.5\ \text{ft})} = 10.37\ \text{lbf/in.}^2$$

The size of the building main is determined from system demand, which is 30 gal/min, and the friction loss per 100 ft in the longest run. Keep in mind that while the building main is delivering the maximum anticipated demand, the velocity must remain within the limitation of 8 ft/sec. Entering Fig. 7–6, using a demand of 30 gal/min and friction loss per 100 ft of 10.37 lbf/in.^{2}, we find that the indicated tube size is nearly $1\frac{1}{4}$ inch, and that size will be used for the first piece of pipe out of the water meter. No tube in the system will be larger.

d. Each of the principal cold and hot water branches is sized beginning at the fixtures and working backward toward the building main. The pipes can be sized conveniently by constructing a sizing table from Fig. 7–6 for the various sizes of tube using the friction loss value of 10.37 lbf/in.^{2} per 100 ft (up to the corresponding velocity limitation of 8 ft/sec), which was computed for the longest run. Constructing such a table for the present example, we have

Nominal Tube Size	*Flow Rate gal/min @ 10.37 lbf/in.^{2} loss per 100 ft*	*Approximate Fixture Unit Capacity*
$\frac{1}{2}$ inch	3.7 gal/min	4
$\frac{3}{4}$ inch	9.5 gal/min	12
1 inch	19 gal/min	30
$1\frac{1}{4}$ inch	33 gal/min	60

Sizing the hot and cold water supply lines begins with the upstairs bath; the cold water demand of 6.75 cwsfu will require a $\frac{3}{4}$-inch riser. The hot water demand of 3.75 hwsfu is just within the capacity limitation of a $\frac{1}{2}$-inch riser, but a $\frac{3}{4}$-inch pipe will be recommended. One-half inch supply pipes will be used from the riser to the individual fixtures. The first floor bathroom has the same demand as the second floor bathroom, so $\frac{3}{4}$-inch tube will be used to the first fitting, after which $\frac{1}{2}$-inch tube will be used to supply the individual fixtures. The principal branch feeding both bathrooms has a demand of 13.5 cwsfu and 7.5 hwsfu. Since the cwsfu demand is very nearly within the capacity of the $\frac{3}{4}$-inch line and the hwsfu demand is easily within that capacity, two $\frac{3}{4}$-inch lines will be used (notice that a 1-inch line would raise the capacity to 30 wsfu, or more than twice the demand).

The principal branch to the kitchen supplies the kitchen sink, dishwasher, and wet bar. Both the demand of 1.75 cwsfu and 3.25 hwsfu can be supplied through $\frac{1}{2}$-inch lines.

The laundry room branch feeds the automatic washer and laundry tray. Both the demand of 3.75 cwsfu and 3.75 hwsfu can be supplied through $\frac{1}{2}$-inch lines.

Each of the three sill cocks has a demand of 5 gal/min (18.9 l/min) and will be supplied by a $\frac{3}{4}$-inch supply line. It should be noted at this point that the number of sill cocks in this installation have greatly increased the total demand and thereby the size of the system. It is also common practice to supply sill cocks with a $\frac{1}{2}$-in. tube, although this limits their usefulness.

Now the individual branches are connected, starting at the fixtures farthest from the building main. The hot water line is sized first to the water heater and then its load is added to the demand on the water main. Where the hot water branches from the kitchen and laundry room join, the demand is 3.25 + 3.75 = 7 hwsfu, and the line is increased to $\frac{3}{4}$-inch. Adding both bathrooms to the hot water demand generates a total of 7 + 3.75 + 3.75 = 14.5 hwsfu, and this number is used to size both the water heater and the hot water supply lines. A $\frac{3}{4}$-inch line will be used to continue the hot water line to the heater, because its capacity is very near the required demand (a 1-inch line capacity would deliver more than twice the demand necessary). A $\frac{3}{4}$-inch line will also be used to connect the water heater to the building main, and at that point the main must be sized to deliver the hot water demand of 14.5 wsfu.

Connecting the cold water supply branches starting where the farthest sill cock connects with the supply to the laundry room, the demand is 6 + 3.75 = 9.75 cwsfu, and the $\frac{3}{4}$-inch line can continue. Where the kitchen and wet bar join the cold water line, the demand is 9.75 + 1.75 = 11.50 cwsfu, and the $\frac{3}{4}$-inch line can continue. Where the cold water branches from both bathrooms and the water heater join the main, the demand is 11.50 + 13.50 + 14.50 = 39.50 cwsfu, and the line should be increased to $1\frac{1}{4}$ inch. (If either the bathrooms or the water heater were connected separately, the line would have been increased to 1 inch, which from the table has a capacity of 30 wsfu.) Where both sill cocks join the water main the demand increases to 39.5 + 6.0 + 6.0 = 51.5 cwsfu, and the line remains as $1\frac{1}{4}$ inch to the meter.

Larger systems used for commercial buildings, including those in multiple story structures such as apartment buildings and office buildings, are designed and sized by the mechanical engineer or plumbing engineer for the project. Isometric layouts are made from the mechanical drawings, and these are used to generate the list of materials and specifications. The plumber then installs the system for the general contractor from the mechanical drawings and isometric drawings, closely adhering to the materials and sizes specified. The engineer is responsible for the operation of the system within the capability of the design.

8-5 Water Hammer and Shock Control

Water hammer is a noise in the pressure piping that results from a pressure wave or surge traveling back and forth within the system. It is caused by abruptly stopping the flow of fluid. When the water is abruptly stopped, it compresses (reducing its volume slightly) against the obstruction, and shock waves are generated backward and then forward in the system at the speed of sound until the imbalance finally comes to rest and equilibrium. Figure 8-12 illustrates the concept of water hammer caused by a quick-closing valve and how the fluid compresses just before the resounding pressure wave is generated backward in the system from the place where the fluid comes to an abrupt stop. The noise is caused by the nearly incompressible fluid pounding the piping system as it tries to follow the shock

Figure 8-12

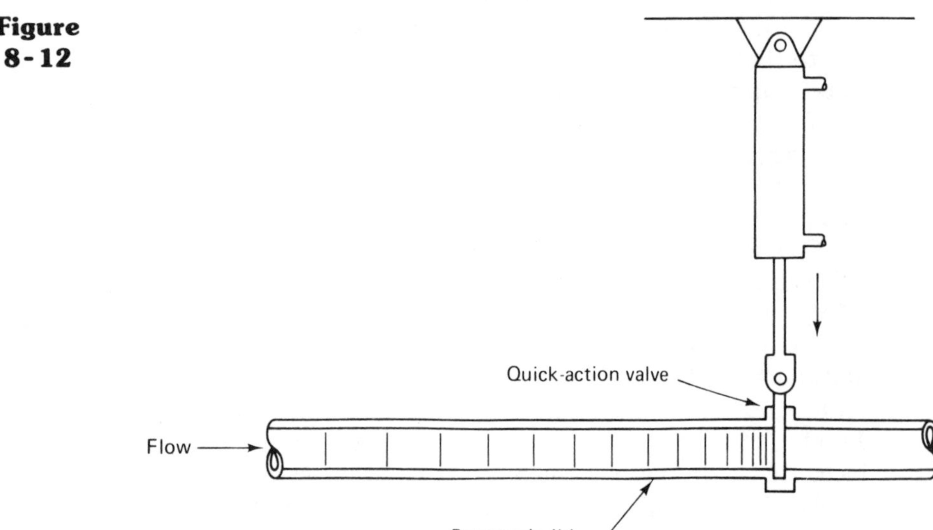

Water Hammer Concept

waves, causing the pipe to stretch and expand. If the shock causing the pressure surges is sufficiently strong, it will rupture the system.

If a valve closes almost instantaneously, the maximum pressure that can be developed in a rigid pipe is

$$p_{max} = \frac{(0.433)cv}{g} \tag{8-2}$$

where c is the celerity (velocity) of the pressure wave and v is the flow velocity of the fluid in ft/sec just before cut off. For purposes of simplification, the value of c can be estimated to be 4750 ft/sec without introducing appreciable error, and Eq. (8–2) becomes

$$p_{max} = \frac{(0.433 \text{ lbf/in.}^2 \text{ per ft})\,(4750 \text{ ft/sec})\,(v)}{(32.2 \text{ ft/sec}^2)}$$

or

$$p_{max} = 64v \tag{8-3}$$

where p_{max} is the pressure added to the operating pressure of the system in lbf/in.².

Example 8-3 A quick-acting valve that controls the flow of water in a line that has an operating pressure of 30 lbf/in.² and a flow velocity of 5 ft/sec is suddenly closed. Estimate the maximum pressure surge in the line.

Solution

The maximum pressure is estimated from Eq. (8–3)

$$p_{max} = (64)\,(5) = 320 \text{ lbf/in.}^2$$

This pressure is added to the line pressure, bringing the maximum pressure surge to

$$p_{surge} = p_{line} + p_{max} \tag{8-4}$$

and

$$p_{surge} = 30 + 320 = 350 \text{ lbf/in.}^2$$

A pressure surge such as the foregoing would be dangerous to the system if it were permitted to occur and continue as described. However, manual valves typically do not close as fast as that described unless they are faulty, and other provisions are made to reduce the possibility of water hammer if they should. Closing the valve more slowly, for example, reduces the possibility of water hammer. It is also noticed from Eq. (8–3) that the magnitude of the peak pressures can be reduced by lowering the velocity of the water in the line, and this is accomplished by increasing its size. Other methods used to reduce water hammer include check valves placed in the line where shocks are expected to occur, mechanical shock alleviators, diaphragm-separated air chambers, and simple air chamber shock tubes

Figure 8-13

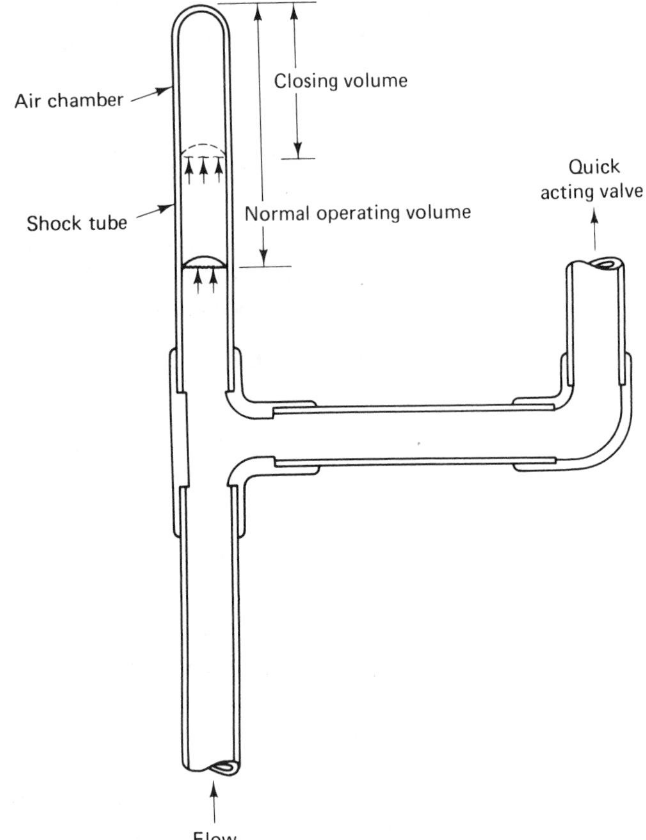

Shock Tube

placed near the fixture or valve which is prone to cause the problem. Figure 8-13 illustrates the use of a simple air chamber constructed from a prefabricated stub-out.

8-6 Summary of Practice

The plumber has the responsibility for installing a distribution system for hot and cold water which is both safe and workable. By "workable" is meant that there is sufficient pressure and volume at all fixtures to make them operate as they were intended by design. The water supply is safe if it conforms to standards established and enforced by health departments, and if the system is installed in such a manner as to protect it from pollution and contamination. Although the plumber typically will not see the system operate for any extended time after it is installed, it is still necessary to anticipate the possible conditions that could cause the potable water supply to become unsafe due to cross connections or back-siphonage through neglect or misuse. Many of these conditions are specified by model and local codes. Perhaps the worst offenders are hoses connected to laundry sinks and sill cocks, and hand-held showers mounted in tubs. Both of these should be connected with vacuum breaker anti-siphon devices as a matter of routine to protect the occupants of the dwelling and others who receive water from the same source of supply. Remember, the contaminated water may be drawn back into the main and distributed some distance from the source of pollution.

The design of the distribution system is based upon demand at the fixtures, that is, through an examination of the use of sinks, lavatories, water closets, and other uses. Demand is specified in water supply fixture units, which are summed and converted to demand in gallons per minute (or liters per minute). The sizes of the pipes necessary to supply the required demand to the building and branches are determined from the available water pressure and losses through the meter, those due to elevation, friction in the longest developed length, and flow through the fixtures. Individual branch lines for hot and cold water are sized similarly. The hot water branch is sized first and added to the cold water demand where it joins the building main.

Water hammer, which can damage distribution piping, is influenced by system pressure, the flow velocity, and the time it takes to close valves that stop the flow. Decreasing the velocity (using larger pipes) and increasing the time to close valves that stop the flow reduce the possibility of water hammer. Controlling devices, such as pressure regulators on the supply, and in-line check valves, water hammer arresters, and air chamber shock tubes near the quick-closing valve, alleviate the problem by stopping or cushioning the pressure wave that is generated backward in the line when the valve abruptly stops the flow.

REVIEW QUESTIONS AND PROBLEMS

1. What is *potable* water?

2. What is meant by the term *fixture unit?*

3. What is a *cross connection?*

4. What is the difference between an *inlet* cross connection and a *direct pressure* cross connection?

5. What is meant by the term *back-siphonage?*

6. What is the difference between *pollution* and *contamination* of the potable water supply?

7. Name and define the five types of *backflow preventers.*

8. How is total demand for a building water supply determined?

9. Why would there be a difference between the demand in wsfu's and demand based on the sum of the flow rates necessary for each fixture?

10. What losses are considered (starting from the street main) when one is sizing a system?

11. What is flow friction loss?

12. What is the difference between *developed length* and *equivalent length?*

13. Where do you begin when determining the sizes of the principal branches?

14. Record an interview with a local plumbing establishment for replay to the class. Ask the person who sizes the systems that they install how it is done in "their business."

15. What is water hammer, and how is it stopped?

REFERENCES

[1] American Public Health Association et al., *Standard Method for Examination of Water and Wastewater* (Washington, D.C.: 1965) 12th ed., pp. 570–71.

[2] American Public Health Association, Report of the Subcommittee on Communicable Disease Control, Committee on Research and Standards, "The Control of Communicable Diseases," *Public Health Reports* 50:1017 (Aug. 9, 1935).

[3] *Journal of the American Medical Association,* February 3, 1934, p. 369.

[4] Watts Regulator Co., Lawrence, Massachusetts 01842.

9 Waste, Vent, and Drain Systems

9-1 Introduction

Waste, vent, and drain systems are constructed to remove waste water and soil containing human fecal matter from buildings, and storm drainage from roofs, paved areas, and yards. Waste piping is constructed so that the contents from each fixture flows through a trap, whereas storm drain water that flows into storm sewers does not. Where the storm drain system flows into a combined storm drain/sanitary drain system, the main leader feeding the system is trapped, as are floor drains carrying sanitary waste water. Storm water and waste water are not drained into sewers intended for sewage only.

The capacity of waste, vent, and drain systems is determined by the load in drainage fixture units (dfu). Each fixture emptying into the system is assigned a value reflecting its trap size and probable frequency of usage, as are other areas such as roofs, paved yard areas, and yards receiving rainwater, which is usually taken at the maximum of four inches per hour. Individual branch pipes and the main stack or leader are sized accordingly by the number of fixture units they carry.

Systems carrying waste and soil in buildings use a trap at each fixture to prevent odor and sewer gas from escaping into the room, and incorporate a vent system protecting each trap from positive and negative pressures of more than one inch of water. Traps subjected to either positive or negative pressures exceeding one inch can lose their seal and become ineffective. Trap seal loss will cause noxious sewer gas to enter the room and render the plumbing system unsafe.

9-2 Flow in Horizontal Branches, Drains, and Vertical Stacks

A *horizontal* drain or branch is one that is sloped at less than 45 degrees from the horizontal. For practical purposes, the flow in horizontal drain and waste pipes can be assumed to be steady and uniform. That is, the pipe is partially filled along its length by liquid that flows smoothly and at uniform depth. Thus, the slope of the liquid surface and the bottom of the

Table 9-1

Flow Velocity and Capacity of Horizontal Drains at Various Slopes

	Slope = $\frac{1}{8}$ in/ft			*Slope = $\frac{1}{4}$ in/ft*			*Slope = $\frac{1}{2}$ in/ft*		
Pipe Size Diameter (in.)	*Velocity Half Full or Full (ft/sec)*	*Flow Half Full (gpm)*	*Flow Full (gpm)*	*Velocity Half Full or Full (ft/sec)*	*Flow Half Full (gpm)*	*Flow Full (gpm)*	*Velocity Half Full or Full (ft/sec)*	*Flow Half Full (gpm)*	*Flow Full (gpm)*
$1\frac{1}{4}$	1.25	3.42	6.9	1.77	4.85	9.7	2.50	6.86	13.7
$1\frac{1}{2}$	1.31	4.10	8.2	1.85	5.80	11.6	2.62	8.20	16.4
2	1.40	6.79	13.6	1.98	9.60	19.2	2.80	13.57	27.1
$2\frac{1}{2}$	1.63	12.30	24.6	2.30	17.40	34.8	3.25	24.60	49.2
3	1.83	19.97	39.9	2.59	28.25	56.5	3.66	39.95	79.9
4	2.07	40.19	80.38	2.93	56.85	113.7	4.14	80.39	160.8
5	2.23	67.73	135.5	3.16	95.80	191.6	4.47	135.46	270.9
6	2.53	110.47	220.9	3.58	156.25	312.5	5.06	220.93	441.9
8	2.87	222.78	445.6	4.06	315.10	630.2	5.74	445.55	891.1
10	3.33	403.66	807.3	4.71	570.95	1141.9	6.66	807.32	1614.6
12	3.76	556.56	1313.1	5.32	928.65	1857.3	7.52	1313.11	2626.2
15	4.37	1191.54	2383.1	6.18	1685.35	3370.7	8.73	2383.08	4766.2

pipe are the same and parallel, and the potential energy lost as the stream flows down the slope is just sufficient to maintain the flow. The minimum velocity of the stream is established at 2 ft/sec (0.61 m/s) so as to promote a scouring action that will keep the pipe clear of dirt and particles, which would otherwise lie in the bottom.

Table 9–1[1] gives the velocity and flow capacity of horizontal drain and waste pipes flowing half full, and full. It is used to:

1. Establish the slope of drain pipes of the various sizes so that the desired stream velocity is maintained.
2. Determine the load-carrying capacity of pipes pitched at common slopes.
3. Size building drains and horizontal branches.

Notice, for example, that a 3-inch pipe must be pitched at a slope of $\frac{1}{4}$ in./ft to impart a velocity greater than 2 ft/sec, whereas for a 4-inch pipe a slope of $\frac{1}{8}$ in./ft is sufficient to impart a velocity greater than 2 ft/sec. Also notice that having a horizontal drain run between half and two-thirds full at capacity requires a pipe size and slope such that the value selected in the table will be greater than half full, but less than full.

Example 9-1

What size drain sloped at $\frac{1}{4}$ in./ft would be required to carry 200 gal/min?

Solution:
With reference to Table 9–1, at a slope of $\frac{1}{4}$ in./ft, a 6-in. pipe would deliver 156.25 gal/min flowing half full, and 312.5 gal/min flowing full. Notice that a 5-inch pipe flowing full would carry only 191.6 gal/min.

Every building with installed plumbing must have at least one main stack that runs undiminished in size and as directly as possible from the building drain through the roof to the open air. The flow of water down vertical stacks varies by the extent to which the pipe is filled. The term *stack* is used to mean a vertical pipe that conveys soil or waste, or vents the system. Flows that occupy less than $\frac{1}{4}$ of the cross-section area of the pipe (Fig. 9–1) travel in sheets attached to the inside wall of the stack for the most part, and the core of this cylindrical flow is filled with air. As flow increases, the cross-section area of flow occupies more of the available space in the stack and frictional resistance causes the water to "drop away" and form slugs that travel short distances before breaking up again and attaching to the flow on the inside wall of the pipe. This phenomenon is called diaphragming and accounts for the pressure oscillation and attendant noises associated with flow in vertical stacks. The upper limit capacity of vertical stacks is reached when they run $\frac{1}{3}$ full and for practical purposes the limitation is reached at $\frac{1}{4}$ to $\frac{7}{24}$ full. Beyond this point diaphragming and overloading are very likely to occur. There is the popular belief that the flow in vertical stacks continues to "free-fall" with increased velocity and

Figure 9-1

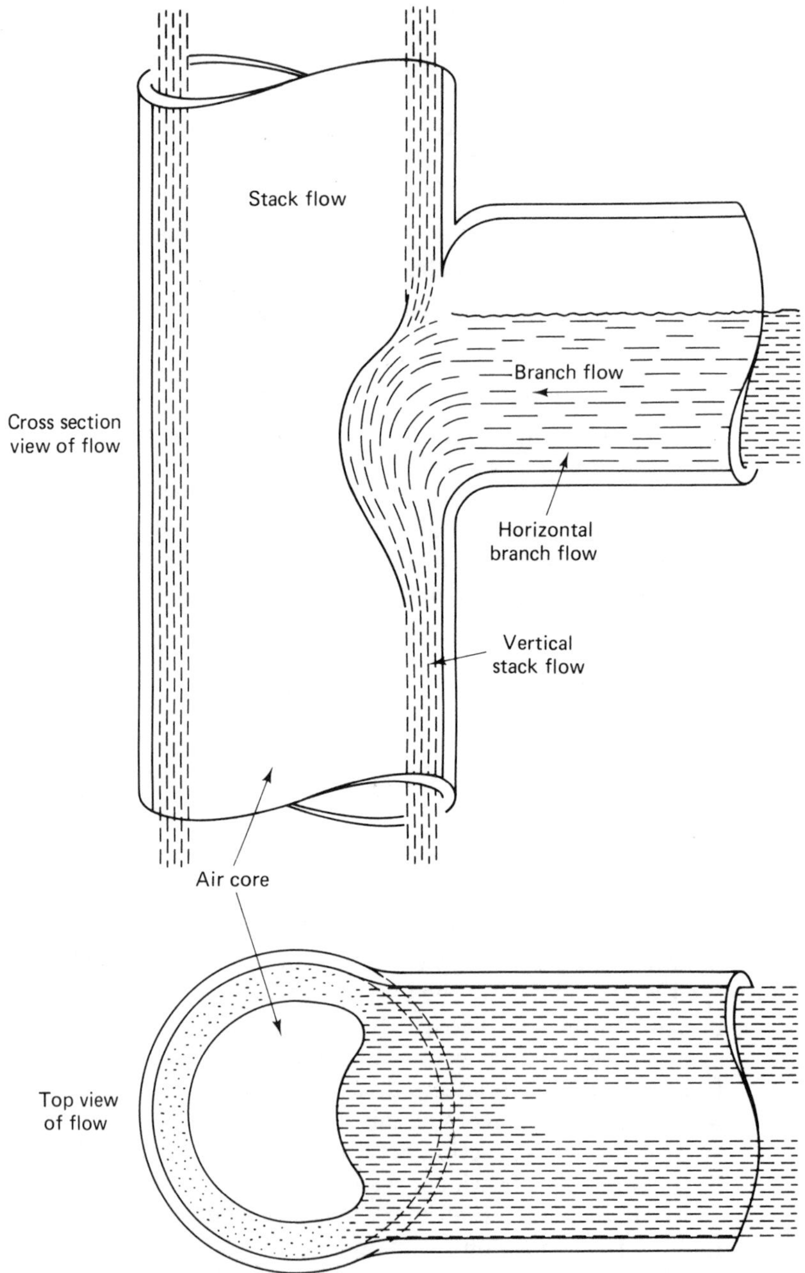

Cross Section of Flow in the Stack and Horizontal Branches

thus to pose a serious danger to the fitting at the base of the stack in tall buildings. This is not the case. Rather, the maximum velocity, called "terminal velocity," is reached within 10–15 feet (3–4.6 m) down the stack from the place of entry at a terminal velocity of 10–15 ft/sec (3–4.6 m/s).

9-3 Sanitary Drainage Systems

The drainage system is covered by all plumbing codes and consists of piping within the premises which transports liquid wastes to the public sewer system itself. The schematic in Fig. 9-2 illustrates the building drain, soil stack, horizontal branches, and stack vent.

The sanitary drainage system transports sewage, that is, animal, vegetable, and chemical matter in suspension, but excludes storm, surface, and ground water. Pipes and fittings that transport wastes must be smooth inside, must join without burrs or obstructions, and must provide for a smooth change in direction. This is accomplished by using *sanitary* and

Figure 9-2

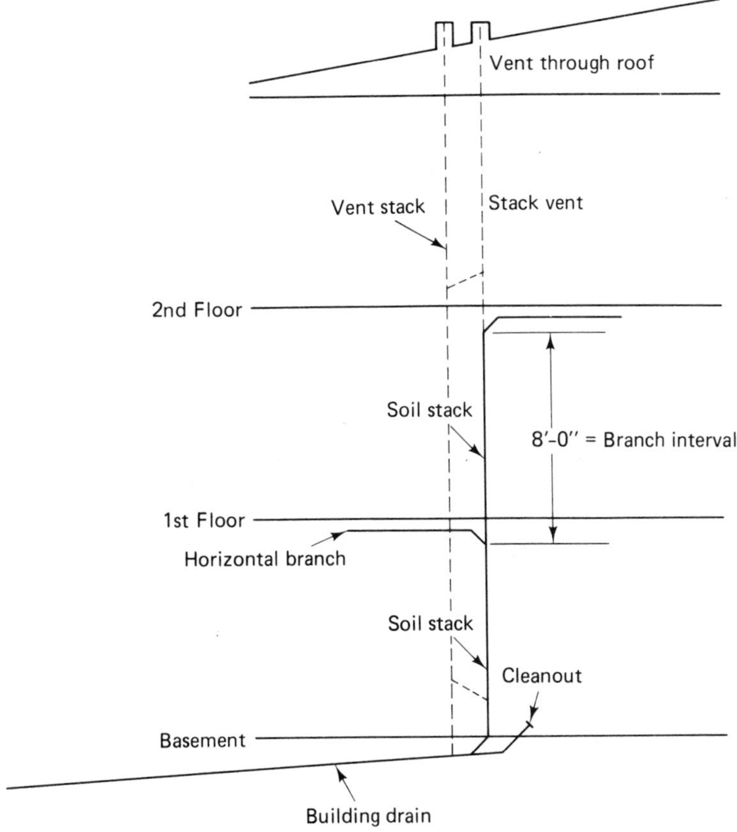

Sanitary Drainage and Vent System

drainage fittings that promote flow and join in a recess that prevents obstruction. Drainage fittings also pitch slightly upward at $\frac{1}{4}$ inch per foot from the horizontal so as to give horizontal pipes and branches the correct fall to promote gravity flow. Materials used for soil and waste pipes include cast iron, galvanized steel, copper tube, and plastic. Only drainage fittings are used on threaded pipe. Pipes to $2\frac{1}{2}$-inch diameter are pitched at not less than $\frac{1}{4}$ inch per foot, whereas pipes over 3 inches in diameter may be pitched at $\frac{1}{8}$ inch. This will ensure that the flow velocity will not be less than 2 ft/sec, which is required for proper scouring action.

Sanitary and drainage pipes feeding into fixture branches, stacks, and mains are sized from the load each carries in drainage fixture units (dfu). The various kinds of fixtures and the load that they impose on the system are listed in Table 9-2. Where a fixture is not listed, a similar fixture corresponding to its size can be used to estimate the dfu value. For equipment that pumps or drains a continuous flow into the drainage system, such as

Table 9-2

Values for Various Fixtures in Drainage Fixture Units

Fixture	*Load in Drainage Fixture Units*
Automatic washer	3
Bathroom group consisting of: Bathtub/shower stall, Lavatory, Water closet	6 (Tank water closet) 8 (Flush-valve water closet)
Bathtub (with or without shower head)	2
Bidet	1
Combination kitchen sink	3
Combination kitchen sink with food waste grinder	4
Drinking fountain	$\frac{1}{2}$
Dishwasher (domestic)	2
Lavatory ($1\frac{1}{4}$-in. waste)	1
Laundry tray (one- or two-compartment)	2
Shower stall (domestic)	2
Urinal (pedestal, syphon jet, blowout)	8
Urinal (wall lip, washout)	4
Wash sink (circular)	2
Water closet (tank operated)	4
Water closet (flush-valve)	8

air conditioning equipment, two fixture units are allowed for each gallon per minute discharge.

The fixture drain or trap size is also used to determine the load capacity; these values are given in Table 9–3.

The maximum number of drainage fixture units that can be carried by a horizontal building drain or building sewer is determined from the diameter size of the pipe and its horizontal pitch, which is given in inches of fall per running foot of pipe. However, in metric units it could also be given in centimeters of fall per running meter of pipe. Table 9–4 gives the load-carrying capacity in drainage fixture units for the various size building drains and sewers pitched at $\frac{1}{16}$ inch, $\frac{1}{8}$ inch, $\frac{1}{4}$ inch, and $\frac{1}{2}$ inch per foot. This information is used to size the main building drain fed by branch drains and stacks throughout the building. An example will clarify how the building drain table is used.

Table 9-3

Fixture Drain or Trap Size and Load in Drainage Fixture Units

Fixture Drain/Trap Size (in.)	*Load in Drainage Fixture Units*
$1\frac{1}{4}$	1
$1\frac{1}{2}$	2
2	3
$2\frac{1}{2}$	4
3	5
4	6

Example 9-2

What size building drain would be required at $\frac{1}{4}$-inch pitch to carry all the fixtures listed in Table 9–2? Assume that there are two bathroom groups, one with a load of 6 dfu and the other with 8 dfu.

Solution:

Summing the values in Table 9–2, $60\frac{1}{2}$ dfu must be carried by the building drain. Entering Table 9–4 using $\frac{1}{4}$-inch pitch, we notice that a 3-inch line will carry 42 dfu, which is too small, and the next size 4-inch line will carry up to 216 dfu. Also notice that the footnote at the bottom of the table specifies that not over two water closets can be connected to a 3-inch line. Thus the proper size drain to carry the fixtures listed in Table 9–2 would be a 4-inch pipe.

Horizontal branch drains receive the discharge from fixtures on the same floor and conduct it to the waste or soil stacks feeding the building main drain.

Table 9-4 **Maximum Loads for Horizontal Drains**

Diameter of Drain (in.)	Building Drain or Building Sewer Slope (inches per foot) $\frac{1}{16}$ (dfu)	$\frac{1}{8}$ (dfu)	$\frac{1}{4}$ (dfu)	$\frac{1}{2}$ (dfu)
$1\frac{1}{4}$				
$1\frac{1}{2}$				
2			21	26
$2\frac{1}{2}$			24	31
3		36*	42*	50*
4		180	216	250
5		390	480	575
6		700	840	1000
8	1400	1600	1920	2300
10	2500	2900	3500	4200
12	3900	4600	5600	6700
15	7000	8300	10 000	12 000

* Not over two water closets or bathroom groups, with computed velocity to be less than 2 ft/s.

There are code restrictions limiting the load in fixture units that can be connected to the stack within each branch interval. A *branch interval* is one story vertical stack height, or eight feet, whichever is larger, within which the horizontal branch drains from that story are connected; Fig. 9–3 illustrates the requirements for a branch interval. The distance between the two horizontal branches in the basement (7 ft 6 in.) does not constitute a branch interval. The total number of drainage fixture units that can be connected to any horizontal fixture branch is shown in the second column of Table 9–5. The third column gives the maximum load for a stack of three branch intervals or less. The fourth and fifth columns are used to size stacks with more than three branch intervals. The following examples will make the use of Table 9–5 clear.

Example 9-3 What is the maximum load in dfu that can be connected to a 4-inch horizontal fixture branch of (a) a three-story building and (b) a building of more than three stories?

Solution:

(a) For any single horizontal branch connected to a 4-inch stack the maximum load is 160 dfu, read from the second column of Table 9–5. The total for the single stack of a three-story (three branch intervals) building is 240 dfu, read from the third column.

(b) For a building of more than three stories, the maximum load for any single branch interval is 90 dfu, read from the last column, and 500 dfu total for the stack read from the fourth column.

Figure 9-3

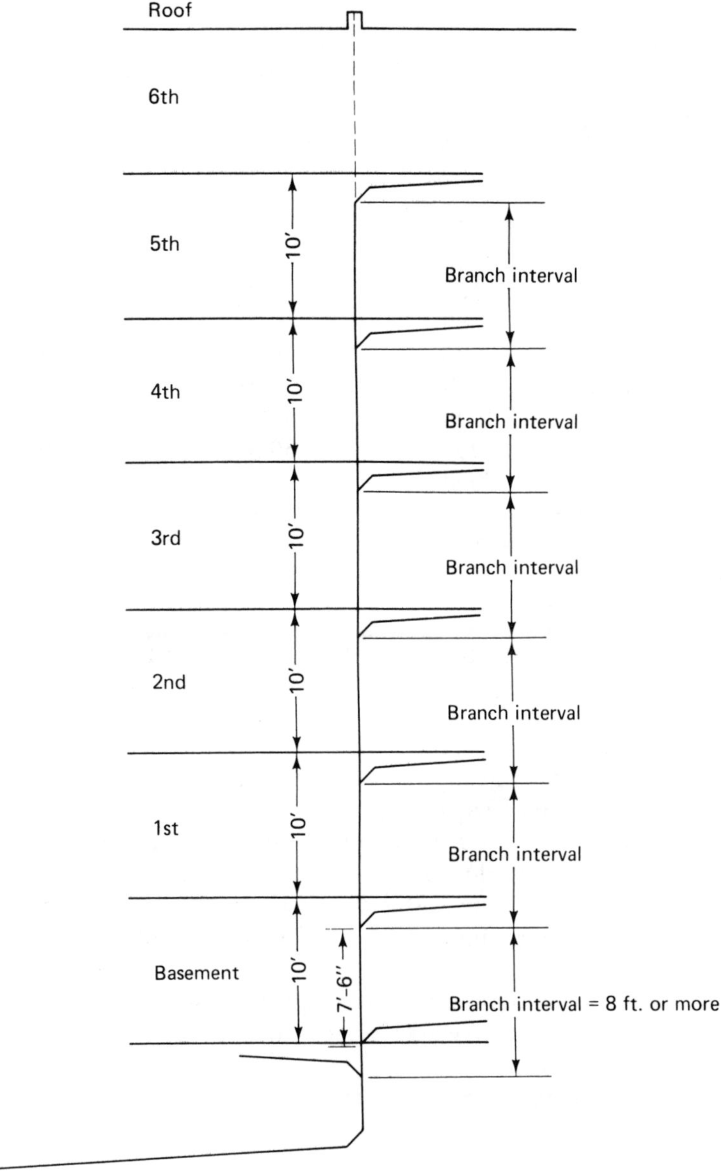

Branch Interval

Table 9-5 **Maximum Loads for Horizontal Fixture Branches and Stacks**

Diameter of Pipe (in.)	*Any Horizontal*[1] *Fixture Branch* (dfu)	*One Stack of Three Branch Intervals or Less* (dfu)	*Stacks with More Than Three Branch Intervals* *Total for Stack* (dfu)	*Stacks with More Than Three Branch Intervals* *Total at One Branch Interval* (dfu)
$1\frac{1}{4}$	1	2	2	1
$1\frac{1}{2}$	3	4	8	2
2	6	10	24	6
$2\frac{1}{2}$	12	20	42	9
3	20[2]	48[2]	72[2]	20[2]
4	160	240	500	90
5	360	540	1100	200
6	620	960	1900	350
8	1400	2200	3600	600
10	2500	3800	5600	1000
12	3900	6000	8400	1500
15	7000			

[1] Does not include branches of the building drain.
[2] Not over two water closets.

There are additional code requirements for sanitary drain systems, and these must be given attention in the overall design:

- Drains:
  1. Drainage pipe installed underground or below basement floors shall be no less than 2 inches.
  2. Horizontal drainage piping will be installed at uniform alignment, and at slopes not less than $\frac{1}{4}$ inch per foot for 3-inch diameter and less piping and, $\frac{1}{8}$ inch per foot for 4-inch-diameter or more piping.
  3. The minimum computed velocity (Table 9–1) will be not less than 2 ft/sec.
  4. The minimum size drain for a water closet is 3 inches, and not more than two water closets will be connected to a 3-inch horizontal branch.
  5. Three or more closets require a 4-inch drain.

- Stacks:
  1. No soil or waste stack shall be smaller than the largest horizontal branch connected to it, except that a 4 × 3 water closet connection shall not be considered as a reduction in pipe size.
  2. The minimum size stack conveying discharge from a water closet is 3 inches.

3. Not more than two water closets are permitted on a 3-inch stack at one story or branch interval.
4. Six branch intervals and six water closets are the maximum that can be served by a three-inch stack.

Stacks are sometimes offset to negotiate the building structure. If the offset is made at a 45-degree angle, the offset is considered to be the same as the stack, and the same rules and tables apply to sizing. If the offset is greater than 45 degrees, the pipe is considered to be in the horizontal plane and the following rules are used to size the stack (*Note:* These do not apply to stacks that offset above the highest fixture. Here the stack is part of the vent system and will be covered in a later section.):

1. That portion of the stack above the offset remains unchanged.
2. The offset length of the stack is sized as a horizontal drain (Table 9–4).
3. The stack below the offset must be at least as large as the offset or sized for the *total* number of fixture units the entire stack serves, whichever is larger.
4. No horizontal branches shall be connected to the stack within two feet above or below a horizontal offset.
5. The offset shall be relief vented, if necessary, as specified (in a later section).

The following procedure is used to size the sanitary drainage system:

1. Size each of the horizontal branches, starting farthest from the building drain, using Table 9–5.
2. Size the stack, starting from the place where the farthest branch drain connects, and proceeding toward the building drain, using Table 9–5.
3. Size offsets as they occur, starting with the stack above the offset, sizing the offset itself, and finally sizing the stack below the offset.

Following are examples to clarify the procedures and calculations used to size stacks and offsets. Refer to Fig. 9–4.

Example 9-4 Figure 9–4(a) illustrates a three-story building with branch intervals including a basement. The stack is fed by four branches. From the information given in the figure, size the branches and the stack.

Solution:
Starting with the branch farthest from the drain and using Table 9–5, we have

Third floor branch 20 dfu = 3 inch (not more than 2 water closets)
Second floor branch 40 dfu = 4 inch
First floor branch 70 dfu = 4 inch

Figure 9-4

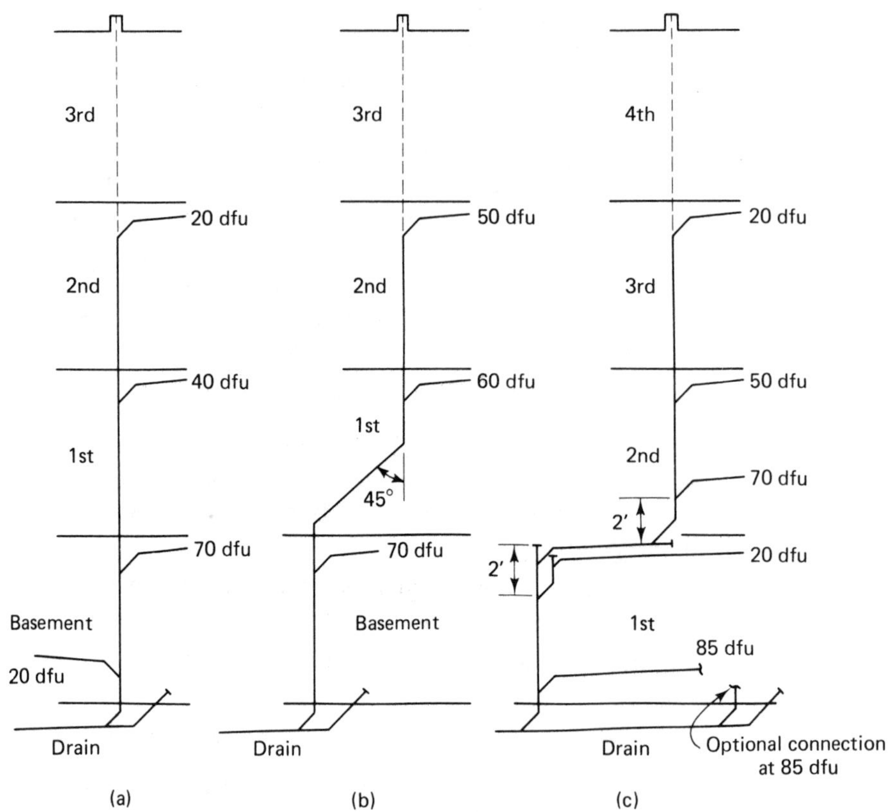

Basement branch 20 dfu = 3 inch (not more than 2 water closets)
Total 150 dfu = 4-inch building drain from Table 9–4 using either $\frac{1}{8}$-inch or $\frac{1}{4}$-inch slope.

Sizing the stack using Table 9–5 and starting at the place where the branch connects from the third floor, we have

At third floor branch = 20 dfu = 3 inch
At second floor branch = 20 + 40 = 60 dfu = 4 inch (entire stack is increased)
At first floor branch = 60 + 70 + 20 = 150 dfu = 4 inch

Notice that the basement is included with the first floor in the first branch interval.

Example 9-5 Figure 9–4(b) illustrates a three-story building with three branch intervals including a basement. The stack is offset at 45 deg. From the information given in the figure, size the branches and the stack.

Solution:
Starting with the branch farthest from the drain and using Table 9–5, we have

Third floor branch = 50 dfu = 4 inch
Second floor branch = 60 dfu = 4 inch
First floor branch = 70 dfu = 4 inch
Total 180 dfu = 4-inch building drain from Table 9–4, using either $\frac{1}{8}$-inch or $\frac{1}{4}$-inch slope.

Sizing the stack, using Table 9–5 and starting at the place where the branch connects from the third floor, we have

At third floor branch = 50 dfu = 4 inch
At second floor branch = 50 + 60 = 110 dfu = 4 inch
At first floor branch = 110 + 70 = 180 dfu = 4 inch

Notice that because the offset is at 45 degrees, no change in stack size is necessary.

Example 9-6 Figure 9–4(c) illustrates a four-story building with three branch intervals. The stack is offset 90 degrees. From the information given in the figure, size the branches and stack, including the offset.

Solution:
Starting with the branch farthest from the drain and using Table 9–5, we have

Fourth floor branch = 20 dfu = 3 inch (not over two water closets)
Third floor branch = 50 dfu = 4 inch
Second floor branch above the offset = 70 dfu = 4 inch
Second floor branch below the offset = 20 dfu = 3 inch (not over two water closets)
First floor branch = 85 dfu = 4 inch
Total 245 dfu

Sizing the stack, using Table 9–5 and starting at the place where the branch connects from the fourth floor, we have

At fourth floor branch = 20 dfu = 3 inch
At third floor branch = 20 + 50 = 70 dfu = 4 inch (increased)
At second floor branch = 70 + 70 = 140 dfu = 4 inch

Since the offset is at 90 degrees, it is sized as a drain from Table 9–4.

Offset = 140 dfu = 4 inch at either $\frac{1}{8}$ or $\frac{1}{4}$ inch/ft slope.

Sizing the stack below the offset, it is seen that the total load on the stack will equal 245 dfu, and the sizing table indicates that the stack would have to be increased to 5 inch. The building drain would also be sized to carry

245 dfu, and from Table 9–4 this would also be 5 inch. Notice that, if the lowest branch on the first floor were piped directly into the drain rather than into the stack, the load on the stack below the offset would be 245 − 85 = 160 dfu, and the stack could continue as 4 inches rather than increased to 5 inches. This would result in a sizable savings, considering that the stack must continue, undiminished in size, through the roof.

9-4 Venting Requirements

A *venting system* provides for the flow of air to or from a drainage system, or within a drainage system, to protect trap seals from back-pressure and siphonage. When a pipe runs full, even for a short interval, air is pushed ahead of the flow and siphoned behind it. Thus, to be effective, the venting system must allow free circulation of air as the soil and waste flow through the drainage system. The *stack vent* is the extension of the soil or waste stack above the highest horizontal drain connection. The term "stack vent" is also synonymous with waste vent or soil vent. The *vent stack,* which connects to the base of the soil and waste stack, is a separate vertical vent pipe installed for the purpose of circulating air to and from parts of the drainage system. Individual fixture vents, continuous vents, circuit vents, battery vents, loop vents, relief vents, and other specific types of vents are installed to protect trap seals and connect to the vent stack, which passes through the roof and into the open air.

Any structure that has a building drain must have a *stack vent* or *vent stack* of not less than 3-inch diameter, or not less than the size of the building drain if it is less than 3 inches, carried full size (without reduction) through the roof. In a one-story building, branch vents may connect directly to the stack vent. A vent stack or main vent is required, in addition to the soil or waste stack, in a building that has vents at two or more branch intervals. That is, buildings of two or more stories require a separate vent stack. This vent can terminate independently above the roof or can be connected to the stack vent and can be carried through the roof, so long as it connects to the stack at least 6 inches above the flood rim of the highest fixture. Where the vent stack (or stack vent) goes through the roof, it must be made weathertight to prevent the roof from leaking. The vent terminates at least 6 inches above the high point of the roof and is sealed with a flashing. Figure 9–5 illustrates the termination of a stack vent to which the vent stack is connected, and the installation of the vent flashing over the vent extension.

Vents serve two main purposes:

1. They vent the main stack, relieving positive and negative pressures within the stack and promoting air circulation within the stack.
2. They vent branches and individual fixture traps, thereby allowing the system to function properly while preventing trap seal loss at individual fixtures.

Figure 9-5

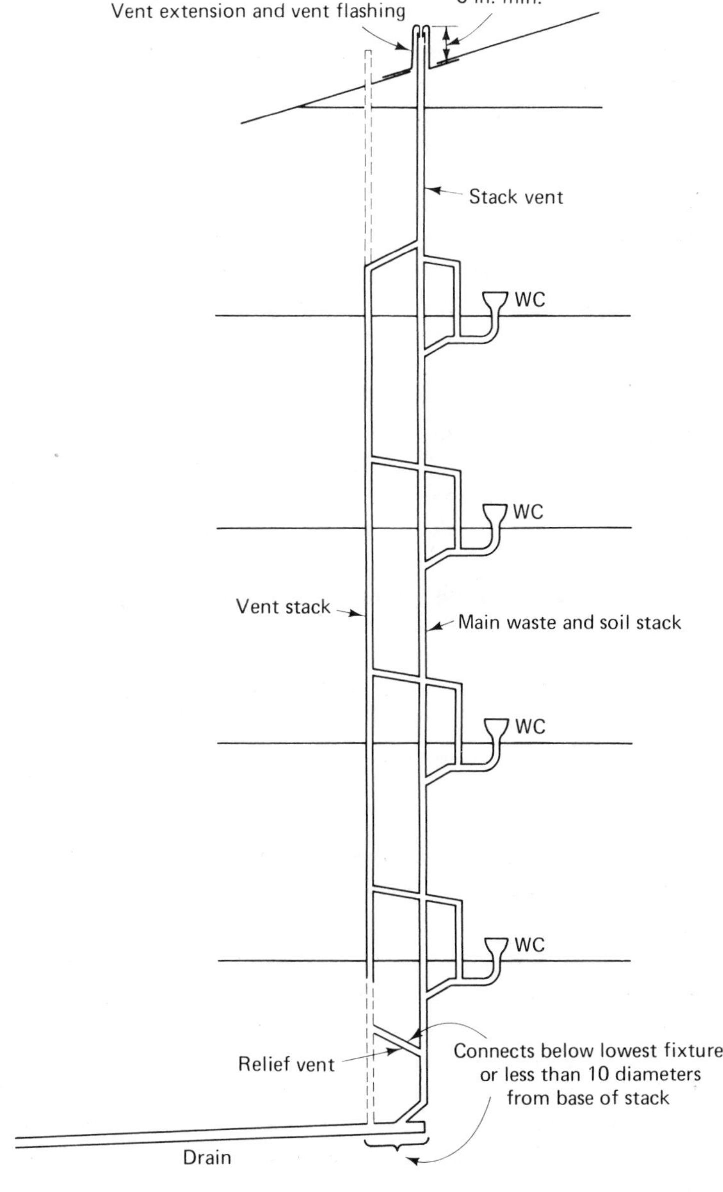
Vent extension and vent flashing
6 in. min.
Stack vent
WC
WC
Vent stack
Main waste and soil stack
WC
WC
Relief vent
Connects below lowest fixture
or less than 10 diameters
from base of stack
Drain

Vent Stack

Vent stacks, circuit vents, and individual fixture vents are sized on the basis of the number of fixture units served and the developed length of the vent piping. In no case should relief vents (that connect to the main stack) be less than half the diameter of the stack. The minimum size for individual vents is $1\frac{1}{4}$ inch or one-half the diameter of the drain to which it is connected, whichever is larger. This means that for a 4-inch stack, the relief vent would have to be at least 2 inches, and for a branch drain of 2 inches, the circuit vent would have to be $1\frac{1}{4}$ inches. A vent pipe less than $1\frac{1}{4}$ inches is not adequate to convey air to vent fixtures because of the high velocity and attendant friction that accompany the necessary air flow. It is also prone to clogging. The developed length of a vent pipe is the farthest distance measured from the place where the vent is connected to the base of the stack to the end of the vent extension (through the roof). The same holds true whether the vent stack extends separately through the roof, or connects to a header that connects to the stack vent, which then proceeds through the roof.

Buildings that have more than ten branch intervals in the soil or waste stack must have a (yoke) relief vent installed at each tenth interval, counting from the top floor. The relief vent connection to the stack at the bottom must be through a wye below the horizontal branch serving that floor, and the connection at the top must be three feet or more above the level of the top floor.

Vent stacks, stack vents, and vent headers (the horizontal branch connecting the vent stack to the stack vent) are sized from the total number of fixture units connected to the waste or soil stack, and the developed length of the vent piping. Table 9–6 gives the size of vent stacks based upon the size of the waste or soil stack, number of fixture units, and the maximum developed length of the vent in feet. It is read across starting from the left.

Example 9-7 A vent stack with a developed length of 80 ft serves to vent a plumbing system carrying 256 dfu. What size vent stack is required?

Solution:

Referring to Table 9–6, and starting at the left using a soil or waste stack size of 4 in., we notice in the second column that 200 dfu is less than the connected load and the next value, 500 dfu, will have to be used. Reading across the table (past 70 ft to the next size larger) to 180 ft, and then up, we find that a vent pipe size of 3 in. is indicated. Notice, however, that if the 3-in. vent stack connects to the stack vent to be carried through the roof, the size will have to be increased to 4 in. because the stack size cannot be reduced.

Horizontal vent circuits are also sized by using the load connected to the horizontal branch in drainage fixture units and the length of the circuit

Table 9-6

Vent Sizing Table

Size of Soil or Waste Stack (in.)	Fixture Units Connected	Diameter of Vent Required (in.)								
		$1\frac{1}{4}$	$1\frac{1}{2}$	2	$2\frac{1}{2}$	3	4	5	6	8
		Maximum Length of Vent (ft)								
$1\frac{1}{4}$	2	30								
$1\frac{1}{2}$	8	50	150							
$1\frac{1}{2}$	10	30	100							
2	12	30	75	200						
2	20	26	50	150						
$2\frac{1}{2}$	42		30	100	300					
3	10		30	100	200	600				
3	30			60	200	500				
3	60			50	80	400				
4	100			35	100	260	1000			
4	200			30	90	250	900			
4	500			20	70	180	700			
5	200				35	80	350	1000		
5	500				30	70	300	900		
5	1100				20	50	200	700		
6	350				25	50	200	400	1300	
6	620				15	30	125	300	1100	
6	960					24	100	250	1000	
6	1900					20	70	200	700	
8	600						50	150	500	1300
8	1400						40	100	400	1200
8	2200						30	80	350	1100
9	3600						25	60	250	800
10	1000							75	125	1000
10	2500							50	100	500
10	3800							30	80	350
10	5600							25	60	250

vent. In addition, it must be remembered that the vent pipe cannot be less than one-half the size of the horizontal branch or the diameter of the vent stack, whichever is smaller. Table 9-7 lists the sizes for horizontal vents based upon the size of the horizontal waste or soil pipe, the maximum number of fixture units to be carried by the line, and the maximum developed length of the vent in feet. The table is read starting from the left column. For example, reading across line 4, a 3-in. horizontal branch line carrying 10 dfu would require a $2\frac{1}{2}$-in horizontal vent if the developed length of the vent were 40 feet.

Table 9-7 **Circuit and Loop Vent Sizing Table**

Line	Soil or Waste Pipe Diam (in.)	Fixture Units (max number)	Diameter of Circuit or Loop Vent (in.) $1\frac{1}{2}$	2	$2\frac{1}{2}$	3	4	5
			Max Horizontal Length (ft)					
1	$1\frac{1}{2}$	10	20					
2	2	12	15	40				
3	2	20	10	30				
4	3	10		20	40	100		
5	3	30			40	100		
6	3	60			16	80		
7	4	100		7	20	52	200	
8	4	200		6	18	50	180	
9	4	500			14	36	140	
10	5	200				16	70	200
11	5	1000				10	40	140

9-5 Fixture Traps and Trap Vents

The vent trap provides a means of discharging wastes into the sewer while providing a water seal that prevents sewer gases from entering the dwelling. Figure 9–6 illustrates the parts of a fixture P-trap, so called because it is shaped like the letter P. Waste water flows vertically from the fixture through the trap and over the crown weir and then horizontally at a slope to the vertical waste pipe. When the trap works properly, a water seal remains in the trap, blocking the return of sewer gases. P-traps are available in sizes from $1\frac{1}{4}$ in. to 6-in.

The integrity of the seal maintained by the trap depends upon the depth of seal and the amount of water maintained in the trap. The 2-in. water seal depth (between the crown weir and top dip) is the most common, but deep-seal traps with a 4-in. seal are sometimes used where the trap will be subjected to either excessive back-pressure or -siphonage. Inadequately vented floor drains, for example, may require a deep seal trap to prevent siphonage.

There are a number of traps that are illegal in some jurisdictions, including the "S" trap (which is a P-trap that turns down), traps with moving parts such as check-balls and flappers; traps with baffles; traps with internal obstructions; traps with crown vents; and traps made from fittings. "S" traps are illegal because they promote self-siphoning. That is, the downward turn after the crown weir promotes both the flow velocity and an extended seal in the downward leg with the result that the trap self-siphons. Traps with moving parts tend to become clogged and inoperative with grease, residue, hair, and other foreign matter, which first foul the moving part and then stream out from it, causing a capillary siphoning ac-

Figure 9-6

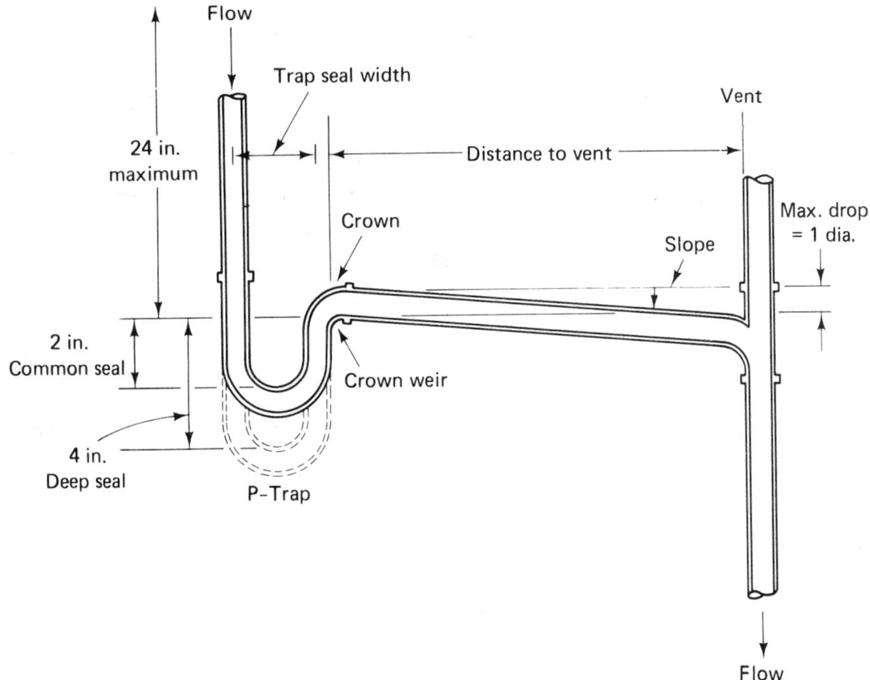

Fixture P-trap

tion over the weir of the trap. Traps with baffles and sharp obstructions suffer the same failures. The P-trap is effective because it offers a smooth and unobstructed passage for the flow, which also promotes a scouring action of the trap, keeping it clear.

The dimensions of the P-trap and how it is installed are also important. As stated previously the depth of trap seal is 2–4 in. In addition, the width of the seal, the distance between the center of the vertical leg from the fixture and the center of the vertical upward leg which approaches the crown weir, is not over 6 in. Narrower widths would unnecessarily obstruct the trap, whereas wider widths would interfere with the scouring action, which keeps the trap clear of material that might otherwise settle out and become an obstruction to the flow. In the installation, it is important that the horizontal leg from the trap which connects to the waste and soil pipe slope downward from the weir but not more than one diameter of the pipe (Fig. 9-6). A negative slope (upward) would impede the flow, whereas too much downward slope would increase the velocity of flow in the horizontal leg unnecessarily. Both conditions of extreme slope cause the horizontal leg to run full somewhere along its length and turn the P-trap into an S-trap. This causes the final vertical drop to siphon the trap.

Each fixture is provided with one and *only one* trap, and each trap is normally provided with a vent. The distance that the trap is located from the vent is measured from the crown weir to the outside of the vent pipe, and this is dependent on the size of the fixture drain. In no case can the trap vent be located closer than two pipe diameters to the crown weir; if it is, the trap is considered to be crown vented and illegal. Crown-vented traps tend to clog the vent as the flow rushes up and over the crown weir, and the momentum deposits waste in the vent pipe. The maximum distance that the trap can be located from the vent is given in Table 9–8, and is taken as the developed length of the pipe rather than the straight horizontal distance. For example, the maximum distance that a $1\frac{1}{2}$-inch trap can be located from the vent is 42 inches (108 cm).

Table 9-8

Maximum Distance from Trap Weir to Vent

Diameter of Trap Arm (in.)	*Distance from Trap Weir to Vent (ft-in.)*
$1\frac{1}{4}$	2 ft 6 in.
$1\frac{1}{2}$	3 ft 6 in.
2	5 ft
3	6 ft
4	10 ft

Traps can lose their seal for a number of reasons, including self-siphoning, capillary action, blowout due to back-pressure on the system, evaporation, wind effects, and leaks. Self-siphoning is caused by the improper construction or installation of the trap. Capillary action can occur in an S-trap, for example, if a string or an absorbing rag hangs over the weir and acts as a wick, pulling the contents of the trap over the weir. Blowout is caused by back-pressure in the system, usually generated on lower floors or at the base of stacks. When this occurs, pressure that develops at the base of the stack is not relieved, and compressed air in the stack is not relieved, and compressed air in the stack blows the trap seal. Traps located in heated areas where the fixtures are seldom used can evaporate the contents, thereby rendering the trap ineffective. A leak, of course, also causes the trap to lose its contents and allows the passage of sewer gas and vermin. Wind effects ripple the contents of the trap. This is due to pressure fluctuations in the vent stack which are brought about by gusting wind that alternately increases and decreases the atmospheric pressure in the vent stack. If the problem becomes serious, the vent should be redesigned or relocated.

The trap for water closets is in the fixture, which is made from vitreous china to ensure a smooth and unobstructed flow of fecal matter and waste into the soil pipe. The basic types are: (1) the siphon-jet, which is the most popular for residential use; (2) the washdown; and (3) the blowout, which is used for commercial applications. A fourth type, the reverse trap design, is very similar to the siphon jet except that the water surface, depth of seal and size of the trapway are smaller, thus requiring less water for efficient

Figure 9-7

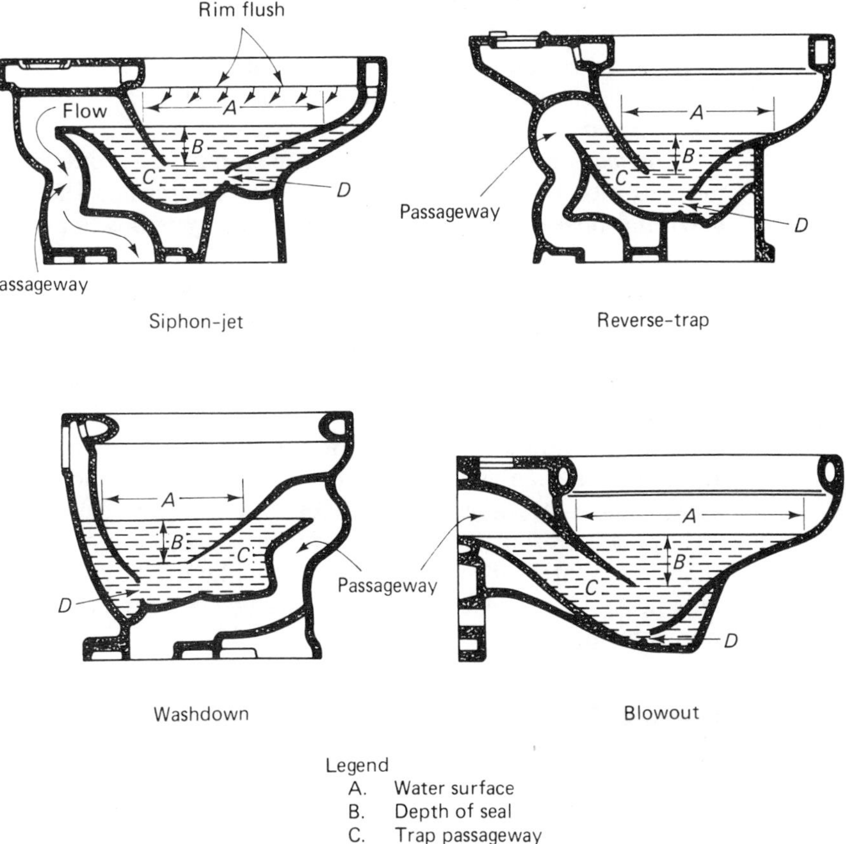

Water Closet Designs

operation. Consequently, the reverse trap design is considered to be a less expensive version of the siphon jet water closet.

A comparison of the designs of the four water closets is shown in Figure 9-7. Their operations are described as follows: The *siphon-jet* flushes when water from the flush tank is released to the bowl. A jet of water is directed into the up-leg to promote the flush and cleansing action at the base of the bowl while water is simultaneously released from the rim at an angle to supply water for the flush and promote swirl. The passageway is designed with offsets to fill and develop a head build-up in the fixture. This eliminates water from the passage; and as the rapid flow of water from the fixture and the atmospheric pressure on the water surface increase the siphonic action, the waste with the water is drawn through the trap efficiently and quietly. Notice that the design has a relatively large water surface, which retards the attachment of soil and waste to the surface of the bowl.

The *reverse trap* design resembles the siphon-jet but has less water surface, a shallower trap seal, and a smaller passageway. Since it uses less water and because it is essentially a smaller fixture, it is also less expensive. Reverse trap water closets are used when economy is an important consideration.

The *washdown* type bowl has the trap at the front; the jet is directed forward into the up-leg. The washdown design has the simplest type of bowl and operates efficiently. Washdown type bowls are also used when economy is important. Both the reverse-jet and washdown types operate satisfactorily with either a flush valve or low tank.

The *blowout* water closet directs the jet up the slightly tilted up-leg and drives the soil and waste over the weir of the trap, rather than relying upon siphonic action. Because of this, it also requires a higher water pressure and flow capacity to ensure that a sufficient amount of water is available to force the flush and cleansing action in the bowl. Blowout water closets are much noisier than those that work primarily on siphonic action. Blowout water closets are available in floor and wall hung models. They receive widespread use in commercial facilities, including schools and office buildings. Water closets intended for public use have elongated bowls and split seats to prevent splashing, soiling the seat, and transferring communicable diseases.

9-6 Types of Vents

There are a number of ways of properly venting fixture traps, and the soil and waste stack. These include:

- Trap Vents:
  1. Individual, back, or continuous vent
  2. Common vent
  3. Branch circuit and loop vent
  4. Wet vent
  5. Stack vent
  6. Combination waste and vent
- Stack Relief and Yoke Vents:
  1. Relief vent
  2. Yoke vent
  3. Suds vent
  4. Ejector and sump vent
  5. Frost closure (at the termination of the vent)

The *individual,* back, or continuous vent serves a single trap. The terms *back vent* (in back of the fixture) and *continuous vent* mean the same thing. By continuous is meant that the vent is a continuation of the vertical drain. The trap in Fig. 9-6 is equipped with a continuous vent. Individual continuous venting is probably the most effective means of protecting a trap seal from the effects of siphonage and backpressure.

Figure 9-8

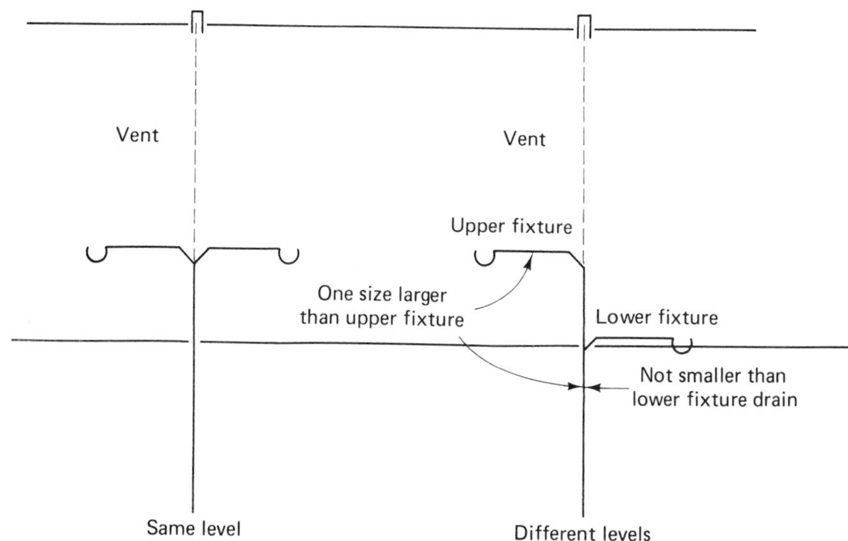

Common Vents

A common vent serves two fixture traps connecting at the same level (Fig. 9–8), and the vertical extension is a continuation of that double connection. A common vent may connect two fixtures at different levels but on the same floor, provided that the vertical drain is one pipe size larger than the upper fixture drain and not smaller than the lower fixture drain, whichever is larger.

Branch circuit and loop vents serve two or more fixtures in a circuit. If the vent connects to a separate vent stack, it is called a *circuit vent.* If it "loops" or connects back to the main stack above the highest fixture (where the soil stack becomes the stack vent), it is called a *loop vent* (Fig. 9–9). Notice that the vent connects downstream from the last fixture trap (to keep the vent clear), and that a loop vent is used on the top floor, since it connects back to the main stack. The horizontal pipe of both the loop vent and circuit vent slope to drain condensate from the vent piping to the main stack.

A *wet vent* serves two functions: It vents one fixture trap while it serves as the drain for another. That is, waste runs down some portion of the vent. Figure 9–10 illustrates several wet vent circuits used for bathroom groups. A single bathroom group of fixtures may be installed with the drain from a back-vented lavatory serving as the wet vent for a bathtub or shower stall and for the water closet, if it meets the following conditions:

1. No more than one dfu drains into a $1\frac{1}{2}$-inch wet vent, or more than four dfu drain into a 2-inch wet vent.

Figure 9-9

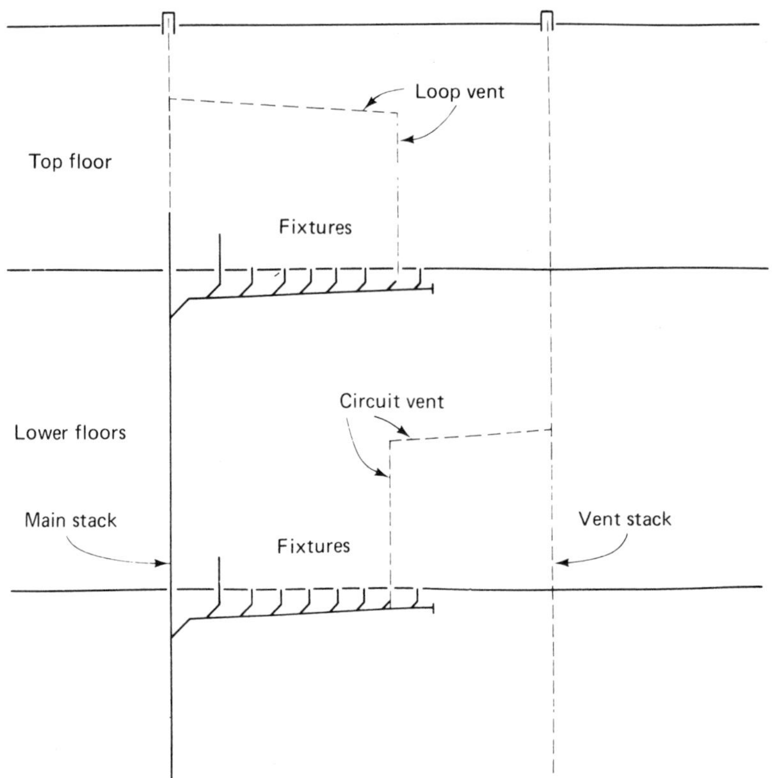

Circuit and Loop Vent

2. The horizontal branch must be at least 2 inches and connect at or below the level of the water closet when it is installed on the top floor. It can also connect to the water closet bend.

A common wet vent can serve back-to-back bathroom groups on the top floor consisting of two lavatories and two bathtubs or shower stalls, provided that the wet vent is 2 inches and the length of the fixture drains conforms to Table 9-8. On lower floors the wet vent and its extension to the vent stack must be 2 inches and water closets must be individually back-vented, unless the 2-inch wet-vented waste pipe connects to the upper half of the horizontal water closet drain at an angle of no greater than 45 degrees from the horizontal in the direction of flow, or if it connects to a special stack fitting for that purpose.

Stack venting uses the soil and waste stack as the vent for individual fixture traps as well as for the system. It is used primarily for single bathroom groups in one-story buildings or on the top floor of multistory buildings. Typically, the fixtures are clustered around the stack, as shown

Figure 9-10

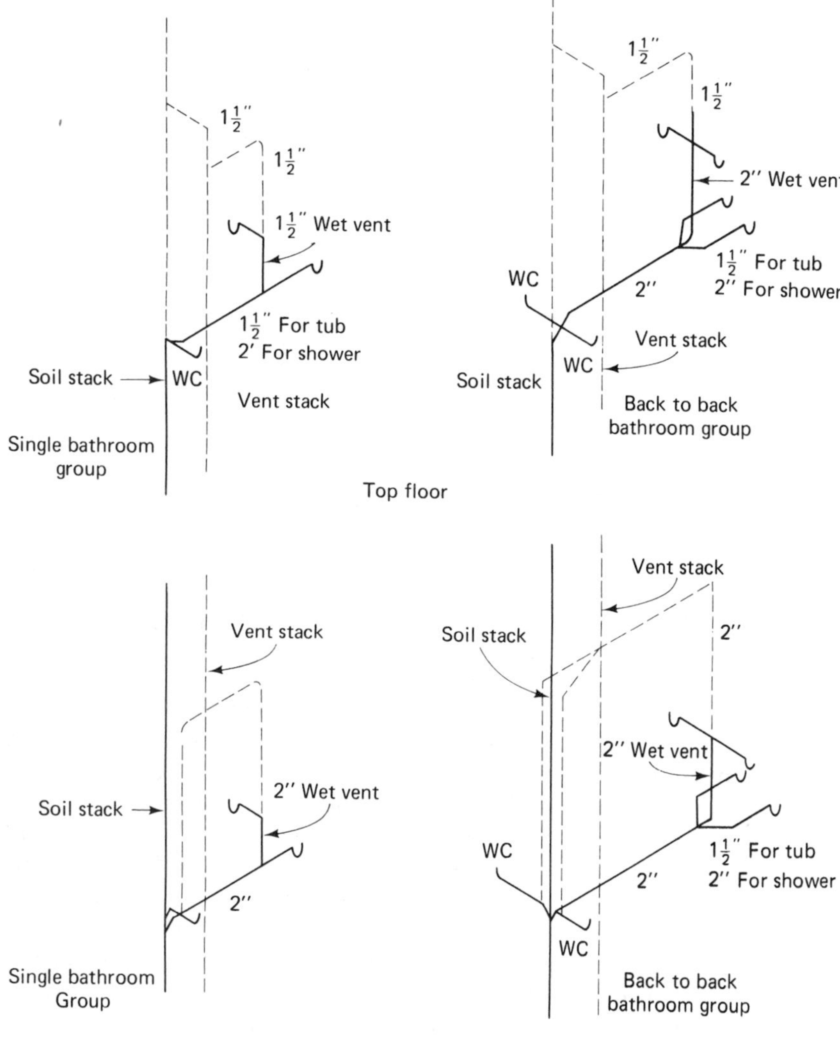

Wet Venting

in Fig. 9-11. However, the distance of each trap from the stack must meet the limitations of Table 9-8, or must be individually vented back to the stack vent, thus, stack and loop venting are used in combination to accommodate the bathroom layout.

Combination waste and vent system is sometimes permitted for use on floor drains and sinks that cannot be vented by ordinary means. Its use is limited to floor drains and sinks. It consists of a wet-vented installation of waste piping without individual fixture vents. Where combination waste and vent pipes are used, the horizontal runs must be at least two pipe sizes

Figure 9-11

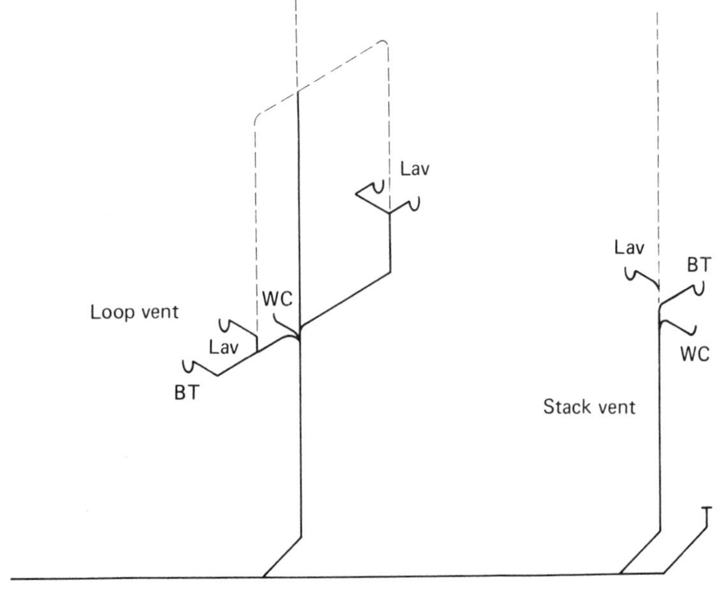

Loop and Stack Vents

larger than the values given in Table 9–8. Oversizing the horizontal run permits the flow of air above the flow line of the liquid in the same pipe. The horizontal run is then relief-vented at both ends, as shown in Fig. 9–12.

Stack relief and yoke vents prevent both pressure buildup and excessive pressure fluctuations in the main stack. This is particularly true in multistory buildings where fixtures discharge simultaneously on different floors. While the vent stack which attaches to the main stack near the base can relieve pressure at the base and prevent trap siphonage by the building drain, pressure fluctuations within the stack above require relief vents for stack offsets and at specified branch intervals.

A *relief vent* is required where the stack offsets more than 45 degrees from the vertical. Abrupt changes in direction cause surging and generate high pressures in the stack, which affect the operation of the system adversely unless they are relieved. Offsets are relief-vented as two separate soil or waste stacks. The section above the offset is provided with a relief vent that is sized for the load carried by the stack above the offset. The section below the offset is vented by installing a relief vent as a continuation of the lower section of the stack, or as a side vent connection to the lower section between the offset and the next lower fixture. The section below the offset is sized for the total load connected to the stack. The diameter of both relief vents must be as large as the main vent or the soil or waste stack, whichever is smaller. Figure 9–13 illustrates relief venting at stack offsets.

Figure 9-12

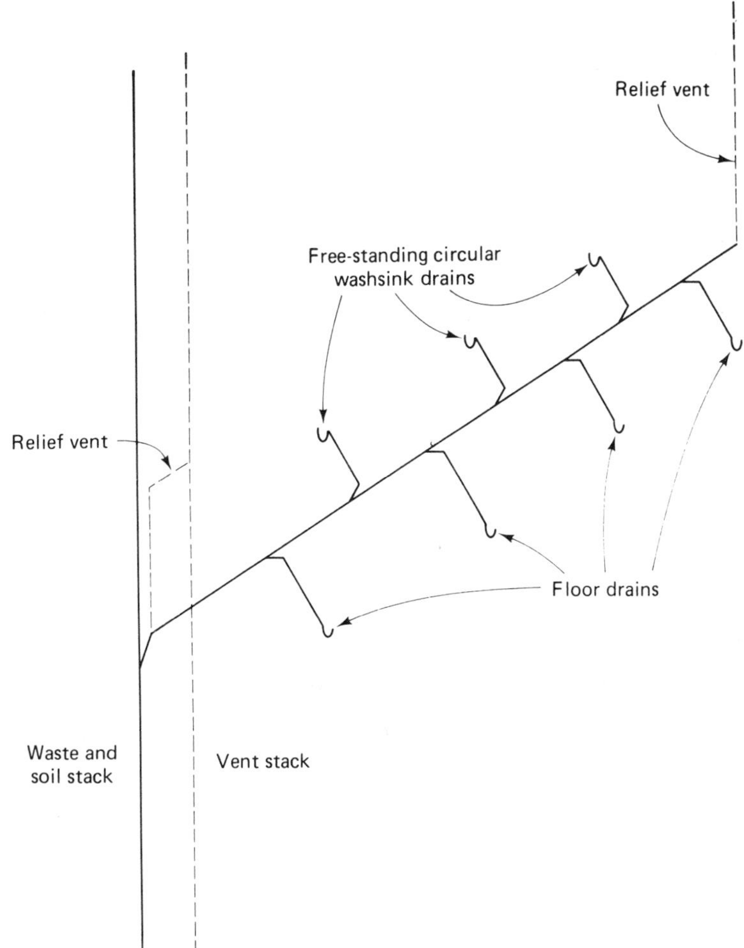

Combination Waste and Vent System

Yoke vents connect the main stack with the vent stack and are required in buildings having more than ten branch intervals. The connection is made at each tenth interval, counting from the top. Yoke vents are the same size as the vent stack to which they connect. The connection of the yoke vent to the main stack is made with a wye fitting below the horizontal branch serving that floor, and the connection to the vent stack is to be at least three feet above the floor. Figure 9–14 illustrates the requirement and connection of a yoke vent.

Suds pressure zones are caused by the mixing of liquid wastes with high-sudsing detergents at areas where the soil and waste pipe changes direction. While the liquid continues to flow down the soil and waste pipe,

Figure 9-13

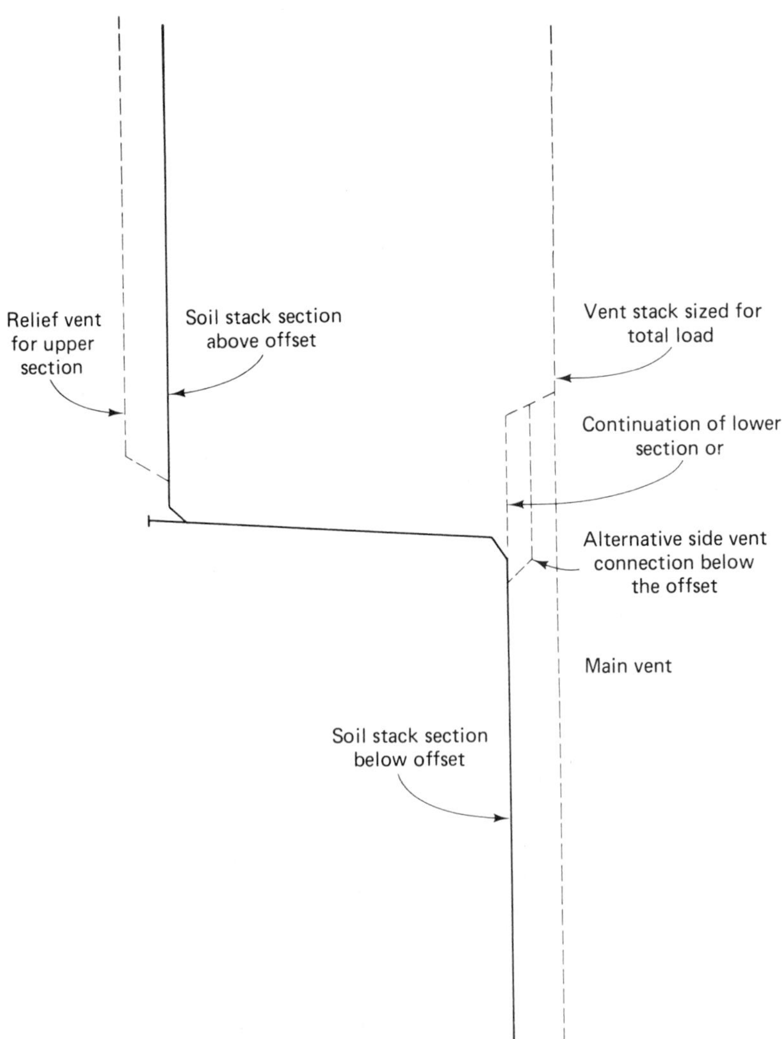

Relief Venting Offsets

the suds remain in zones near the offsets to restrict the flow of air in the system. Failure of the vent system from suds in the soil and waste stack, particularly on lower floors, can cause suds to be backed up in branches and individual fixture drains and vents, and even appear in fixtures. Consequently, accepted practice is to avoid connections to the stack at suds pressure zones and if necessary to provide separate suds relief vents for them in buildings receiving sudsy wastes from fixtures on two or more stories. Suds pressure zones are considered to exist in the following areas (Fig. 9–15):

Figure 9-14

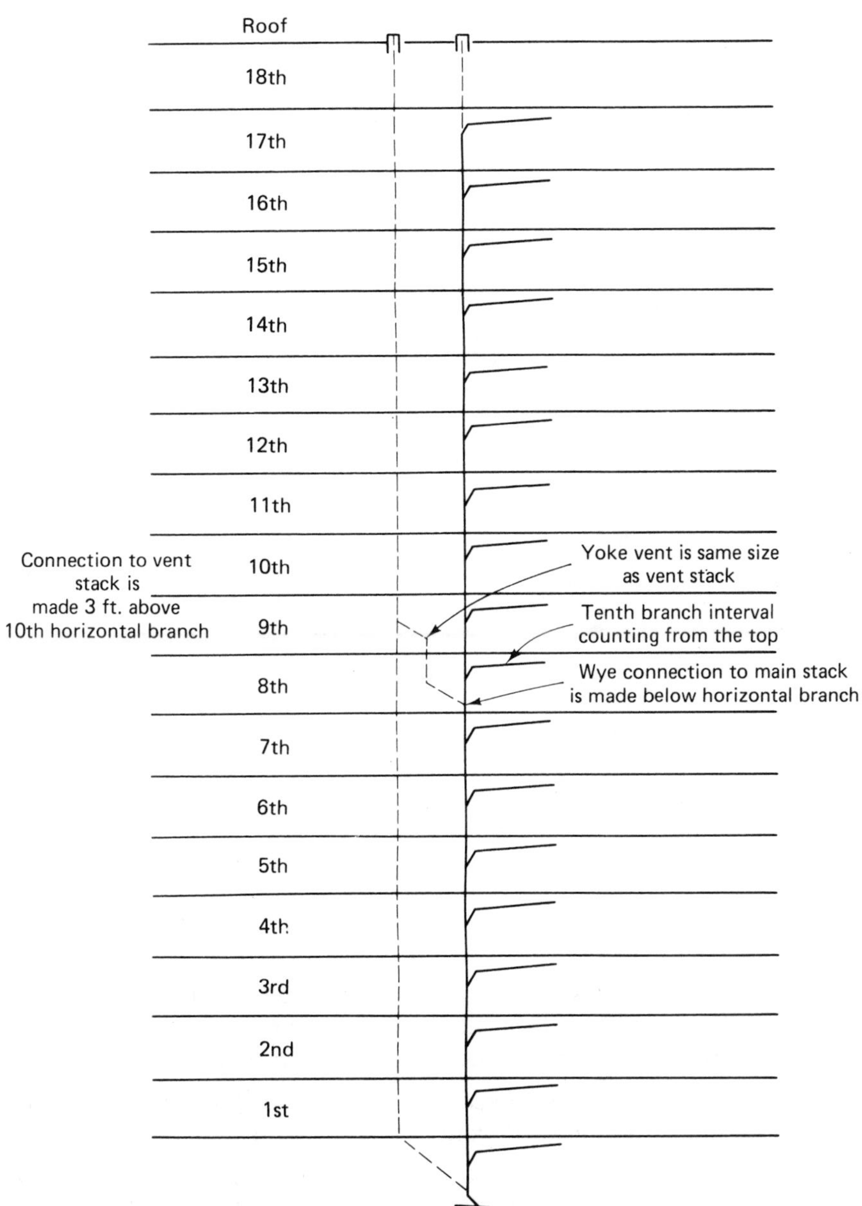

Yoke Relief Vent for Stack of More Than Ten Branch Intervals

Figure 9-15

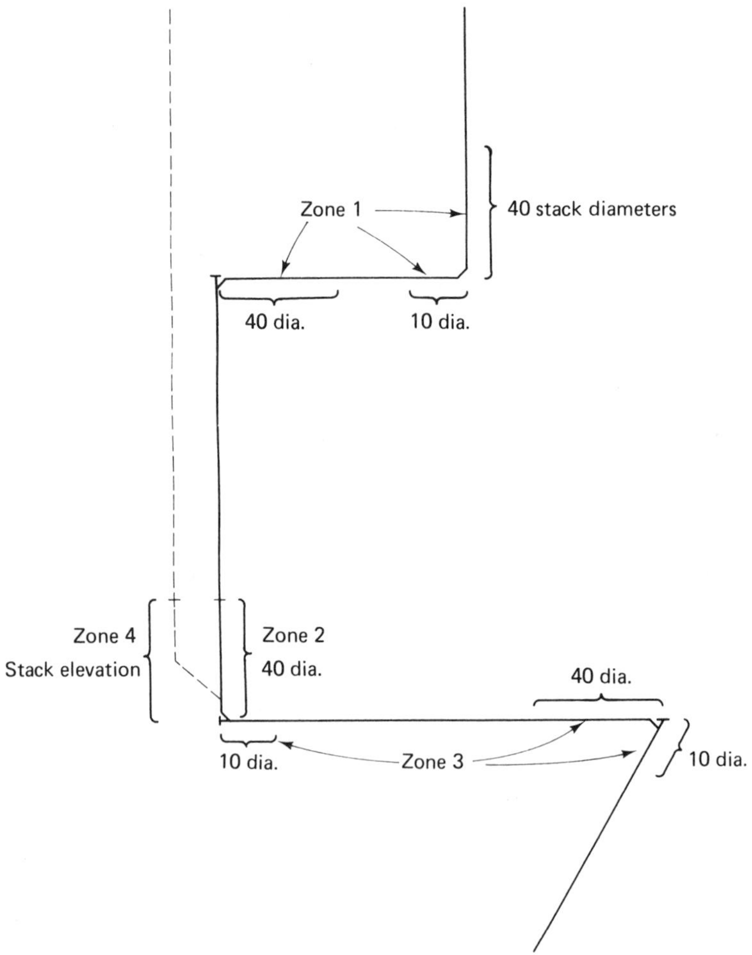

Suds Pressure Zones

Zone 1. At stack offsets of more than 45 degrees, 40 stack diameters upward, 10 stack diameters horizontally, and 40 stack diameters horizontally preceding the top fitting of the lower vertical stack.

Zone 2. At the base of the soil stack extending upward 40 stack diameters upward from the base fitting.

Zone 3. In the horizontal drain, extending 10 stack diameters following the stack fitting, and if a subsequent offset of more than 45 degrees occurs farther down the drain, then 40 diameters upstream and 10 stack diameters downstream from that offset fitting.

Zone 4. If the stack vent connects to a suds pressure zone at the base of the stack, then upward in the vent stack to the same suds level as in the soil and waste stack.

Ejectors are used when the level of the fixture is below the level of the public sewer. Those of the sump pump variety are vented in the same way as gravity systems, with the air requirement determined by the amount of sewage drained in and pumped out. If the ejector is of the pneumatic type, the vent must be carried separately to the open air as a vent terminal to prevent blowing trap seals in the gravity system.

Frost closure. In colder climates where the temperature remains below -20°F (-6.7°C), there is the possibility of vent terminal closure, as the inside air condenses and forms frost on the inside of the vent terminal. Other factors affecting frost closure include the wind velocity, humidity in the inside air, the length and diameter of the vent terminal, and air velocity in the vent stack. The common method of preventing frost closure is to increase the diameter of the vent terminal, beginning the diameter increase one foot below the roof. The minimum vent terminal diameter to prevent frost closure is 3 in. Another method used is to install a special jacketed vent terminal that uses heat from the building to warm the vent pipe. Local code authority and approval are required for the method used to prevent frost closure.

9-7 Storm Water Plumbing

Storm water systems convey precipitation from roofs, foundation walls, and areas adjacent to buildings, to the storm sewer. In all cases the storm water plumbing and sanitary plumbing are run separately inside the building. Outside the building some have combined sanitary and drainage sewers. Most municipalities now have separate storm water sewers, and where they do, the two systems are not interconnected. Where the municipality uses a combined system, the storm water plumbing connects to the combined system outside the building, preferably in a manhole so that future attachment to the separate storm sewer can be made without alterations to the building. In a combined system, where the storm water plumbing is connected to the sanitary plumbing, the connection is made through a trap and downstream at least 10-ft (3 m) from any branch or the soil stack.

Storm water plumbing consists of the drain in the collection area, leaders, which are usually run vertically and are called *downspouts*, and the storm drain itself, which is comparable to the horizontal building drain in sanitary systems. Storm water systems are installed both inside and outside the building. Inside the building the leaders run vertically to the roof supported by columns or free-standing in chases, and terminate through the roof deck at the roof drain fitting. Just under the roof deck the pipe is offset to allow for expansion of the leader, which otherwise could force the vertical leader up through the roof when it expands. An alternate system is to use slip expansion joints, which will permit vertical movement without disturbing the roof drain fitting. Figure 9–16 illustrates an inside and out-

Figure 9-16

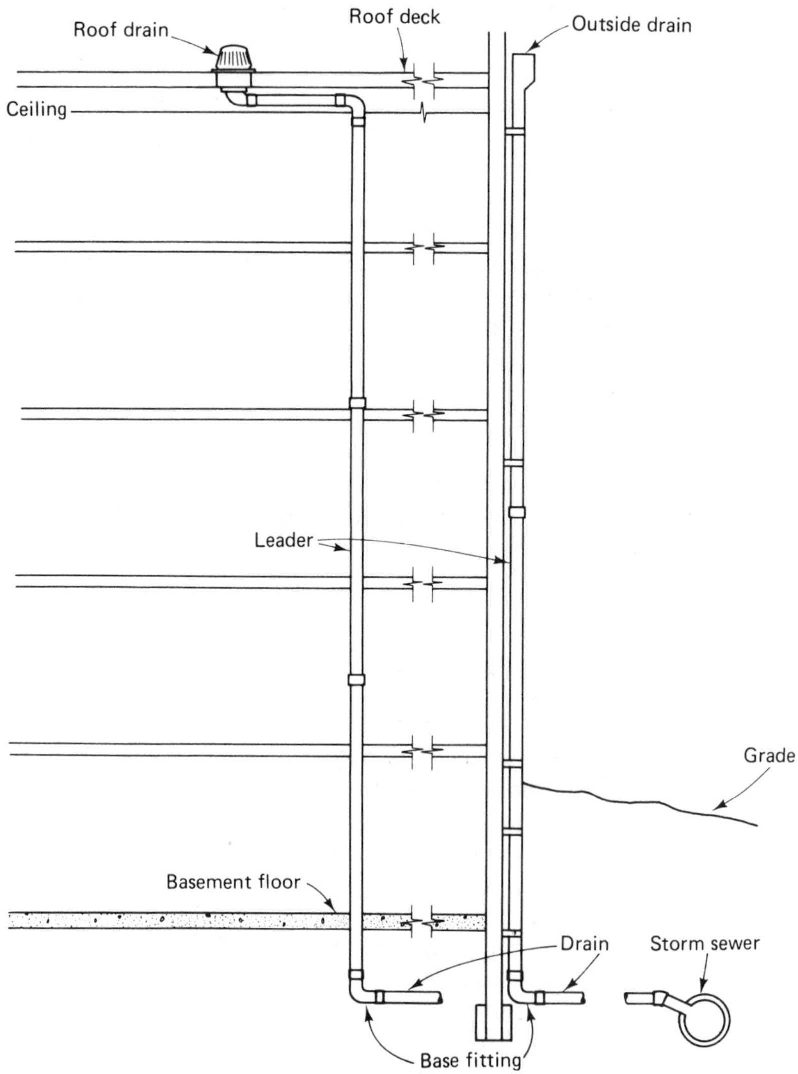

Inside and Outside Roof Drain System

side roof drain system. Figure 9-17 illustrates typical roof drain and grate drain fittings and their component parts. When the system is mounted to the outside of the building, the downspout is attached to the building and terminates at ground level in the storm water system. Above ground the outside leader can be pipe, sheet metal, or plastic, but below grade level either cast iron or other material that conforms to the code in the local jurisdiction must be used.

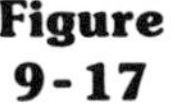

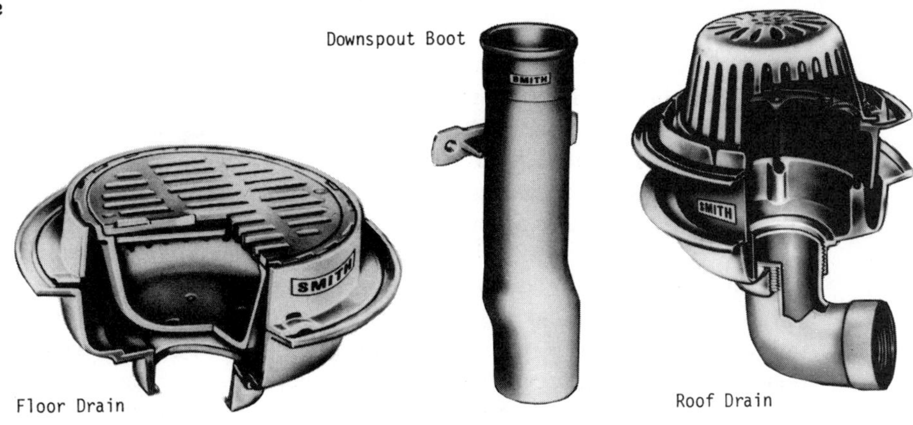

Drainage Fittings (J.R. Smith Mfg. Co.)

Roof gutters, storm drains, and leaders are sized from the projected area being drained and the maximum rate at which the rain is expected to fall. For most areas the sizing tables are calculated for a maximum rainfall of 4 in. per hour. The projected area equals the ground square footage the building occupies plus any roof overhang. It is independent of the pitch of the roof. Where the roof steps and is on two levels, it is customary also to add in 50 percent of the vertical wall connecting the two levels, although the validity of this calculation has been questioned. It is based upon the assumption that a driving rain will strike the wall portion and accumulate on the lower roof, but observations of accumulations along sidewalks adjoining large buildings that pose a similar circumstance do not bear this out.

Example 9-8 A square building 100 ft × 100 ft with no overhang receives rainwater at a maximum rate of 4 inches per hour. At what rate in gallons per minute will the water accumulate on the roof?

Solution

The projected area is computed from:

$$\text{Area} = \text{length} \times \text{width} = (100\text{ ft})(100\text{ ft}) = 10{,}000\text{ ft}^2$$

At a rainfall rate of 4 inches per hour, the accumulation will be

$$Q = \text{area} \times \text{depth/hr} = (10{,}000\text{ ft}^2)(\tfrac{1}{3}\text{ ft/hr})(1728/231\text{ gal/ft}^3)$$

$$Q = 24935\text{ gal/hr }(94\ 389\text{ l/hr})\text{ or }416.6\text{ gal/min }(1573.2\text{ l/min})$$

The same would be true for a courtyard or other projected ground area.

Gutters are sized by load in a manner similar to horizontal branches. Table 9–9 gives the maximum permissible load (in ft^2 of surface area) for semicircular gutters pitched at $\frac{1}{16}$ in./ft, $\frac{1}{8}$ in./ft, and $\frac{1}{4}$ in./ft, and based on a 4-in./hr rainfall. If the roof were pitched, then the gutters serving each side of the roof would receive their respective amounts, based upon the projected areas associated with each side. If a roof were perfectly flat and the building were square, then guttering around the building would receive approximately one-fourth of the water at each side.

The maximum permissible loads for the various sizes of storm leaders and horizontal piping delivering water to the storm sewer are given in Table 9–10. The tabled values are for round pipe. If the leaders are rectangular, which is often the case when they are mounted to the outside of residential and light commercial buildings, they can be sized about ten percent smaller to compensate for the gain in circular area of the pipe cross section.

Table 9-9

Maximum Load for Semicircular Gutters

Gutter Diameter (in.)	*Drained Area (ft^2)* $\frac{1}{16}$ *in./ft*	$\frac{1}{8}$ *in./ft*	$\frac{1}{4}$ *in./ft*
3	170	240	340
4	360	510	720
5	625	880	1250
6	960	1360	1920
7	1380	1950	2760
8	1990	2800	3980

Table 9-10

Maximum Load for Storm Leaders and Drains

Gutter Diameter (in.)	*Leaders Drained Areas (ft^2)*	*Horizontal Drains Drained Area (ft^2)* $\frac{1}{8}$ *in./ft*	$\frac{1}{4}$ *in./ft*
2	720	—	—
$2\frac{1}{2}$	1300	—	—
3	2200	822	1600
4	4600	1880	2652
5	8650	3340	4720
6	13 500	5350	7550
8	29 000	11, 500	16, 300

Example 9-9 Size the roof gutter, vertical leader, and horizontal drain fed from a roof with a projected area of 1800 ft^2 with a maximum rainfall of 4 in./hr. Assume that the gutter is pitched at $\frac{1}{8}$ in./ft.

Solution:

First, sizing the gutter from Table 9-7, we see that a gutter with a 7-inch diameter would handle 1950 ft^2 if it were pitched at $\frac{1}{8}$ in./ft.

Sizing the vertical leader and horizontal drain from Table 9-8, we see that a 3-in. circular leader would handle 2200 ft^2, and a 4-in. diameter circular horizontal drain would handle 1880 ft^2 if it were pitched at $\frac{1}{8}$ in./ft.

9-8 Copper Sovent Single Stack Drainage System

The single stack plumbing system is the invention of Fritz Sommer of Bern, Switzerland, 1959. It is used in multistory office buildings, in apartment buildings, and in hotels in heights from 3 to 30 stories, and for loads to 400 dfu. In the United States alone there have been over 40,000 installations. It is available in copper, and most of the development in this country has been done under the leadership of the Copper Development Association Inc.[2] More recently, sovent fittings have also become available in cast iron.[3] Sovent single stack systems are engineered for specific applications. A comparison of the traditional two-pipe system and sovent system is illustrated in the schematic shown in Fig. 9-18. Installation of the system requires approval by the local code authority.

The copper sovent system uses DWV fittings, a copper DWV stack and horizontal branches; and two basic "sovent" fittings to accomplish the self-venting feature of the system:

1. An *aerator* fitting at each floor level.
2. A *deaerator* fitting at the base of the stack and for horizontal offsets.

Figure 9-19 illustrates the action of the aerator and deaerator fittings. The aerator functions to limit the velocity of both liquid and air in the system, prevents the cross section of the stack from filling with a plug of water (which would effectively upset the pressure balance in the stack and threaten trap seals), and mixes the waste entering the branch and waste inlets of the fitting with the air and vertical flow in the stack. At floors that do not have waste connections requiring the aerator fitting, a double inline offset is used. The deaerator fitting consists of an air separation chamber with a pressure relief outlet. It is used to relieve the pressure at the bottom of the stack and at horizontal offsets. At the base of the stack the deaerator fitting connects the pressure relief line back into the top half of the building drain. At horizontal offsets, the pressure relief line connects into the vertical portion of the stack below the offset through a wye fitting. The arrangements for connecting the deaerator fitting at the base of the stack and at horizontal offsets are shown in Fig. 9-20.

Copper sovent drainage stack design features are given in the system shown in Fig. 9-21. They are sized similarly to conventional two-pipe systems. Notice that the stack must also be carried through the roof full

Figure 9-18

SOVENT's cost-saving potential can be seen by considering a 12-story stack to serve this apartment grouping. The material saving is shown graphically in the schematic riser diagrams for two-pipe and SOVENT systems.

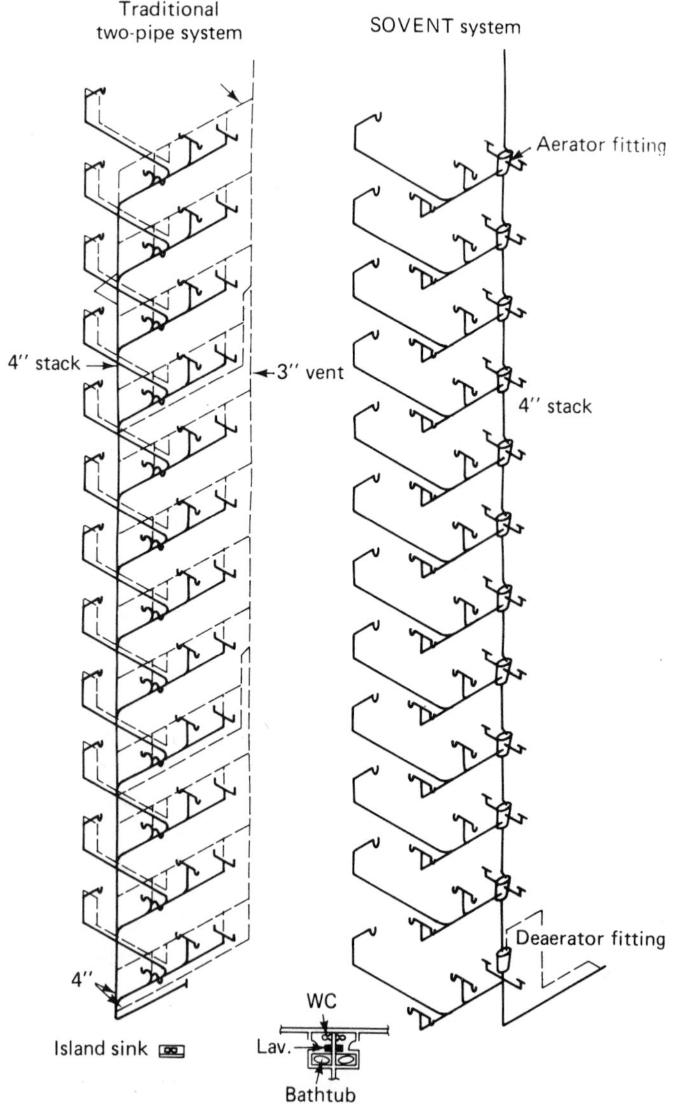

Comparison of Conventional Two-Pipe Sanitary Drainage System with Single Pipe Sovent System

Figure 9-19

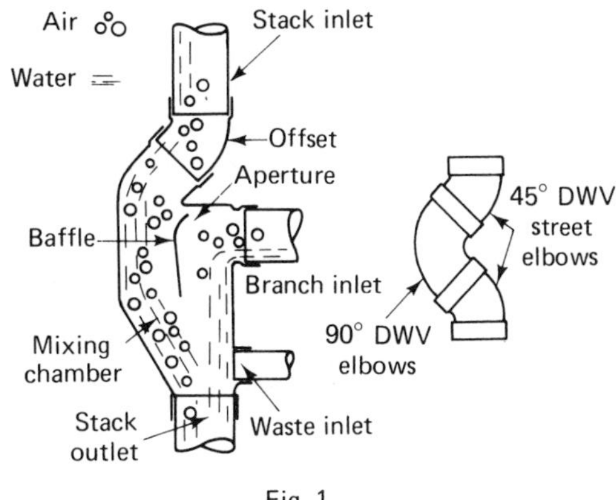

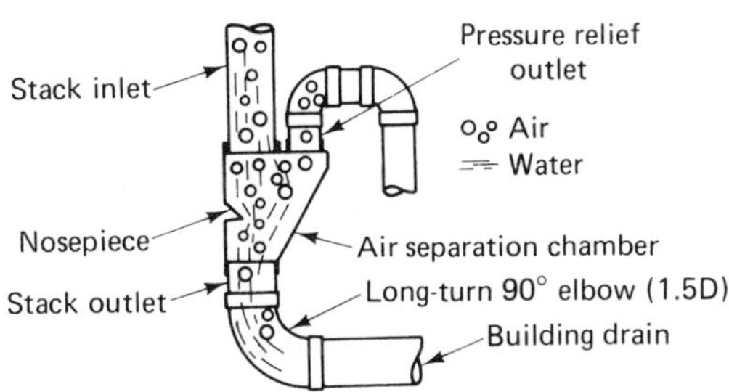

Action of Aerator and Deaerator Fittings

size. It is increased one diameter if the distance between two stacks connected by a header is over 20 ft long. There are a number of other novel design features that permit soil and waste lines to connect to horizontal offsets without separate venting, waste connections to the horizontal pressure relief line, and soil and waste branches connecting immediately below the deaerator fitting at the bottom of the stack. These are illustrated in Fig. 9-21.

9-9 Summary of Practice

The sanitary soil and waste system removes the liquid containing fecal matter from water closets, and waste water from sinks, lavatories, bathtubs, and showers. Each fixture is equipped with a liquid trap seal to prevent sewer gas and vermin from entering the room and building. Soil and waste water exit the building through horizontal branches, which lead to the vertical soil stack and building drain. Both the stack and individual fixtures are equipped with a vent pipe to reduce pressure fluctuations in the stack and thus allow the liquid soil and waste to flow properly. Vents also

Figure 9-20

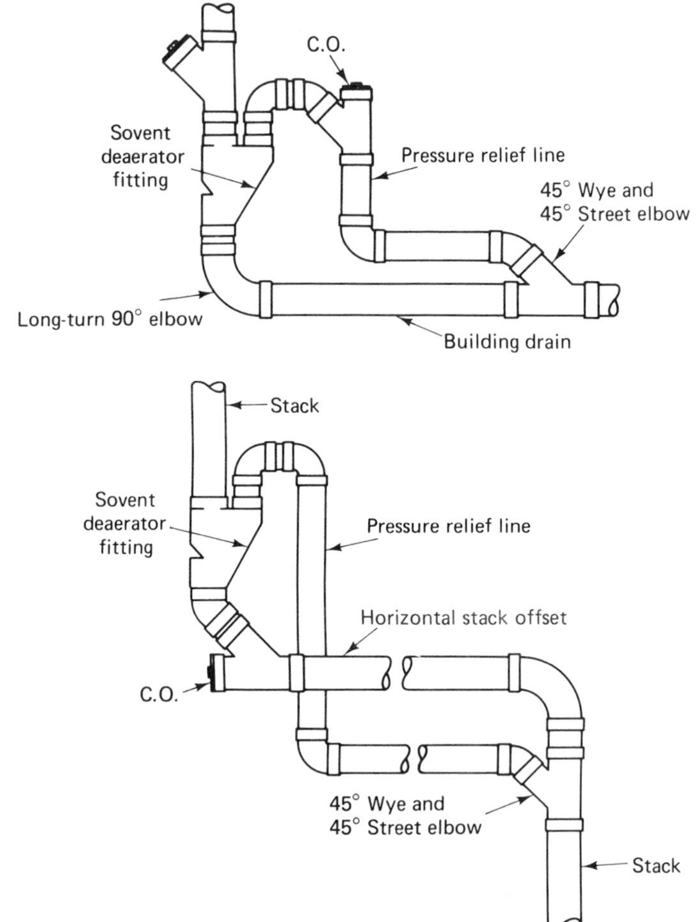

Installation of Sovent Deaerator Fitting at Stack Base and Stack Offset

protect trap seals at individual fixtures which otherwise could be blown by positive back-pressure or siphoned by negative pressures. The "sovent" single stack plumbing system combines the stack and fixture vents with the waste and soil pipe.

The storm drain system conveys water from precipation and runoff from roof gutters, down vertical leaders (downspouts), to storm sewers. Where they are separate from the sanitary system, traps are not used. Where they are combined, individual leaders entering the sanitary system are trapped, although in well-designed systems justification for this requirement is questionable.

Sanitary drain and vent systems are sized from tables giving loads in drainage fixture units, whereas storm water systems are sized from tables

Figure 9-21

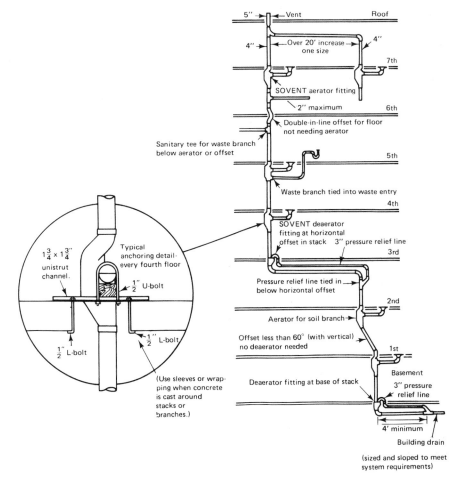

Sovent Sanitary Drainage Stack Design

giving the load in square areas and maximum expected rainfall runoff, which is 4 inches/hr for most parts of the country. Increasing the pitch of horizontal branches, gutters, and drains can increase the capacity of a system for a given diameter size.

REVIEW QUESTIONS AND PROBLEMS

1. What piping is included in the sanitary drainage system?
2. How does liquid flow in horizontal pipes differ from liquid flow in vertical pipes?

3. What purposes do the waste and vent stacks serve?

4. What basic sizing requirement applies to both waste and vent stacks?

5. Why is a minimum velocity of 2 ft/sec (0.6 m/s) chosen for the flow in horizontal branches and drains?

6. How are offsets of more than 45 degrees in the soil and waste stack sized?

7. What is the purpose of the trap?

8. Which traps are illegal?

9. List requirements governing the sizing and installation of traps.

10. List and describe five ways of venting traps.

11. List and describe two venting requirements that apply to the soil and waste stack.

12. What means is used to prevent frost closure?

13. What arrangements can be made for a floor drain that cannot be vented by ordinary means?

14. Size the gutter, downspout (vertical leader), and horizontal drain for a storm drain system receiving water at the rate of 4 in./hr from a roof with an area of 1200 ft^2 (111.5 m^2).

15. What is a "sovent" system?

REFERENCES

[1] Table 9–1 was developed by using the Chezy-Manning equation for turbulent flow:

$$v = KR^{2/3}S^{1/2}$$

where v = flow velocity in ft/sec or m/s
K = dimensional constant which accounts for the size and roughness of the pipe
R = hydraulic radius of the pipe in units of ft or m, defined as

$$R = \frac{\text{Cross section area of flow}}{\text{Whetted perimeter}}$$

for pipes running half-full and full

$$R = \frac{D}{4}$$

S = slope of the pipe (dimensionless)

The dimensions of v in English units are ft/sec, and in SI metric units m/s. Approximations for the recommended values of K in both English and metric units for the various pipe sizes used for sanitary drains and storm sewers are given below.

Pipe Size (inches)	*SI units K-value K ($m^{1/3}$/sec)*	*English units K-value K ($ft^{1/3}$/s)*
$1\frac{1}{4}$ and $1\frac{1}{2}$	83	124
2 through 3	77	114
4	71	106
5 and 6	67	99
8 and larger	63	93
storm sewers	69	103

The flow rate is computed by substituting the Chezy-Manning equation in the flow equation $Q = Av$.

$$Q = KAR^{2/3}S^{1/2}$$

where A = cross-section area of flow in ft^2 or m^2. The student of history will be interested to read Robert Manning's original contributions describing the flow coefficient: "On the Flow of Water in Open Channels and Pipes," *Transactions of the Institution of Civil Engineers of Ireland,* Vol. 20 (1891), p. 161, and Vol. 24 (1895), p. 179.

[2] *Copper Sovent Single Stack Plumbing System Design Handbook,* Copper Development Association Inc., 405 Lexington Avenue, New York, N.Y. 10017

[3] Sovent Corporation, P.O. Box 6561, Tyler, Tx 75711, c/o Mr. Virgil Conine.

10 Installing and Testing a System

10-1 Introduction

The plumbing system is designed to serve the needs of fixtures installed throughout a building. That is, the location and type of fixture determine where the pipes must terminate to deliver water from the distribution system, and return it to the building drain. The design is typically developed by an engineer or specialist, and then is installed and tested by the plumber. The materials list is made from isometric drawings, developed by the plumber from the engineer's plans. That is, the plumber constructs an isometric of the overall building or fixture groups within the building, and then figures the list of materials and labor necessary to complete the job. The actual running of the pipes often requires some slight deviation from the isometric drawing to clear obstructions and building supporting members.

The installation is done in several steps and begins with establishing the exact location of the fixtures within the building. This allows the plumber to begin roughing-in the system. The drain, waste, and vent system is installed first, because the pipes are larger and less flexible. The water distribution system is run next because it can be plumbed around the drain, waste, and vent system. The system is then tested and inspected by the local inspector to ensure that the piping meets the plumbing code for the local jurisdiction, and can pass tests rendering it free from leaks and safe to operate. The water heater and fixtures are installed and adjusted next. This completes the installation. A final test of the system is then made to make sure that it operates satisfactorily, and that the fixtures are free from defects. The owner or contractor is given the literature and warranties for each of the fixtures to file for future reference when repairs or adjustments might be necessary.

10-2 Permits

Every plumbing installation is covered by a plumbing code, and local permits are required before an installation is begun. The plumbing code used is often the state code with additional requirements imposed by an ordinance of the city, municipality, or local community government. Larger cities have their own code. In smaller cities, the code is often covered by an ordinance that is appended to the state code. Beginning a plumbing job without a permit or covering the pipes (in the walls, etc.) and putting the system into operation without an inspection and approval of the local inspector is prohibited. There are usually penalties and sometimes fines associated with violation of the plumbing and inspection ordinance.

Permits to install a plumbing system and connect it to a municipality's water system are obtained by the plumbing contractor through the local county or municipal government office. A fee is charged for each permit to recover a portion of the cost of administering the code and inspecting the job. The permit should be kept in a separate file with other materials for each installation so that information about the job will be available from a single source. A sample permit is shown in Fig. 10-1.

10-3 Roughing-in Drawings and Materials Estimates

The engineer, architect, or general contractor gives the plumber a set of plans for the building that is to be plumbed with the responsibility to estimate the materials and labor to complete the job. These are incorporated in the final bid. The price must be competitive enough to get the job, but at the same time must still allow for a profit. Although different methods are used to arrive at the final bid, each one must account for the materials needed to complete the work before any estimate can be made. Materials lists are usually taken from an isometric drawing that the plumber makes from the floor plans that locate the fixtures in their respective places. The pipes in the system are sized on the isometric drawings, and then each pipe and fitting is checked off the drawing as it is added to the materials list.

Similar layouts for fixture groups recur in several kinds of buildings. For example, a typical bathroom or kitchen layout can be used as a standard, and with some modifications it can be used again in other buildings. As drawings and materials lists are made to plumb groups of fixtures, they can be saved and refined. This not only saves time in making future drawings and materials lists, but improves the accuracy of subsequent bids and increases the chances of getting the job. Additions have to be made for the pipes necessary to connect the standard groups together, but this is much simpler than starting over for each job. Figure 10-2 illustrates several common groupings for water piping fixtures and the plumbing associated with each. The layout of the drain, waste, and vent piping would have another layout that matches it. That is, they would match if one were to be laid over the other. Each one is sized and then the materials to complete the job are checked off the drawing and listed. After all the fixture groups have been estimated, the pipes used to connect them to each other are added to the list. Labor is estimated for the number of hours necessary to install the fix-

Figure 10-1

PLUMBING PERMIT

№ 01600 Date ____________

____

Issued To ____________

Property Owner ____________
and
Address ____________

Street & No. of Job ____________

Lot No. ____________ Block No. ____________

Class of Bldg. ____________

Sewer Assessment Paid Amt. ____________

Water Assessment Paid Amt. ____________

Size of Lateral ____________

No. of Fixtures ____________

Permit Fee $ ____________

Said permit is valid providing said work is done in accordance with all codes and ordinances of The City of ________, ______, and the laws and rules of the State of ______.

Refer to Bldg. Permit No. ____________

Final Inspection ____________

Plumbing Inspector ____________

Sample Plumbing Permit

ture groups. The bid is then made by adding together the costs for material, labor, overhead, and profit.

Figure 10–3 illustrates the floor plans for a standard two-story, conventional house with a basement. The drain, waste, and vent system pipes are the largest and are shown on the first and second story floor plans. The hot and cold water distribution system pipes are plumbed around the drain waste and vent system. They are shown for the basement layout. The basement contains the washer (and dryer), laundry tray, water heater, floor drain, and building drain. The first floor contains a bathroom group, kitchen sink with dishwasher, and bar sink. The second floor has a bathroom group. The isometric drawing for the drain, waste, and vent system for all three floors is shown in Fig. 10–4, as is the materials list.

Figure 10-2

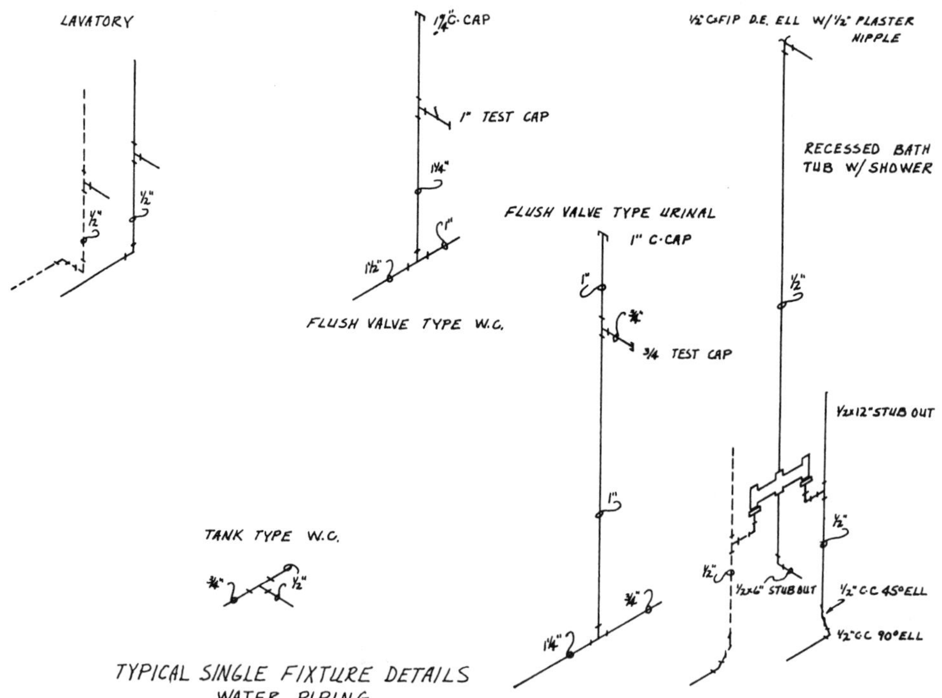
LAVATORY
½"
½"
1¼" C. CAP
1" TEST CAP
1¼"
1"
1½"
FLUSH VALVE TYPE W.C.
½" C×FIP D.E. ELL W/½" PLASTER NIPPLE
RECESSED BATH TUB W/SHOWER
FLUSH VALVE TYPE URINAL
1" C·CAP
1"
¾"
¾ TEST CAP
½"
½×12" STUB OUT
1"
½"
TANK TYPE W.C.
¾"
½"
½"
½×6" STUB OUT
½" C.C 45°ELL
¾"
½" C.C 90°ELL
1¼"
TYPICAL SINGLE FIXTURE DETAILS
WATER PIPING

Typical Single Fixture Details
Water Piping

Lavatory

3'	1/2" Type L Copper Pipe
3	1/2" C x C 90 Ell
2	1/2" C x C x C Tee
2	1/2 x 6" Type L Stub-outs
2	1/2 x 12" Type L Stub-outs
2	1/2" Copper-coated Tube Strap with Nails
1	Misc. Solder, Flux, and Gas

Tank Type Water Closet

2'	3/4" Type L Hard Copper Pipe
1'	1/2" Type L Hard Copper Pipe
1	3/4 x 1/2 x 1/2" C x C x C Tee
1	1/2 x 6" Type L Stub-out
1	Misc. Solder, Flux, and Gas

Flush Valve Type Water Closet

3'	1-1/4" Type L Gard Copper Pipe
1'	1" Type L Hard Copper Pipe
1	1-1/4 x 1-1/4 x 1" C x C x C Tee
1	1-1/4" Copper Cap
1	1" Copper Test Cap
1	1-1/4" Copper-coated Tube Strap with Nails
1	Misc. Solder, Flux, and Gas

Flush Valve Urinal

5'	1" Type L Hard Copper Pipe
1'	3/4" Type L Hard Copper Pipe
1	1 x 1 x 3/4" C x C x C Tee
1	1" Copper Cap
1	3/4" Copper Test Cap
1	1" Copper-coated Tube Strap with Nails
1	Misc. Solder, Flux, and Gas

Recessed Bathtub with Shower

10'	1/2" Type L Hard Copper Pipe
5	1/2" C x C 90 Ell
2	1/2" C x C 45 Ell
2	1/2" C x C x C Tee
5	1/2" C x M Adapter
1	1/2" C x F Drop Ell
1	1/2 x 6" Stub-out
2	1/2 x 12" Stub-out
4	1/2" Copper-coated Tube Strap with Nails
1	1/2 x 6" Black Steel Plaster Nipple and Cap
1	Misc. Solder, Flux, and Gas

Figure 10-3

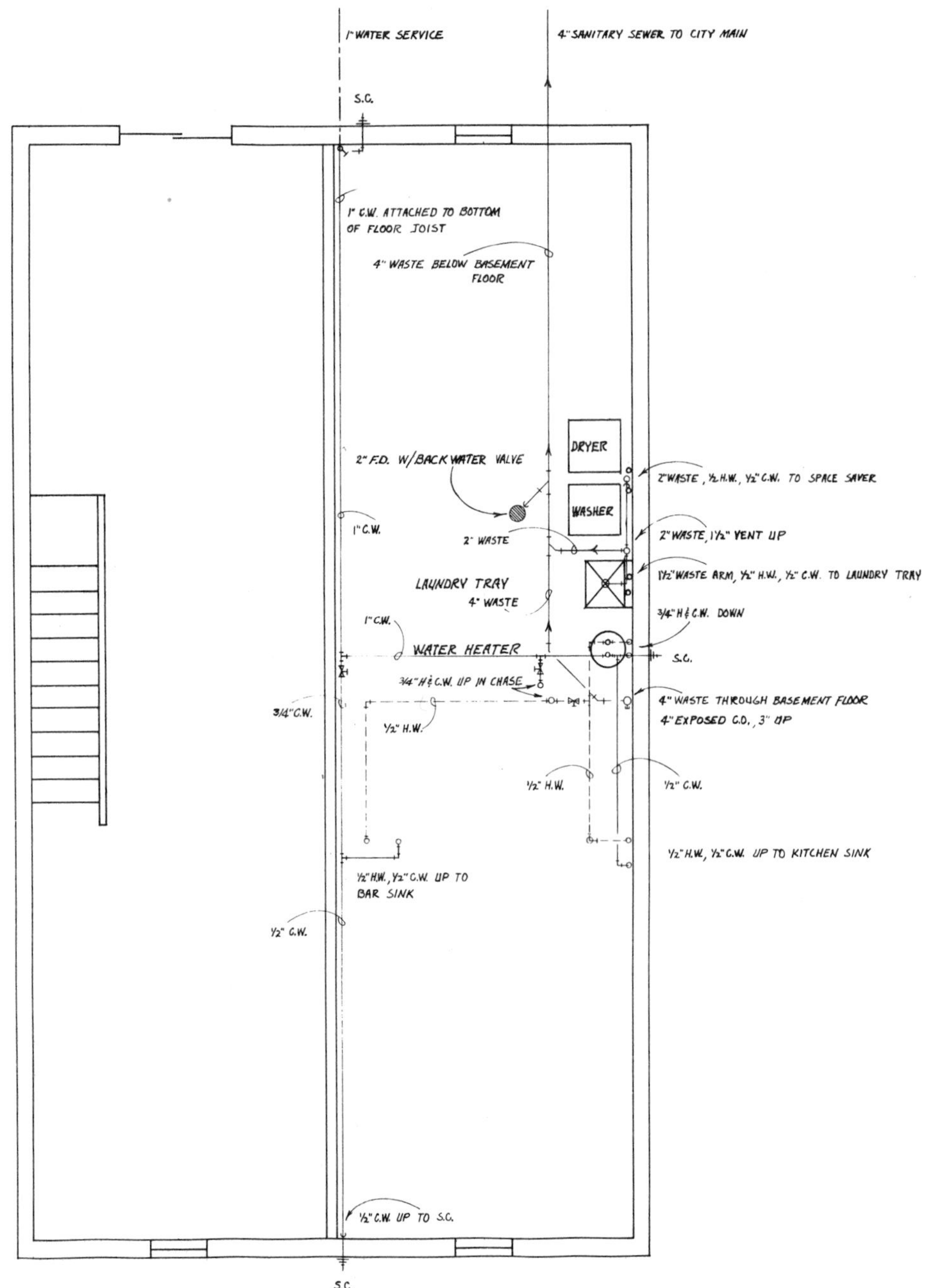

BASEMENT

Floor Plans for Standard Two-Story Ranch House

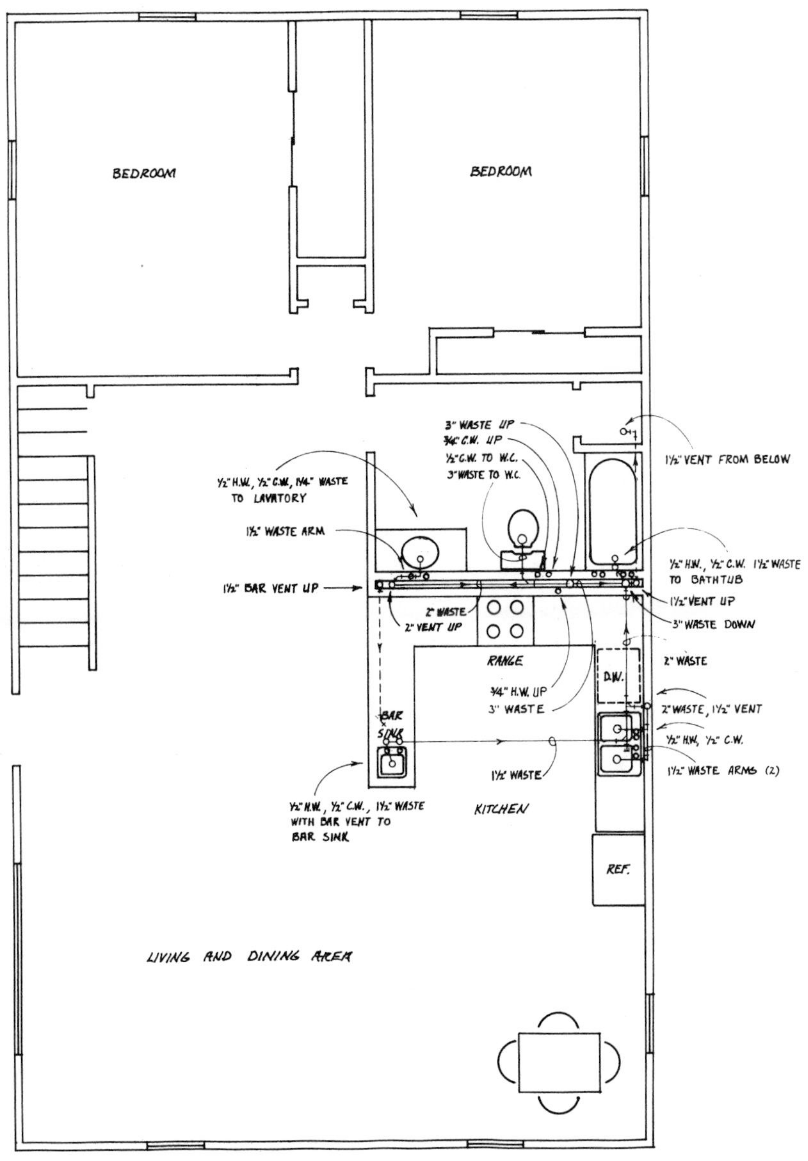

1ST FLOOR

(Continued on following page.)

Figure 10-3 (Cont'd)

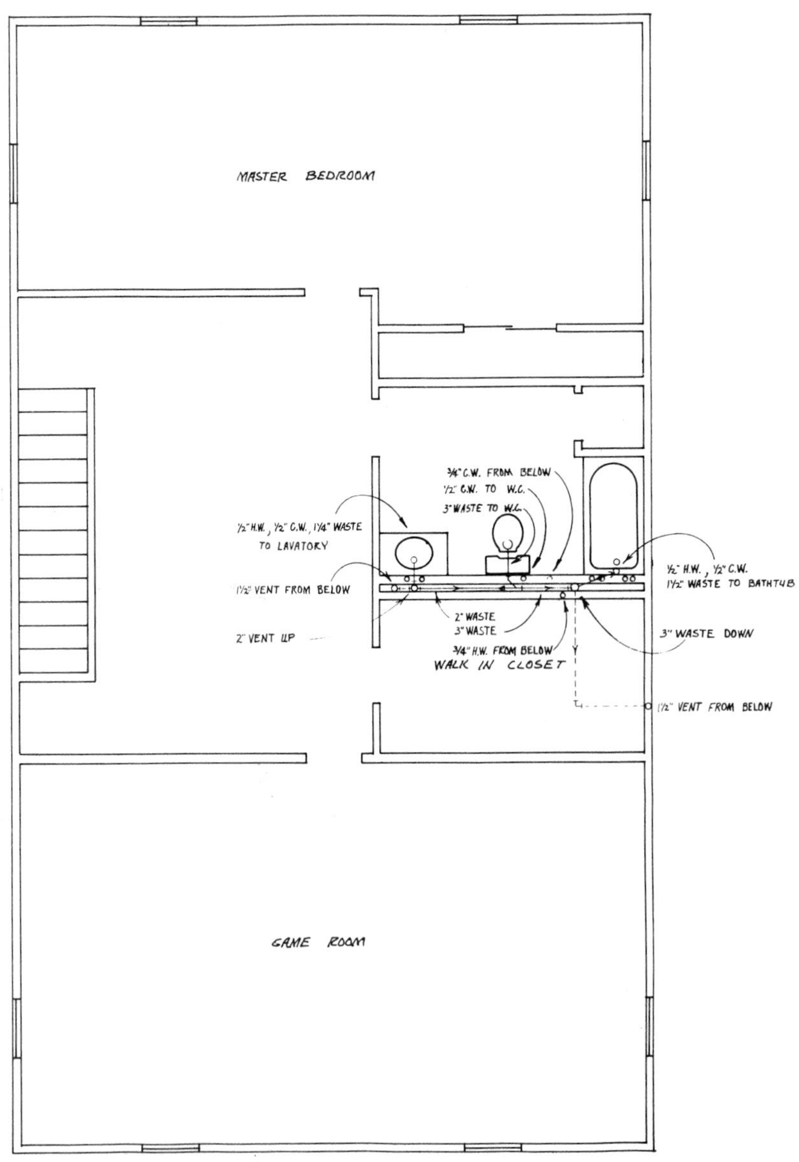

2ND FLOOR

Figure 10-4

Above Ground Waste and Vent Piping
Plastic Pipe and Fittings

Quantity	
1'	4" Schedule 40-ABS-DWV Plastic Pipe
40'	3" Schedule 40-ABS-DWV Plastic Pipe
72'	2" Schedule 40-ABS-DWV Plastic Pipe
84'	1-1/2' Schedule 40-ABS-DWV Plastic Pipe
1	4" Single 45° Wye
1	3 x 3 x 2" Single 45° Wye
1	4" Spigot x FPT Cleanout Adapter
1	2" Spigot x FPT Cleanout Adapter
1	4" Cleanout Plug
1	2" Cleanout Plug
1	4 x 3" Spigot x Hub Bushing
2	3 x 2" Spigot x Hub Bushing
1	3 x 3 x 2" ¼ Bend w/Low Heel Inlet
3	3" Single Comb. Wye and 1/8 Bend
1	2" Single Comb. Wye and 1/8 Bend
1	2 x 2 x 1-1/2" Single Comb. Wye & 1/8 Bend
2	3" 1/4 (90°) Bend
5	2" 1/4 (90°) Bend
15	1-1/2" Bend
2	4 x 3" Hub Close Flange
1	3" Sanitary Tee
1	3 x 3 x 1-1/2" Sanitary Tee
5	2 x 2 x 1-1/2" Sanitary Tee
2	2 x 1-1/2 x 1-1/2" Sanitary Tee
2	2" Sanitary Tee
2	1-1/2" Sanitary Tee
1	3 x 3 x 1-1/2" Sanitary Tee Street
1	2 x 2 x 1-1/2" Sanitary Tee Street
1	1-1/2" Sanitary Tee Street
1	3 x 3 x 2 x 2" Double Sanitary Tee
1	2" 1/8 (45°) Bend
1	1-1/2" 1/8 (45°) Bend
1	2" P Trap
6	1-1/2" P Trap w/Union
1	2" H x MPT Adapter
6	1-1/2" H x SJ Pipe Trap Adapter
2	1-1/2 x 1-1/4 Spigot x SJ Fitting Trap Adapter
1	1-1/2 Coupling
1	3" Lead Vent Flashing
3	3" Clevis Hanger
4	2" Clevis Hanger
6	1-1/2" Clevis Hanger
13	3/8" Coach Screw Rods
26	3/8 Hex Head Nuts with Washers
1	Quart ABS Solvent Cement

(Continued on following page.)

Drain, Waste, and Vent System Isometric

Figure 10-4 (Cont'd)

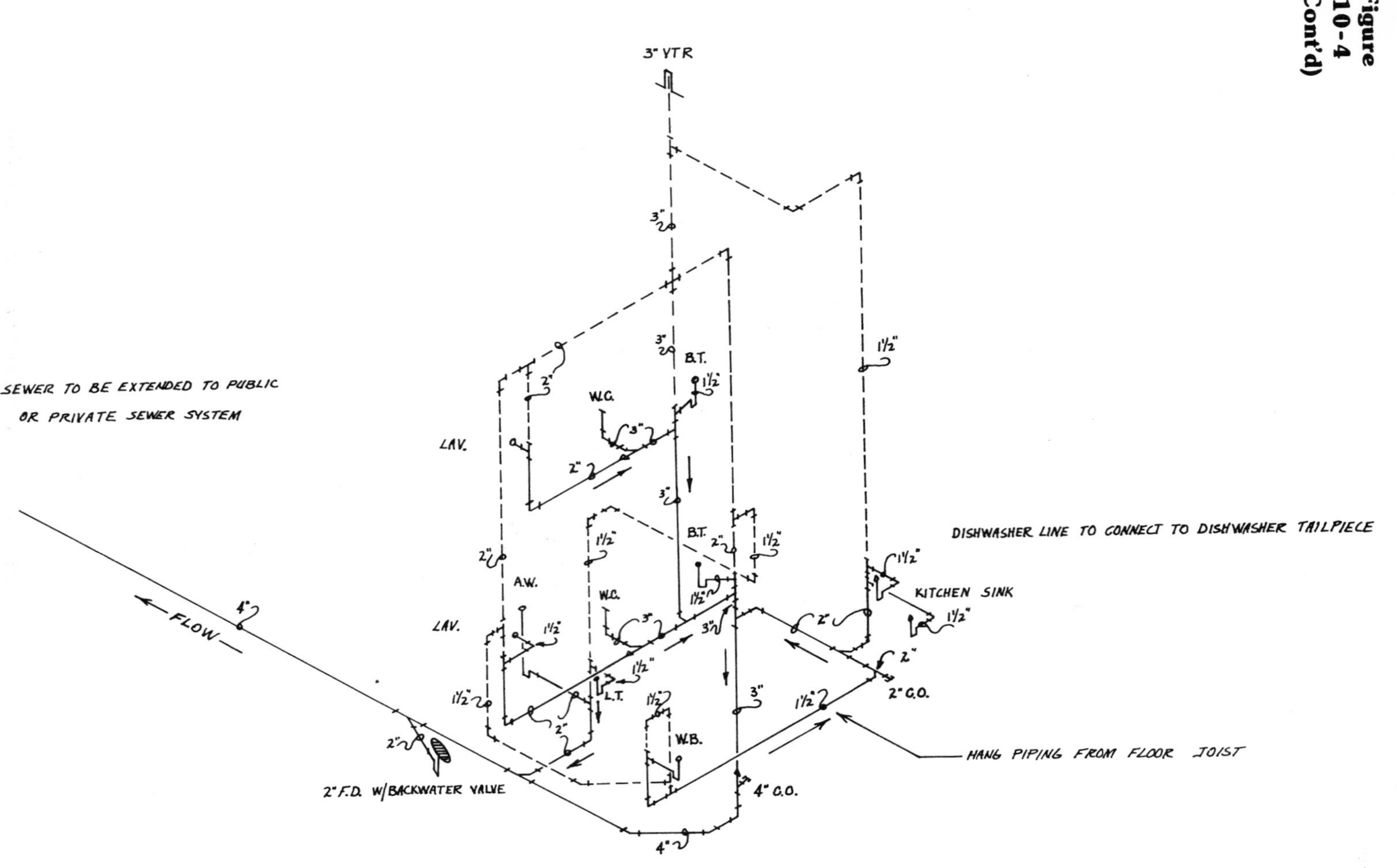

PIPE SCHEDULE

———————	WASTE PIPING
— — — — — — —	VENT PIPING

SYMBOLS

3"VTR	3" VENT THROUGH ROOF
2" F.D.	2" FLOOR DRAIN
C.O.	CLEANOUT
B.T.	BATHTUB
LAV.	LAVATORY
W.C.	WATER CLOSET
A.W.	AUTOMATIC WASHER
W.B.	WET BAR
▷—	REDUCER
←	DIRECTION OF FLOW

Figure 10-5

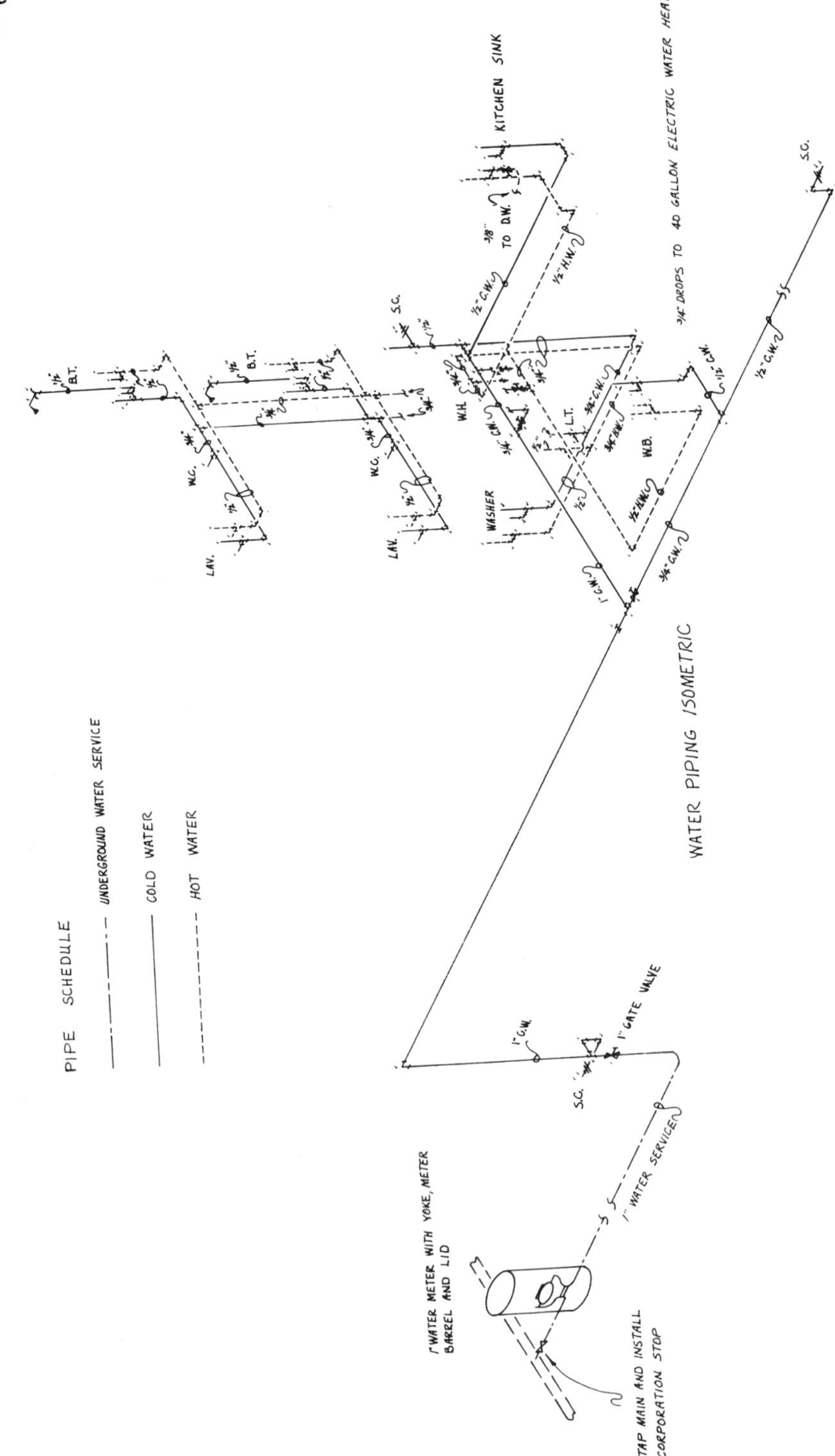

Hot and Cold Water Isometric

Water Piping Takeoff

Below Grade and Exterior Water

100'	1" Type K Soft Copper Water Tubing
1	1" Corporation Stop
1	1" Copper Meter Yoke with Plain Stop and Copper Coupling
1	18" Meter Box Cover
1	18 x 24" Fiber Meter Box
1	1" Frostproof Water Meter

Above Grade Water Piping

39'	1" Type L Hard Copper Pipe
71'	3/4" Type L Hard Copper Pipe
105'	1/2" Type L Hard Copper Pipe
8'	3/8" Type L Soft Copper Pipe
1	1" Gate Valve
3	3/4" C x C Stop
1	3/8" C x C Stop
3	1/2" x 12" Frostproof Sillcock
3	3/4" HT Backflow Preventer
1	Washing Machine Supply and Drain
1	1" C x C 90 Ell
6	3/4" C x C 90 Ell
40	1/2" C x C 90 Ell
3	3/4" C x C x C Tee
14	1/2" C x C x C Tee
1	1 x 1 x 1/2" C x C x C Tee
4	3/4 x 3/4 x 1/2" C x C x C Tee
1	1 x 3/4 x 1" C x C x C Tee
1	1 x 3/4 x 3/4" C x C x C Tee
5	3/4 x 1/2 x 1/2" C x C x C Tee
4	3/4 x 1/2 x 3/4" C x C x C Tee
1	1/2 x 1/2 x 3/4" C x C x C Tee
1	1/2 x 3/8 x 1/2" C x C x C Tee
1	1" C x C Coupling
8	1/2" C x M Adapter
13	1/2 x 6" Stub-outs
16	1/2 x 12" Stub-outs
2	1/2" C x F Drop Ell
1	3/4" C x C Crossover
1	1/2" C x C Crossover
6	1" Copper-coated Tube Strap with Nails
10	3/4" Copper-coated Tube Strap with Nails
15	1/2" Copper-coated Tube Strap with Nails
4	3/4" Copper-clad Van Hangers
8	1/2" Copper-clad Van Hangers
2	1/2" x 6" Black Steel Plaster Nipple and Cap
1-3/4#	50-50 Solder
4 oz.	Solder Flux

Notice that the isometric drawing shows each fitting that will be used in the system and that these are checked off as the list is made. Figure 10–5 illustrates the isometric drawing of the hot and cold water distribution piping for the same system. If the two drawings were constructed on transparent material, they could be superimposed one over the other and match perfectly.

10-4 Roughing-in the Drain, Waste, and Vent System

The drain, waste, and vent system is installed in steps and the plumber is expected to work by the contractor's schedule. This allows the pipes to be roughed in and tested after the framing is done, rough walls are up, and the roof is on. However, if there is to be a concrete floor in the basement, the drain must be installed before the floor can be poured. Or, if a residence is built on a slab, such as is common in some housing developments, then both the waste and fresh water must be roughed-in under the slab before the building framing is begun. (In some installations, the hot and cold water distribution system is run in the attic or ceiling and fed down between the studs in the walls to the plumbing fixtures.) The important point to remember is that although there is a desirable sequence of steps to follow in roughing-in the system, the plumber does not work in isolation from the other trades, and the sequence must be established to fit both the contractor's schedule and accepted plumbing practice.

In commercial structures the vertical drain, waste, and vent pipes are usually installed in chases. These are vertical passages that have walls around them. In effect, they allow the plumbing to be installed and then later for a wall to be built around them. Toilet rooms, for example, that have several water closets on one wall will require a chase for the vertical pipes as well as ample space behind the wall to mount closet carriers, braces, and connecting pipes. In residential plumbing, it is common to run both the vertical and horizontal pipes in the vertical walls, bracing them where they are weakened structurally or making them thicker if the pipe diameter is too large. A wall thickness of 6 in., for example, is adequate to support and cover vertical 3-in. or 4-in. stacks, and to structurally support horizontal branch lines connecting them. And while the plumber is not responsible for insulating drain and waste pipes in walls which generate objectionable noises, having extra room within a wall permits this added feature. Sound suppression is particularly beneficial with plastic pipe installed in first floor walls which carries soil and waste from the upstairs bathroom.

If there is no immediate need to pour the concrete floor, the following sequence is recommended to rough-in the drain, waste, and vent piping:

1. Locate the fixtures in their respective rooms, using the floor plans and fixture rough-in dimensions.

2. Run the vertical waste and vent stacks (with branch fittings connected).
3. Install and connect the drain to the stack.
4. Rough-in each fixture group.
5. Connect the fixture groups to the stacks.

The layout and rough-in for the fixtures are made after the walls are up, the roof is on, the subfloor is laid, and the outside walls have been covered with sheathing. The inside walls have not been covered, however, and the plumber and electrician are usually working at the same time to install the rough plumbing and wiring for the building.

The fixture layouts are located where they will be installed from the room floor plan. This does not mean that the actual fixture is put in the room where it will be located, but rather that the layout is made for the room to determine exactly where each fixture will be placed when the job is completed. The floor plan is used for this purpose. After the fixture location has been determined, then the roughing-in dimensions from the fixture manufacturer are used to make measurements on the subflooring, wall sills, wall studs, and sill plates at the top of the wall.

A typical bathroom fixture layout is shown in Figure 10-6. This is the floor plan for the room, which will have dimensions indicating where the fixtures should be placed when the job is completed. The fixtures are placed on one wall to simplify the explanation. Figure 10-7 illustrates each of the three fixtures that are to be placed in the bathroom and the roughing-in dimensions supplied by the fixture manufacturer. Notice that the bathtub

Figure 10-6

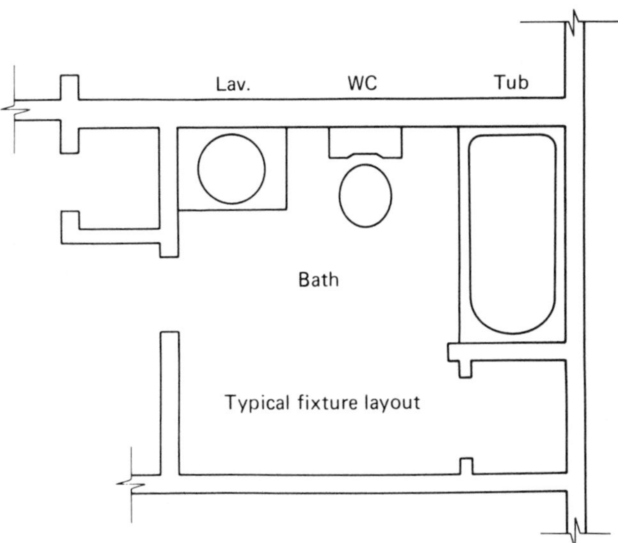

Bathroom Fixture Floor Plan and Layout

Figure 10-7

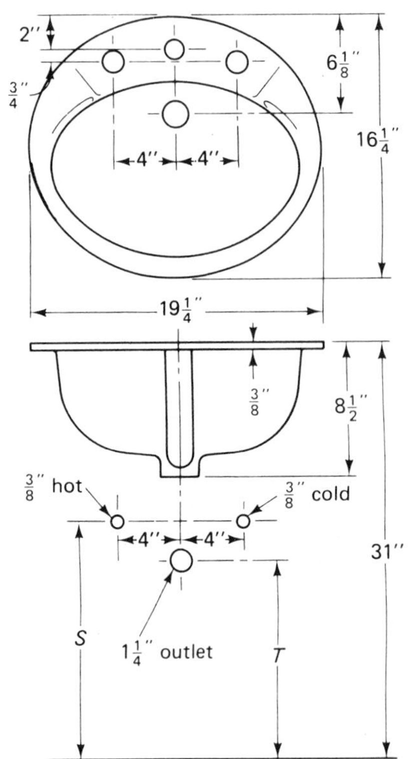

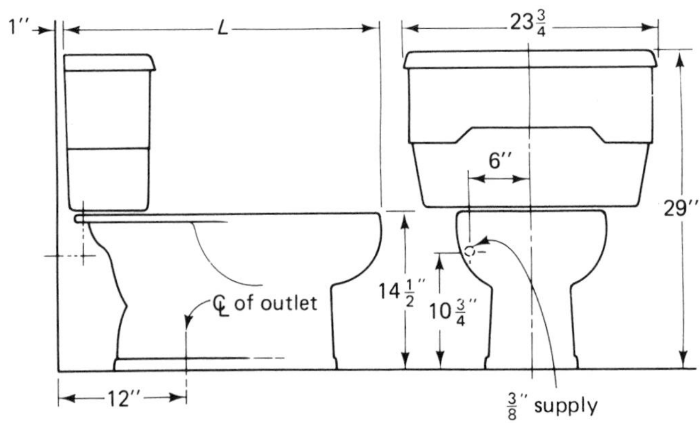

Bathroom Fixture Rough-in Dimensions

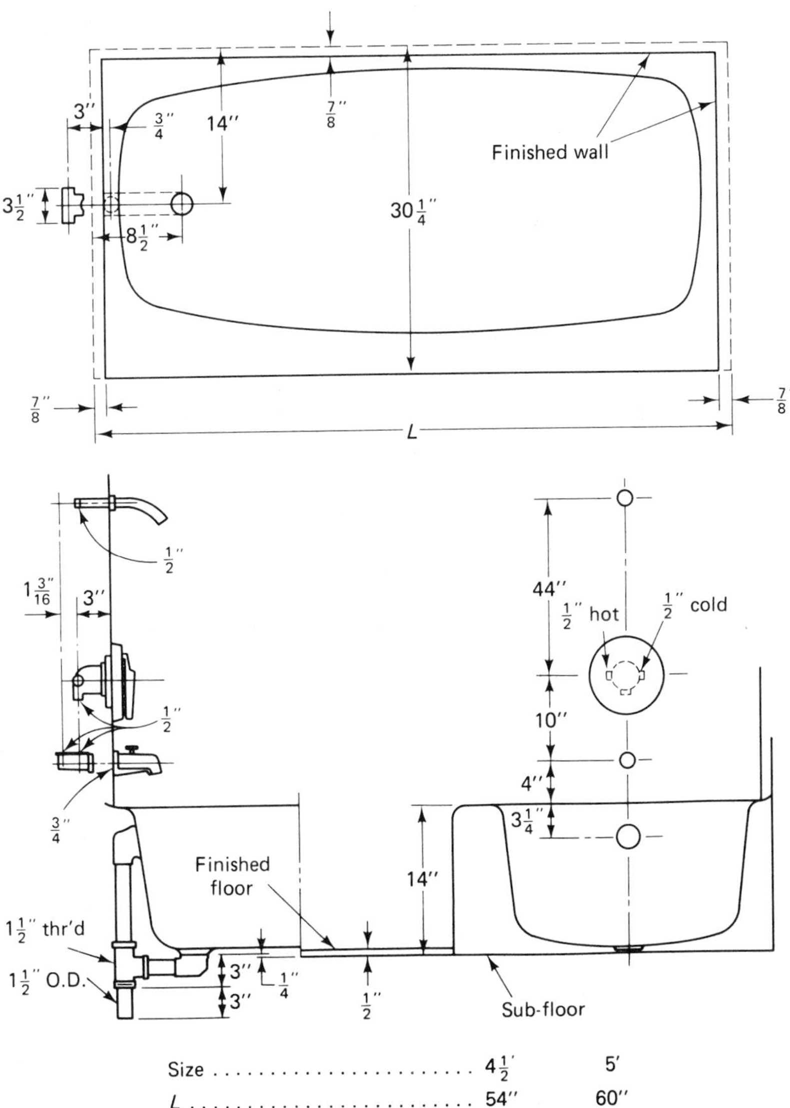

Size	4 1/2′	5′
L	54″	60″

No change in measurements if connected drain and overflow.

is "left hand", that is, the drain is at the left end when viewed from the front. The dimensions given for the fixtures are from the finished wall and subfloor. Thus, $\frac{1}{2}$-inch allowance is made for the thickness of the wall covering when the dimensions for the water closet are taken. Notice that the dimensions for the bathtub are taken from the subfloor and an allowance is made for the thickness of the finished floor. The actual dimensions that are marked on the subfloor, wall sill, and sill plate are shown in Fig. 10-8. These dimensions locate where the holes will be made for the drain, waste, and vent piping, and identify the size that the holes should be to accommodate the pipes.

The waste stack with stack vent is run from below the first floor through the roof with the branch fittings installed to receive the branch drains. Figure 10-9 illustrates the main stack for the residence, which attaches in this case to the bathroom group previously described. The waste height openings are shown for the bathroom group, as well as the hot and cold water distribution. The water closet opening is shown above the wye facing forward. Installation of the vent flange is shown in Fig. 10-10.

As the main stack and the stack vent (vent stack) are run vertically, some means of centering the vertical openings through which the pipe runs is necessary. This is done with a plumb bob and string. After the rough-in hole for the stack is cut in the first floor, the plumb bob can be hung from the hole to the basement area below to establish where it will intersect with the building drain. Above, where the vent stack or stack vent must go through the roof, a plumb bob is hung from the roof to intersect with the

Figure 10-8

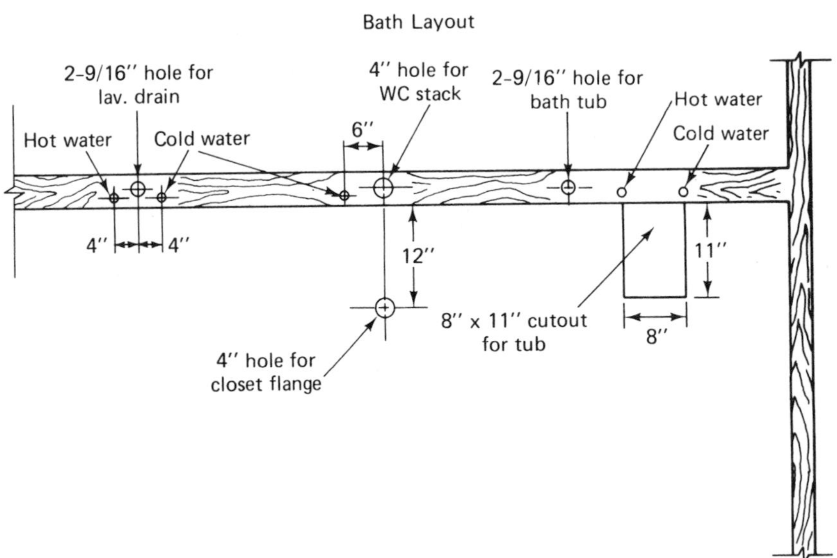

Marking the Rough-in Dimensions for the Bathroom

Figure 10-9

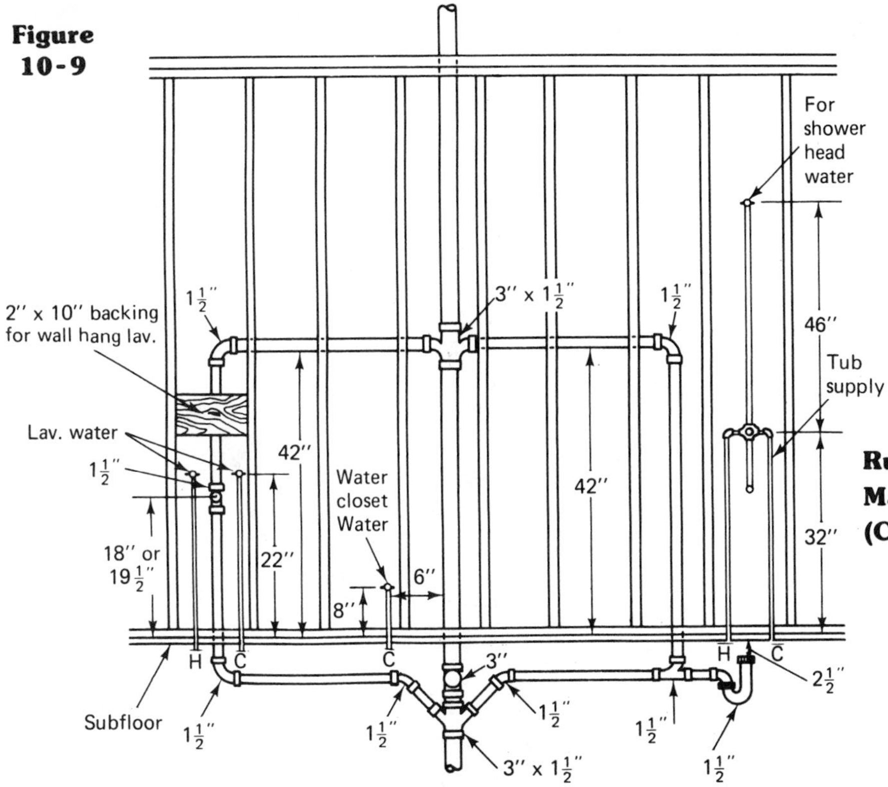

For shower head water

2" x 10" backing for wall hang lav.

$1\frac{1}{2}$"

3" x $1\frac{1}{2}$"

$1\frac{1}{2}$"

46"

Tub supply

Lav. water

$1\frac{1}{2}$"

42"

Water closet Water

42"

32"

18" or $19\frac{1}{2}$"

22"

6"

8"

H C C H C

3"

$2\frac{1}{2}$"

Subfloor

$1\frac{1}{2}$"

$1\frac{1}{2}$"

$1\frac{1}{2}$"

$1\frac{1}{2}$"

3" x $1\frac{1}{2}$"

$1\frac{1}{2}$"

Running the Main Stack (Capp Homes)

Figure 10-10

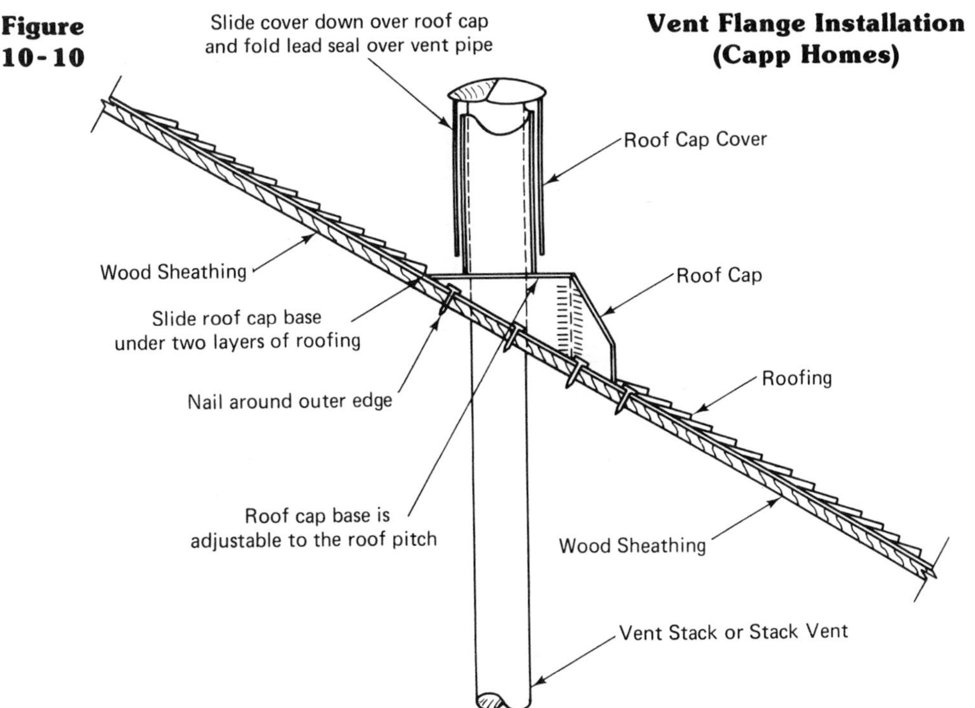

Vent Flange Installation (Capp Homes)

Figure 10-11

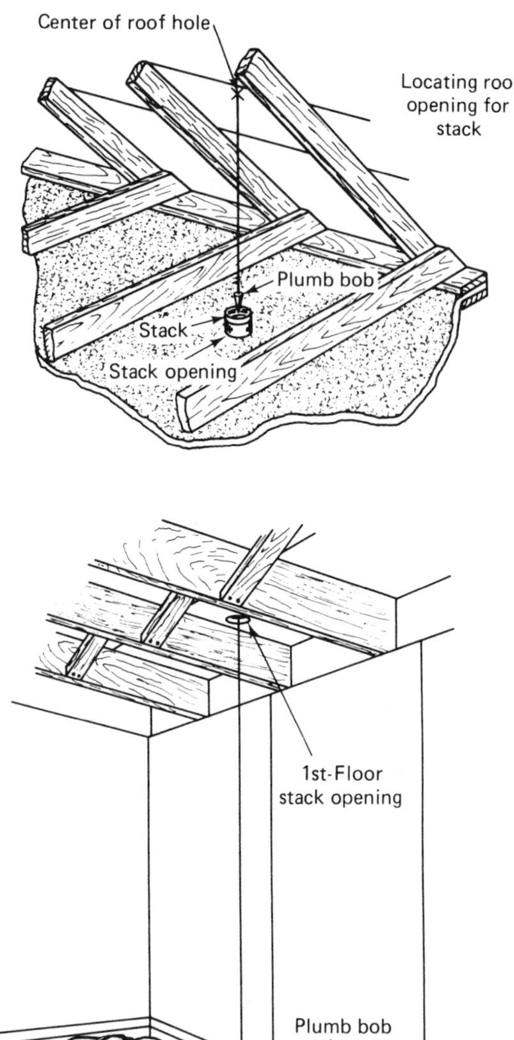

Locating stack base

Centering Stack Openings

stack opening. The use of the plumb bob to locate the building drain as well as the stack vent hole through the roof is shown in Fig. 10–11.

The building drain begins at the base of the stack and extends across the building, where it goes under the foundation and connects to the sewer outside the building (Fig. 10–12). If there is to be a basement floor, or if the house is built on a slab, the drain will be installed first and must be supported adequately at the correct height to have the cleanout located above the floor and the floor drain with the correct pitch below floor height. The drain has a uniform pitch, usually $\frac{1}{4}$ in. per foot, starting at the base of the stack. This means that the depth of the trench must be computed starting from the base fitting to the place where the drain extends under the foundation. For example, if a 4-in. drain starting at the stack base fitting is to be covered by a 4-in. concrete floor, the minimum depth the trench could be would be just over 8 in. If the drain extends 30 ft to the opposite wall before going under the foundation, the depth of the trench would have to be 8 in. plus 30 ft $\times$ $\frac{1}{4}$ in. per ft = $15\frac{1}{2}$ in. Thus the pipe would have a total drop of $7\frac{1}{2}$ in., and the lowest portion of the pipe at the foundation wall would be at $15\frac{1}{2}$ in. The lowest portion of the inside drain where it leaves the building is called the *invert elevation.* Sometimes it occurs that the minimum depth of the trench computed from the depth of the stack base will not permit the drain to clear the underside of the foundation. In that case, the usual procedure is to increase the depth of the trench uniformly, starting at the foundation wall and extending back to the stack base fitting, until the drain will clear the foundation. Outside the building, the drain extends to the sewer.

Figure 10-12

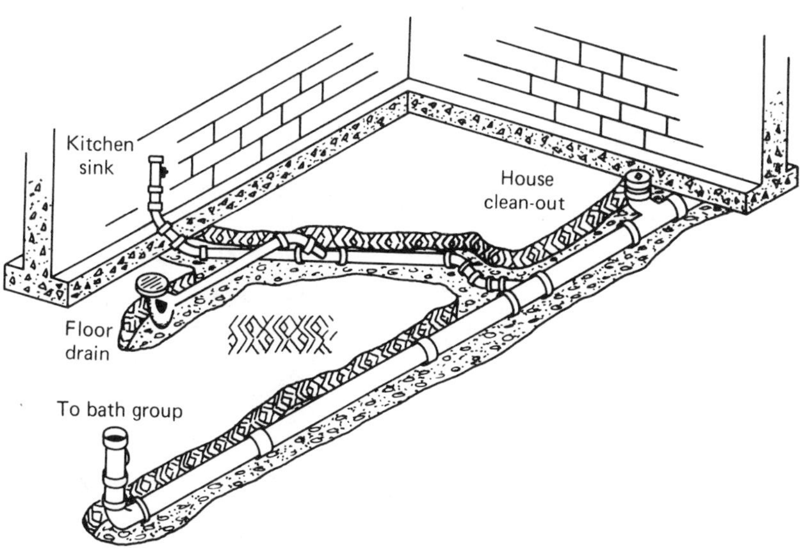

Building Drain

Allowance will have been made to have all houses laid out in a subdivision above the sewer, or to drain into lift stations to accomplish the same purpose. If the drain is several feet above the sewer, then its pitch cannot be uniform. In this case the drain is laid in stages, running several feet at a uniform pitch of $\frac{1}{4}$ in. per foot, and then is dropped to the next stage with $\frac{1}{8}$th (45 deg) fittings. This staging continues until the drain can be connected to the sewer. It is good practice to draw the elevation first and compute the drop of each stage before attempting to install the line with cleanouts.

The drain inside the building is assembled starting with the cleanout and first piece of pipe which extends under the foundation wall (Fig. 10–13). The end of the pipe extending under the wall is capped to prevent it from being plugged with dirt. With the pipe and fitting at the proper elevation, successive sections of pipe and fittings are connected back to the stack base fitting. If the trench is dug using a line, then only the pitch of each piece needs to be checked with a level during assembly. In lines of 100 feet or more, the drain trench and pipe outside are pitched using a tripod and transit bevel, or cold beam laser, to establish a uniform and accurate grade.

Figure 10-13

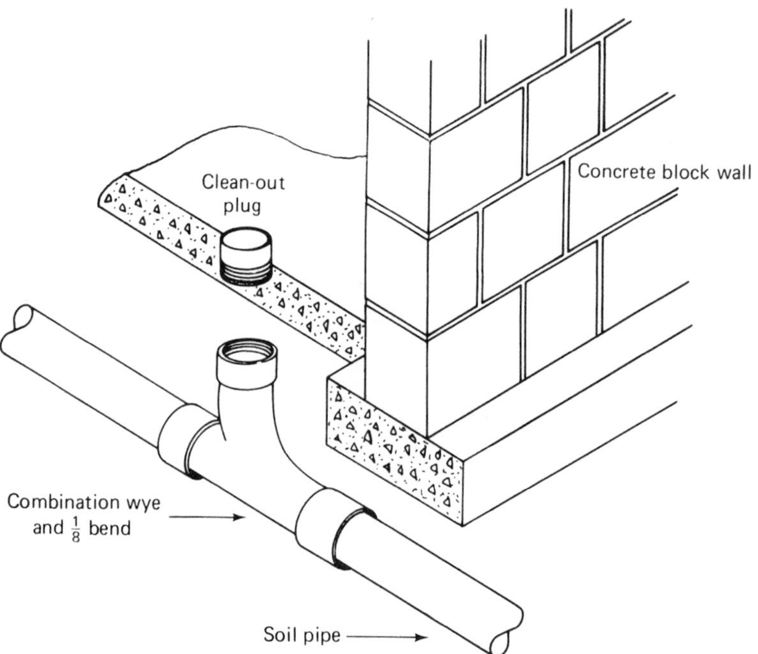

Building Drain Cleanout Installation (Capp Homes)

When one is making the layout and actually digging the trench for the building drain, it is good practice to stake out the trench and pull a string along the path that the main drain and branch drains will take. This serves several purposes: First, it permits the drain to be dug on a straight line. Second, a string can be pulled and tied between the stakes at the proper grade, a string level being used, and then uniform measurements can be taken from the string to the bottom trench. This is preferred to laying a level in the trench every few feet to check the grade. Finally, if strings are used, the layout can be checked to make sure that branch drains can be made to intersect the main drain at angles requiring the minimum number of fittings. Notice, for example, the extra fittings used to connect the branch to the main drain in Fig. 10–12. Some of these might have been eliminated by staking out the system first and checking the plan before the trench was dug. If the dirt is not fine or if it contains rocks, allow room below the pipe for up to 6 inches of fill sand to provide a proper bedding.

The connection between the main stack and the building drain is made between the base fitting and the first floor. One method of taking this measurement is shown in Fig. 10–14. If plastic pipe is used, this is relatively simple, but if the stack is cast iron, the pipe and base fitting should

Figure 10-14

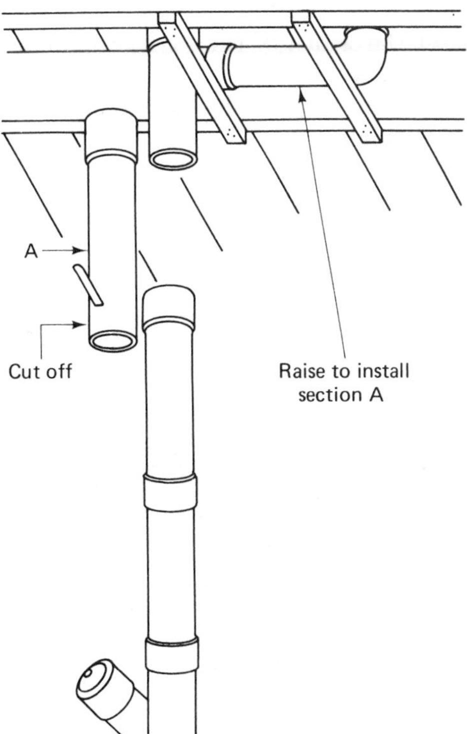

Connecting the Stack to the Building Drain

be braced first, because the lower pipe and fittings in the system are not rigid enough to support themselves.

After the main stack is connected to the building drain, each fixture group is roughed-in and connected to the stack at the fittings already provided. Before actually roughing-in the fixture groups (and for that matter when the vent pipes are run), some thought must be given to insuring the structural integrity of the horizontal supporting members of the dwelling. Drilling holes and cutting away material in the horizontal floor joists weakens them. The maximum size hole that can be cut in the center section of a joist is equal to one-third the width of the joist, and must be in the center one-third section of the joist. (This does not hold true if several adjacent joists were drilled, since it would seriously weaken the floor.) Only one such hole is permitted in the center section, and no hole may be cut in the upper or lower thirds of the joist. For a 2-in. × 10-in. floor joist, this would be a 3-in. hole. For the same joist in the end sections, the maximum size hole is 2 in., with the minimum distance from the hole to the bottom of the joist being 2 in. The holes must be at least 6 in. from the end of the joist. The requirements for cutting holes in floor joists are shown in Fig. 10–15. Remember that a 3-in. hole is required for clearance of a 2-in. pipe, and a 2-in. hole is required for a $1\frac{1}{2}$-in. pipe. Notching to provide access for pipes presents a more serious problem. Notching should never be done in the center half of the joist. It should be done only near the ends, up to one-quarter the width of the joist. Figure 10–16 illustrates the notching rule. When notches are cut in joists, thought should be given to bracing to reinforce the place where they are weakened. Depending upon the depth of the notch, nailing 2 × 4's on one or both sides is common practice. Sometimes

Figure 10-15

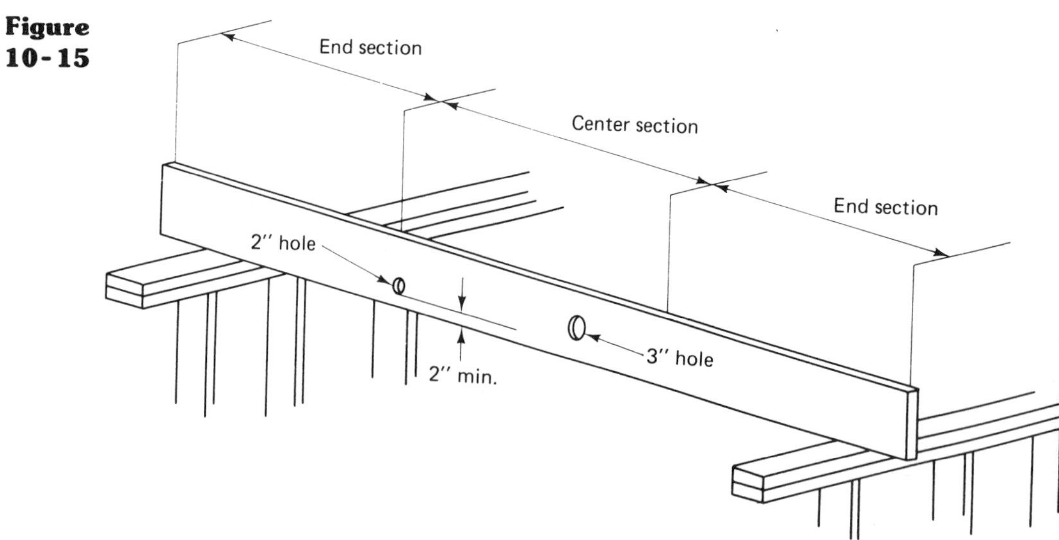

Cutting Holes in Horizontal Joists

steel straps are used. The important thing to remember is that not only must the joist remain safe, but the bracing will undergo the scrutiny of everyone who sees it, including the owner, and this will reflect on the workmanship of the plumber. If there is some question about bracing, talk to the carpenter *before* making the notch or hole in the joist. Where a closet bend runs across the joists, cutting a hole or notch to give access is usually not possible, and would weaken the joist to the extent that it is not advised. Rather, the joists are cut off and boxed with headers to maintain their structural integrity. The common method used is to double up the headers using wood and nail construction, although metal strap hangers are also used for this purpose. The details of these two methods of boxing the floor joists around the closet bend are shown in Fig. 10–17.

After the main stack and vent stack (if one is used) have been connected to the building drain, each fixture group is roughed-in. That is, the remainder of the holes, notches, and cutouts necessary to provide access to the fixtures are made. The closet connection and pipe from the tee at the stack are braced if the pipe is cast iron. If plastic or copper is used, the strength of the joints is sufficient to support it when the closet flange is attached to the subfloor. Figure 10–18 illustrates the single and back-to-back installation of the closet flange. If the upstairs bath does not access a chase or plumbing wall leading directly downward, an offset such as that shown in Fig. 10–19 may be necessary. In this case 6-in. plumbing walls are required, and insulating the pipes should be considered. The bathtub is roughed-in before the plasterboard is put on and will require some carpenter work. That is, if the access is small, or if a combination tub/shower fixture is specified, a portion of the wall must be left out to allow for the installation. Figure 10–20 illustrates a first-floor bathtub waste installation that uses a P-trap from the waste and overflow. The

Figure 10-16

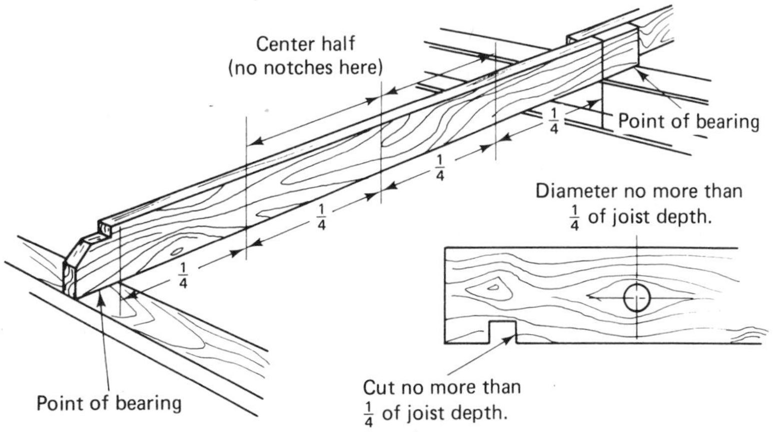

Notching Rule

Figure 10-17

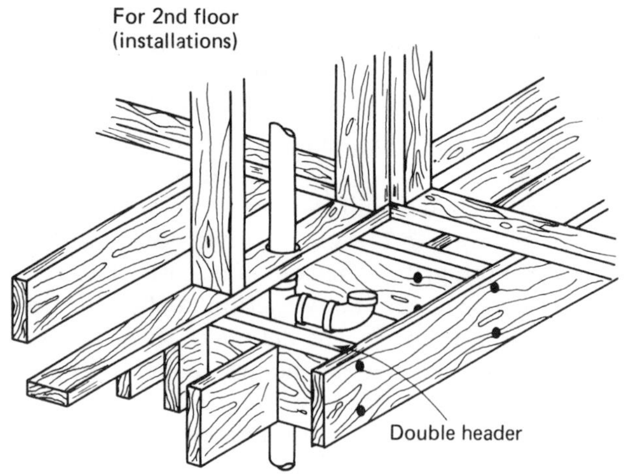

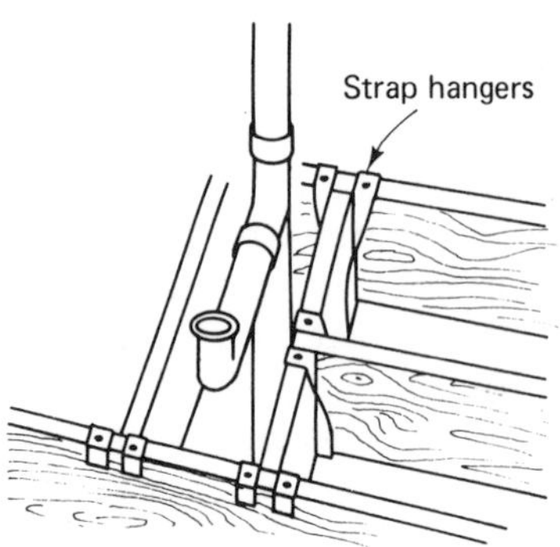

Boxing Floor Joists

roughing-in detail for the bathtub is shown in Fig. 10–21. Notice the access panel at the fixture end of the tub and brace across the back wall to support the tub. The connection and future maintenance to the tub are made through the access hole if the installation is upstairs, and from the basement for a first-floor bathroom. Be sure that the bathtub is level end-to-end and front-to-back so that it will drain properly. A typical layout for a kitchen sink with washer connection in the basement is shown in Fig. 10–22. The waste line must proceed up the basement wall and break over before proceeding up inside the 2-in. × 4-in. outside wall to the vent terminal. That is, the pipe makes an offset with two one-eighth bends. In the basement the standpipe connects to the drain through a P-trap. If it is anticipated that a laundry tub will be installed later, the 2-in × $1\frac{1}{2}$-in. sanitary tee should be replaced with a 2-in × $1\frac{1}{2}$-in. sanitary cross, which will be capped on one side.

Figure 10-18

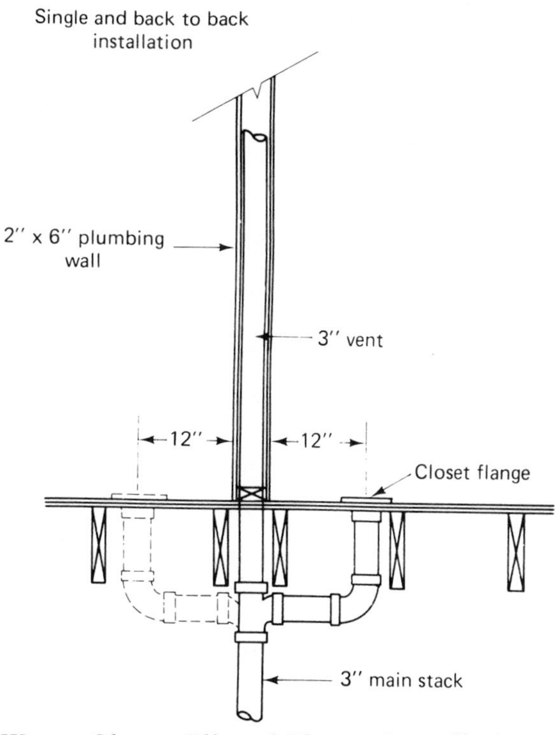

Water Closet Ell and Flange Installation (Capp Homes)

Figure 10-19

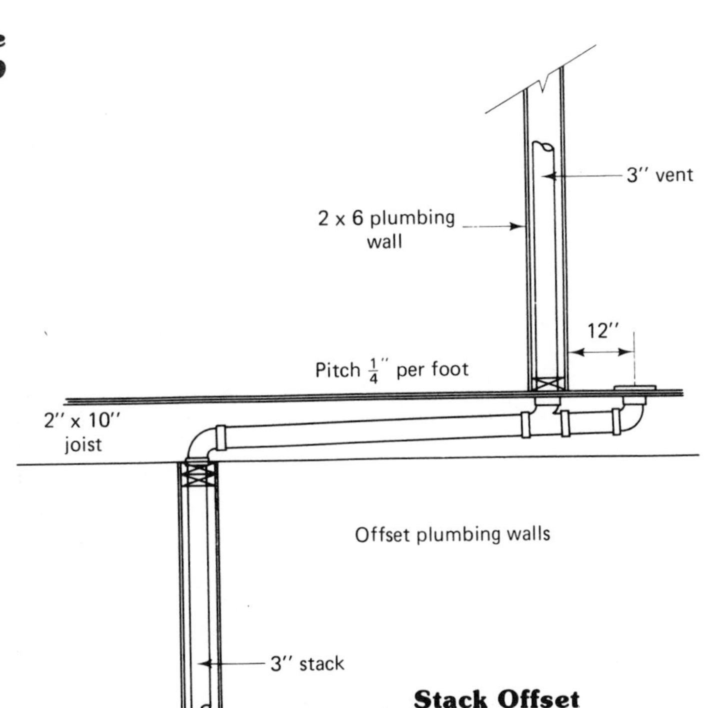

Stack Offset

Figure 10-20

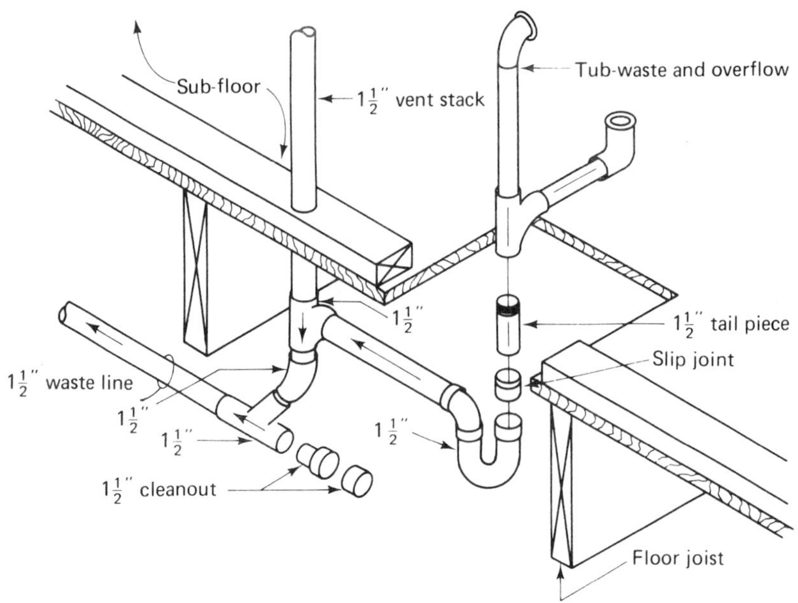

Bathtub Waste Installation (Capp Homes)

Figure 10-21

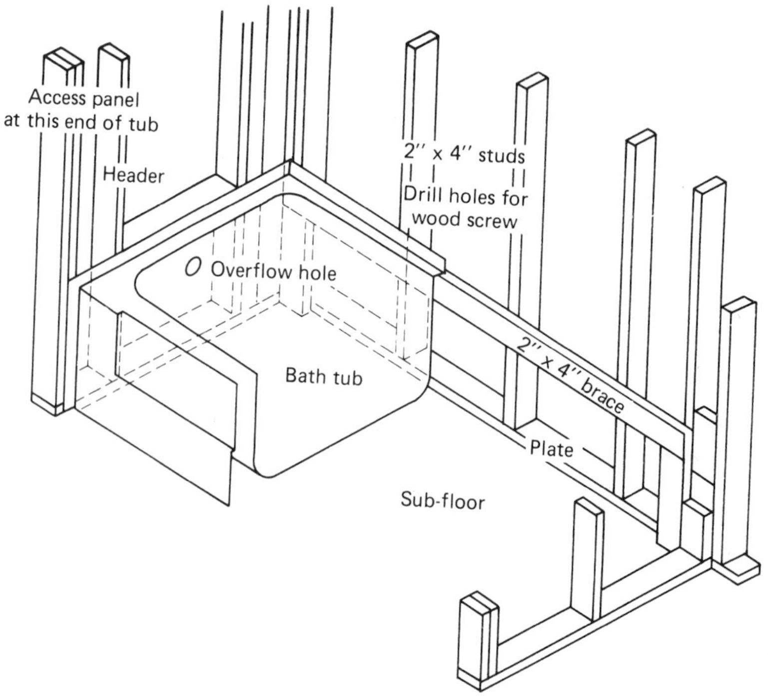

Bathtub Rough-in

Figure 10-22

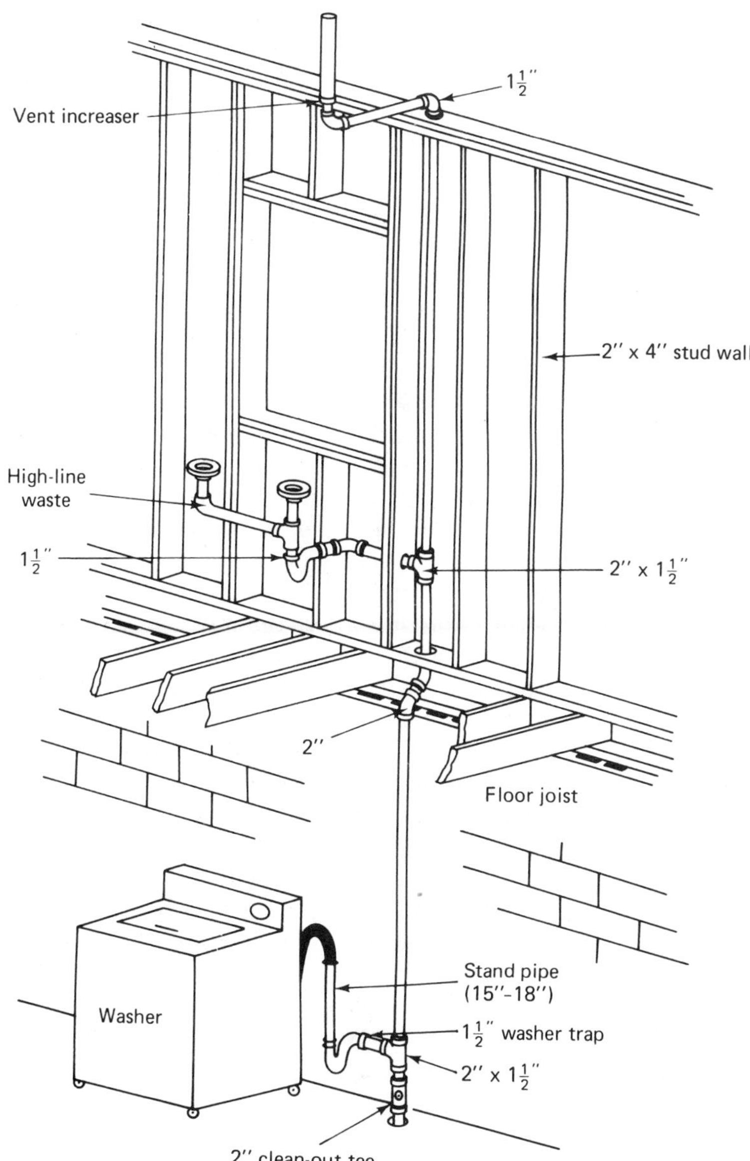

Kitchen Sink Waste Water Layout

10-5 Roughing-in the Hot and Cold Water Distribution System

The water distribution system is roughed-in starting at the service entry. If the meter is outside, as shown in Fig. 10-5, a stop and waste valve should be just inside the wall to turn the system off and drain it. For a typical home, and depending upon the demand and corporation policy, the meter and water entry pipe will be sized at $\frac{3}{4}$ in. or 1 in. It is good practice to secure the pipe near the valve to ensure that future operation will not damage or loosen the water pipe and joints. Then run the cold water directly to the hot water tank and to the farthest connection. Install the fittings for the branches using the dimensions shown on the floor plan.

Before proceeding to cut holes in floor joists, sill plates, and studding, it is important to refer to the floor plan layout drawing and isometric drawing to determine where the hot or cold water pipes, which run parallel, roll over each other as they change directions. That is, one pipe must cross the other, and this often requires an offset. One method used to reduce the number of offsets is to drill one set of holes in the joists an inch higher than the other. Remember that the layout is probably the most important consideration in installing an efficient and professional-looking water distribution system, so use care in this part of the plan. The more fittings that are used, the less efficient is the system.

While the water distribution system can be roughed-in and installed one piece at a time starting where the water service enters the building, it is probably more efficient to rough-in each of the fixture groups and then tie the system together. In this manner the plumber can be sure where the horizontal pipes running through the walls and under the floor must be and thereby eliminate the possibility of having to use extra fittings to connect misaligned fixture groups to the branches and main. Another time-saving technique is to use a chalk line and level to mark the location of all the holes for a fixture group first, or for branch lines, to be sure that the layout is the most efficient, and that the holes will line up when they are drilled or cut. It may seem to require more planning and layout time at the outset, but ultimately it reduces the total labor cost of the job and improves the quality of workmanship.

Another method to reduce installation time and improve the workmanship of the job is to cut several pieces of pipe and assemble them without soldering the joints. In this way larger portions of the system can be cut and fit, followed by soldering all the joints at the same time. This is known as prefabrication. It also provides one more chance to correct an oversight in the design or error in the installation. Fluxed copper joints should not be left overnight.

When actually cutting and connecting the pipe, be sure to remove the burrs by reaming if a tubing cutter is used, or with a knife if a saw is used, and then remove the cuttings from the pipe. It must be remembered that cuttings and debris left in the pipes may stay there rather than being flushed out, reducing the efficiency of the system as well as increasing the possibility for damaging valves, faucets, and fixtures. After the end of the pipe is cleaned and assembled, it must be bottomed in the fitting, or else solder (or solvent/glue when plastic is used) enters the tube, disrupting the smooth inside surface and flow path, causing friction and a pressure drop in the line.

Because the water system will be tested before the fixtures are installed, preformed stubouts are commonly used to extend the pipe from the wall and cap it off in one operation. This eliminates the necessity of capping each one separately. Later, after the system has been tested, these stubouts are cut off to leave the proper length of tubing extended, and then are used for air chambers at other jobs. In this way, several fittings are eliminated and nothing is wasted.

Piping for the water closet is roughed-in 6 in. left of center as you face it, and at an elevation of 8 in. above the subfloor measured to the center of the eared ell. The pipe will be finish cut to 4–6 in., so a 1-ft preformed stubout would leave enough remaining for use later as an air chamber. The pipes for the lavatory will come through the sill plate and into the wall for either vanity style or wall-hung lavatories. The vertical pipe lengths are 24 in. for the water closet and 30 in. for the lavatory. This allows sufficient length to drop through the floor inside the wall. They are stubbed out 4 to 6 in. Kitchen sink supplies are brought through the floor and cabinet base, in line with the faucets. The hot water heater is installed with unions, and a gate valve is placed in the cold water line.

The supply pipes for bathtubs, lavatories, and water closets come through the sill plate typically and turn 90 degrees, using an eared ell. The ears are used to attach the pipe in the wall, and a 1-in. by 8-in. board or 1-in. × 6-in. board is nailed between the studs at the proper height for this purpose. Wood screws are used to attach the eared ells. Wall-hung lavatories also require a 2-in. × 10-in. brace mounted at the front between the studs for the mounting bracket, centered 30 in. above the subfloor.

10-6 Testing the System

The plumbing is pressure-tested before the pipes are covered by the inspector having jurisdiction over the area. This includes the building sewer, sanitary drainage and vent piping, the water service, and the hot and cold water distribution system. Covering the pipes before they pass inspection is prohibited and usually carries a penalty. Depending upon the size and complexity of the system, it may be inspected in parts or all together. Where the water service and sewer trenches are extensive, they may be inspected separately so that they can be covered.

Inside the building the drain, waste, and vent system is given a low-pressure test of 5 lbf/in.^{2} (34 kPa) or water column test; and the water distribution system is given a high-pressure test of $1\frac{1}{2}$ times the working pressure, or 150 lbf/in.^{2} (1034 kPa). In large buildings or several stories, it may be necessary to block off and test successive floors. It must be

Test Plug

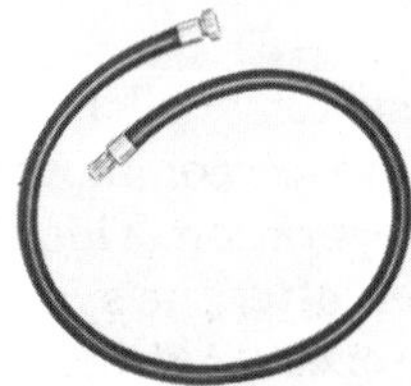

Extension Hose

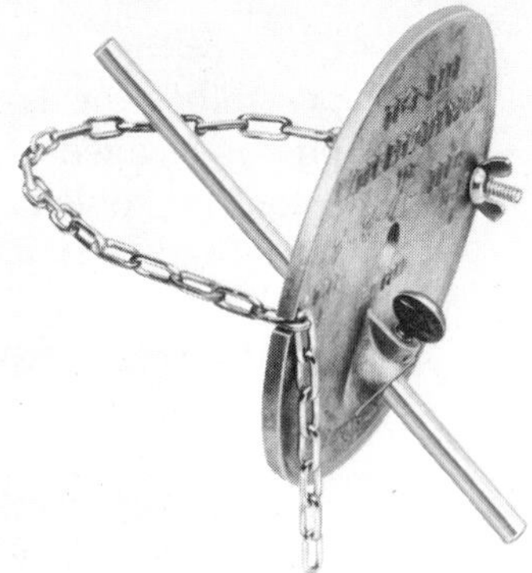

Water Removal Plate

Figure 10-23

Petcock Assembly

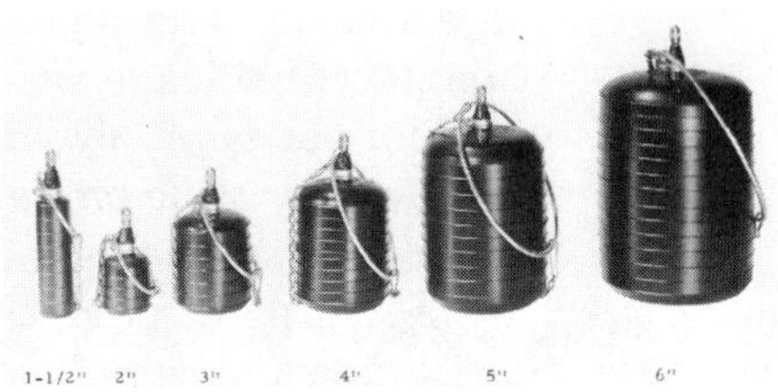

Test Balls

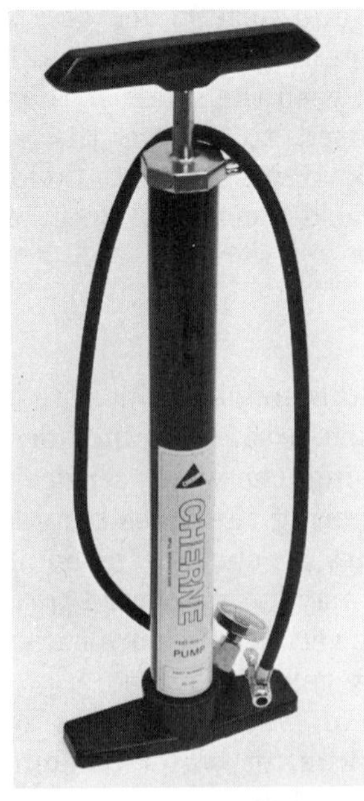

Test Pump

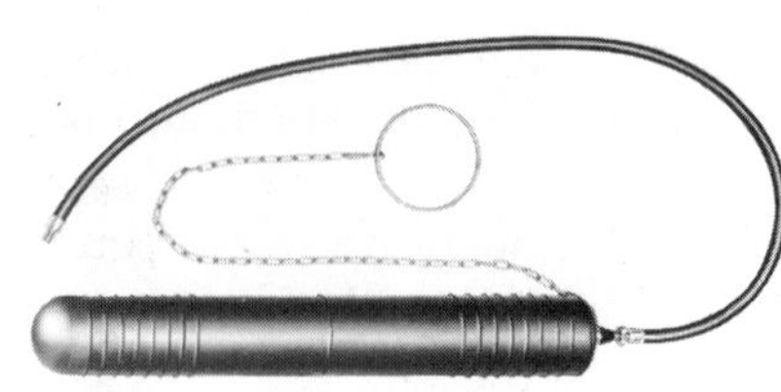

Long Test Ball

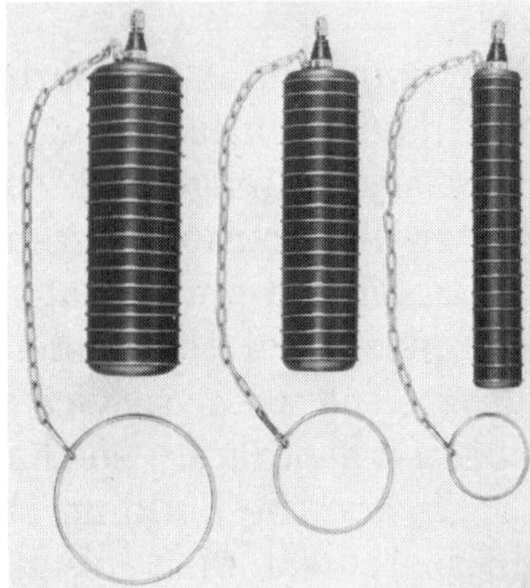

High-Rise Test Ball

Testing Equipment

remembered that other workmen are in the building continuing their work during the time when the plumbing work is being installed, and their work cannot continue past a certain point, e.g., covering the walls, until the plumbing is inspected.

Before the plumbing inspector is called, it is important that the plumber already have conducted the pressure tests to be sure that the system will pass. There will be obvious embarrassment if the inspector is called a day or two in advance as required, with the anticipation that the system will be completed, and then it is not completed or will not pass the required tests. The best method is to have the system ready, test it, and then work on other parts of the system or at other jobs while waiting for the inspector.

The standard methods of test for drain, waste, and vent systems are the 10-ft (3.048-m) water test or the 5-lbf/in.^{2} air pressure test. Equipment used to plug and pressurize the system is shown in Fig. 10–23.

The heavy-duty air pump is equipped with a brass check valve and break resistant pressure gauge. The hose has a quick disconnect coupler, which permits instant release from the test plugs. The extension hose is used when it is necessary to reach farther inside the pipe and still operate the pump in the normal position. The petcock assembly (valve in hose) is used to close off the line at the desired pressure before the hose from the pump is released.

The water test is done as follows:

1. Plug all openings in the system except the vent terminal. Care must be taken not to overinflate the test plugs. On plastic plumbing this could break the pipe. The methods of installing the test plugs is shown in Fig. 10–24.[1]
2. Fill the system with water through the vent terminal. All soil and waste stack plumbing must be tested to 10 feet of water.

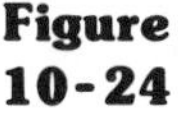
Figure 10-24

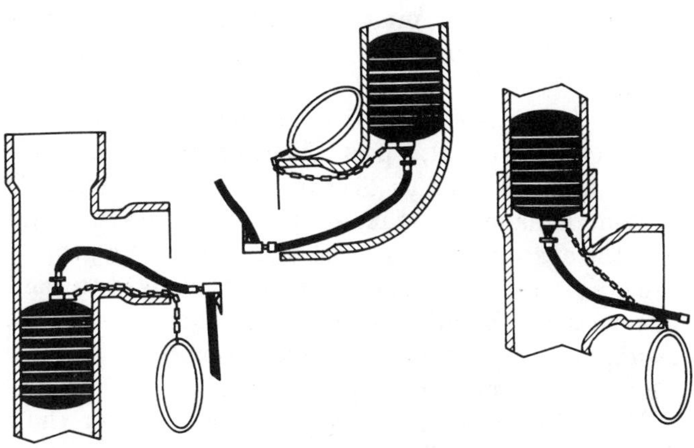

Test Plug Installations

3. The water should stand in the system 15 minutes before the inspection begins, and remain tight and leak-free to pass.
4. To remove the water, the test plug is deflated at the test tee, as shown in Fig. 10–25. Notice that the water removal plate serves two purposes: First, it has a support rod that holds the deflated plug while the test water drains, usually with vigorous action. Second, it reduces the amount of water that would otherwise exit out the test tee if it were not present.

Figure 10-25

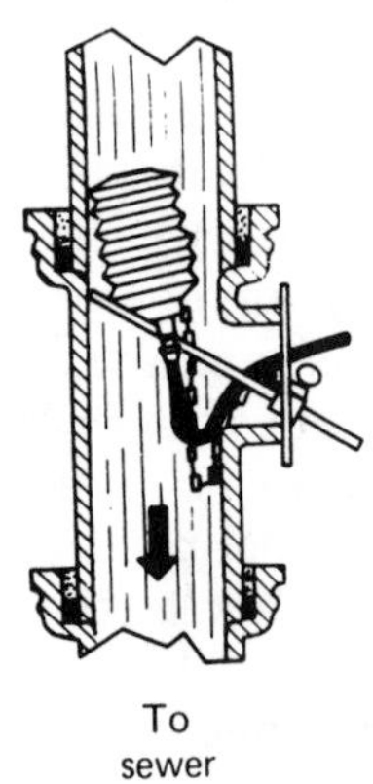

Releasing the Water at the Test Tee

The air test is appropriate in freezing climates or where it is not practical to use the water test. It is conducted in the same manner except that all the openings are plugged and the air pump is used to pressurize the system. The system must hold air under a pressure of 5 lbf/in.^{2} (34 kPa) for a period of 15 minutes without the introduction of additional air by the pump.

For larger systems with volumes that make filling the system with air by hand impractical, a portable compressor can be used. Care must be exercised, however, not to overpressure the system. This can result in both system damage by breakage, as well as personnel injury should a test plug exit the system under pressure and become a projectile. If a compressor is used, a pressure regulator such as that used to regulate low-pressure fuel gas systems should be installed on the pressure line, and the compressor system should be monitored and attended at all times in person during the time the system is being pressurized.

The final test of the system is made after the fixtures have been installed, and this will be discussed later.

10-7 Sizing and Installing the Water Heater

The storage water heater consists of a storage tank and heater. The heater may be gas or electric. If it is gas, additional plumbing is required to pipe in the gas, and a vent system is necessary to vent the fumes.

Water heaters pipe cold water in from the top through the dip tube to the bottom where the water is cooler, and draw it off near the top at the hot water outlet, where it is hotter. Typical gas and electric storage water heaters are shown in Fig. 10-26. The components include the immersion heaters with controls, the glass-lined and insulated tank, which is guaranteed from 5 to 10 years, cold water supply dip tube, hot water dip tube, immersion heater elements, combination T-P (temperature and pressure) relief valve, and drain valve and lines. In addition, water heaters are equipped with a magnesium anode rod, or this may be incorporated with the hot water outlet tube, to inhibit electrolysis action associated with water stored at elevated temperatures.

Residential water heaters are sized by the capacity of the tank, which ranges from 20 to 100 gallons, by recovery rate measured as 100°F (56°C) temperature rise for the number of gallons per hour specified, and by the input wattage (electrical) or heat units (BTU's if gas fired). A typical tank would have a capacity of 40 gal (151 l), with a recovery rate of 34–45 gal/hr with temperature rise of 100°F. That is, water entering the tank at 50°F (10°C) would be heated to a temperature of 150°F (65.5°C) at the rate of

Figure 10-26

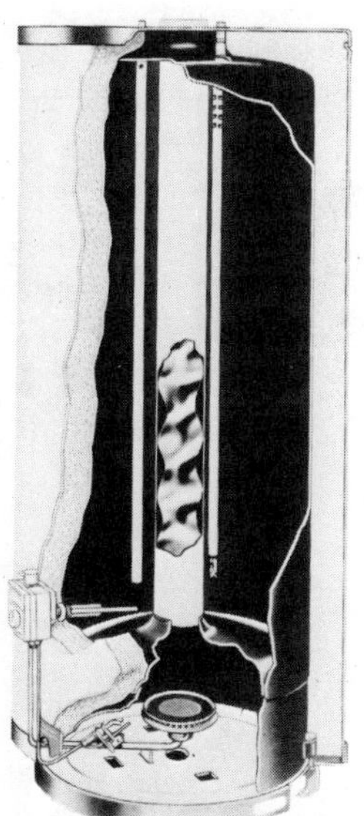

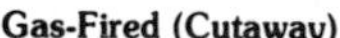

Gas-Fired (Cutaway)

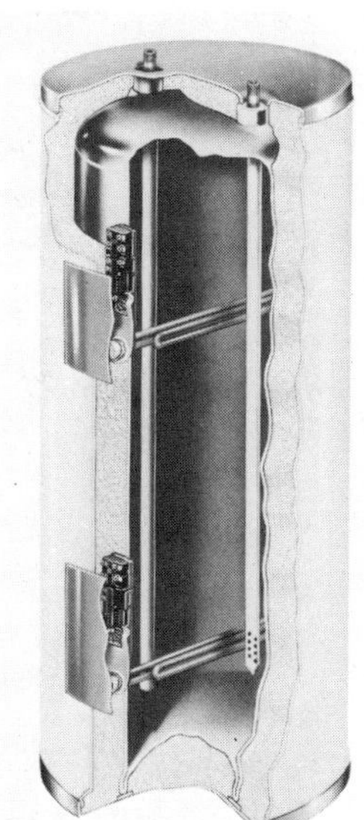

Electric (Cutaway)

Storage Water Heaters (A.O. Smith)

35–45 gal/hr. Only about 70 percent of the capacity of the tank is usable before the incoming cold water at the bottom of the tank dilutes the hot water remaining and drops its temperature. Remember that the tank remains full at all times and when hot water leaves the tank at the top, cold water must enter at the bottom. That is, it is the pressure of the cold water inlet which moves the water through and out of the tank when a hot water faucet is opened.

The combination temperature/pressure (T–P) relief valve guards against the conditions of overheating caused either by an improperly set or faulty temperature control, and overpressure caused by high water line pressures. It is sized at $\frac{3}{4}$ inch minimum inlet and outlet. Although it is true that overheating is also accompanied by pressure buildup in the tank, the chance that the pressure relief valve will not operate makes it necessary to have this dual control. The danger occurs in that while an overpressure condition may cause the tank to rupture and leak, overheating generates steam, and this compressible gas can cause an explosion of the tank, hurling it as a projectile through the building. The condition would be similar to a boiler explosion. A combination T–P relief valve and cutaway are shown in Fig. 10–27. The pressure relief function responds to the water pressure, which exerts a force against a spring to unseat the valve. Water passes through the valve and exits at the threaded connection, which must be connected to a drain pipe extending down the side of the heater to within 18 inches of the floor. The pressure relief valve is set at 125 lbf/in.^{2} (862 kPa), and is equipped with a test lever, which is used to check the valve and clear

Figure 10-27

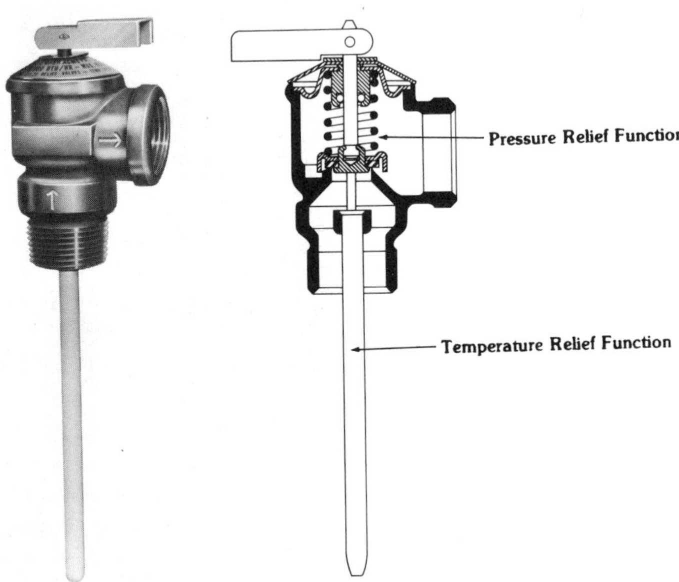

Combination Temperature-Pressure Relief Valve (A.W. Cash Manufacturing Corp.)

it periodically of scale and rust. A faulty or corroded leaking pressure relief valve will drip. The temperature relief function of the valve consists of an extension thermostat installed with the temperature-sensing element that extends about 6 in. down into the water. Overheating causes the thermostat piston to extend and lift the valve from the seat. The temperature relief valve is set just below boiling at 210°F (98.8°C). At 212°F and atmospheric conditions (zero gauge pressure), water boils, vaporizes, and becomes steam. When it is under pressure, however, the boiling temperature is increased, as shown in Table 10-1. The important point to be made is that under normal operating pressures of the system, the water in the tank would not be boiling, but if it were released through a hot water faucet it would be, and could result in a scalding injury.

Figure 10-28 illustrates the installation of both gas-fired and electric water heaters. Notice that the cold water inlet pipe is equipped with a shutoff valve.

Table 10-1 **Relationship between Pressure and Boiling Temperature**

Gauge Pressure		*Temperature*	
lbf/in²	kPa	°F	°C
0	0	212	100
30	207	274	134
60	414	307	152
90	621	331	166

10-8 Installing Plumbing Fixtures

Installation of the plumbing fixtures and appliances is called *finishing.* This must be done carefully, for it is here that errors can result in breakage, which requires costly delays and replacement. Before starting the installation, uncrate and inspect the fixtures and appliances to be sure that they are in satisfactory condition. This is usually done when they arrive at the job or warehouse. If they are found to be unsatisfactory, this should be discovered as soon as possible so that replacements can be secured from the supplier. After the inspection, it is important that the fixtures be put back in their cartons and adequately protected so that they are not knocked over and broken at the site. This is an unfortunate circumstance, but it occurs.

The bathtub and shower stall are built-in fixtures, since they must be installed before the wall covering can be applied to the studding around them. The important thing to remember for these fixtures is that they can be damaged both during installation as well as afterward by other craftsmen, either by a hammer blow or dropped tool, or by standing in them, which abrades the surface with the debris under foot. The best method to protect the tub and shower stall is to clean them after the installation and

Figure 10-28

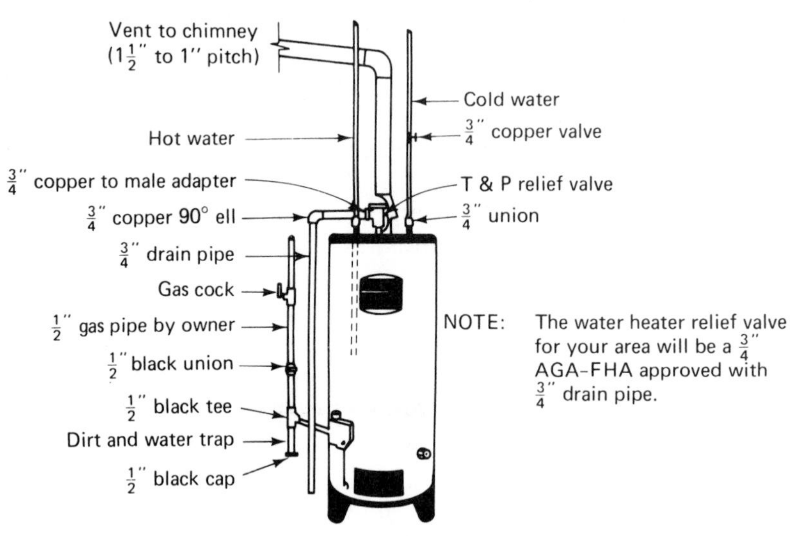

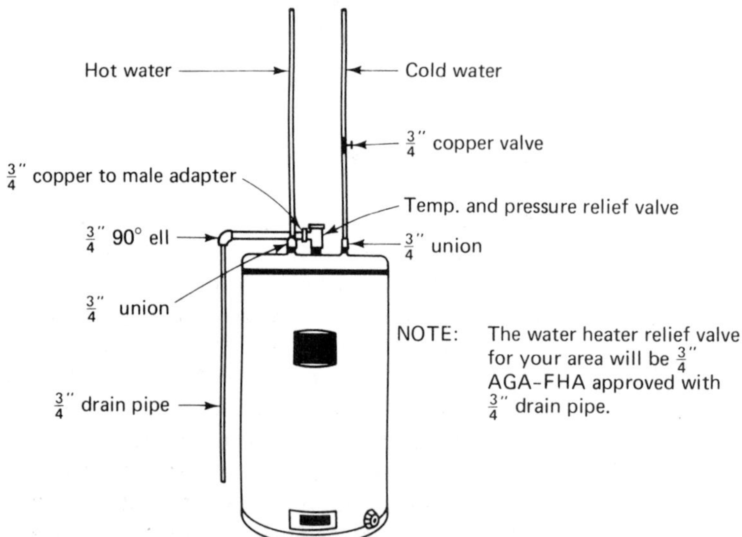

Installation of Gas and Electric Storage Water Heaters (A.O. Smith)

then install a tub and shower stall cover to protect the surfaces. They are available commercially. The next best method is to have the job foreman inspect them after the plumbing work is finished to witness that they were left in good condition and without damage.

The water closet is held in place by the closet bolts. It is sealed at the closet flange with a wax gasket ring, and between the subfloor and the base of the bowl with caulking. To begin the installation, first cut the stubout off, leaving $1\frac{1}{8}$ in. protruding from the wall. Next, clean the area around the closet flange and place the closet bolts in the slots provided in the flange. Place the bowl in position over the closet flange and draw a line around it. Now the final assembly can begin. With the bowl removed, run a bead of caulking around the inside of the pencil mark that indicates the outline of the bowl. Remove the wax ring from the box and place it over the closet flange. Place the closet bowl over the bolts and rock it gently until the bolts are exposed through the holes. Place the washers and nuts on the bolts and tighten gently but firmly until the closet bowl rests evenly on the floor. Remove the excess putty from around the base of the closet bowl. The mounting detail for the closet is shown in Fig. 10-29. The closet tank is mounted to the bowl next. It is a good idea to sit on the closet bowl during the installation to seat the fixture as well as to support the tank. Place the tank over the bowl with the gasket in place, insert the bolts and nuts with their gaskets and washers, and tighten evenly until the tank sits level and firm on the closet bowl. Next, mount the toilet seat to the closet bowl. Complete the installation by cutting, bending, and installing the supply tube. The compression nut and washer for the shutoff valve are now slipped over the supply pipe, and it is tightened in position. Finally, the flexible closet supply is adjusted, cut to final length, and assembled with the compression nuts and 4 rings. The last step is to recheck and tighten all closet bolts and nuts, adjust the water level with the float mechanism, and place the cover on the closet tank.

A typical installation for the kitchen sink and dishwasher is shown in Fig. 10-30. To begin the installation, mount the strainers in the sink. Apply a $\frac{1}{8}$-in. bead of plumber's putty to the flange and place the strainer body through the opening. Press down firmly so that the putty spreads evenly. To secure the strainer, place the rubber washer and metal washer over the basket strainer, as shown in the detail of Fig. 10-31, and then screw on the locknut hand tight. To prevent the strainer from turning while the locknut is being tightened, the handles of a pair of pliers may be placed in the crosspiece of the strainer and held with a screwdriver. Be careful not to scratch the sink during this operation. The garbage disposal attaches to the sink in place of one of the basket strainers and empties through a trap into the waste pipe through a wye, as shown in Fig. 10-32. The installation is similar to that previously described, except that the flange is held in place with three bolts instead of a locknut. These bolts must be tightened evenly to seal and align the disposal unit. Restraint must be exercised so as not to overtighten the mounting bolts and thus damage the sink or disposal unit.

Figure 10-29

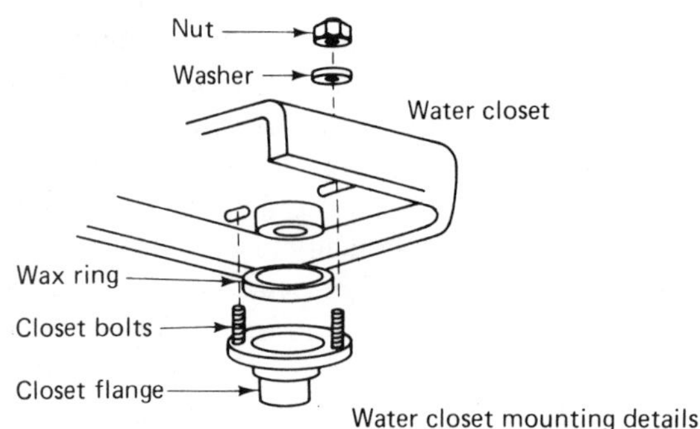

Bottom view

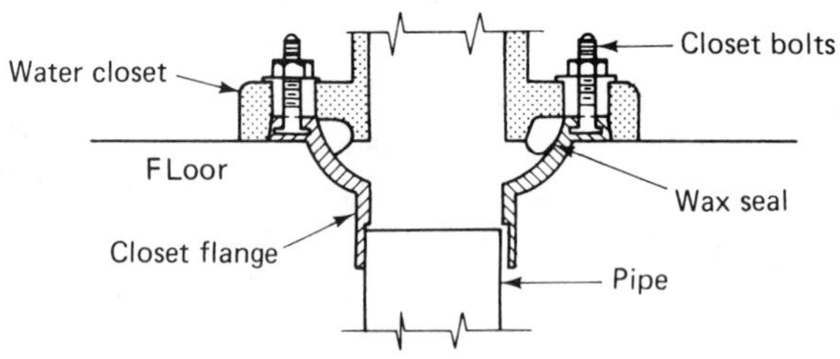

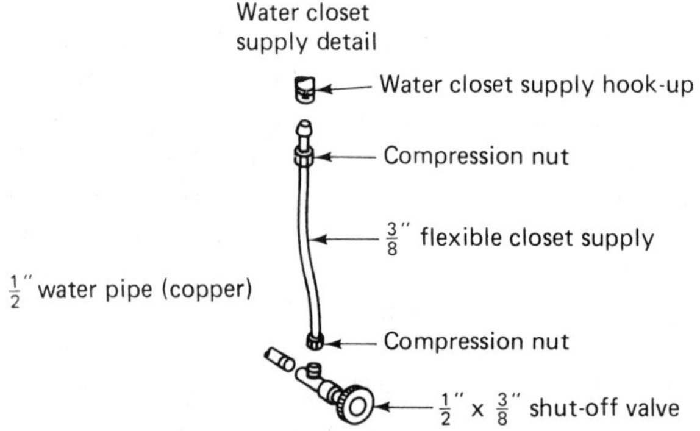

Mounting Detail for Water Closet

Figure 10-30

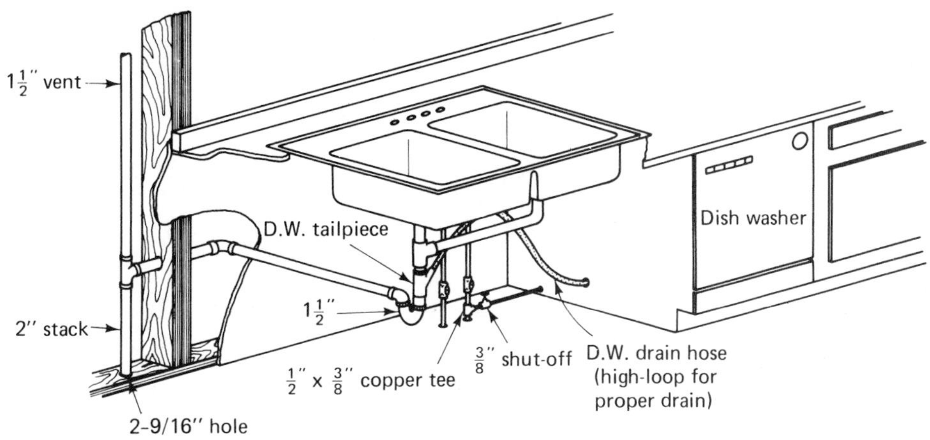

Installation for Kitchen Sink and Dishwasher (Capp Homes)

Figure 10-31

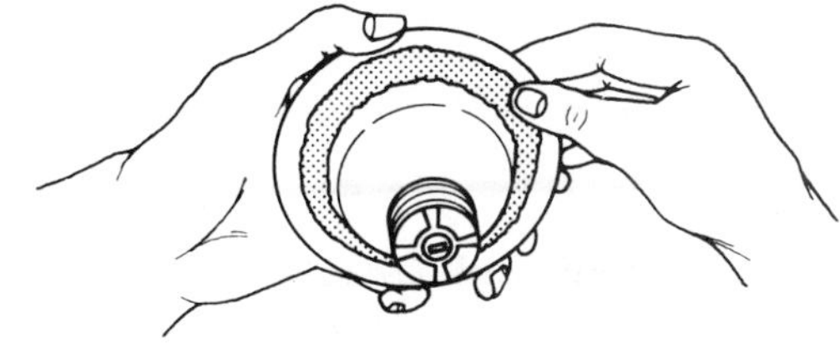

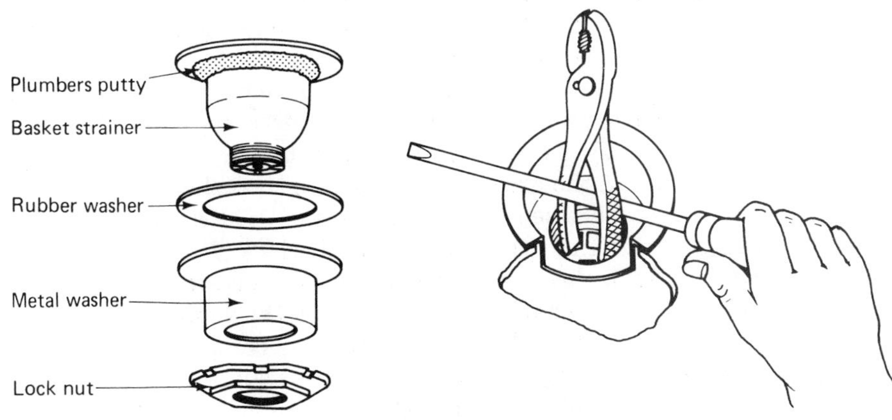

Installing the Strainer (Capp Homes)

Figure 10-32

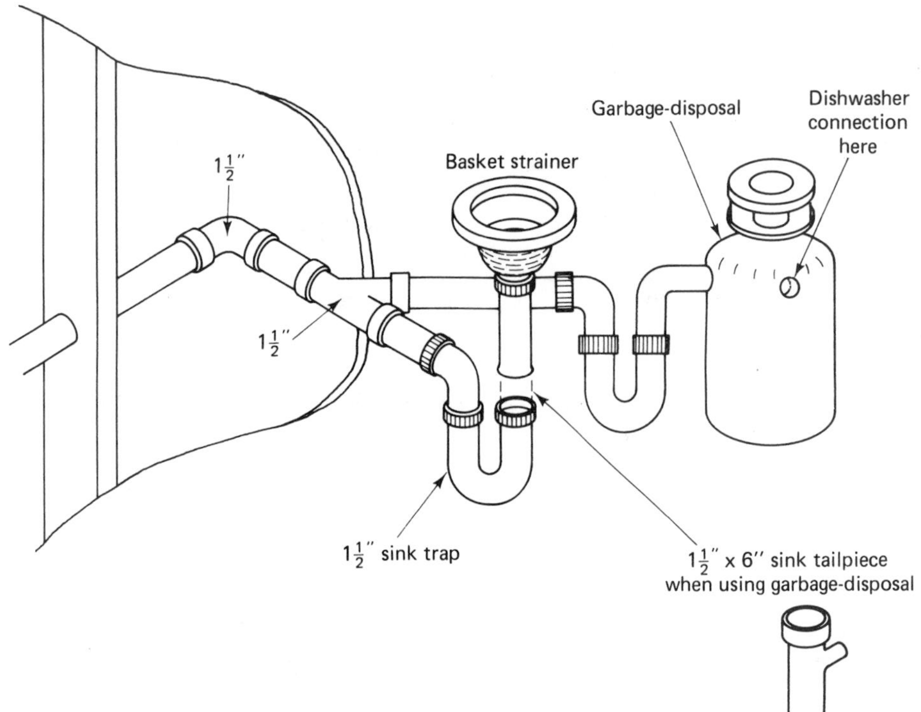

Kitchen Sink Detail with Garbage Disposal (Capp Homes)

A cutaway view of a typical garbage disposal unit is shown in Fig. 10-33. Notice that the bolts exert an upward pressure on the metal washer to seal the unit in the sink. The rubber washer (which goes between the basket and metal washer) is not shown.

A typical laundry tub installation is shown in Fig. 10-34. To place the tub in the correct position, it may be necessary to cut the supply pipes to length first. If this is the case, mount the faucet assembly to the tub first, as is shown in the assembly figure, and then slide the tub near the wall to take the measurements to cut the pipes. An alternate method is to take the measurement from the floor (to both the faucets and the pipes), but irregularities in the floor may cause this dimension to be faulty. Be sure to allow sufficient length for the unions on the supply pipes. The assembly of the supply pipes to the faucet is done by compression unions, and this requires that the large nut be placed on the supply pipe first, followed by the attachment of the union. If copper tubing is used, it would be soldered to the supply pipe. The final installation requires tightening the union nuts and securing the pipes (and the sink if necessary) to the wall so that accidental movement of the sink will not break the installation and seal.

Figure 10-33

Garbage Disposal Cutaway (In-Sink-Erator)

10-9 Summary of Practice

Installing and testing a plumbing system is an activity that requires the coordination of several steps. The following sequence is typically followed to plumb a house:

1. Develop a successful bid.
2. Secure necessary permits.
3. Arrange the schedule of installation with the general contractor.
4. Rough-in the system.
5. Test the roughed-in system and have it inspected.
6. Install the fixtures and appliances.
7. Test the final installation and have it inspected.

The successful bid is developed from the floor plans, which are used to develop the isometric drawing and materials list. The job is bid on the basis of the time to install the required materials. Some plumbing establishments sell both the appliances as well as the installation, and in this case the bid will be for both. Permits are required by local authority to ensure that plumbing within the jurisdiction is safe. The schedule for the installation is

Figure 10-34

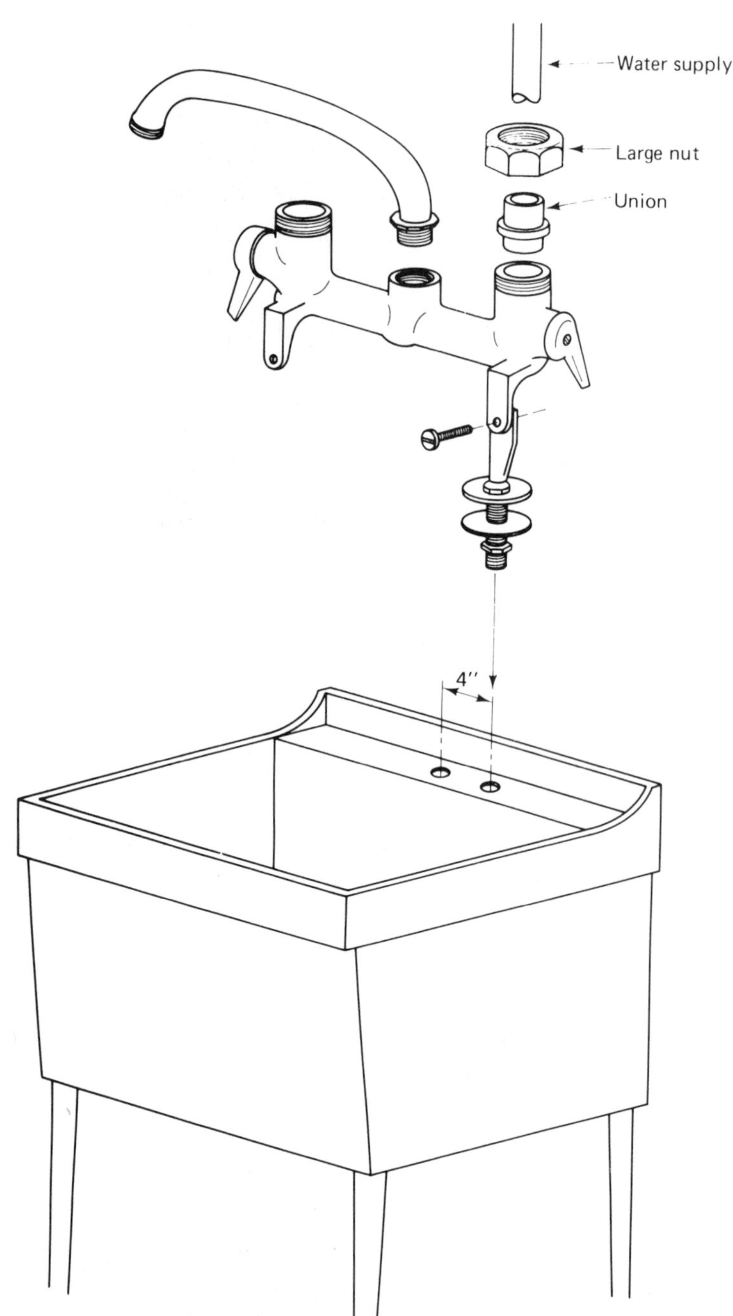

Laundry Tray Installation

arranged with the general contractor, who must coordinate all the work of the various crafts to complete the building. Scheduling, and then updating the schedule every few days, is necessary in case some jobs progress slower or faster than expected. That is, because the plumber must work within the general contractor's schedule, it is necessary to check the progress of the jobs from time to time so that a full and orderly work schedule can be maintained. The plumbing system is roughed-in after the subfloor is down, walls are up, and the roof is on. Typically, the plumbing is installed at the same time the electrical wiring is run. Carpenters are usually on the job to assist with structural alterations, which may be necessary to accommodate the pipes. The bathtub and shower stall are built in with the rough-in plumbing. After the installation is roughed-in, it is pressure-tested according to weather conditions, accepted practice in the area, and ordinances in the local jurisdiction. The finish plumbing is installed after the drywall and paneling are up and the kitchen cabinets have been installed. The installation of the fixtures requires care to make sure that they are not damaged or broken.

The final test of the system consists of two parts. First, the plumber adjusts, cleans, and inspects each fixture and appliance to be certain that it operates properly as one part in the total system. Then the final test is made to certify that the system will operate all together, for example, that several water closets can be flushed simultaneously, without siphoning or blowing trap seals. This is required in multistory buildings. The final test of the drain and vent system requires that traps that have been filled with water can hold a pressure of 1 inch of water for 15 minutes. Either smoke or oil of peppermint is introduced in the system (2 oz of oil is poured down each stack), after which the vents are capped. The absence of the smell of smoke or oil of peppermint assures that the traps and connections for water closets, sinks, and drains are absolutely sound and gas- and watertight.[2]

REVIEW QUESTIONS AND PROBLEMS

1. What does the plumber need to submit a bid for an installation?

2. What is the purpose of the permit? For your area, what is the penalty for beginning a job without a permit?

3. What purposes does the isometric drawing serve?

4. Make an isometric drawing and materials list for a dwelling from a set of blueprints.

5. Why is the local code important when one is estimating the cost of a plumbing installation?

6. Why is the drain, waste, and vent system installed first?

7. What sequence of steps is followed when one is roughing-in the DWV system? Knowing that each plumber has an individual style of doing a job, ask one local plumber to explain that preferred sequence. Record the interview on tape, if possible, to share the experience with others in a group discussion. A specific building plan can be used to structure both the interview and the group discussion.

8. Why is the placement of the hot water heater important?

9. How are hot water heaters rated?

10. Explain one test of the roughed-in DWV system.

11. Explain the test of the water distribution system.

12. Review the sequence of steps to install a water closet.

13. Why is it important to have the general contractor or the foreman for the job inspect the plumbing after the fixtures and appliances have been installed?

14. Who pays for the repair if a carpenter drives a nail through a drain, waste, or vent pipe after the inspection is made? Refer the question to several local contractors and compare answers.

15. Explain the final test.

REFERENCES

[1] The plumber should be aware that test plugs are extremely dangerous when used improperly. For example, overinflating the plug or system can cause the plug to come out with enough force to kill or maim a person. Refer to *Safe & Proper Use of Pneumatic Plugs* for safe operating procedures (Cherne Industries, Edina, MN 55436).

[2] It must be remembered during the peppermint test that the person who handles and pours the oil of peppermint down the stack is not permitted in the building until after the test is completed, since the scent from the oil would spoil the test.

11 Plumbing Systems Components

11-1 Introduction

There are a number of components that are installed and maintained as part of the plumbing system by the plumber. These include pumps, valves, meters, and gauges, in addition to the piping systems and fixtures that they serve.

Pumps are used for fresh water supply, ejection or discharge of soil and waste water, and recirculation of water and other fluids that must be replenished. For example, pumps bring fresh water from wells and cisterns, as well as boost the supply in tall buildings. They are also used to empty sumps of drain water and sewage when drains are below sewer levels. In hot water circulating systems and solar systems, they are used to circulate water to maintain constant outlet temperatures or transfer heat to storage for later distribution.

Valves are used to control the direction, flow rate, and pressure of fluids. They stop, start, and regulate the direction, amount, and pressure in response to system and operator requirements. Temperature control valves are used to regulate fluids at a preset temperature. Temperature relief valves release fluids at preset temperatures to protect equipment and prevent injury. Pressure relief valves have the same purpose, but respond to pressure rather than temperature. Simple control valves open and close manually to regulate flow between the limits of full-open and closed. Pressure regulating valves are designed to limit the pressure in the line on the downstream side. For example, they are used to reduce the pressure from the water main to a value below 80 lbf/in.^{2} (551 kPa) for use in a building. Flow control valves limit the rate of flow rather than pressure, and may or may not be temperature- and pressure-compensated. They are used extensively for water conservation.

Water meters record the cumulative amount of water used, usually in gallons. The amount of water used determines the water utility charge. They are also useful sometimes in discovering leaks.

The plumber uses several gauges to monitor and test the operation and safety of piping system, fixtures, appliances, and components such as pumps and valves. Although the pressure and temperature gauges are most common, multimeters are also used to check electrical switches, circuits, appliances, and related equipment such as water heaters, pumps, and thermostat controls.

11-2 Pumping Principles

A fluid is pumped when its volume is displaced and transferred from one place to another. The pumping action can be positive, as when an exact amount of fluid is displaced for each stroke or revolution of the pump, or it can be displaced and transferred using the inertia of the fluid in motion. When the pumping action displaces a specified amount of fluid for each stroke or revolution, it is referred to as a *positive displacement* pump. Gear, vane, piston, and screw pumps are positive displacement pumps. Pumps using the inertia principle to propel the fluid are of the *nonpositive displacement* type and include centrifugal propeller and mixed flow impeller designs. Positive displacement pumps are used where the primary consideration is one of pressure, whereas nonpositive displacement pumps are used where high volume is most important. Because of their simple design and fewer number of moving parts, nonpositive displacement pumps cost less to install and operate with less maintenance.

Essentially, a pump consists of a low-pressure inlet, a pumping chamber with a drive mechanism, and a high-pressure outlet (Fig. 11-1). Energy is added to the fluid in the pumping chamber by the action of the mechanical drive. In positive displacement pumps, the high- and low-pressure portions of the pumping chamber are separated so that the fluid being pumped cannot leak back and return to the low-pressure source. The chamber in nonpositive displacement pumps is connected so that as the

Figure 11-1

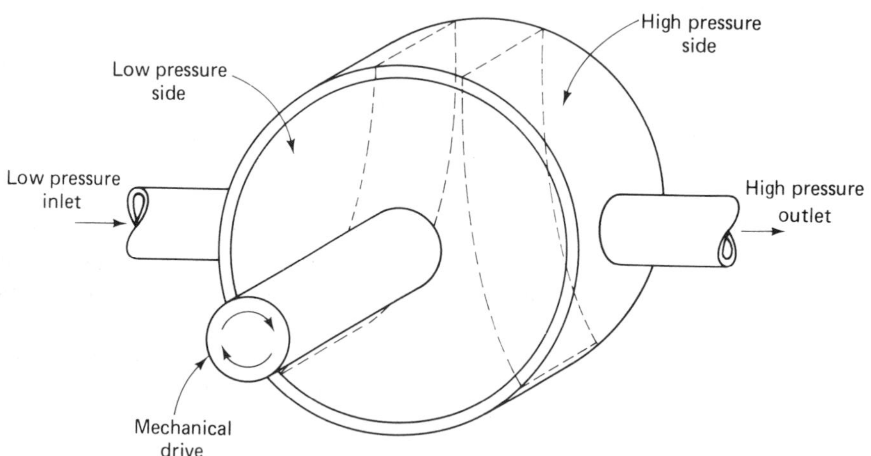

Basic Parts of a Pump

pressure increases to overcome resistance to flow at the outlet, fluid within the pumping chamber circulates.

Nonpositive displacement pumps make use of Newton's first law of motion to move the fluid against system resistance (pressure). This law states, in effect, that a body in motion tends to stay in motion. The action of the mechanical drive in the pumping chamber speeds up the fluid so that its velocity accounts for its ability to move against the resistance of the system. In centrifugal pumps, rotational inertia is imparted to the fluid (outward from the axis of the shaft), whereas in propeller pumps, the inertia imparted to the fluid is translational (in line with the shaft). For a given pump size, the volume of fluid pumped is influenced by the speed of the rotating member, resistance to flow (pressure) at the outlet, and internal clearances.

Positive displacement pumps cause the fluid to move by varying the physical size of the pumping chamber in which the fluid is sealed. As the low-pressure inlet side of the pumping chamber increases in volume, atmospheric pressure causes the fluid to enter the pump. That is, the increases in the size of the chamber cause a vacuum, and fluid is sucked in. As the fluid moves through the pump and to the high-pressure outlet side, its volume is reduced, causing it to be expelled. The action in positive displacement pumps, then, is to alternately increase and decrease the volume of the pumping chamber to accomplish the pumping action. Since the volume for each stroke or rotation is fixed by the positive displacement characteristic of the pumping chamber, the volume of fluid pumped for a given pump size is influenced only by the number of cycles made by the pump. Leakage is a factor, of course, but this is considered as a loss in volumetric efficiency rather than a characteristic of the pump.

There are some other characteristics that should be remembered about positive displacement and nonpositive displacement pumps. These can be summarized as follows:

Positive Displacement Pumps	*Nonpositive Displacement Pumps*
• high pressure	• low pressure
• self-priming	• usually primed
• by-pass valve required to protect pump against high pressure	• internal circulation protects pump against high pressure
• some form of lubrication usually required because of meshing parts	• impeller requires no lubrication
• medium volume	• high volume
• high efficiency	• medium efficiency
• wear occurs in pumping action	• little or no wear in pumping action

11-3 Pump Types

The nonpositive displacement pumps most often encountered in general plumbing applications are the centrifugal, propeller, and mixed-flow type. In positive displacement application, the piston, vane, and gear pumps are the most common.

Centrifugal pumps can transfer large volumes of fluids. The construction of a centrifugal pump is shown in Fig. 11-2. The parts include an inlet port, pumping chamber shaped as an involute, the drive shaft and impeller, and the outlet port. In operation, rotation of the impeller imparts a centrifugal force to the fluid as it enters at the eye, causing it to be directed outward from the center of the impeller. The blades are curved opposite to the direction of rotation to increase efficiency. Defuser blades are sometimes attached stationary to the pump housing to redirect the fluid in such a way as to reduce the velocity and internal clearances. This increases the ability of the pump to develop pressure against the resistance to flow. Defuser action is shown in Fig. 11-3. Although centrifugal pumps are normally used to transfer large volumes at relatively low pressures up to 50

Figure 11-2

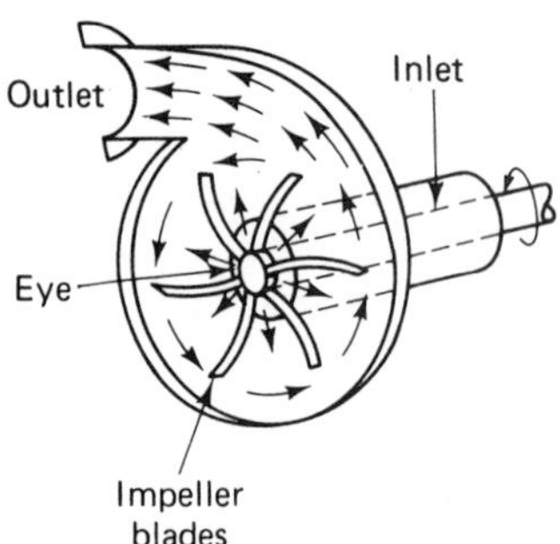

Centrifugal Pump Construction

Figure 11-3

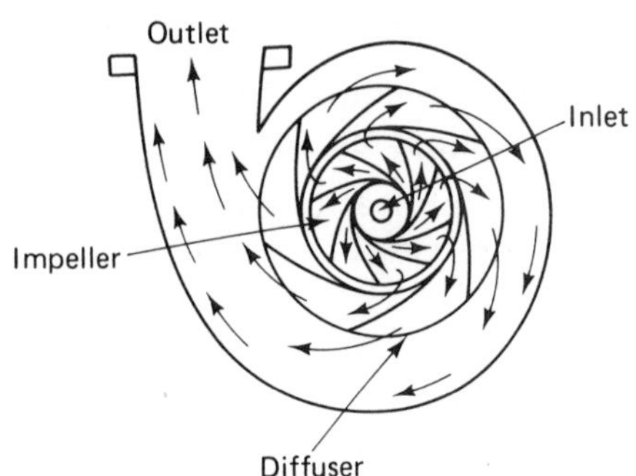

Centrifugal Pump Diffuser Action

lbf/in.^{2} (345 kPa), staging has the effect of raising the pressure. As with other series circuits, the pressure adds from stage to stage, while the volume flow rate is the same through each of the stages. Centrifugal pumps have several advantages, including low production cost, simplicity of operation, high reliability, low maintenance factor, low noise level, and the ability to pump nearly all fluids without clogging or damage to internal parts. Because the inlet and outlet passages are connected hydraulically, centrifugal pumps are not self-priming and must either be positioned below the level of the fluid or primed to start the pumping action. Most have relatively large internal clearances between the impeller and the housing, although some water pumps use flexible impeller blades that run close to the chamber wall to reduce internal clearances and increase pumping efficiency.

Centrifugal pumps have wide application as ejector pumps, circulating pumps, and booster pumps. They are trouble-free and easy to maintain, and they have a long life, since the only moving part subject to the fluid is the impeller. Failure, when it does occur, usually begins at the seal that protects the bearing next to the impeller or in the bearing itself. Once the seal or bearing fails, the fluid enters the motor portion. A good indicator of the condition of the pump is the impeller bearing. If it turns smoothly without roughness by hand and there is little or no play in the bearing, chances are the pump is still serviceable. Knowing the number of hours on the pump, or the age of the pump, is also a good way to determine condition, since both seal and bearing life can be predicted from these indicators. That is, both bearing wear and lubricant dry-out cause failure.

Figure 11–4 illustrates a typical submersible centrifugal pump, which is used for pumping water from basements, lifting raw sewage and effluent,

Figure 11-4

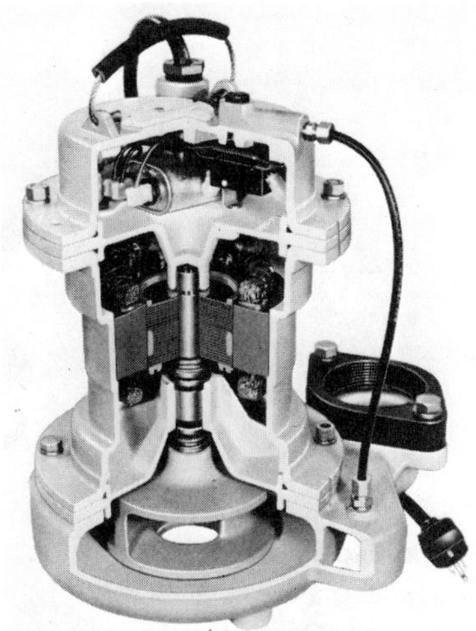

Submersible Centrifugal Pump (Kenco)

and light industrial applications. The pump is completely sealed and submersible. The motor chamber is filled with oil. Pumps of this type will pass semi-solids with diameters up to 1.5 in. and have lifting heads of 20 ft. Notice also that these pumps are equipped with a pressure-actuated diaphragm control switch, which is activated by pressure at the outlet. When the pump is submersed at heads of 8 to 20 ft (2.4 to 6.1 m), the pressure in the tube leading from the outlet activates the switch to start the motor. Because the tube is connected to the outlet, which is under pressure while the pump has a supply of water, the motor will continue to operate until it pumps down. Once the water at the outlet is gone, the diaphragm switch turns the motor off, and the pump remains at rest until it again becomes submersed to the required depth.

A typical maintenance free single-stage (one impeller) centrifugal pump used for booster and circulating pump applications is shown in Fig. 11–5. Many of the internal components of the pump are stainless steel. The multispeed motor runs sealed and submersed in oil. Having the ability to change motor speeds provides a means to eliminate velocity noise that is sometimes associated with a booster having a higher flow rating than the application requires. The alternate method is to add spacers to increase the size of the pump chamber, thus reducing efficiency, or to change or alter the impeller to reduce pump capacity. These measures require disassembly of the pump,

Figure 11-5

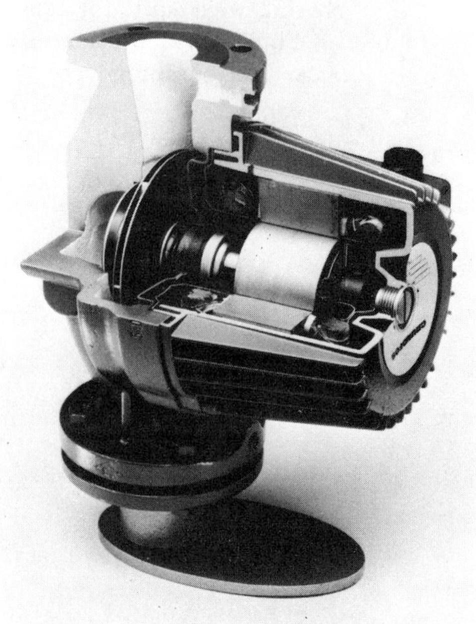

Centrifugal Booster/ Circulating Pump (Grundfos Pumps Corp.)

Propeller pumps are also nonpositive displacement pumps. Figure 11–6 illustrates the essential parts consisting of the low-pressure inlet, higher-pressure outlet, propeller, and drive. In operation fluid is swept along by the action of the close-fitting propeller in the pump housing. The fluid moves along the axis of the propeller drive shaft longitudinally. Propeller pumps are also staged to raise their pressure capability. They are used in underground water lines to boost pressures as an alternative to elevated storage tanks, and are sometimes called torpedo pumps. The propeller principle is also used in some water meters.

Mixed-flow pumps impart both a longitudinal and an axial thrust to the fluid. In single stages they will handle relatively large volumes at low pressures, but with staging, mixed-flow pumps can build the pressure to 400 lbf/in.^{2} (2759 kPa) for deep well pump and irrigation applications. The construction of the pump and impeller is shown in Fig. 11–7. Another advantage to mixed-flow pumps over the straight centrifugal pump is reduction in the diameter of the housing and an increase in overall efficiency. The impeller in mixed-flow pumps is sometimes called a Francis vane impeller after the inventor.[1]

The piston pump uses the action of a close-fitting piston in a cylinder barrel to first draw fluid through the inlet into the cylinder and then expel it from the outlet. There are several readily available examples of piston pumps, including the pitcher pump (excluded by most plumbing codes), the hand tire pump, and the reciprocating shallow well pump. The action of the pump requires it to have two check valves, one at the inlet and one at the outlet. The parts of a piston pump are shown in Fig. 11–8. On the intake stroke the piston increases the volume of the cylinder, and fluid enters through the inlet check valve. The outlet check valve is closed. When the piston returns on the pumping stroke, it reduces the volume in the cylinder and expels the fluid through the outlet check valve. During this time the inlet check valve is closed. Repeated cycling causes the pump to alternately intake and expel fluid. Because the parts are close-fitting, only relatively clean fluids can be pumped, and the valves, springs, and internal parts must be noncorrosive. The displacement (V) of piston pumps is computed from the volume expelled during each cycle. That is,

Figure 11-6

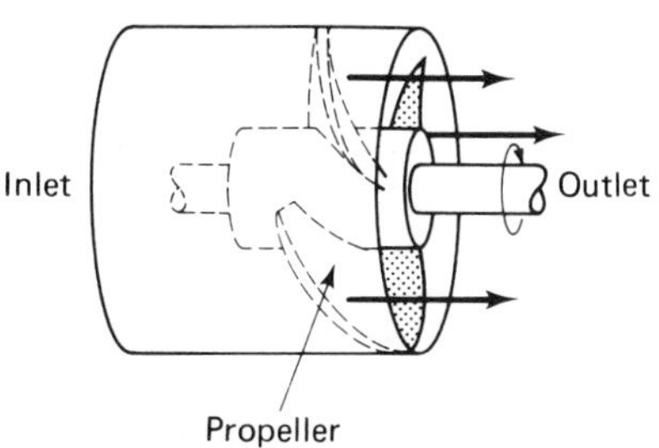

Propeller Pump

Figure 11-7

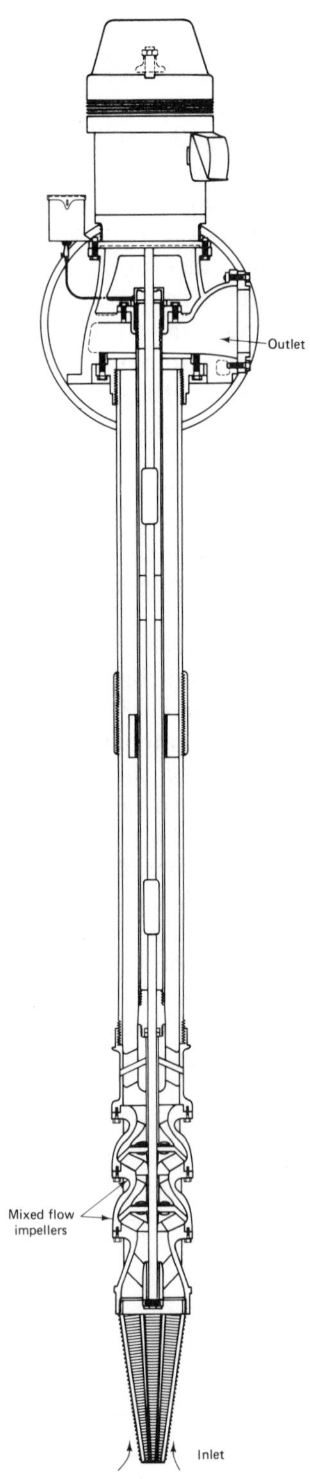

Mixed Flow Irrigation Pump

Figure 11-8

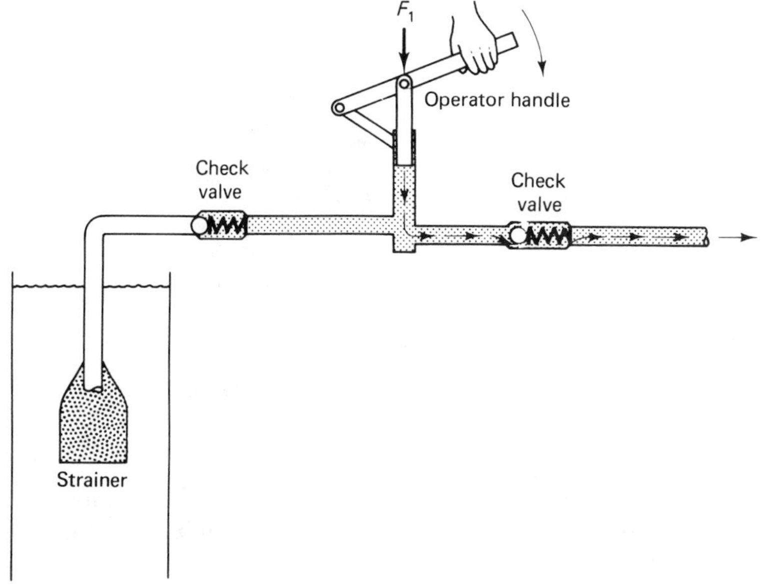

Parts of a Piston Pump

$$V = \frac{\pi D^2 l}{4} \tag{11-1}$$

where D is the diameter of the cylinder bore and l is the stroke. The delivery rate of the pump is given by the displacement V times the cycle rate N. That is,

$$Q = VN \tag{11-2}$$

Example A single-acting piston pump cycles at a rate of 60 cycles/minute. If the bore is 2 in. (5.08 cm), and stroke is 1.5 in. (3.81 cm), compute the displacement and discharge in gal/min. Note that 1 gallon = 231 in.^{3}.

Solution:
From Eq. (11.1) the displacement of the pump is

$$V = \frac{(3.14)(2 \text{ in.})^2(1.5 \text{ in.})}{4} = 4.71 \text{ in.}^3 \ (77 \text{ cm}^3)$$

And from Eq. (11-2), the delivery is

$$Q = (4.71 \text{ in.}^3)(60/\text{min.}) = 282.6 \text{ in.}^3/\text{min.} \ (4630 \text{ cm}^3/\text{min.})$$

Knowing that a gallon contains 231 in.^{3}, we obtain

$$Q = \frac{282.6 \text{ in.}^3/\text{min}}{231 \text{ in.}^3/\text{ gal}} = 1.22 \text{ gal/min} \ (4.63 \text{ l/min})$$

Figure 11-9

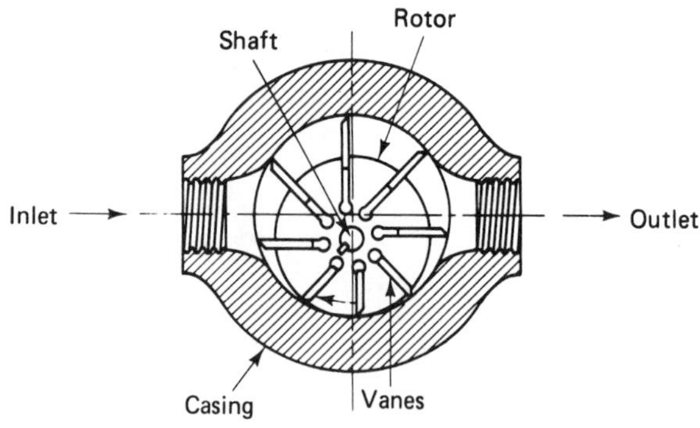

Parts of a Vane Pump

Vane pumps are positive displacement pumps. The essential components of vane pumps include the inlet and outlet ports, driven rotor, slide vanes, and stationary cam ring. The rotor, vanes, and cam ring, and sometimes the end wear plates, are replaceable as a cartridge. The components of a vane pump are shown in Fig. 11-9. Pumping action occurs when the sliding vanes in the rotor alternately increase and decrease the crescent-shaped space between the rotor and the cam ring. The increasing volume up to the halfway or *crossover point* causes suction at the inlet port and fluid to enter the low pressure inlet cavity. Notice that the inlet and outlet ports are isolated from each other by the spacing and seal of the vanes in the case. After the crossover point, the fluid volume is reduced in the outlet cavity at system pressure. As the rotor turns, the vanes in the rotor slots alternately extend and retract through each half-revolution, trapping fluid from the inlet port and transferring it to the outlet port. Vane pumps are used for high-pressure applications that require pressures of several hundred lbf/in.^{2}.

Gear pumps are positive fixed displacement rotary pumps. They are of both the external and internal gear type. They are simple in design and relatively inexpensive to manufacture and maintain. The pumping action in the external gear pump shown in Fig. 11-10 occurs when the input drive causes one gear to turn the other. As the teeth part on the suction side of the pump near the low-pressure inlet port, increases in the volume of the inlet chamber cause a slight vacuum. The rotating gear teeth transfer fluid drawn in and trapped in the gear teeth around the outside periphery of the gears to the high-pressure outlet chamber. Meshing of the teeth in the outlet chamber reduces the cavity by the amount equal to that displaced between the teeth as they mesh. This forces fluid from the outlet cavity and port at system pressure. Figure 11-10 shows a cutaway of an external gear pump.

Internal gear pumps operate similarly to external gear pumps. The internal spur gear drives the outside ring gear, which is set off center. Be-

Figure 11-10

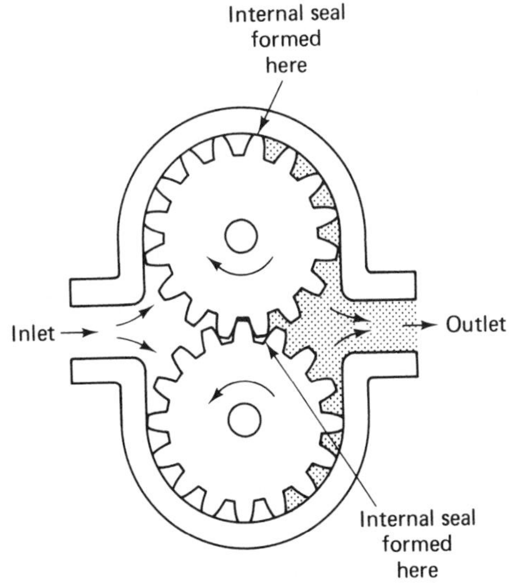

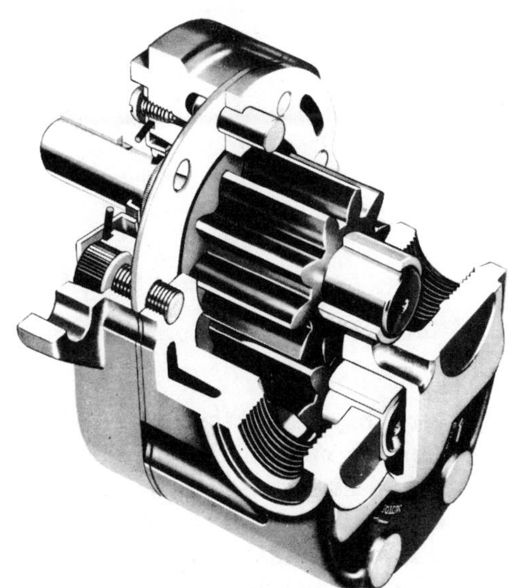

External Gear Pump

tween the two gears on one side is a crescent-shaped spacer around which the fluid is carried. The inlet and outlet ports are located in the end plates between where the teeth mesh and the ends of the crescent-shaped spacer. The pumping action shown in Fig. 11–11 shows fluid carried from the inlet to the outlet port in the following manner. The internal gear drives the ex-

Figure 11-11

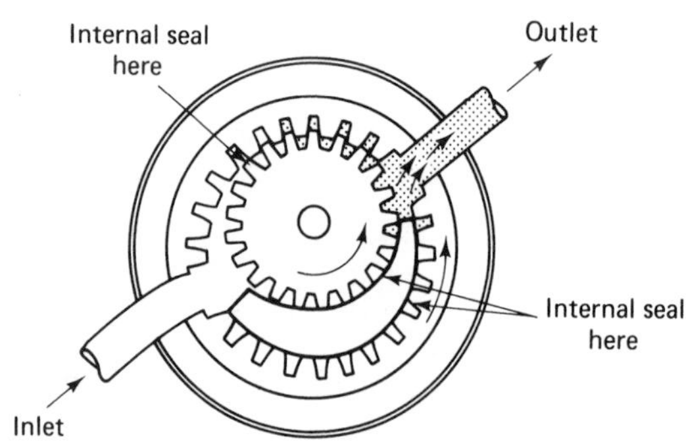

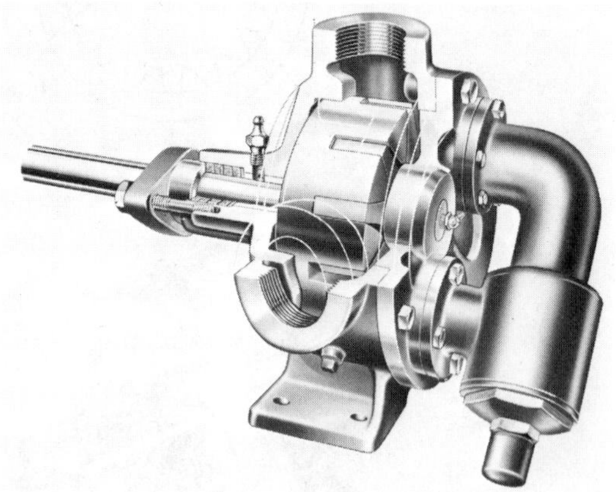

Internal Gear Pump

(Viking Pump Division, Houdaille Industries, Inc.)

ternal ring gear. They seal at the bottom where they mesh. Rotation causes the teeth to unmesh near the inlet port, increasing the cavity volume and creating a suction. Fluid entering the pump is trapped between both the internal and external gear teeth on both sides of the crescent-shaped spacer, and is carried from the inlet to the outlet cavity of the pump. Meshing of the gear teeth near the outlet reduces the volume, and the fluid is expelled from the pump at system pressure. A cutaway view of a general-purpose external gear pump is shown in Fig. 11-11(b). Because they have a positive displacement, gear pumps are self-priming at lifts up to several feet.

11-4 Directional Control Valves

Directional control valves are used to stop, start, check, and throttle the flow of fluids under pressure in pipes. In plumbing, the fluid is usually water, but in other trades liquids, gases, vapors, and even solids in suspension are also controlled by valves. The most common control valves used in plumbing are

- Gate valves
- Globe valves
- Ball valves
- Butterfly valves
- Check valves

Gate valves are so named because they resemble ancient flood gates, which are raised to open and lowered to close. The parts of a gate valve are identified in Fig. 11–12. Gate valves use a wedge-shaped disc, which may be either solid or split, to seal by downward force against the seats. That is, downward force by the valve stem wedges the disc between the two seats until it seals. The stem may rise when the valve is open, or remain stationary while the seat rises. The advantage of a rising stem is that just by looking it is obvious when the valve is open or closed. A nonrising stem, however, gives the valve a lower profile when the valve is in the open position, because the stem and hand wheel remain in the same position.

Gate valves are used in the full-open or full-closed position. Water service valves, for example, are gate valves (or plug valves). When they are full-open, they offer little resistance to flow and reduce pressure drop to a minimum. When they are left partially open, however, the throttling occurs mostly at the bottom of the disc and seats, subjecting them to erosion (called wire drawing). It is also common to hear chattering from vibration because the disc is not rigidly supported, and this can cause damage to the seating surfaces. Thus, gate valves should never be used for throttling or left in the partially open position.

Globe valves receive the name because they are generally globe-shaped. The parts and action of a globe valve are shown in Fig. 11–13. They are also designed to be used for regulating flow (throttling), and are specified when the valve must be opened or closed frequently. A globe valve causes a pressure drop because the flow must change direction. Balanced against this factor, however, are several advantages. First, flow takes place around the entire periphery of the disc, resulting in even wear. Then, too, the discs are replaceable and the seats can be reconditioned much more easily than in a gate valve. There is also a wide choice of metal and nonmetal discs available to accommodate a variety of applications, including hot and cold water, air, oil, gas, steam, and even volatile liquids.

Ball valves are generally used when a quick-acting, low-profile valve is necessary to fit into a confined space. A 90-deg turn of the valve handle quickly opens and closes the valve, and the direction of the handle indicates whether the valve is open or closed. The parts and action of a ball valve are shown in Fig. 11–14. When a standard ball valve is fully open, it offers little resistance to flow. If the ball has what is known as a full internal port, then it offers almost no resistance in the open position. Under noncritical conditions, ball valves may be used to regulate flow. A major advantage of the

Figure 11-12

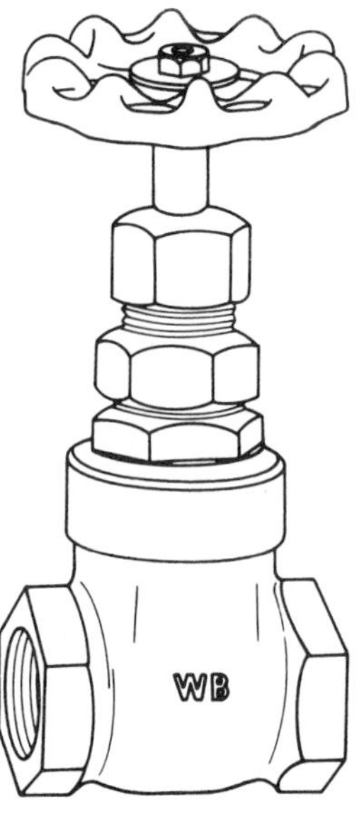

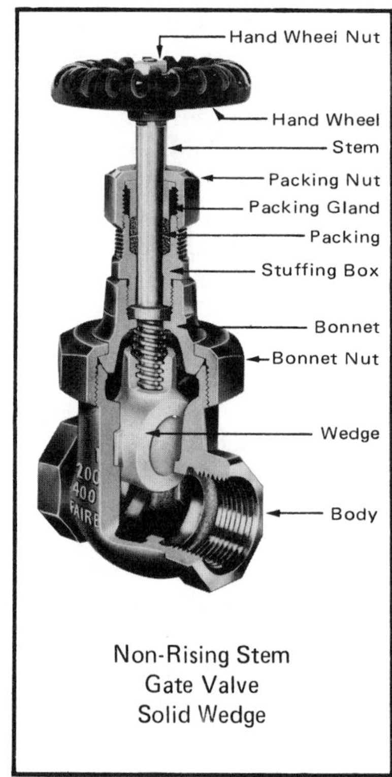

Non-Rising Stem
Gate Valve
Solid Wedge

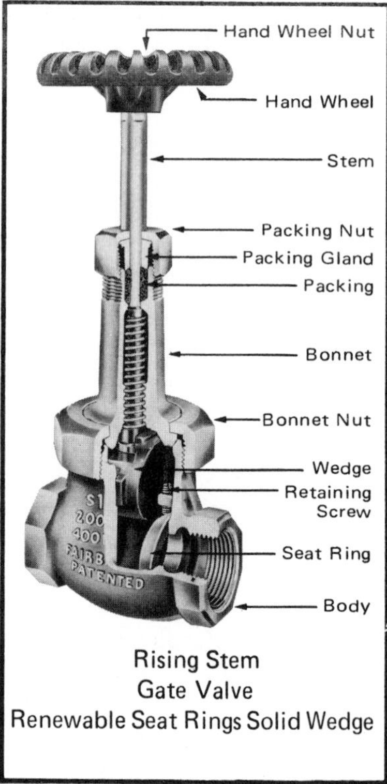

Rising Stem
Gate Valve
Renewable Seat Rings Solid Wedge

Gate Valve
(Wolverine Brass Works and Fairbanks Company)

Figure 11-13

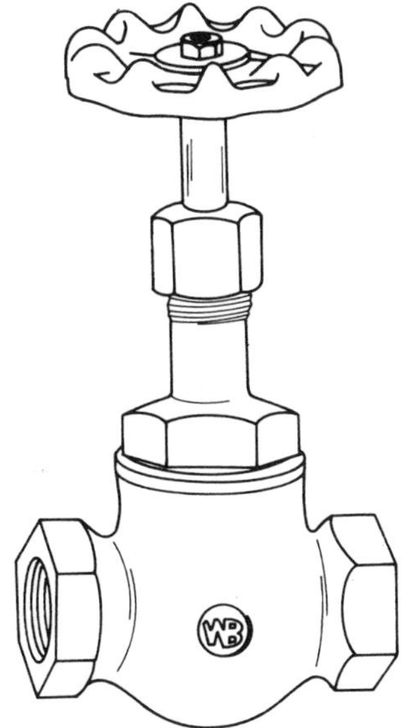

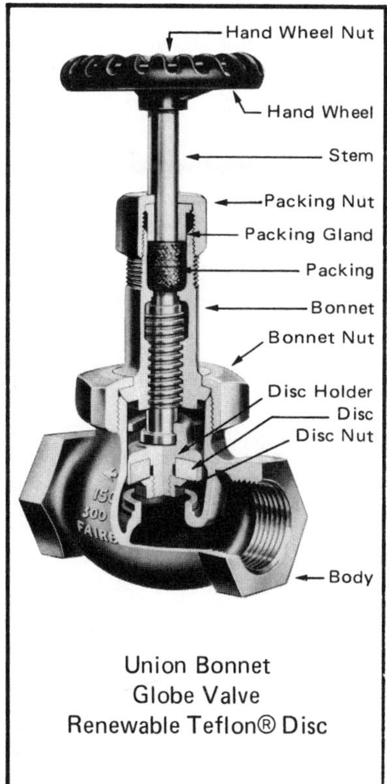

Union Bonnet
Globe Valve
Renewable Teflon® Disc

Globe Valve
(Wolverine Brass Works and Fairbanks Company)

Figure 11-14

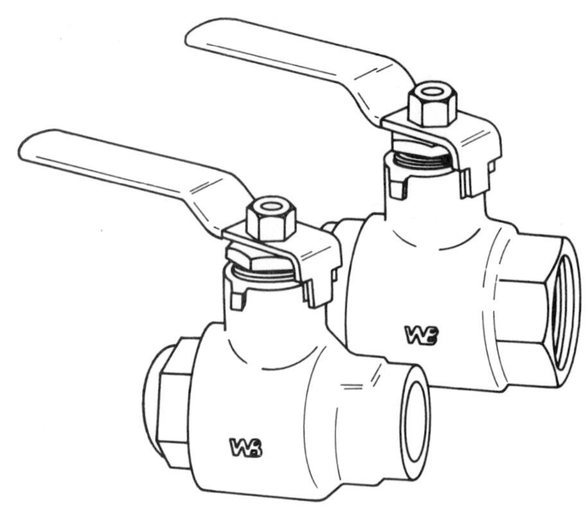

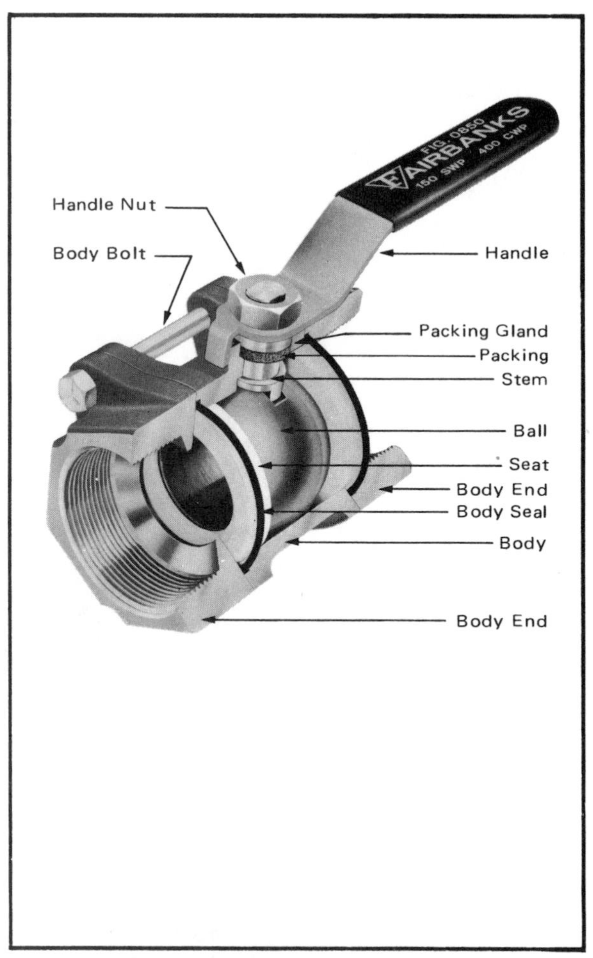

Ball Valve
(Wolverine Brass Works and Fairbanks Industries)

ball valve is the ease with which the design lends itself to repair. Ball valves, which were derived from their ancient counterparts, the plug valve, are illustrated in Fig. 11–15, and are used as meter and corporation stops.

Butterfly valves such as that shown in Fig. 11–16 are receiving increased use in the plumbing industry for several reasons. The design is mechanically simple and requires about 30 percent of the space needed for a similar size gate valve. They are also versatile, requiring only a 90-deg turn of the handle for flow in either direction with equal efficiency. They can be used full-open, full-closed, or for throttling applications. Finally, they are more economical because of material savings and lend themselves to automation.

Butterfly valves close by seating the streamlined alloy disc against a reinforced resilient seat, which is often made of E.P.T. Ethylene Propylene Terpolymer (designated EPTM), a synthetic for cold water service. The inset in Fig. 11–21 illustrates the reinforced elastomer seal, which also acts as its own flange gasket, isolating fluids from the valve body, shaft, and external parts of the valve. The liner is easily replaceable and restores the valve to like-new condition.

Figure 11-15

Two-Way

Three-Way

Plug Valves
(A.W. Cash Manufacturing Corp.)

Figure 11-16

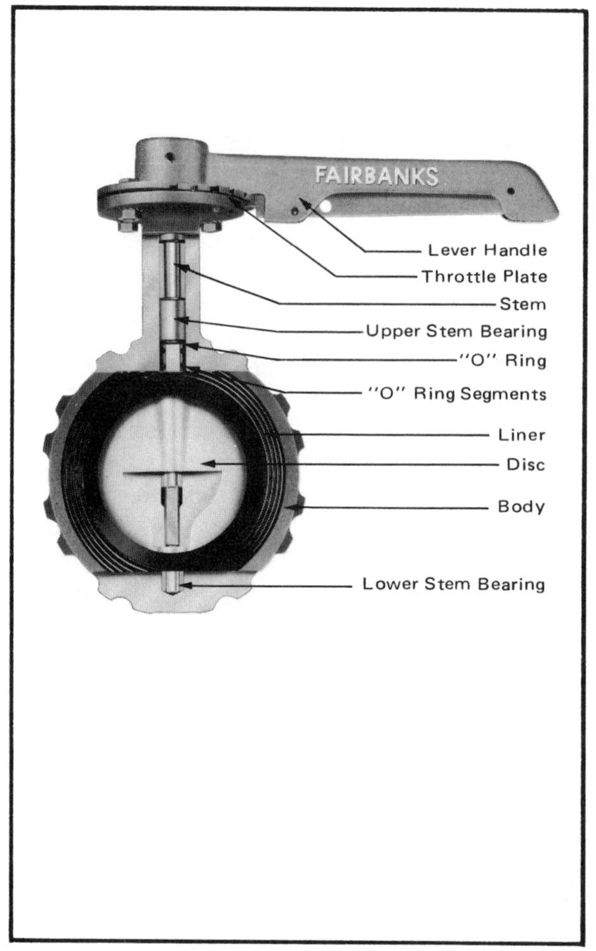

Butterfly Valves
(Fairbanks Company)

Check valves permit free flow in one direction and stop the flow in the opposite direction. Check valves are installed in plumbing systems to prevent reverse flow. Well pumping systems, for example, require them normally to prevent the pump from losing its prime. Piston-type water pumps require two check valves permitting flow in one direction, one on the inlet and the other at the outlet, for the pump to function properly. Check valves may be installed in the cold water inlet of water heating systems to prevent hot water from backing up in the cold water line. Other arrangements of check valves in water piping systems prevent water hammer, reverse flow, and give direction to the flow in mains and branch lines.

There are a number of mechanisms used to cause the valve to check, including swing discs mounted on a hinge pin, lift checks, which rise off a seat, and ball checks, which usually have a light spring to hold the ball against the seat. A typical swing check valve is shown in Fig. 11-17. Flow to the right causes the disc to swing free from the seat and give unobstructed flow. When the flow attempts to reverse, the disc swings against the seat to stop the flow. The cap (or bonnet) at the top serves to inspect and recondition the valve. The plug at the side gives access to the hinge pin.

Figure 11-17

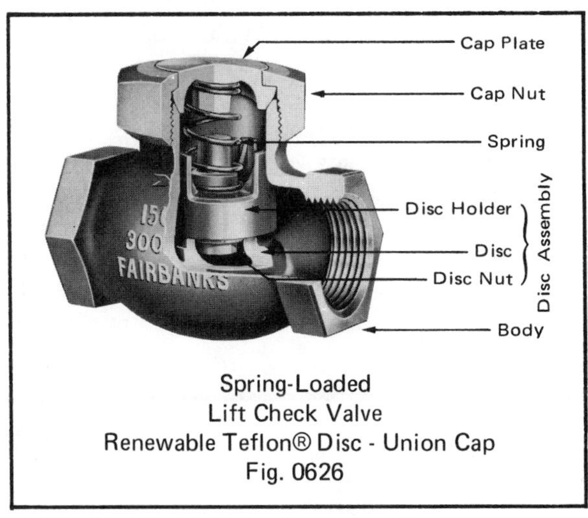

Spring-Loaded
Lift Check Valve
Renewable Teflon® Disc - Union Cap
Fig. 0626

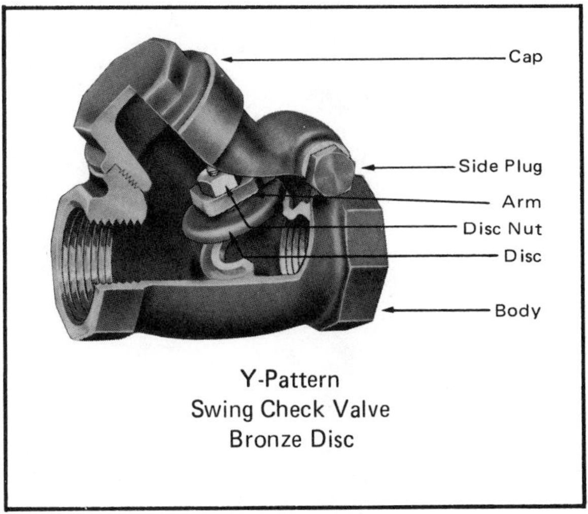

Y-Pattern
Swing Check Valve
Bronze Disc

Check Valve
(Fairbanks Company)

11-5 Pressure, Flow, and Temperature Control Valves

Pressure control valves release or limit the pressure downstream from the valve. Relief valves are the most common pressure control valves. They are located near pump outlets and function to provide the system plumbing and components with protection against pressure overloads. Figure 11–18 illustrates the operation of a direct-acting spring-type ball relief valve. Spring pressure acts to keep the ball element against the seat and the valve in the closed position. System pressure at the inlet port acts against the exposed (projected) area of the ball. When the force of the fluid (pressure × area) becomes greater than the opposing resistance offered by the spring, the ball is forced from its seat, the valve opens, and fluid is directed to an area of low pressure through the outlet port. The pressure at which the valve opens is called the *cracking pressure.* The pressure at which the rated flow passes through the valve is called the *full-flow pressure.* The pressure at which the valve ceases to pass fluid after being opened is termed the *closing pressure.* Adjustment within the pressure range of the valve is made with the adjustment screw, which acts to compress the check valve spring.

True pressure regulating valves are designed to automatically reduce higher inlet pressures to lower outlet pressures. Those designed for plumbing systems usually do not divert or "dump" the fluid, but rather lower the

Figure 11-18

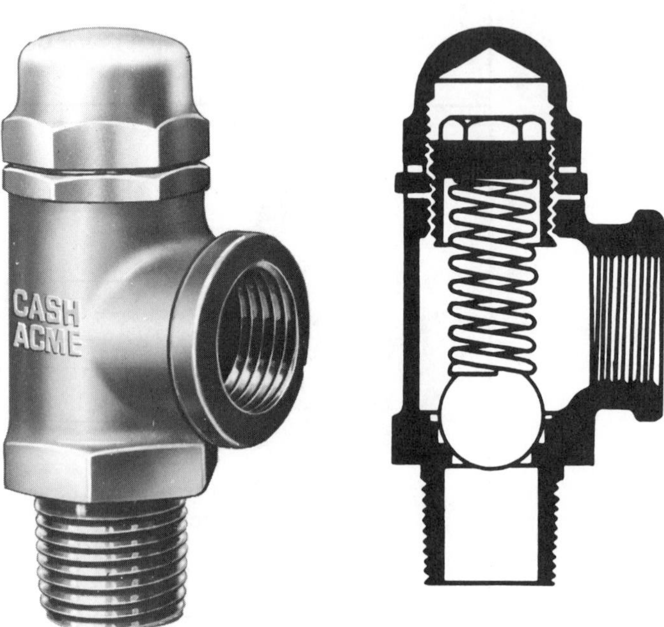

Direct Acting Pressure Relief Valve (A.W. Cash Manufacturing Corp.)

pressure at the outlet to acceptable limits. Many plumbing codes require pressure reducing valves whenever water pressure exceeds 70 to 85 lbf/in.^{2} (483 to 586 kPa). The pressure regulator shown in Fig. 11–19 is designed for residential or commercial applications, is available in sizes $\frac{1}{2}$ in. through 2 in., and has a maximum inlet pressure capability of 300 lbf/in.^{2} (2969 kPa). In operation, spring pressure against the diaphragm keeps the valve open until the pressure at the outlet becomes sufficiently great to act upward against the diaphragm and compress the spring. This causes the plug at the bottom to lift against the valve seat and the valve to close or throttle the flow sufficiently to reduce the pressure at the outlet. The pressure at the outlet is determined by the setting of the spring. Notice that both the valve disc and seat are replaceable.

A pressure regulator correctly installed in a water line will normally give years of satisfactory service. Such an installation is shown in Fig. 11–20. The piping should first be thoroughly flushed out to ensure removal of all foreign matter. It is best to anchor the valve and adjacent piping. Notice that a shutoff valve is installed ahead of the regulator, as is a strainer to protect it from dirt and silt. It is particularly important to keep the valve clean for proper regulation.

Regulators make a closed system, since water cannot flow back through them. The pressure in a closed system may rise due to thermal expansion of water from the water heater. Thus a relief valve is necessary on these hot water devices or at other places in the system where this is likely to occur to protect it from overpressure. Another option is to install a bypass check valve around the regulator, but if the inlet line into the regulator

Figure 11-19

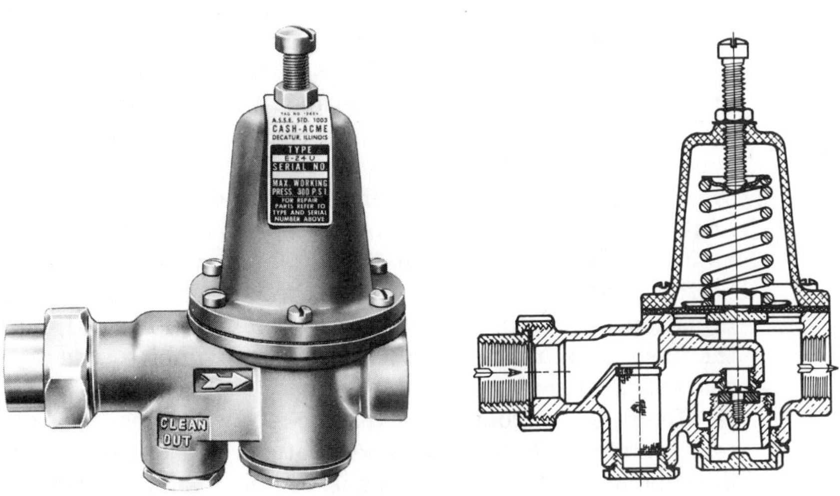

Pressure Regulating Valve
(A.W. Cash Manufacturing Corp.)

Figure 11-20

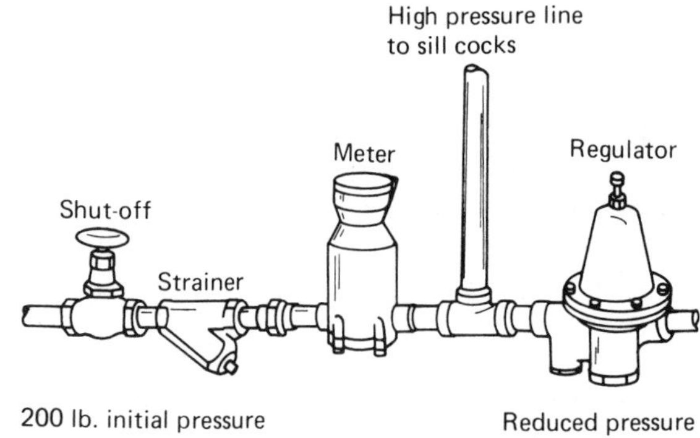

Pressure Regulator Installation

carries water at sufficient pressure to damage the system anyway, this is not acceptable. The best installation, then, will place the pressure relief valves in conspicuous locations, where they can be monitored and checked periodically to be sure that they operate properly. A permanently installed pressure gauge on the low-pressure side can serve as a quick check on system pressure.

Flow control valves limit the fluid volume flow rate rather than the system pressure. In plumbing applications their purpose is usually to conserve water, and at the fixture outlets they accomplish this by passing the water through a fixed or adjustable restriction. Whereas faucet flow control valves, usually called flow limiters, attach to the threaded end of the spout and limit the flow by passing the water through an orifice, some shower flow limiters have been designed so that the orifice is used to mix air with the water, still limiting the flow to approximately 3 gal/min (11.4 l/min). This has the effect of providing a satisfying shower without the reduction in water volume becoming noticeable.[2] Other devices to reduce water consumption have been applied to water closets. The adjustable pressure set type valves shut off the flow when the depth and volume of water in the tank reaches the established limit. The traditional method is to adjust the float rod to lower the fill level. A recent innovation incorporates two flush valves at different levels, one for a full volume flush of fecal matter at the bottom, and the second one at a higher level for a partial volume flush for urination waste. Another commonly publicized method, placing bricks in the flush tank, has been criticized by the plumbing industry as an ineffective means of saving water.[3]

Temperature control valves are of two types: those that limit the temperature, and those that regulate it. An example of a temperature limiting valve used in plumbing is the relief valve used on hot water tanks and explained in Chapter 10. Their purpose is to release and open when the water temperature reaches a predetermined limit. For water heaters, this is the boiling temperature at atmospheric pressure. Water temperature

Figure 11-21

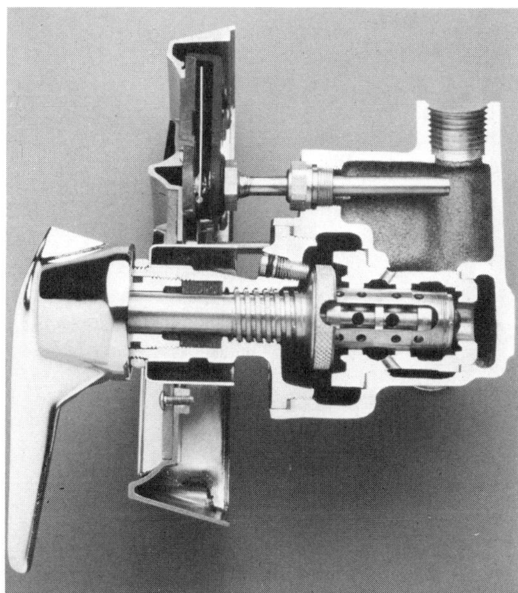

Pressure Balancing Temperature Control Valve (Symmons Industries Inc.)

regulating valves, on the other hand, regulate the temperature by tempering the hot water with cold water. The valve is sensitive to changes in the water flowing through it and adjusts internally to maintain the temperature setting.

Figure 11-21 illustrates enclosed and cutaway views of a single-handle pressure-balancing mixing valve with integral thermometer. In the cutaway view, the thermometer is shown inserted in the well near the water outlet at the top. This bi-metal sensor actuates the thermometer hand that reads the outlet water temperature. The valve is adjusted until the thermometer indicates that the desired temperature on the dial has been reached. Thereafter, the valve will maintain this temperature regardless of pressure fluctuations in the hot and cold water.

The action of the pressure compensation feature is shown in the four illustrations of Fig. 11-22. When the pressure of the hot and cold water is equal, as shown in Fig. 11-22(a), the compensating piston is centered, allowing equal inlet for the hot and cold water. When the hot water pressure is high, and the cold water pressure is low, as shown in Fig. 11-22(b), the piston moves toward the cold end, reducing openings at the hot end. This reduces the hot pressure to equal that of the cold, and the temperature remains at the original setting. When the hot water pressure is low and the cold water pressure is high, the opposite occurs [Fig. 11-22(c)]. That is, the compensating piston moves toward the hot end and reduces the pressure of the incoming cold water. Should either the cold or hot water

Figure 11-22

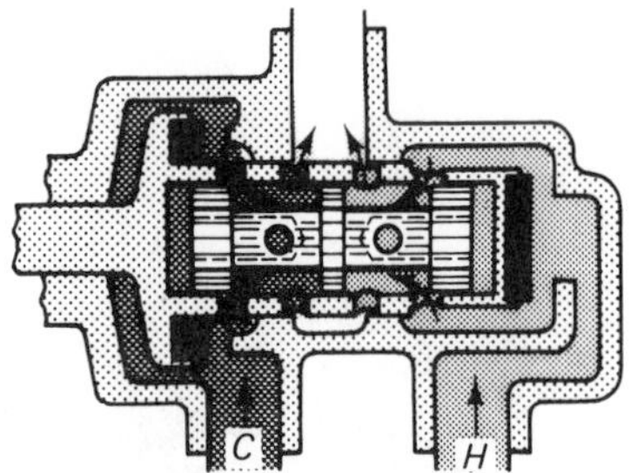

(a) Equal hot and cold pressure supply; piston is in center position allowing equal inlet openings for hot and cold.

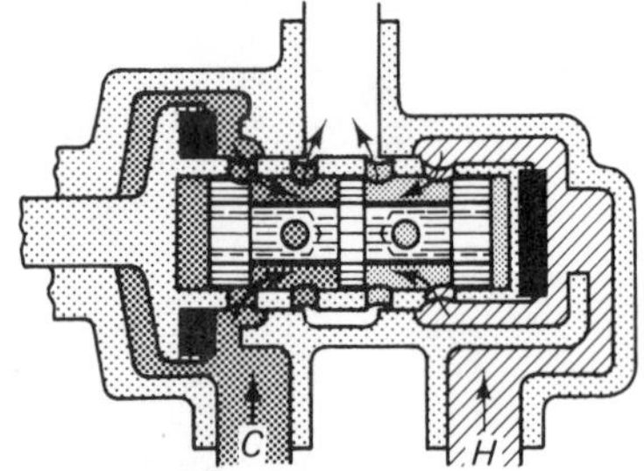

(c) Low hot pressure and high cold pressure supply; piston moves toward hot end and reduces openings at cold end, thus reducing cold pressure to equal hot. Temperature remains at original setting.

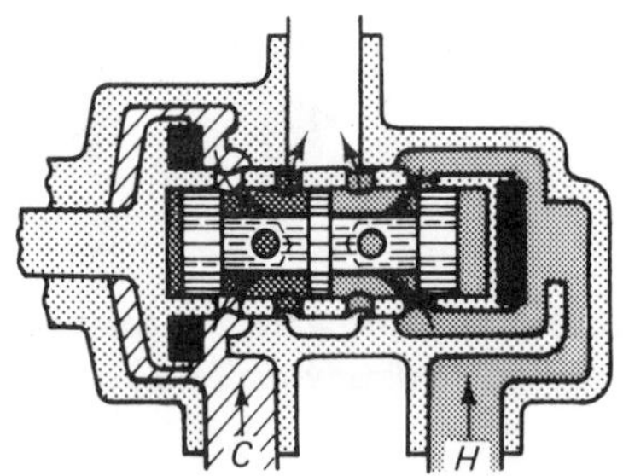

(b) High hot pressure and low cold pressure supply; piston moves toward cold end and reduces openings at hot end, thus reducing hot pressure to equal cold. Temperature remains at original setting.

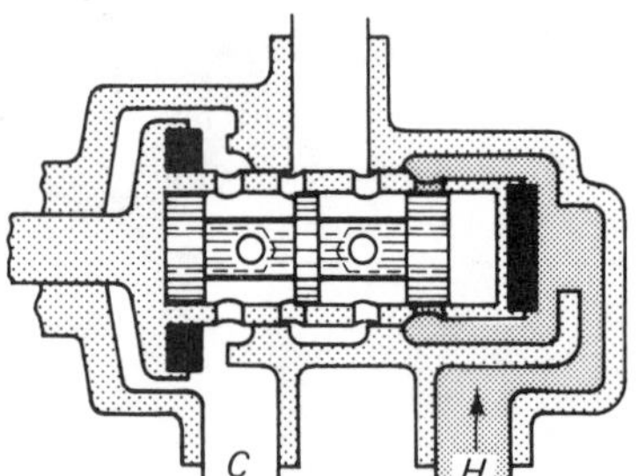

(d) Cold water supply failure; piston moves to extreme position at cold end closing hot openings completely, thereby reducing hot pressure in valve to zero; result—no water flow through valve. In the event of hot water supply failure the opposite sequence occurs.

Pressure Compensation Temperature Control Valve Operation

supply fail, as shown in Fig. 11-22(d), the pressure compensating piston moves to the extreme position, shutting off the other supply. That is, if the cold water supply pressure fails, the compensating piston shuts the hot water off. The opposite will occur if the hot water supply pressure fails.

11-6 Plumbing Valve Applications

There are a number of applications in plumbing that make use of valves in other than the traditional sense where the valve simply looks like a gate, globe, or check valve with a handle on it. However, they are valves just the same, since they stop, start, control, or divert the flow of water or other fluids. For example, the ballcock valve and flush valve[4] in the water closet

Figure 11-23

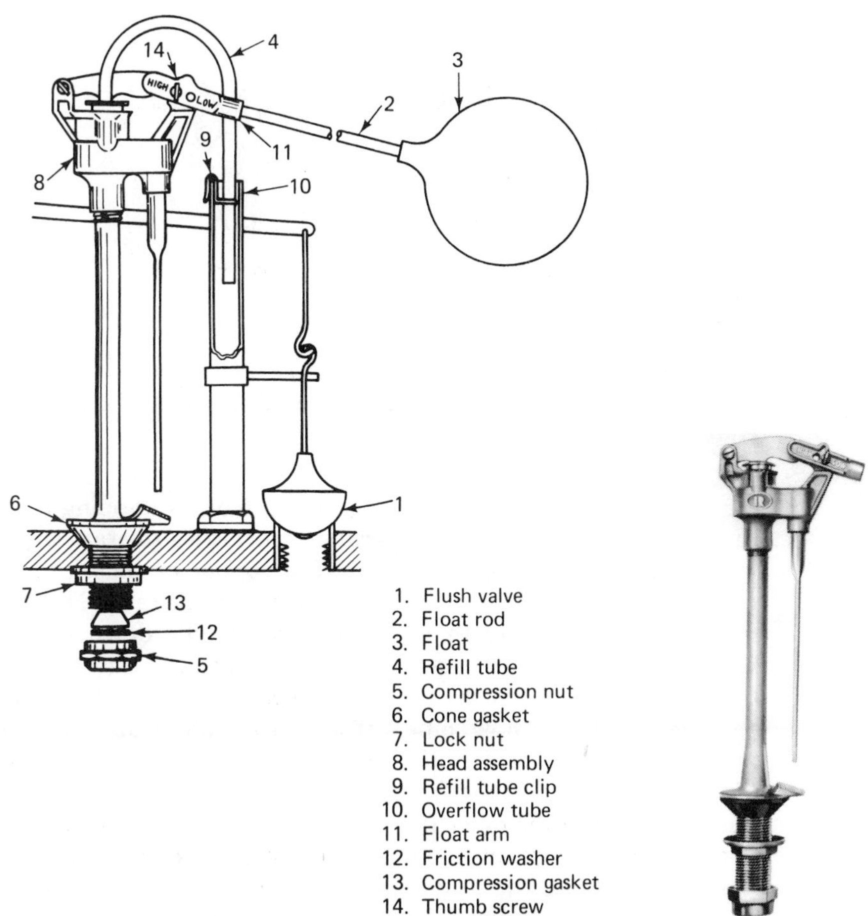

Ballcock and Float Valve Assemblies (Rockford-Eclipse Division of Eclipse, Inc.)

tank are control valves, as are the faucet and mixing valves for the lavatory and shower, and the sink and tub drain valves. They are unusually simple devices, and that is why they work so well, but because they have mechanical parts, adjustment, maintenance, and repair are still required occasionally.

The ballcock valve and flush valve shown in Fig. 11-23 are two of the oldest devices used for inside plumbing. The ballcock valve shuts the water off when the tank is filled by the upward action of the float, which forces the plunger assembly down onto the valve seat. The level of the water in the tank and thus the amount of the flush is determined by the float setting, which is adjusted by bending the float rod. The flush valve,

Figure 11-24

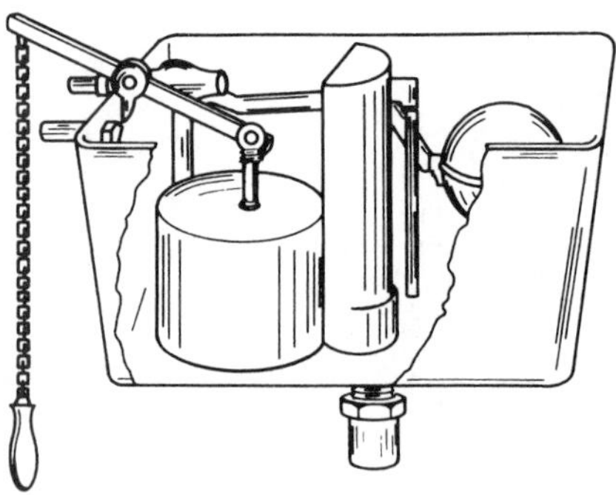

Crapper Flush Valve Circa 1872

sometimes called the flapper valve because of its flapping action, initiates the flush when the handle is operated, and closes to start the filling action when the tank has been emptied. The rubber or plastic hollow ball-like valve floats on top of the water as it recedes in the tank. This eliminates the requirement of having to hold the flush handle down during the flush. When the tank empties, the flush ball settles on the flush valve seat to stop the flow of water from the tank. Meanwhile, water continues to enter both the tank and overflow pipe through the refill tube. This provides an afterflush action which both aides the transport of solid waste in the soil pipe and refills the bowl trap, which was siphoned by the action of the flush. The refill tube must flow into the overflow tube above the water line to prevent siphonage of the tank. The parts of the ballcock valve and its action are illustrated in Fig. 11-25. Ballcock valves and flush valves are repairable. A number of replacement types, including the diaphragm and float-cup types, are available as alternatives to repairing those that are badly worn and obsolete.

Some installations, particularly commercial ones, use flushometer valves instead of flush tanks, the advantage being the elimination of the waiting period while the tank fills between flushing and reduction of the water required per flush by a gallon or more. They do, however, require a higher flow rate with a supply pipe size of $\frac{3}{4}$ in. to $1\frac{1}{2}$ in.

Flushometers discharge a predetermined quantity of water to fixtures and are actuated by direct water pressure. The two main types are the diaphragm type, easily recognized by the large top cover; and the piston type, which is narrower at the top. Both of these are shown in Fig. 11-26.

The check valve shown to the right in both valves prevents backflow and is necessary to the controlled operation of the valve. Should the water pressure drop in the system, it also prevents all the other flushometers in the building from operating automatically when the water service is restored. The cross section of the check valve is also shown in Fig. 11-26.

The diaphragm-type flushometer uses the diaphragm (1) to divide the

Figure 11-25

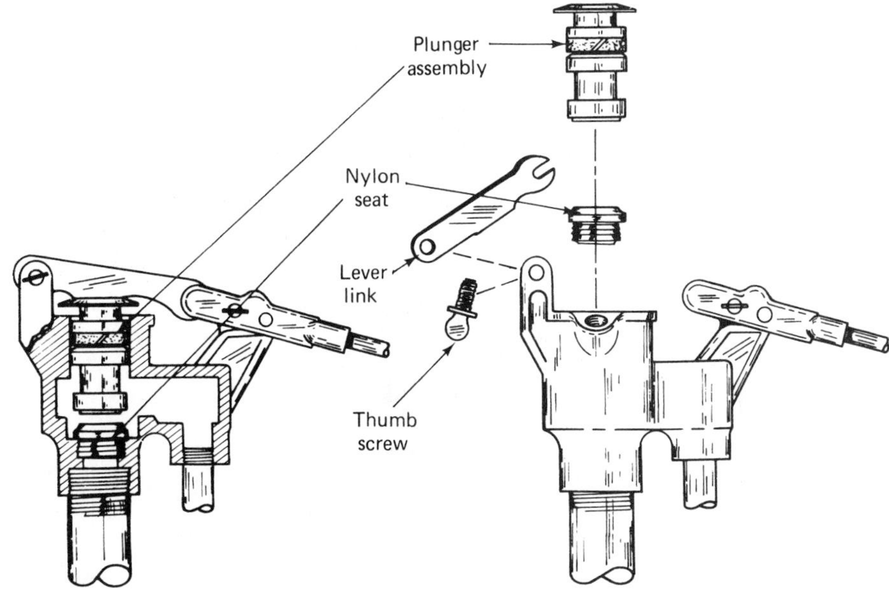

Ballcock Valve Parts

valve into an upper and lower chamber with equal water pressures in both chambers when the valve is in the closed position. The greater pressure area on the top of the piston (2) holds the valve closed on the main seat (3). A slight rocking movement of the handle (4) pushes the plunger *in* (5), which tilts relief valve (6) and allows water to escape from the upper chamber. The water pressure in the lower chamber below the cap (1) now becomes greater, and this raises the piston (2) from the main seat (3), allowing water to flow down through the valve outlet (7) to flush the fixture. While the valve is operating, a small amount of water flows upward through the by-pass (8) and gradually fills the upper chamber. This causes the piston (2) to return to its seat (3) and close the valve. Notice from the figure that inward movement of the plunger will tilt the relief valve, but that because it lifts when the flush begins that holding the handle will not continue the flush indefinitely. Rather, when the diaphragm and valve return to the closed position, the tilting mechanism will telescope, leaving the valve stem sitting on top rather than at the end of the plunger. Adjustment of flow from the diaphragm-type flushometer is accomplished by adjusting the check stop valve adjustment screw out one or two turns so that the toilet or urinal flushes quickly but does not flood, and refills the bowl to the normal level.

Piston-type flushometers operate similarly to the diaphragm-type, but their adjustment is different. On these valves, remove the cap that covers the adjustment at the top of the piston valve assembly and turn the screw first full-in, and then back it out one or two turns. The farther it is turned out, the longer the flush.

Figure 11-26

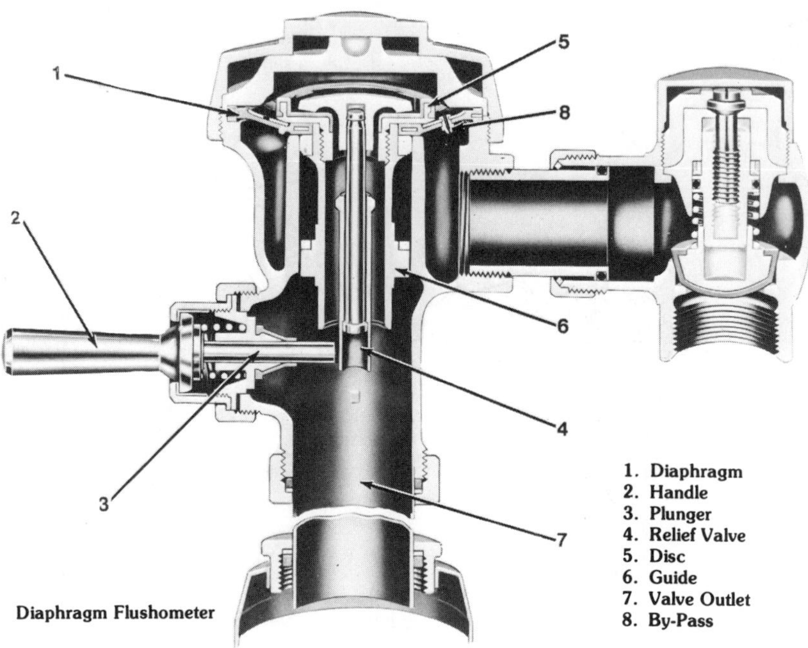

Diaphragm Flushometer

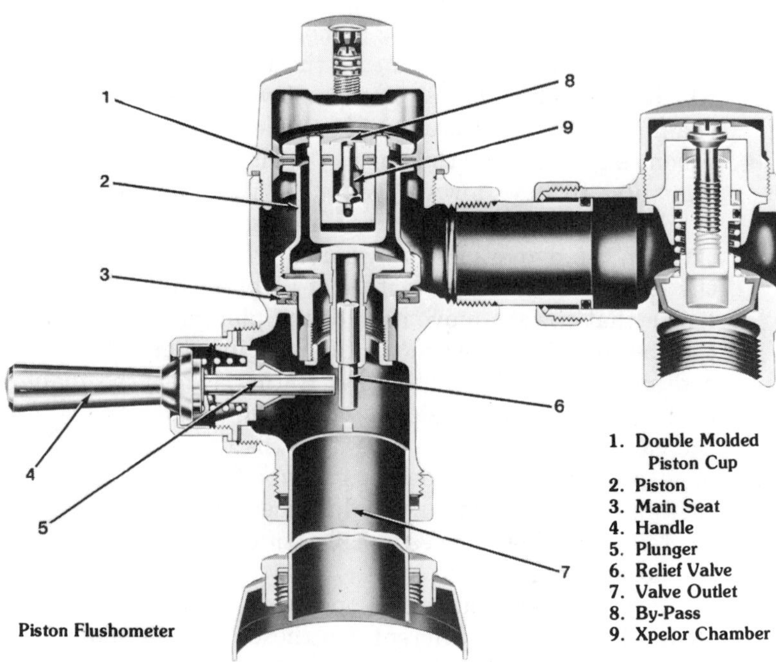

Piston Flushometer

Flushometers (Sloan Valve Company)

Figure 11-27

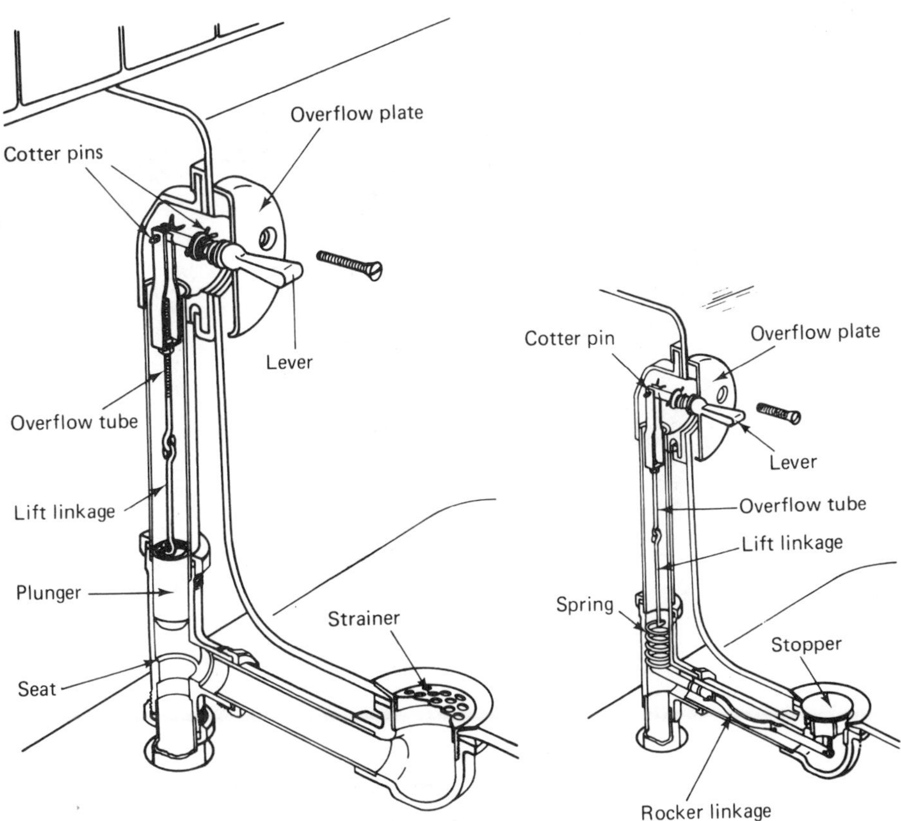

Plug and Pop-Up Bathtub Drain Valves

The two most common types of valves that are used to plug bathtub drains are shown in Fig. 11-27. The plug type lifts or falls by the action of the lever. Notice that the drain is plugged at the base of the overflow pipe. If adjustments are required, the valve is removed through the overflow plate by simply removing the plate and pulling the flexible linkage out through the pipe. The adjustment is made by turning the brass yoke on the threaded rod. The pop-up type lifts the stopper by an upward movement of the lever. This pushes down on the linkage and spring, which causes the rocker linkage to act like a rocking horse and lift the stopper. Notice that there is no screen in the drain and consequently hair and other material may sometimes cause the stopper plug seal to fail. The stopper assembly may be removed by first removing the overflow plate and lifting the linkage assembly, and then by simply working the stopper with the rocker linkage out through the drain hole. Typically, the stopper assembly can be restored to service by cleaning it and replacing the gasket on the stopper. For reassembly, the stopper and rocker linkage are installed first, followed by the lifting linkage through the overflow plate.

Figure 11-28

Hot and Cold Water Valves (Gerber Plumbing Fixtures Corp.)

A hot and cold water faucet set with shower diverter is shown in Fig. 11-28. These are modified globe valves with composition disc washers to resist the effects of hot water. The diverter valve at center is used to select either the shower or tub spout as the direction for the water flow. The valve stems, packing glands, bonnets, washers, and seats are removable for repair and replacement.

Figure 11-29 illustrates a typical supply stop used for the water closet (angle stop), lavatory, and kitchen sink. They are a modified globe valve with replaceable stem packing and compression washer disc.

11-7 Water Meters and Gauges

Meters and gauges are used to monitor the flow rate and volume, line pressure, and temperature of water in the system. The water district or corporation sells water based upon usage in gallons (or liters). In large systems (2 in. to 12 in.) it is also common to measure the volume flow in main lines in such units as cubic ft, acre ft, cubic meters, miners' inch hours, barrels, and other standard units. Pressure gauges read line pressure at several

Figure 11-29

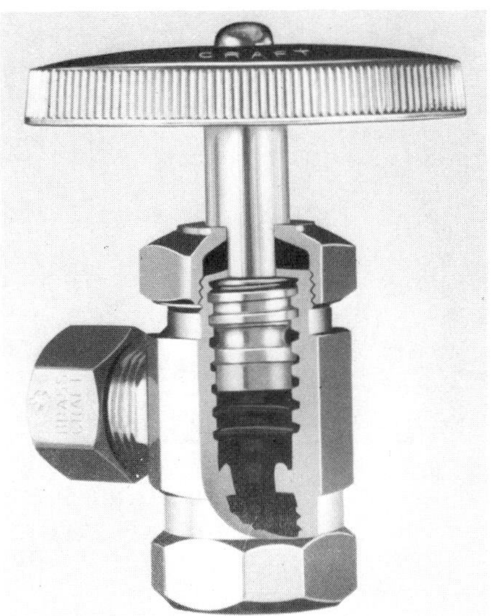

Supply Stop (Brass Craft)

places in the system. They tell whether the pressure is too high to be safely contained by the pipes and equipment, too low to satisfy supply demands, or within the normal operating range. Fluctuations in pressure can also indicate the condition of pumping machinery, surging, and the effects of intermittent demand. Common pressure scales include the lbf/in.^{2} (sometimes written psi), Pascal (Pa), and other units peculiar to individual countries and applications, such as the kg/cm^2. Temperature gauges monitor hot water systems, including hot water supply heaters, boilers for heating domestic water and hydronic heating, and of course solar hot water systems, where their permanent installation to determine system operation and efficiency is a necessity. Temperature gauges are scaled in degrees Fahrenheit (°F) and degrees centigrade (°C).

Water meters are installed in a meter box, as shown in Fig. 11–30. The fixture that holds the meter is called a *linesetter;* its purpose is to provide a rigid mounting that is in proper alignment for the meter when it is installed. The brace bar across the bottom holds the ends of the linesetter in place. Water does not flow through it (inset). Even when service pipes are shallow, raising the meter above the service line level has the advantage of keeping it out of the dirt and mud, making it easier to read and extending its life.

Water meters are available in standard sizes of $\frac{5}{8}$ in., $\frac{3}{4}$ in., 1 in., $1\frac{1}{2}$ in., 2 in., 3 in., 4 in., and 6 in. with flow capacities from 1 gal/min to 1000 gal/min. Basically there are two types, positive displacement and turbine type.

Figure 11-30

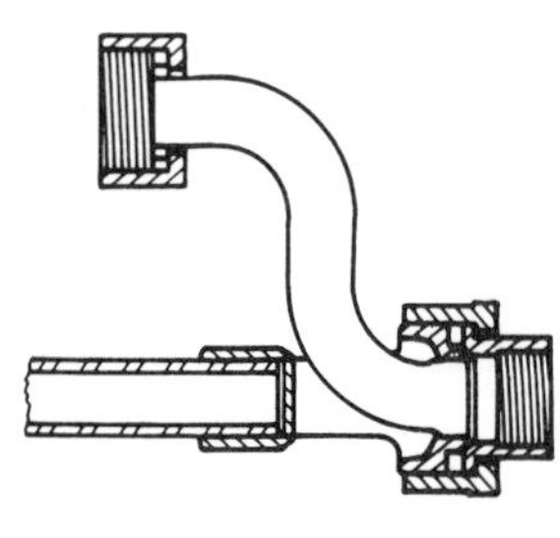

Water Meter Installation (Ford Meter Box Company)

Positive displacement water meters measure the flow by using a nutating (oscillating while it rotates) disc or rotating piston that registers the amount of water displaced as it flows through the meter. The driving mechanism from the rotating disc or piston can be either through a direct-connected shaft or through a magnetic coupling. The magnetic coupling allows the register, which is the portion of the meter containing the reading mechanism, to be hermetically sealed, which assures a longer service life. The two types of positive displacement meters are shown in cross section in Fig. 11–31. The cases have a replaceable bottom plate to protect the meter movement should it freeze.

The action of the nutating disc meter is caused by water passing through the chamber where the disc is located, which causes it to nutate as it rotates about its central spherical bearing. It resembles the spiral action of a wobbling top turning on its end. Each rotation of the disc is recorded by the register, which is geared so that this translates to the appropriate measure, gallons, cubic feet, liters, etc. The register is read directly like the odometer on an automobile, with a larger rotating hand on the register dial indicating that water is flowing through the meter. Two direct-reading registers are shown in Fig. 11–32, indicating units in gallons × 1000 and cubic feet × 100 for each rotation of the hand. The register accumulates starting from the right, with each successive digit to the left being ten times greater than the one preceding it.

In the rotating piston-type water meter there are two primary elements, the hermetically sealed register and the measuring chamber with the piston. Permanent magnets transfer the movement of the piston to the register. This magnetic coupling operates between the driver magnet in the piston and a follower roller magnet sealed inside the cylindrical portion of the register case which extends into the measuring chamber. The motion of the roller magnet turns a crank attached to the first pinion of the register

Figure 11-31

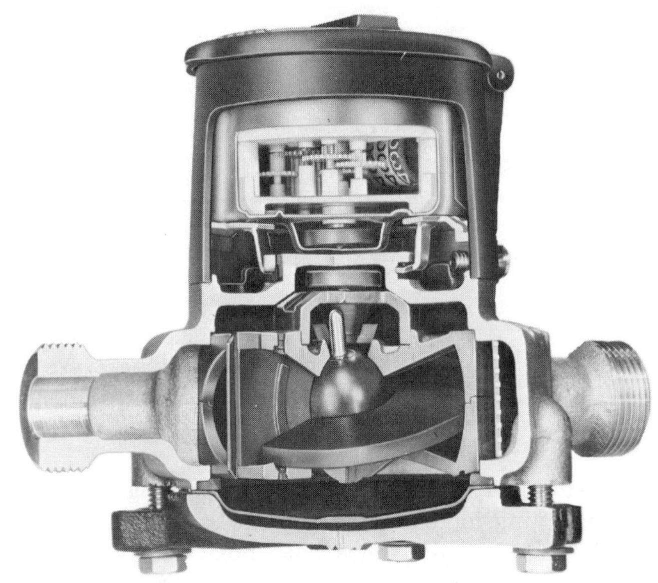

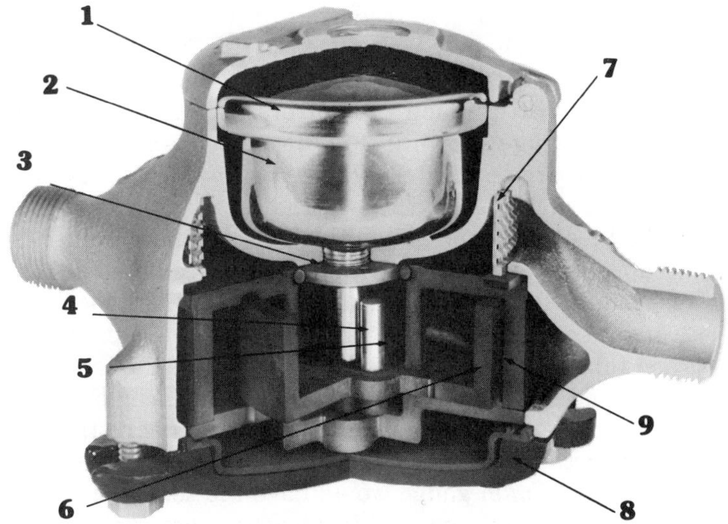

Rotating Piston Meter (Rockwell International)

1. Register Glass
2. Sealed Register
3. Register Retainer
4. Magnetic Drive
5. Magnetic Shield
6. Oscillating Piston and Piston Roller
7. Cylindrical Strainer
8. Interchangeable Bottom Plate
9. Cast Bronze Case

Positive Displacement Water Meters

Figure 11-32

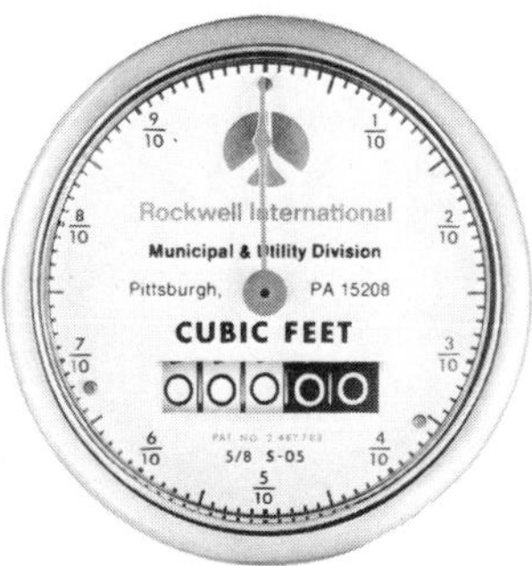

Direct Reading Registers (Rockwell International)

gear reduction. The instrument-type gearing requires only a small portion of the driving force available in the magnetic coupling.

Pressure gauges sense both positive and negative pressures from a datum or standard pressure which is taken to atmospheric pressure (14.7 lbf/in.^{2} or 101 kPa). If the pressure is above atmospheric pressure, the gauge reads positive (for example, at a pump outlet), and at less than atmospheric pressure it reads a vacuum (for example, at a pump inlet). At atmospheric pressure it reads zero. Pressure gauges are calibrated to read most accurately in the center of their scale. A 100-lbf/in.^{2} gauge would thus be the most accurate at a half-scale reading of 50 lbf/in.^{2}.

The most common type of pressure gauge for plumbing work is the bourdon gauge shown in Fig. 6–6. Its operation is quite simple, and the readings are accurate. When positive pressures are applied through the hole at the threaded end, the air within the shaded circular bourdon tube becomes compressed. The tube is hollow and its cross section is egg-shaped. Compressed air within the tube causes it to straighten a bit. That is, it uncurls, and pulls on the connecting link that drives the hand pointer through a simple gear mechanism. When the pressure is released, the bourdon tube returns to its original position, and a light coil spring returns the hand and driving mechanism, keeping the lash (clearance) out of the gears to keep the gear train snug. Some gauges have the case filled with oil to dampen the movement. Negative pressures have the opposite effect. A pressure/vacuum gauge would have the zero part way up the scale so the hand could move and read in a negative pressure (vacuum) direction.

While pressure gauges are most often seen in commercial applications, for example, in testing backflow preventers, there are a number of applications to all systems that warrant their use, particularly since they are inexpensive. Low-pressure readings at the service main can account for improper operation of fixtures, heavy water usage outside the building, and leaks or breaks in the water line. Remember that the customer typically pays for water usage on the building side of the meter should a leak occur. High pressures can give the owner an indication of whether or not a pressure reducing valve would be desirable at the meter to protect pipes and fixtures within the building. A maximum reading hand on the face will also indicate if this has occurred since the last reading was taken.

Temperature gauges are used to monitor the water temperatures of domestic hot water and heating systems. They may be used periodically to calibrate the water temperature thermostat in such applications as water heater storage tanks, or may be installed permanently in tanks and lines to give a constant readout. Because they are inexpensive and there is increased consciousness toward energy conservation, more permanent installations will be seen. They are a must for solar plumbing systems to determine the temperature and efficiency of solar panels through which water and other fluids circulate.

There are a number of accurate temperature gauges. Those that connect directly into the system through a fitting or tank connection are usually of the bimetal type. Mercury gauges are also inexpensive and accurate, but precaution is necessary to ensure that corrosive action does not deteriorate the sensing end and allow contamination of the system with mercury. Remote gauges use extended pressure sensing bulbs, which are attached to the gauge mechanism by the pressurized tube filled with a liquid which expands in response to a temperature increase. Remote gauges connected by electrical wires consist of a resistance sensing element and a gauge connected by a wire. They generally require low voltage power.

Flow sights and flow meters are used to determine if fluid is flowing in the system and how much is displaced or transported from one place to another. For example, if a circulating pump moves water through a conventional hot water system, or if it delivers water or a transfer fluid through solar panels on the roof, it is necessary to know the flow rate to determine if the pump and system are working properly. The pressure shown at the pump outlet may be within the limits specified, but restrictions in the piping system can still reduce the flow below that required for it to operate satisfactorily. Only with a flow meter can the discharge of the pump and flow rate circulating through the system under pressure be determined accurately. Where flow sights will give an indication that there is flow in a system by either direct observation of the fluid moving through the gauge or by observing a displacement spinner in the sight glass, flow meters give an indication of the actual amount flowing in gal/min or liters/min. Flow meters of the direct reading type typically measure the flow as a pressure drop across the orifice. Electrically driven flow meters measure the change in resistance caused by a wafer element that is suspended in the flowing stream. These are slightly more accurate than the direct reading type and offer less resistance to flow, but for plumbing applications their additional cost is prohibitive. Figure 11-33 illustrates several flow sights and flow meters.

Figure 11-33

Direct Reading Flow Meter
(RCM Industries)

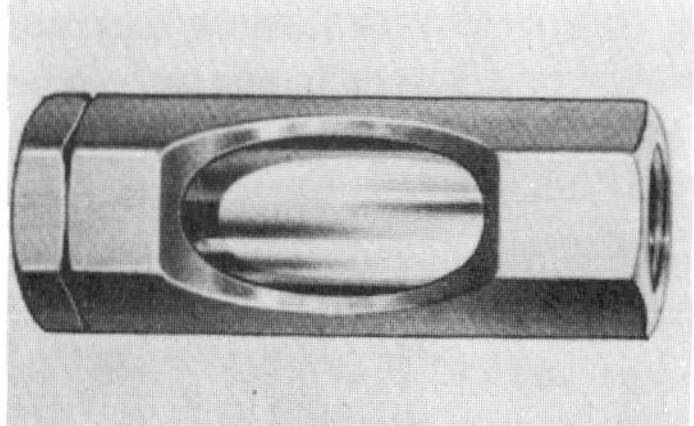

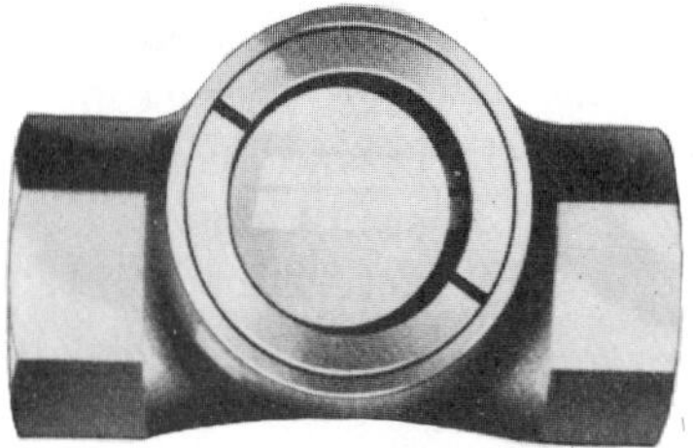

Flow Sights
(Lube Devices, Inc. (top)
and W. E. Anderson, Inc.)

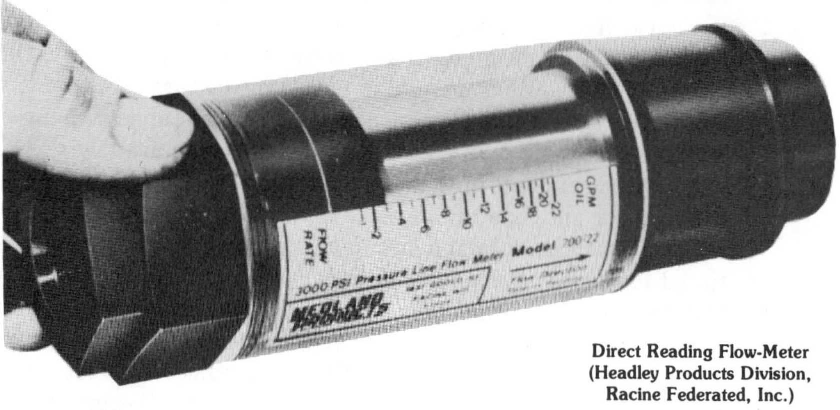

Direct Reading Flow-Meter
(Headley Products Division,
Racine Federated, Inc.)

Flow Sights and Flow Meters

11-8 Summary of Practice

The plumber must be familiar with many related components used in plumbing systems, in addition to piping systems and installing fixtures. In fact, most failures other than stoppage in the waste and drainage system do not occur in the pipes themselves, but with such components as pumps, valves, and control devices. For example, the dripping faucets and leaking ballcock and flush valves can account for more wasted water and expense than any other source, since it is estimated that nearly half the water used is flushed through urinals and water closets.

Pumps have always been important to the plumbing market. They are taking on increased importance in the domestic market now that they are being used for recirculating solar-heated water and transfer fluids. Swimming pools and aeration septic systems, of course, have always been equipped with recirculating pumps. Well and sump pumps are also used extensively. Installing and repairing pumps at the job site and in the shop have the potential for improving the services available from the plumber and increasing product sales and profits as well. Nearly every plumbing shop markets and repairs one or more pump lines.

Valves control the flow of water and other fluids under pressure. The drain system has no valves, except the back check system provided by fixture traps. Shutoff and fixture valves are the more visible ones that are installed and serviced, but others, such as pressure and relief valves, which are less conspicuous, safeguard the system and prevent personal injury. Their workings and repair must also be understood so that they can be checked for proper operation and renewed for continued service. The pressure and temperature relief valve installed in the water heater tank is an important example. The pressure regulator sometimes installed in the meter box to reduce line pressures to the building is another. Installing a valve backwards, for example a pressure regulator or check valve, defeats its purpose and can damage the system and threaten the safety of building occupants.

Traditionally, plumbing has been a craft, but extensive application, monitoring, and testing procedures qualify many of the skills as technical. Calculating system pressures, temperatures, and flow rates, for example, and then installing gauges to monitor and check these require understandings that usually have not been associated with the plumber's collection of skills. Current practice, however, requires these competencies if one is to be competitive in the field. For example, the plumber who installs a circulation system, according to specifications, that does not function correctly or deliver the required flow rate then must be able to determine the cause of failure from readings taken from electrical, pressure, temperature, and flow-rate gauges to solve the problem. Storing test meters and gauges safely in a special tool box is the best way to have them available when they are needed. Many plumbing systems can be improved by permanently installing pressure gauges, temperature gauges, and flow sights at places where this information will indicate if the system and components are operating properly. Line pressures, pump pressures, water temperatures, and circulation or flow rate indications can be monitored with permanent instrumentation without imposing a significant increase in system cost, particularly when the benefits of improved operation and speed of repair are considered.

REVIEW QUESTIONS AND PROBLEMS

1. What is the difference between a positive displacement pump and a non-positive displacement pump?

2. How does a centrifugal pump operate? Why must it be primed?

3. What are the three basic parts of a pump?

4. How do positive displacement pumps operate?

5. Name six types of pumps and where they are used.

6. A single-acting piston pump with a 2-in. bore and 3-in. stroke cycles at the rate of 480 cycles/min. Compute the volume flow rate and discharge for one hour of operation.

7. What is the purpose of directional control valves and what types are there?

8. What is the difference between a rising stem valve and a nonrising stem valve?

9. Why shouldn't a gate valve be used for regulating flow?

10. What is the difference between the operation of a pressure relief valve and a pressure reducing valve?

11. What functions does a hot water tank water temperature (and pressure) control valve serve? Does it limit the temperature of the water?

12. Survey one or more local plumbers to determine:
   a. Which ballcock valves are used in most local houses?
   b. Which ballcock valve do they prefer to install in a new job, and in a retrofit job?
   c. What causes most ballcock valves to fail?

13. What are the two basic types of water meter, and when are they used?

14. How does a bourdon gauge operate?

REFERENCES

[1] Initially the Francis impeller was conceived as an inward flow reaction turbine, and was perfected by James B. Francis at Lowell, Massachusetts, around 1855. Francis turbines are used, for example, at the Grand Cooley Dam (Washington State) with outputs as high as one million horsepower.

[2] Reducing the flow increases the temperature span of hot and cold water changes caused by sudden use of water elsewhere in the plumbing system. For an explanation of this condition see: Albert G. Fehrm, "Water Can Be Conserved in Showers, but Consider the Safety Factor," *Domestic Engineering*, July, 1977.

[3] An entertaining account of the controversy surrounding placing bricks in toilets is written in "Bricks and Toilet Tanks Provoke Nationwide Battle," *Scoreboard,* Fluidmaster, Inc., P. O. Box 4264, 1800 Via Burton, Anaheim, Ca. 92803.

[4] The English counterpart to the American flush valve shown in Fig. 11-24 is attributed to the inventiveness of Thomas Crapper, 1872. It uses siphonic action to initiate the flush and earned for Crapper the position of Royal Plumber, an appointment made by the Lord Chamberlain.

12 Introduction to Solar Plumbing

12-1 Introduction

Solar energy comes from the sun. Solar plumbing transfers the heat available from fluids passed through collectors exposed to the sun to storage, and then distributes it within the system to places where it is needed. The most common application of solar plumbing is domestic hot water, although space heating and cooling are becoming more common. The field is growing, propelled by a conservation-minded public, and the scarcity and cost of traditional fossil fuels such as oil and gas. In 1977 there were about 66,000 installations in America, with about 95 percent of them used for domestic hot water heating, and the remainder for space heating. Currently, the estimate places this number at over 200,000.[1]

Solar systems are supplemental to the domestic hot water and space heating requirements of a building. This means they require a backup system to supply that portion of the heat which cannot economically be realized from the sun, and for times when the sky is dark or cloudy. Typically they will supply 50–80 percent of the heating requirements. Oversizing a solar system causes it to be less efficient and so is to be avoided.

The heat loss and heat gain requirements of a system must be accurately determined if efficiency and energy savings are to be realized. Using rules of thumb is discouraged, although they are useful to check calculations for heating domestic hot water.

The present discussion is intended to give an overview of solar energy systems, and to give the plumber useful information needed to install those for domestic hot water heating.

12-2 Solar Language

The basic language of solar systems is derived from solar geometry, that is, how the earth faces the sun as it rotates about it and the amount of energy this provides. Figure 12–1 illustrates the relationship of the movements of the earth to the sun. The earth rotates about its own axis once each 24-hour day, and around the sun once each solar year. We have different seasons

Figure 12-1

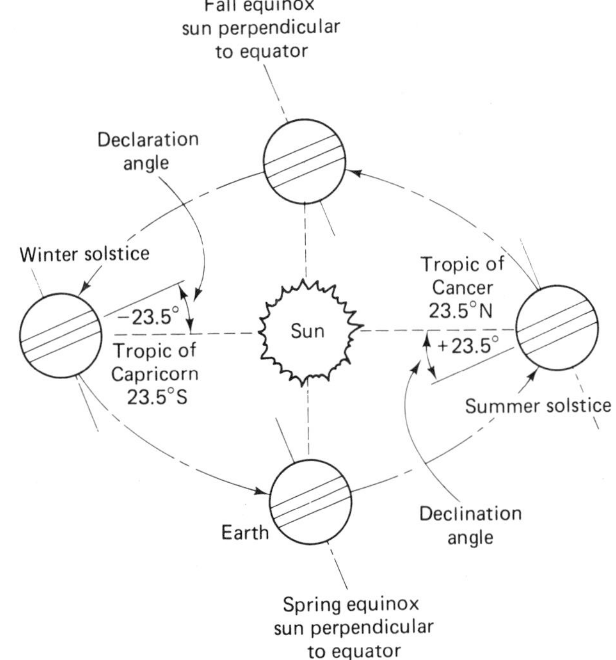

**Solar Geometry
(Copper Development Assn.)**

because the sun faces us from different angles. For example, during the *winter solstice* the sun is perpendicular to a line drawn around the earth 23.5 degrees down from the equator called the *tropic of Capricorn,* and during the *summer solstice* it is perpendicular to a line drawn around the earth 23.5 degrees up from the equator called the *tropic of Cancer.* During the *fall equinox* and *spring equinox,* it is perpendicular to the equator. This means that at different seasons the angle at which the sun shines against the portion of the earth where we are standing changes. The angle of tilt between the equator line on the earth and the sun is called the *declination angle,* and accounts for changes in seasons, the amount of daylight and darkness, the length of days, and the intensity of the solar radiation.

The amount and intensity of solar radiation that reaches the earth is influenced by the distance that it must travel through the earth's atmosphere, which acts as a filter. About 30 percent of the radiation is reflected back into space by the cloud cover of the earth. Another portion is reflected and absorbed by particles of water, dust, and pollutants. This scattered radiation is called *diffuse* radiation. The remainder is *direct radiation,* mostly in the form of visible light. This direct radiation and a portion of the diffuse radiation are available for collection by solar systems. The steeper the angle through which the sun strikes the earth, the farther it must travel through the atmosphere before it reaches the earth and the less

Figure 12-2

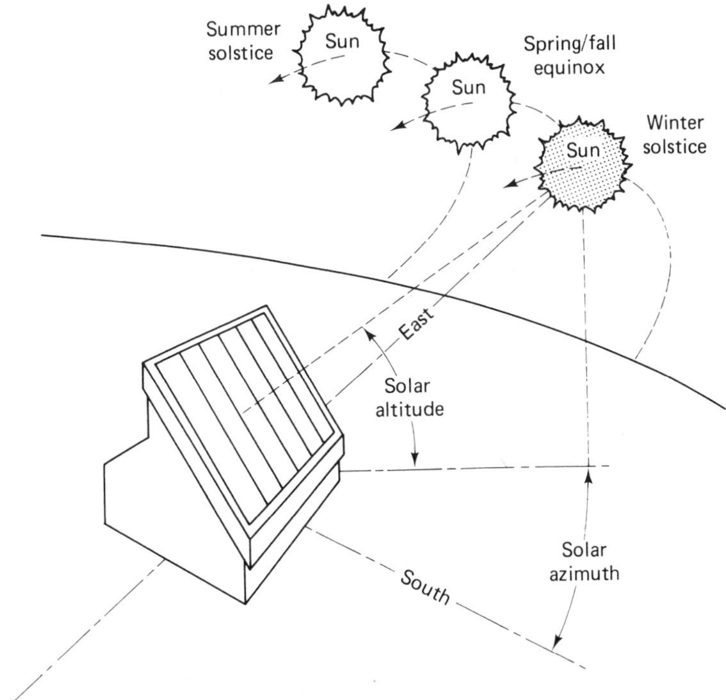

Solar Altitude and Azimuth (Copper Development Assn.)

energy that is available for useful purposes. This explains why the sun is less intense in the winter.

Flat plate solar panels are tilted to collect the most energy for the application used. For example, in North America, while the sun travels from east to west, the flat plate solar panel is always facing toward the south. This is illustrated in Fig. 12–2. The direction of the sun measured by a compass either to the east or west of the panel is called the solar *azimuth* angle. The direction upward in the sky from the earth is referred to as the solar *altitude* angle. Collector panels are mounted to face slightly east or west (azimuth angle), and are tilted upward (solar altitude) to direct the most sunlight against the panel for the specific application. As a general rule, to collect the most solar radiation during the winter the collector is tilted at an angle equal to the latitude plus 15 degrees. To collect the most radiation during the summer, it is tilted at an angle equal to the latitude minus 15 degrees. If it is set at an angle equal to the latitude, it will collect the most radiation averaged over the entire year. These are taken as averages and vary with location, since magnetic compass south is not true polar south, and other circumstances based upon local conditions cause collector panel performance to vary.

A solar system consists of a collector, storage for the heat energy once it has been collected, and a distribution system to convey the heat from storage to places where it can be used later. In a typical domestic hot water system, heat energy is collected by using a solar panel that heats circulating water or an alcohol solution. The heat energy is transferred to water stored in an insulated tank at a temperature of about 135°F (57.2°C). The distribution system consists of the hot water pipes leading from the storage tank to the faucets. When these are opened, the hot water from the tank is available for use. The three components of the system are shown in Fig. 12-3.

Of the two types of solar collector panels, *concentrators* and *flat plate,* the latter is the most common. Concentrators, as the name implies, reflect direct sunlight against a single tube placed at the base of a trough. This single tube gets much hotter but makes use only of the direct sunlight. The diffuse radiation is lost, as is much of the direct sunlight, unless the concentrator is made to follow the sun through the sky. Typically, this is not economically feasible. Flat plate collectors convert both direct and diffuse radiation falling on the plate through a wide range of angles. The construction of a typical flat plate collector is illustrated in Fig. 12-4. They use liquid as the transfer medium. The sun shines on a *flat black* or *selective black* collector surface that is integral with an array of copper tubes. The entire assembly is set in a boxlike frame with insulation at the back and a single or double glazing cover over the front. Notice that the flow length is the same for each tube path through the collector to achieve a balanced flow

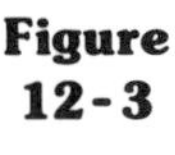
Figure 12-3

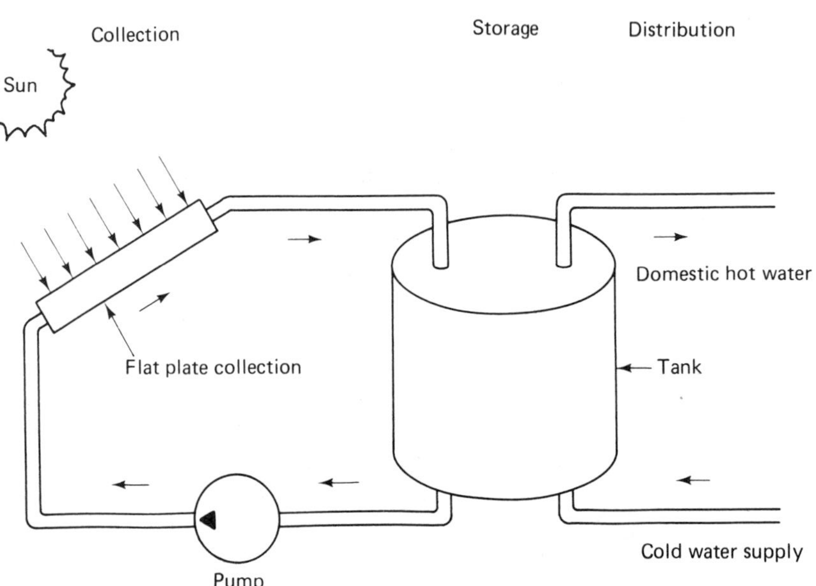

Solar Domestic Hot Water System Components

Figure 12-4

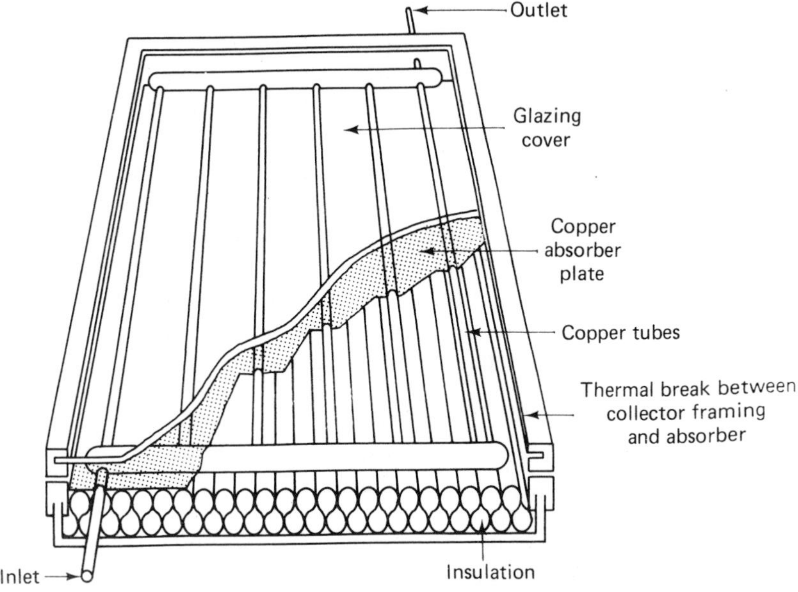

Typical Flat Plate Collector (Copper Development Assn.)

and uniform temperature across the entire collector plate. The insulation at the back of the collector reduces losses once the heat is collected. The single- or double-glazed covering at the front reduces the loss of convective heat and insulates the front against loss of collected heat during cold weather. Selective collector plate surfaces are more effective in collecting radiation, partly because their thickness is approximately equal to the radiation wave length. When used with single or double glazing, these collector panels are the most efficient, although the cost of selective coatings are higher than flat black coatings. Solar collectors are sealed to prevent the formation of condensation, which would reflect incoming radiation. The insulation and materials used must also resist decomposure and *outgassing,* a condition that accompanies high temperatures and clouds the glazing cover, reducing its efficiency. Although an operating temperature of 200°F (93.3°C) is normal, no flow conditions can reach 400°F (204°C). Condensation is commonly controlled with drying agents called *desiccants,* which are placed between the glass panels.

12-3 Hot Water Requirements

Although both solar collector panels and the entire piping system can be engineered and fabricated by the plumber, it is anticipated that at least initially the engineering will have been done by a professional designer. There are also many packaged systems available.[2] The plumber, then, has the responsibility for installing the domestic hot water system, checking its performance, and maintaining it through its useful life. This requires that a number of calculations be performed. Some of these calculations are involved with the heating requirement of the system, while others have to do with the hydraulics of fluid flow through the piping. Both types of calculations are necessary for the plumber to be able to troubleshoot the system once it is installed. A number of design manuals are available that give in-depth treatment of solar systems,[3] as well as automated programs that can perform system calculations with low-cost calculators[4] (Fig. 12–5).

A simplified design calculation for determining the domestic hot water heating requirements can be made by using the following steps:

1. Determine the hot water needs.
2. Calculate the outlet and inlet temperatures.
3. Calculate the heat required.

The hot water needs are computed by multiplying the use per person by the number of occupants. Added to this are the requirements for the washing machine, dishwasher, and any other add-on requirements, such as hot water for washing cars. Each occupant will normally use about 15 gal per day of water heated to 135°F (57.2°C). This would allow for one shower, using about 10 gal, and leave 5 gal for other purposes. If each occupant takes more than one shower per day, then the hot water requirement in-

Figure 12-5

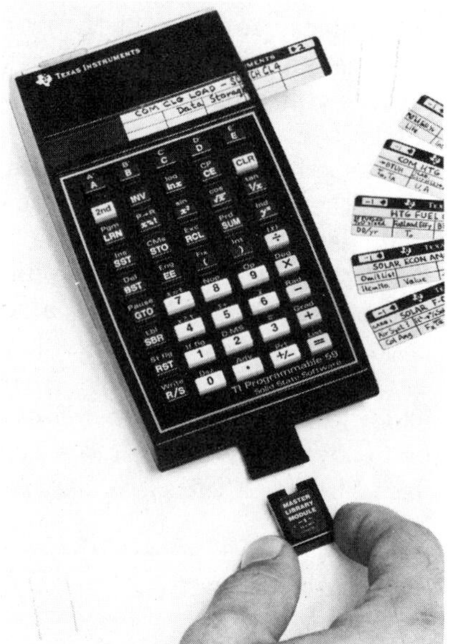

Solar Design Program Calculator (Scotch Programs, Inc.)

creases. Multiple showers are a common practice with teenagers. A washing machine will normally add 4 gal per day for each occupant, and an automatic dishwasher another 15 gal per day for the entire family. Totaling the hot water consumption for a family of four, we have

Hot water for four occupants = (4 × 15 gal/day)	=	60 gal/day
Hot water for washing machine = (4 × 4 gal/day)	=	16 gal/day
Hot water for dishwasher	=	15 gal/day
Total		91 gal/day @ 135%F

Depending upon the individual needs of the family, this minimum might increase as much as 33 percent, to 120 gal per day. This would allow two or more of the occupants to take two showers.

The inlet temperature (t_{hw}) and outlet temperature (t_m) of the domestic hot water are computed to find out how much the amount of hot water required must be heated to meet the demand. This is done by subtracting the inlet temperature of water entering from the service in the ground from the hot water supply leaving the water heater. The procedure is repeated by approximating the average temperature differences each month of the year, although the accuracy is increased for space heating and cooling applications by making the calculation for each day or even each hour throughout the year for a particular location.[5] As an average, 135°F (57.2°C) can be used as the water temperature in the heater and/or storage tank. During the summer months this temperature can be expected to increase to 160°F (71.1°C) or more because of increased sun and decreased heat loss. The temperature of the water entering the system through the service can be assumed to be the same as the ground temperature. This varies with each season of the year. Monthly ground temperatures for fourteen selected cities are given in Table 12–1. This information is also available from the local weather forecasting service.

The heat necessary to raise the temperature of the water from ground temperature at the inlet to 135°F required at the outlet of the hot water heater is computed as follows: First, determine the temperature difference Δt between the inlet and outlet water temperatures for each month. Second, multiply the temperature difference by the heat value necessary to raise water the required number of degrees. Finally, sum the heat values for the entire year to compute the total equivalent cost of heating the domestic hot water for the dwelling. This total value is useful to figure the pay-back time for the hot water system.

The temperature difference Δt is computed by subtracting the temperature of the water entering at ground temperature t_{hw} from the temperature of the water at the outlet of the water heater. For example, if during December in a city like Chicago, the water temperature at the service entrance is 35°F and the hot water requirement is 135°F at the outlet of the tank, the temperature difference $\Delta t = (135 - 35) = 100$°F. This means that during the month of December the domestic hot water heating system would

Table 12-1

Monthly Ground Temperatures for Fourteen U.S. Cities

		J	F	M	A	M	J	J	A	S	O	N	D
Albuquerque	W	72	72	72	72	72	72	72	72	72	72	72	72
Boston	Re	32	36	39	52	58	71	74	67	60	56	48	45
Chicago	L	32	32	34	42	51	57	65	67	62	57	45	35
Denver	Ri	39	40	43	49	55	60	63	64	63	56	45	37
Fort Worth	L	46	49	57	70	75	81	79	83	81	72	56	46
Los Angeles	Ri,W	50	50	54	63	68	73	74	76	75	69	61	55
Las Vegas	W	73	73	73	73	73	73	73	73	73	73	73	73
Miami	W	70	70	70	70	70	70	70	70	70	70	70	70
Nashville	Ri	46	46	53	63	66	69	71	75	75	71	58	53
New York	Re	36	35	36	39	47	54	58	60	61	57	48	45
Phoenix	Ri,Re,W	48	48	50	52	57	59	63	75	79	69	59	54
Salt Lake C.	W,C	35	37	38	41	43	47	53	52	48	43	38	37
Seattle	Ri	39	37	43	45	48	57	60	68	66	57	48	43
Washington	Ri	42	42	52	56	63	67	67	78	79	68	55	46

Source Date From *Handbook of Air Conditioning System Design*, p. 5-41 through 5-46; McGraw Hill Book Company, New York (1965). Abbreviations: C—Creek, L—Lake, Re—Reservoir, Ri—River, W—Well

have to raise the temperature of the water used each day (91 gal/day for the present discussion) by an average value of 100°F.

The heat necessary to raise the temperature of a fluid like water is computed in the English system in British thermal units (Btu's), and in the metric system in calories (cal) or joules (J). One Btu is the heat required to raise the temperature of one pound of water from 39°F to 40°F. This can be expressed as

$$1 \text{ Btu} = 1°\text{F/lb of } H_2O$$

since raising the temperature 1°F at other temperatures does not require appreciably different heat values. A calorie is the amount of energy required to raise the temperature of one gram of water from 4° to 5° on the centigrade scale. This is also approximated to be

$$1 \text{ cal} = 1°\text{C/gm of } H_2O$$

A joule equals a newton-meter of mechanical energy, and when conversions are necessary from Btu's or calories, the following factors can be used:

$$\text{Btu's} \times 1.055 \times 10^3 = \text{joules} \quad (12\text{-}1)$$

and

$$\text{cal} \times 4.18 = \text{joules} \quad (12\text{-}2)$$

Where the solar radiation is given in langleys, the conversion to Btu is made by using

$$\text{langleys/day} \times 3.69 = \text{Btu/ft}^2\cdot\text{day} \qquad (12\text{–}3)$$

The flow of water circulating through a solar system is usually measured in gallons or liters per unit of time, and thus it is more convenient to compute the heat necessary to raise a gallon or liter of water one degree than to use other units. Since one gallon of water weighs approximately 8.33 lb, the amount of heat necessary to raise it a prescribed amount, say 100°F, would be

$$\text{Btu's} = (1\ \text{Btu/lb}\cdot{}^\circ\text{F})(8.33\ \text{lb/gal})(100^\circ\text{F}) = 833\ \text{Btu/gal}$$

Thus the heat value necessary to supply the domestic hot water needs for each month is simply the product of the usage in gallons per month, the average temperature rise required, and the heat required to raise each unit (gallons or liters) one degree Fahrenheit or centigrade. Summing the Btu value for twelve months and multiplying it by the dollar value of an equivalent heat source would show the utility cost of supplying the dwelling hot water for one entire year. This calculation of the heat required to raise the Fahrenheit temperature of gallons of water a specified number of degrees is

$$\text{Monthly heat requirement (Btu's)} = 8.33 \times \text{days} \times \text{gal} \times \Delta t$$

where Δt is the average monthly temperature rise of the water. (12–4)

Example 12-1

Compute the monthly and total Btu requirement to heat 100 gal of domestic hot water each day to a temperature of 135°F.

Solution:
For convenience, Table 12–2 illustrates the steps and gives the average ground water temperature entering the hot water tank. To complete the answer, the necessary calculations have been made in rows 5 and 6. Row 6 was computed by using Eq. (12–4). The monthly Btu requirements were then summed to arrive at the total for the entire year.

To give some idea of the cost of this Btu energy by an equivalent source of heat, consider the case where the power is supplied by the local electrical utility at the rate of 5¢ per kilowatt-hour (kWh). One way this computation can be made is to convert the heat required in Btu's to joules (J) (an SI metric unit), and also to convert the power available from the utility in kilowatt-hours to joules. They then have a common base, and the Btu energy units most familiar for heating water can be converted to kWh from which the utility bill is calculated by using Eq. (12–1). That is,

$$\text{Btu} \times 1.055 \times 10^3 = \text{joules}$$

Table 12-2

Domestic Hot Water Heating Requirement (Load)

Month	*Jan.*	*Feb.*	*Mar.*	*Apr.*	*May*	*June*
1. Days	31	28	31	30	31	30
2. Hot Water Requirements (gal/day)	100	100	100	100	100	100
3. Outlet Temperature (°F)	135	135	135	135	135	135
4. Inlet Temperature (°F)	35	35	37	42	52	56
5. Temperature Difference (Δt°F)	100	100	98	93	83	79
6. Heat Requirements L_d (Btu/day)	83 300	83 300	81 634	77 469	69 139	65 807
7. Heat Requirements Load L_m (Btu/mo)	2 582 300	2 332 400	2 530 654	2 324 070	2 143 309	1 974 210
Month	*July*	*Aug.*	*Sept.*	*Oct.*	*Nov.*	*Dec.*
1. Days	31	31	30	31	30	31
2. Hot Water Requirements (gal/day)	100	100	100	100	100	100
3. Outlet Temperature (°F)	135	135	135	135	135	135
4. Inlet Temperature (°F)	58	61	66	59	51	39
5. Temperature Difference (Δt°F)	77	74	69	76	84	96
6. Heat Requirements L_d (Btu/day)	64 141	61 642	57 477	63 308	69 972	79 968
7. Heat Requirements Load L_m (Btu/mo)	1 988 371	1 910 902	1 724 310	1 962 548	2 099 160	2 479 008

Annual Btu Requirement 26 051 242

and

$$\text{kWh} \times 3.6 \times 10^6 = \text{joules} \tag{12-5}$$

so that in deriving the constant

$$(\text{Btu})\left(\frac{1.055 \times 10^3\ \text{J}}{\text{Btu}}\right)\left(\frac{\text{kWh}}{3.6 \times 10^6\ \text{J}}\right) = \text{kWh}$$

or

$$\text{Btu} \times 0.000\,293 = \text{kWh} \tag{12-6}$$

For Example 12–1, if the utility cost is 5¢/kWh, the value of the heat required for domestic hot water (neglecting losses) would be

$$\text{Heat required (kWh)} \times \text{rate (\$/kWh)} = \text{value}$$

and

$$(26\,051\,242\ \text{Btu} \times 0.000\,293\ \text{kWh/Btu})(\$0.05/\text{kWh}) = 381.65$$

For a typical residence this could approach 25 to 30 percent of the total energy consumed.

12-4 Solar System Capabilities

The solar system capability to generate the heat required to supply the domestic hot water comes from the sun shining on solar collector panels. Water or other fluid circulating through the array is heated and transfers this heat to water in the storage tank. In the design of a system it is common to size the panels and storage tank to provide a reasonable portion of the hot water required rather than all of it. This requires these additional steps:

4. Calculate the solar collector panel output or harvest.
5. Size the collector for 50–80 percent of the requirements.
6. Size the storage tank.

A fixed-angle flat plate solar collector panel facing approximately south recovers only a portion of the solar energy available. Even on a clear day, the sun moves across the sky to change the angle of incidence. So, too, does the change of seasons from summer through the fall, to winter and then to spring. This means that average collection values rather than exact values are used. More sophisticated computer methods make this calculation on an hour-by-hour basis throughout the year for a specific location, particularly when space heating and cooling systems are being designed, but domestic hot water systems do not warrant this level of detail and associated costs. Rather, the average available insolation (radiation) can be determined on a month-by-month basis from weather graphs, and then a collector panel area is selected that will convert this to the level of energy specified, for example, to 50–80 percent of total requirements, when mounted at a fixed angle facing approximately south. The amount of collectible solar energy available from the sun through the atmosphere as well as collector efficiency enter into the calculation of how large the array must be to supply the necessary heat to the fluid pumped through the tube passages.

The fluid is circulated between the collector panel and the storage tank at a rate of 0.5–3 gal/hr per ft^2 of plate area and for best results is heated sufficiently to raise the temperature a few degrees, for example, 5°F on each pass. The collector panel is sized from the recoverable solar energy available to each square foot of plate area.

The earth receives 429 Btu/ft^2 per hour falling on each square foot of surface perpendicular to the direction of the sun's rays at the outer edge of the atmosphere. This rate is known as the *solar constant.* However, only a portion of this radiation is available at the surface, and this amount is influenced by the geographic location of the panel and the time of year. Solar insolation maps for each month of the year are used to show the radiation available; they also indicate what percentage of the radiation available is diffuse or scattered radiation. These maps are given in Appendix L. The top number at each geographic location is the average Btu/ft^2 falling on the horizontal plane each day for that month, and the bottom number is the percentage of the total that is diffuse or scattered radiation.

The average daily radiation I_d available to a collector plate tilted toward the sun is computed as the product of the average radiation falling on a horizontal surface I_h and a factor F_b which accounts for the diffuse radiation, latitude, and tilt angle of the collector plate. That is,

$$I_d = I_h \times F_b \tag{12-7}$$

The values for F_b are given in Table 12–3. To use the table, first consult the insolation maps given in Appendix L for a specific month and location to determine the average daily solar radiation received on a horizontal surface, and locate the percentage of that total which is diffuse radiation. For example, during the month of March at 40 deg latitude in central Illinois, the average daily insolation is given as 1196 Btu/ft^2, with 52 percent being diffuse radiation. Now look at Table 12–3. Since the diffuse radiation is given as 52 percent, the value falls between 40 to 60 percent and will have to be interpolated. Now consider the case where the collector is tilted at an angle equal to the latitude or approximately 40 deg. Reading across the table to March at 40 percent and 60 percent diffuse, we see that the value for 52 percent falls between 1.21 (at 40 percent diffuse) and 1.14 (at 60 percent diffuse). The value of 52 percent diffuse will have to be interpolated. Making the computation, we obtain

$$\frac{52}{100} \times (1.21 - 1.14) = 0.04$$

Subtracting this difference from 1.21 gives a value of 1.17 for the factor F_b. Finally, solving for the average daily radiation available at a collector plate located in central Illinois and tilted at an angle equal to the latitude (40 deg), we have

$$I_d = I_h \times F_b = (1196 \text{ Btu/ft}^2 \cdot \text{day})(1.17) = 1399 \text{ Btu/ft}^2 \cdot \text{day}$$

For the complete month of March the total available energy would be 31 times this average daily amount, or 43 369 Btu/ft^2.

How much energy the collector plate can convert to useful purposes is determined by its efficiency and the difference between the average fluid temperature entering the collector and the ambient temperature of the air surrounding the collector plate. Instantaneous collector efficiency (measured during 15-minute intervals) is given by the formula

$$\text{Collector efficiency } (\eta) = \frac{\text{useful solar energy collected } (q_u/A)}{\text{solar energy incident on the collector } (I_t)}$$

or

$$\eta = \frac{\frac{q_u}{A}}{I_t} \times 100 \qquad (12\text{-}8)$$

where the efficiency η (Greek letter eta) is in percent, the output of the panel q_u is measured in Btu/hr (W/hr), A is the area of the collector (plate aperture) in ft^2 (m^2), and I_t is the average solar energy incident on the collector plate in Btu/ft^2 ·hr (W/hr). The efficiency for typical flat plate collectors with single and double glazing is shown in Fig. 12–6.

It is common practice for manufacturers to plot the efficiency of solar collector plates against the ratio of the temperature difference to the solar energy incident on the collector plate. This ratio is called the collector efficiency factor and is given by the equation

$$\frac{\Delta t}{I_t} = \frac{(t_i - t_a)}{I_t} \quad \frac{°F}{\text{Btu/ft}^2 \cdot \text{hr}} \qquad (12\text{-}9)$$

where Δt is the temperature difference between the inlet fluid t_i °F (°C) and the ambient air surrounding the plate t_a °F (°C). The temperature of the fluid as it enters the collector is influenced by the temperature of the cold water supply, the heating load, and the amount of energy available to the collector panel. In reality, these values vary throughout the day, and from season to season. Estimates of the fluid inlet temperature for calculation purposes can be made by substituting the values of Table 12–1 into Table 12–4.

The ambient temperature near the collector plate is influenced by the amount of radiation received from the sun and the heat loss associated with the construction of the panel. Collector plates covered by a double glass covering separated by an air space lose less heat than those covered by a single glass covering. Those with no covering whatsoever, of course, lose

Table 12-3 Collector Tilt Factors (Copper Development Association Inc.)

Diffuse	Collector Tilt	Latitude	Jan	Feb	Mar	Apr	May	June	July	Aug	Sept	Oct	Nov	Dec
20% Diffuse	*Collector Tilt = Lat. − 15°*	24°	1.13	1.09	1.04	1.00	0.97	0.96	0.97	1.00	1.04	1.10	1.12	1.15
		32°	1.34	1.24	1.11	0.99	0.93	0.94	0.92	0.98	1.11	1.28	1.34	1.42
		40°	1.67	1.41	1.23	1.01	0.90	0.90	0.89	1.02	1.21	1.42	1.73	1.76
		48°	2.18	1.64	1.51	1.10	0.96	0.80	0.95	1.08	1.43	1.73	2.26	2.56
	Collector Tilt = Latitude	24°	1.31	1.21	1.08	0.95	0.87	0.85	0.87	0.94	1.08	1.21	1.28	1.37
		32°	1.56	1.39	1.16	0.97	0.79	0.76	0.78	0.96	1.15	1.45	1.56	1.71
		40°	1.95	1.55	1.28	0.98	0.75	0.75	0.74	1.00	1.25	1.57	2.05	2.10
		48°	2.53	1.80	1.57	1.08	0.90	0.74	0.89	1.05	1.49	1.90	2.63	3.04
	Collector Tilt = Lat. + 15°	24°	1.40	1.25	1.06	0.80	0.70	0.67	0.70	0.79	1.05	1.27	1.36	1.50
		32°	1.69	1.45	1.13	0.89	0.59	0.55	0.58	0.88	1.12	1.54	1.68	1.89
		40°	2.11	1.60	1.25	0.91	0.67	0.65	0.66	0.92	1.23	1.63	2.24	2.30
		48°	2.72	1.85	1.54	1.00	0.79	0.65	0.78	0.97	1.45	1.98	2.84	3.34
40% Diffuse	Collector Tilt = Lat. − 15°	24°	1.10	1.07	1.03	1.00	0.97	0.96	0.97	1.00	1.03	1.07	1.09	1.12
		32°	1.27	1.20	1.09	0.98	0.93	0.95	0.92	0.97	1.09	1.22	1.26	1.33
		40°	1.52	1.33	1.18	0.98	0.89	0.89	0.89	1.00	1.16	1.34	1.59	1.61
		48°	1.93	1.50	1.42	1.06	0.94	0.78	0.93	1.04	1.35	1.57	2.02	2.23
	Collector Tilt = Latitude	24°	1.24	1.16	1.06	0.95	0.88	0.86	0.88	0.94	1.05	1.17	1.21	1.28
		32°	1.44	1.31	1.12	0.96	0.79	0.76	0.78	0.95	1.11	1.36	1.43	1.56
		40°	1.74	1.44	1.21	0.96	0.74	0.74	0.73	0.97	1.19	1.46	1.84	1.87
		48°	2.20	1.61	1.45	1.03	0.88	0.73	0.87	1.01	1.38	1.71	2.31	2.60
	Collector Tilt = Lat. + 15°	24°	1.32	1.19	1.03	0.80	0.72	0.69	0.72	0.79	1.03	1.20	1.26	1.38
		32°	1.53	1.36	1.09	0.89	0.59	0.56	0.58	0.88	1.08	1.42	1.52	1.70
		40°	1.85	1.47	1.18	0.88	0.67	0.66	0.66	0.90	1.15	1.50	1.98	2.03
		48°	2.34	1.64	1.41	0.95	0.78	0.64	0.77	0.93	1.34	1.75	2.46	2.82

Diffuse	Collector Tilt	Lat.											
60% Diffuse	Collector Tilt = Lat. − 15°	**24°**	1.07	1.05	1.02	1.00	0.98	0.97	0.98	0.99	1.02	1.05	1.06	1.09
		32°	1.20	1.15	1.06	0.97	0.93	0.95	0.92	0.96	1.06	1.17	1.19	1.25
		40°	1.38	1.25	1.13	0.96	0.88	0.89	0.88	0.98	1.12	1.26	1.45	1.46
		48°	1.69	1.35	1.33	1.02	0.92	0.77	0.91	1.00	1.27	1.42	1.77	1.90
	Collector Tilt = Latitude	**24°**	1.17	1.11	1.04	0.94	0.89	0.87	0.89	0.93	1.03	1.12	1.13	1.20
		32°	1.31	1.24	1.07	0.94	0.79	0.76	0.78	0.94	1.07	1.26	1.30	1.41
		40°	1.53	1.32	1.14	0.93	0.73	0.74	0.73	0.95	1.12	1.35	1.63	1.65
		48°	1.87	1.42	1.34	0.99	0.86	0.72	0.85	0.97	1.28	1.51	1.98	2.16
	Collector Tilt = Lat. + 15°	**24°**	1.22	1.13	1.00	0.80	0.73	0.71	0.73	0.79	1.00	1.13	1.16	1.25
		32°	1.37	1.26	1.04	0.88	0.59	0.56	0.58	0.87	1.03	1.30	1.36	1.50
		40°	1.60	1.34	1.10	0.86	0.66	0.66	0.66	0.88	1.08	1.37	1.72	1.76
		48°	1.96	1.43	1.29	0.91	0.77	0.64	0.76	0.89	1.23	1.52	2.09	2.31
80% Diffuse	Collector Tilt = Lat. − 15°	**24°**	1.04	1.03	1.01	1.00	0.98	0.97	0.98	0.99	1.01	1.03	1.04	1.05
		32°	1.12	1.11	1.04	0.96	0.93	0.96	0.92	0.96	1.04	1.11	1.12	1.16
		40°	1.24	1.17	1.08	0.94	0.88	0.88	0.87	0.96	1.07	1.18	1.30	1.31
		48°	1.44	1.22	1.24	0.98	0.90	0.75	0.89	0.97	1.19	1.27	1.53	1.57
	Collector Tilt = Latitude	**24°**	1.10	1.07	1.01	0.94	0.90	0.89	0.90	0.93	1.01	1.07	1.06	1.11
		32°	1.19	1.16	1.03	0.93	0.79	0.77	0.78	0.93	1.03	1.17	1.18	1.26
		40°	1.33	1.21	1.07	0.90	0.73	0.74	0.72	0.93	1.06	1.23	1.42	1.43
		48°	1.55	1.25	1.22	0.94	0.84	0.70	0.84	0.93	1.18	1.31	1.66	1.72
	Collector Tilt = Lat. + 15°	**24°**	1.12	1.06	0.98	0.80	0.75	0.73	0.75	0.79	0.97	1.06	1.05	1.13
		32°	1.21	1.17	0.99	0.87	0.60	0.57	0.59	0.86	0.99	1.18	1.19	1.31
		40°	1.35	1.20	1.02	0.83	0.66	0.66	0.65	0.86	1.01	1.24	1.47	1.49
		48°	1.58	1.23	1.16	0.87	0.76	0.63	0.75	0.85	1.12	1.29	1.71	1.79

Figure 12-6

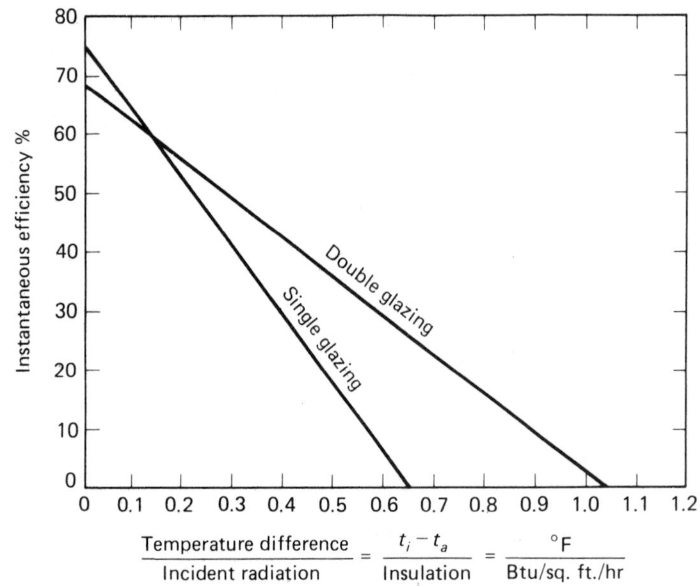

Collector Efficiency Graph

Table 12-4 Collector Inlet Temperatures (Copper Development Association Inc.)

	Domestic Water Heating (Add to Cold Water Supply Temperature)
January	CWS + 20°
February	CWS + 35°
March	CWS + 60°
April	CWS + 75°
May	CWS + 85°
June	CWS + 95°
July	CWS + 100°
August	CWS + 100°
September	CWS + 95°
October	CWS + 70°
November	CWS + 50°
December	CWS + 30°

the most. The amount of insulation behind the collector and the thermal barrier around the edges also influence the extent of the heat loss and drop in the collector plate temperature. In the absence of information about a specific application, the temperature of the outlet water is often used to make preliminary calculations for double-glazed panels. This can be assumed to be the design temperature of 135°F (57.2°C), but in fact temperatures may approach 200°F (93.3°C) under some conditions during the summer months.

To determine the output of the panel q_u/A it is necessary to first establish the operating temperature characteristics for the specific application (or manufacturer's product) and then determine the efficiency for that condition. The efficiency curve and equation can then be used to solve for the energy output.

Example 12-2 Compute the value of the collector efficiency factor, i.e., ratio of the difference between the inlet fluid temperature and ambient temperature, to the incident solar radiation, using the following values: inlet temperature, 135°F (57.2°C); ambient temperature, 70°F (21.1°C); incident solar radiation, 200 Btu/ft^2·hr.

Solution:
Setting up the ratio, using Eq. (12–8), we have

$$\frac{\Delta t}{I_t} = \frac{(t_i - t_a)}{I_t} = \frac{(135°\text{F} - 70°\text{F})}{(200 \text{ Btu/ft}^2 \cdot \text{hr})} = 0.325 \frac{°\text{F}}{\text{Btu/ft}^2 \cdot \text{hr}}$$

Example 12-3 A particular solar collector to be tested by an independent laboratory, using ASHRAE Standard 93–77,[6] is supposed to have an instantaneous efficiency of 50 percent under the conditions given in Example 12–2. Compute the power output per ft^2 of collector panel.

Solution:
Solving Eq. (12–7) for $\frac{q_u}{A}$, we obtain

$$\frac{q_u}{A} = \frac{\eta \times I_t}{100} = \frac{(50\% \times 200 \text{ Btu/ft}^2 \cdot \text{hr})}{(100\%)} = 100 \text{ Btu/ft}^2\text{/hr}$$

The number of daylight hours during each day that the solar panel will generate heat varies by the season, location, time of day, and the collector tilt angle. Since on the average about 90 percent of the sun's energy is received during two-thirds of the daylight hours, a daily output time for the solar collector can be computed by using the calculation

$$\text{Daily output time} = \frac{(2/3 \times \text{daylight hrs/day})}{0.90} \tag{12-10}$$

The number of hours per day is determined from local sunrise-sunset times. In the absence of more accurate information, an average of 12 hr of daylight can be assumed, and this will result in an average output time of 9 hr. per day.

The tank is sized from the square area of the collector area at a ratio of 1.25 to 2.0 gal/ft² (18 to 30 l/m²), with about 1.8 gal/ft² being the maximum for water systems. This is derived from

$$\frac{(15\ \text{Btu/ft}^2\cdot{}^\circ\text{F})}{(8.33\ \text{lb/gal})(1\ \text{Btu/lb}\cdot{}^\circ\text{F})} = 1.8\ \text{gal/ft}^2$$

For glycol systems

$$\frac{(15\ \text{Btu/ft}^2\cdot{}^\circ\text{F})}{(8.80\ \text{lb/gal})(0.83\ \text{Btu/lb}\cdot{}^\circ\text{F}} = 2.05\ \text{gal/ft}^2$$

A sample calculation will be used to demonstrate the sizing procedure.

Example 12-4 Size the collector panel array and storage tank necessary to provide 70 to 80 percent of the hot water needs of a system installed in central Illinois. Use the hot water requirements of Example 12-1 given in Table 12-2, and a double-glazed flat plate collector for which the efficiency can be taken from Fig. 12-7. Estimate the required collector area to be four panels, each with an area of 24 ft² · (2.2 m²).

Solution:

The results of the calculations are given in Table 12-5. Each of the steps is numbered, and they continue in sequence from Table 12-2. Notice that the application, location, latitude, collector type, and tilt angle are listed first. The explanation of the steps is as follows:

8. The horizontal insolation is taken from Appendix L.
9. The percentage diffuse is also taken from Appendix L.
10. The tilt factor is taken from Table 12-3.
11. The daily insolation incident to the collector is computed as the product of the horizontal insolation and the tilt factor.
12. The monthly harvest is the product of the daily insolation and the days in the month.
13. The solar daytime hours are computed from local weather records and Eq. (12-9).
14. The average hourly insolation is computed by dividing the daily insolation by the solar daytime hours.
15. The inlet water temperature is computed from the cold water supply temperature for the nearest city (shown in Table 12-1.), and adding to these values the temperatures shown in Table 12-4. For example, for January, Table 12-1 indicates the temperature for the cold water supply is 32°F, and added to this is 20°F shown in Table 12-4.

Figure 12-7

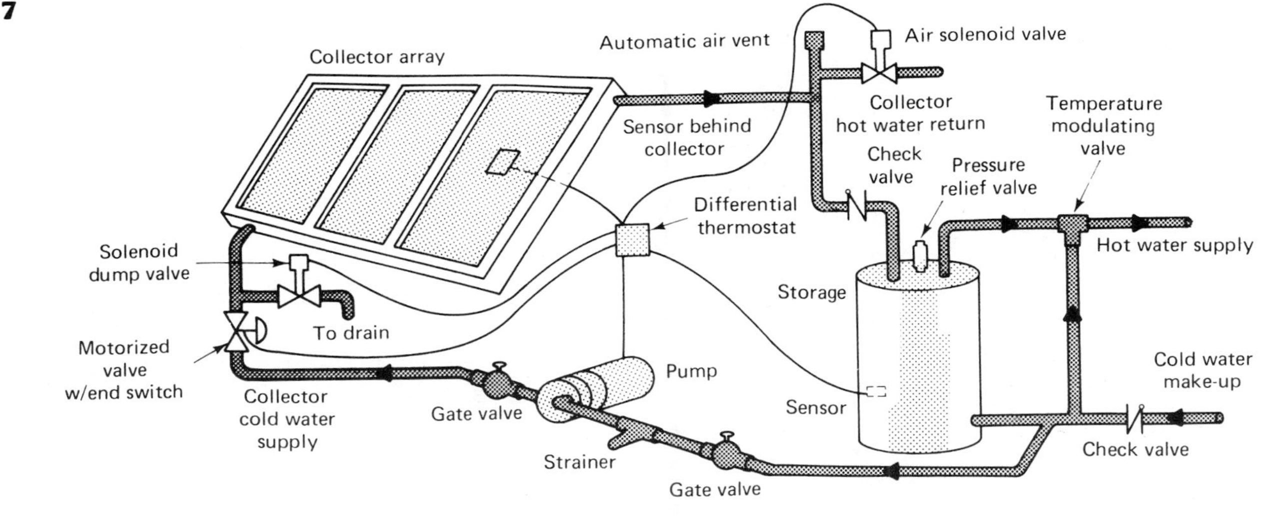

Open Loop System (Copper Development Assn.)

16. The ambient temperature of the air is given by local weather data for the area (not shown) or from climatic references.[7]
17. The temperature difference is computed as the difference between the inlet and ambient temperatures.
18. The collector panel average efficiency factor is derived by dividing the temperature difference by the hourly insolation from Step 14.
19. Collector efficiency is taken from Fig. 12-6; the panel efficiency factor for a double-glazed panel is used.
20. Collector panel output is derived by multiplying the efficiency for the panel by the average hourly insolation I_t times the average solar daytime hours for the month.

Table 12-5

Sizing a Solar Collector System

Application—Hot Water
Location—Central Illinois
Latitude—40°N

Collator Type—Double Glazed
Selective Surface Flat Collator
Collator Tilt—40°N (Latitude)

Month	*January*	*February*	*March*	*April*
Load L_t (Btu/mo)	2 582 300	2 332 400	2 530 654	2 324 070
8 Horizontal Solar Insolation I_h	601	867	1 196	1 546
9 Percent Diffuse	58	54	52	51
10 Tilt Factor F_b	1.55	1.35	1.11	0.94
11 Daily Insolation I_d	932	1 170	1 328	1 453
Days/Month	31	28	31	30
12 Monthly Insolation I_m	28 892	32 760	41 168	43 590
13 Solar Daytime (hr)	6.4	7.9	8.8	9.8
14 Hourly Insolation I_t	146	148	151	148
15 Inlet Temp. t_i (°F)	52	67	94	117
16 Ambient Temp. t_a (°F)	32	36	47	59
17 Temp. Difference $\triangle t$ (°F)	20	31	47	58
18 Collector Panel Factor $\frac{\triangle t}{I_t}$	0.14	0.21	0.31	0.39
19 Collector Efficiency (%)	58	55	48	42
20 Collector Panel Output $\frac{q_a}{A}$ (Btu/ft²·day)	541	644	637	610
21 Collector Area Estimate A (ft²)	96	96	96	96
22 Collector Array Output q_a (Btu/mo)	1 610 016	1 731 072	1 895 712	1 756 800
23 Percent of Monthly Load	62	74	75	76
24 Percent of Annual Load				

21. The collector area estimate is for four panels with an aperture area of 3 ft × 8 ft, for a total area of 96 ft^2.
22. Collector panel array output is the product of the daily output per ft^2, the number of days in the month, and the collector array area (96 ft^2).
23. The percentage of the monthly load supplied by the collector array is arrived at by dividing the output per month by the load per month. In cases where the output is equal to or greater than the load for a particular month, the percentage is 100, and the remaining harvest by the collector array is lost.
24. The percentage of the annual load is derived by dividing the sum of the collector panel output for the 12 months (including only the load for those months where the output exceeds the load) by the annual load. In this case this is 78 percent.

May	*June*	*July*	*August*	*September*	*October*	*November*	*December*
2 143 309	1 974 210	1 988 371	1 910 902	1 724 310	1 962 548	2 099 160	2 479 008
1 956	2 125	2 137	1 908	1 583	1 144	716	539
44	42	40	40	40	44	56	58
0.74	0.74	0.73	0.97	1.19	1.44	1.67	1.67
1 447	1 572	1 560	1 851	1 884	1 647	1 195	900
31	30	31	31	30	31	30	31
44 857	47 160	48 360	57 381	56 520	51 057	35 850	27 900
10.6	10.9	10.9	10.2	9.2	8.3	7.4	6.9
136	144	143	181	205	198	161	130
136	152	165	167	157	127	95	65
69	79	83	81	74	63	47	30
67	73	82	86	83	64	48	35
0.49	0.51	0.57	0.48	0.40	0.32	0.30	0.27
36	35	31	37	42	48	49	52
521	550	484	685	791	791	584	468
96	96	96	96	96	96	96	96
1 550 496	1 584 000	1 440 384	2 038 560	2 278 080	2 354 016	1 681 920	1 392 768
72	80	72	100	100	100	80	56
							(20 240 928 Btu) 78%

Example If the system uses water, the storage tank is sized at 1.25–2.0 times the square footage of the collector array, the tank sizes available being taken into consideration. Making the computation, we have

$$\text{Maximum tank size} = (1.25 \text{ to } 2.0) \times 96 = 120 \text{ to } 192$$

Tanks are available in sizes of 66, 82, 100 and 120 gallons (251, 312, 380 and 455 liters). Depending upon whether or not an existing tank is in place, either a single 120-gallon tank would be recommended (1.25 times the collector array area) or a 100-gallon tank in tandem with the existing water heater storage tank. This could bring the total storage area to 166 gal capacity (if a 66-gal tank were in place), but assumes some interchange between the two tanks, requiring water usage. Some sources recommend connecting the tanks in parallel rather than in series to improve performance during standby when only make-up heat is required.[8]

12-5 Systems Design

There are two basic domestic hot water system types: open loop circulation and closed loop circulation. Open loop circulation systems pump potable water from the lower temperature regions of the storage tank through the solar collector panels and return it to the higher temperature regions near the top of the tank. Closed loop circulation systems pump nonpotable water or a nonfreezing solution from a heat exchanger within or around the base of the hot water storage tank through the collector and return it back through the heat exchanger in a closed loop circuit. There is no interchange or connection between the fluid circulating through the collector panels and the potable water supply in the hot water storage tank. Single-wall heat exchangers are permitted where the antifreeze solution is of the nontoxic type, such as propylene glycol. Double-wall protection is required when the antifreeze solution is toxic.

In an open loop system, as shown in Fig. 12–7, when the differential thermostat indicates that the collector will generate usable heat, the circulating pump delivers water from the lower regions of the storage tank or the cold water supply line to the collector panel array and returns it through the top of the tank. The water continues to circulate from the bottom of the tank to the collector array and back through the top of the tank so long as the differential thermostat indicates that there is usable energy available from the sun. When the control senses that the temperature of the collector panels is dropping, the pump stops circulation. Reverse circulation from the tank through the collector panel array, which would cool the water during the night, is stopped with a check valve in the collector return line. When the differential thermostat senses that the temperature approaches freezing, the system drains the water from the collector array. This occurs as follows: First the motorized valve in the collector supply begins closing. The pump is off. When the valve reaches the closed position, an end switch in the valve energizes both the solenoid dump valve and air solenoid valve. This permits air to enter the system at the top, and the two or three gallons of water in the system drains from the collector array at the bottom. The system is now drained and closed to circulation. Operation is restored when the collector array control signals that it can again supply usable energy. In sequence, the motorized valve begins to open, signaling the dump valve and air solenoid valve to close. When the motorized valve reaches the open position, the pump starts, and water begins to circulate through the collector array and return to the top of the hot water tank. The automatic air vent at the top relieves trapped air from the system as the pump circulates the water.

A closed loop system is shown in Fig. 12–8. Notice that the system uses a heat exchanger immersed in the hot water storage tank, and because of the separate circulation loop to the collector panel array, there must be an expansion tank, pressure relief valve, and place to fill the system. As before, single-wall heat exchangers are permitted only when nontoxic antifreeze solutions are used in the collector loop. When the antifreeze solution is toxic, *double-wall* protection must be provided to put a double barrier between the toxic solution and the potable water supply. The check valve prevents thermo-siphoning and reverse circulation during the night and at other times when the system is not producing heat. The differential thermostat works the same as in the open loop system, signaling the pump when the collector array temperature is sufficiently greater than the line leading from the heat exchanger, indicating the availability of usable energy from the sun.

There are other components that are common to both systems. These include a strainer on the upstream side of the pump to protect it from dirt and sediment, isolation valves, which permit removal of the pump for periodic service, and a temperature modulating valve to temper the hot water coming from the tank. This will prevent scalding when the tap is opened to the hot water supply, particularly during the summer months when the available sunshine and lower heat loss allow water temperatures through the collector to increase. For example, water returning from the collector at 160°F (71.1°C) could easily bring the temperature of the water in the storage tank from 135°F (57.2°C) to 145°F (62.7°C). Another option that is available is to install a high-limit aquastat in the storage tank to limit the circulation of the fluid from the solar array to a maximum temperature, for example, 160°F.

A two-tank open-loop self-draining system that is common in retrofit applications is shown in Fig. 12–9. The solar storage tank is connected upstream of the existing water heater. Water in the solar storage tank is preheated before it enters the existing water heater, thereby supplying the hot water needs up to the capacity of the solar system. The additional hot water needs of the system are supplied by the existing water heater. The system drains down by closing of solenoid valve 5 and opening solenoid

Figure 12-8

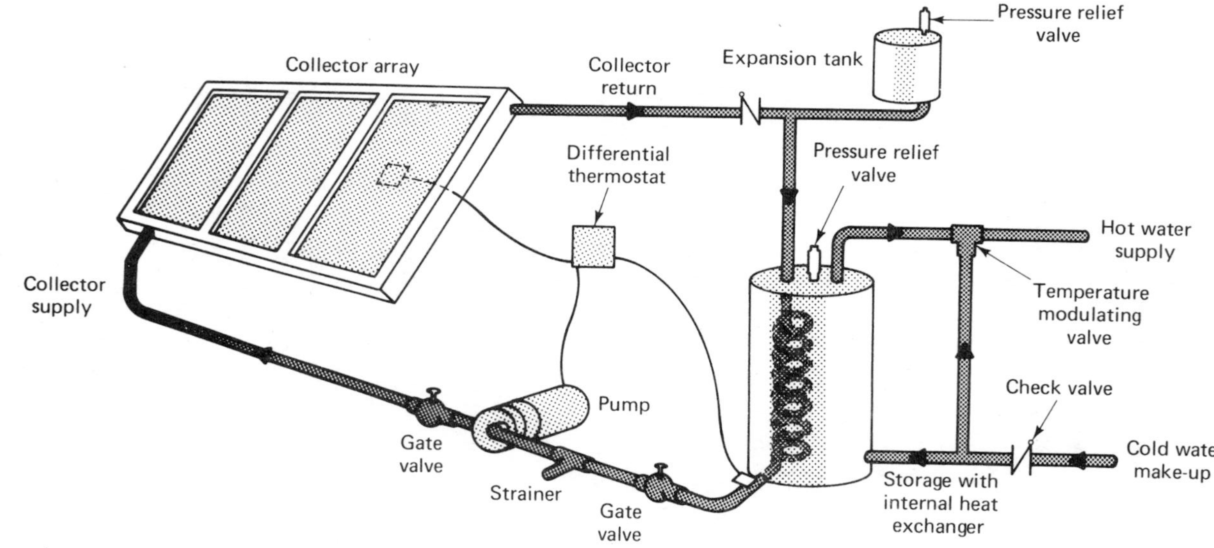

Closed Loop System (Copper Development Assn.)

Figure 12-9

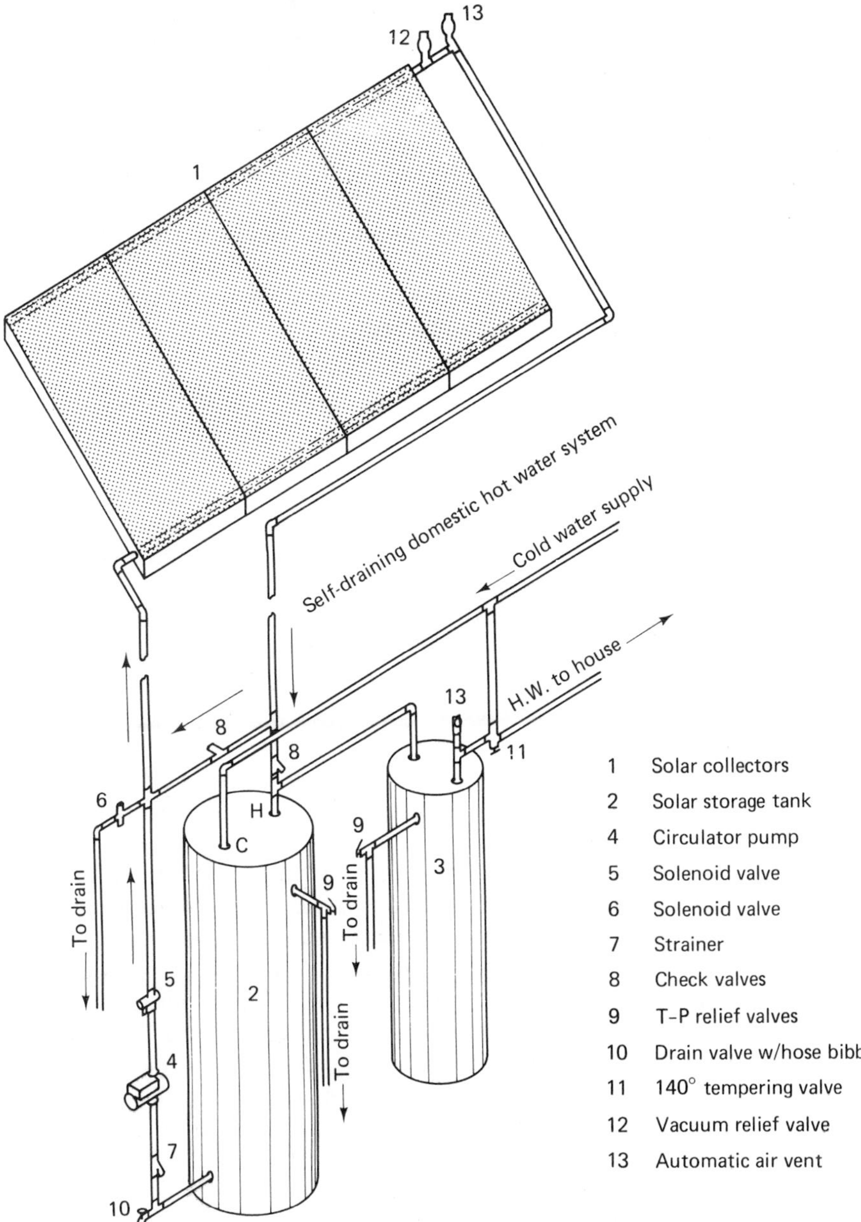

1 Solar collectors
2 Solar storage tank
4 Circulator pump
5 Solenoid valve
6 Solenoid valve
7 Strainer
8 Check valves
9 T-P relief valves
10 Drain valve w/hose bibb
11 140° tempering valve
12 Vacuum relief valve
13 Automatic air vent

Two-tank Open-loop Self-draining System

valve 6, which drains the collectors. Check valve 8 between the two legs allows flow from the return leg to connect to drain when solenoid valve 6 is opened. Check valve 8 in the return water leg just above the tank prevents thermo-siphoning at night and other times when the temperature of the collector panel drops. The collector array is fitted with an automatic vacuum relief valve 12 that activates when the system is drained, and an air vent valve 13 to purge the system during startup. Both tanks are equipped with combination temperature-pressure (T–P) relief valves. Water from the existing tank is connected through a tempering valve 11. Sensors and controls for the system are not shown, nor is the insulation that would cover the hot water pipes to reduce heat loss.

12-6 Solar Hydraulics

A number of factors are known to affect the efficiency of the solar system, including the heat-collecting characteristics of the panel that are built in by the manufacturer and the weather conditions under which the panel must operate. The two are related to each other in part by the panel efficiency factor, which is the difference between the temperatures of the fluid circulating through the collector panel and the ambient air, divided by the incident solar radiation. When the panel efficiency factor has the smallest numeric value, the collector panel has the highest efficiency, and this occurs when the temperature difference between the fluid circulating through the panel and the ambient air is the smallest. For a given weather condition, a temperature difference of 5°F (2.8°C), for example, would be more efficient than 10°F (5.6°C).

The efficiency for a given panel is improved by setting the flow rate so that the temperature difference is minimized, and for a panel with a holding capacity of 0.033 gal/ft^2 of collector surface this occurs in the flow range of 0.5 to 1.0 gal/hr for each square foot of panel. Matching the circulating pump to this system requirement is important to be sure that the necessary flow is maintained while the pumping losses and the power consumed by the unit are kept to a minimum. Although it is routine to specify a pump that will deliver the required flow through the collector array, it cannot be taken for granted that this flow actually occurs in practice, or that the attendant head losses through the system are not excessive.

Initially, in drain-down systems, a static pressure capability of several feet is necessary to lift the fluid from the base of the tank to the roof where the collector array is mounted. As the system is purged of air and the fluid returns to the inlet of the pump, the static losses are recovered. That is, lifting of the fluid is counterbalanced by the falling of the fluid. Thereafter, the losses are associated with the flow through the system, commonly consisting of 200 ft (61 m) of pipe runs; two dozen fittings, which are mostly elbows; a heat exchanger, if one is used; and the collector panels themselves, which usually account for about 1 percent of the total flow losses when plumbed in parallel from a common manifold. Flow losses through the pipe runs commonly account for about 75 percent of the loss, and the remaining 24 percent is attributed to losses through the fittings and valves.[9]

When component systems are built up from off-the-shelf stock items, a major concern should be to keep the system design and routing of the pipes as simple as possible. This will reduce the number of components and the length of the runs, and the number of fittings will be kept to a minimum. Using low-friction components such as ball shutoff valves (to isolate the pump) and gravity swing check valves will also reduce pumping losses.

The friction losses associated with circulating the required flow through the collector array are used to size the pump. These are computed per 100 ft (30.5 m) of developed tube length, including the losses for the equivalent lengths of the fittings. To keep pumping losses to a minimum, the tubing is sized for a flow velocity in the range 0.5 to 1.25 ft/sec (0.15 to 0.38 m/s), which will allow the pressure drop to fall in the range of 0.5 to 1.5 lbf/in.^{2} per 100 ft of developed tube length. Domestic hot water systems using 2 to 5 panels and flow rates of 1–2 gal/min (3.78 to 7.57 l/m) are plumbed with $\frac{1}{2}$-in. or $\frac{3}{4}$-in. tube. Type L or M are used depending upon whether straight lengths or flexibility are required. The velocity equation $v = Q/A$ is used to select the tubing size, and then the head loss associated with that flow rate is used to select the pump, using its characteristic performance curve. Such a curve of head loss vs. flow rate is shown in Fig.

Figure 12-10

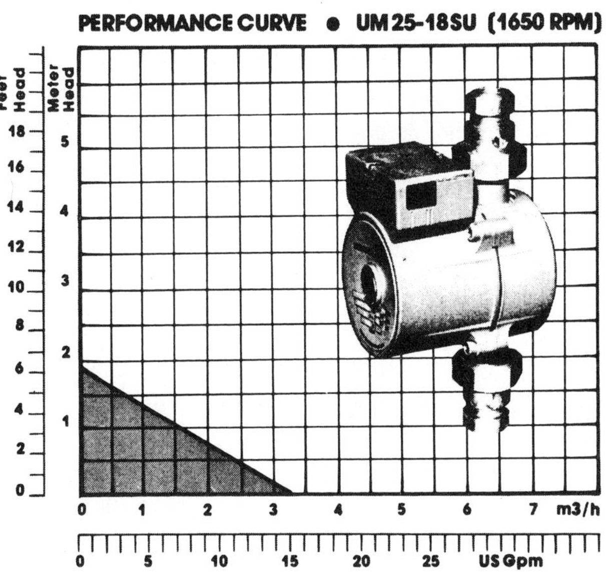

Circulating Pump Performance Curve (Solar Energy Products, Inc.)

12–10. The head loss from friction is given approximately by the Darcy-Weisbach equation:

$$h_f = f(\frac{Lv^2}{D2g}) = 0.07(\frac{Lv^2}{D2g}) \qquad (12\text{–}11)$$

where h_f is the friction loss in ft, $f = 0.07$ is taken as the friction factor, L is the developed length in ft, v is the flow velocity in ft/sec, D is the tube I.D. in ft, and g is the gravitational constant 32.2 ft/sec^2. The value $f = 0.07$ can be used as the maximum allowable friction factor for both water and glycol fluids without introducing appreciable errors in sizing.

After the installation has been made, both the flow rate and pressure loss due to friction can be checked to see if the system is operating properly. What is important to remember is that if the friction losses are high (monitored by a pressure gauge at the pump outlet) or the circulating flow is low, then the system has excessive friction, which will both reduce the efficiency of the solar panels and increase the cost of circulating the fluid through the collector loop. An example will clarify the calculations to size the solar loop tubing and the circulating pump.

Example 12-5 Compute the flow rate and velocity through a $\frac{1}{2}$-in. type L copper tube feeding 75 ft^2 of collector panels with a circulation flow of 0.5 gal/hr for each ft^2 of panel surface.

Solution:
The flow rate is computed from the product of the panel area and the flow rate through each ft^2 of panel.

$$\text{Flow rate} = (75\ \text{ft}^2)(0.5\ \text{gal/hr}\cdot\text{ft}^2)(1/60\ \text{hr/min}) = 0.625\ \text{gal/min}\ (2.37\ \text{l/m})$$

Half-inch type L copper tubing has an inside diameter of 0.545 in., and the flow velocity is computed from

$$v = \frac{Q}{A} = \frac{(0.625\ \text{gal/min})(231\ \text{in.}^3/\text{gal})(1/1728\ \text{ft}^3/\text{in.}^3)(1/60\ \text{min/sec})}{(3.14)(0.545\ \text{in.})^2(0.25)(1/144\ \text{ft}^2/\text{in.}^2)}$$

and

$$v = 0.86\ \text{ft/sec}\ (0.26\ \text{m/s})$$

Example 12-6 Size a pump for Example 12–5 if the system has a developed tube length (including the runs and equivalent lengths of all fittings) of 300 ft.

Solution:
The flow rate through the system is 0.625 gal/min. The head loss can be computed by using the Darcy-Weisbach formula with an assumed maximum value of $f = 0.07$.

$$h_f = \frac{(0.07)(300\ \text{ft})(0.86\ \text{ft/sec})^2}{(0.545\ \text{in.})(\frac{1}{12}\ \text{ft/in.})(2)(32.2\ \text{ft/sec}^2)} = 5.3\ \text{ft}\ (1.6\ \text{m})$$

or, converting this to lbf/in.^{2} (of H_2O), we have

$$h_f = (5.3 \text{ ft})(0.433 \text{ lbf/in.}^2\text{·ft}) = 2.3 \text{ lbf/in.}^2 \text{ (15.8 kPa)}$$

Thus a pump would be selected that could deliver a flow rate of 0.625 gal/min at a head loss of 2.3 lbf/in.^{2}.

Although most systems are not instrumented beyond the sensors and controls necessary for operation, having taps for flow rate indicators, and pressure and temperature sensing gauges to monitor system performance over an extended period of time seems to warrant the added cost. Providing a connection to install a simple flow meter, for example, is almost a necessity to be sure that the rated flow is actually circulating through the system. A permanent tap to monitor pump electrical current will also insure that, when the pump is delivering rated flow, the current draw is within established limits, and available to indicate high friction and pumping losses should failure in the pump occur.

12-7 Summary of Practice

Plumbers install and service domestic solar hot water systems because they pertain to potable water, although local jurisdictions will have to be honored. Most systems will be purchased as a package and installed as a single tank system in new houses, or as a two-tank preheat system in houses that have an existing hot water heater.

The size of a solar domestic hot water system is based upon the water usage of the household and the general climatic condition for that geographic zone. The range of 15–20 gallons of hot water for each occupant is common. The climatic condition for a specific area can be determined from local weather data and experience with existing solar installations in the area. The plumber should keep a reference file with such items as ground water temperatures, ambient air temperatures, daylight hours, and insolation values for that locality. Sizing is done by comparing the available harvest of solar heat using two or more panels with the annual load requirements; 50 to 80 percent of the total requirements is the goal for the system, depending upon the location and preference of the customer. Even with packaged systems a standard chart should be worked out to estimate the capability and performance of the system. This has value not only in checking the performance, but in determining the savings in energy that result from the system. The economic payback for the system is usually within 3 to 6 years, including tax credits,[10] and a number of manufacturers provide a service for analyzing the energy system life cycle costs.[11] Other considerations, such as quality of life, also exert a major influence on the customer to select a solar heating system.

Installation practices suggested by the manufacturer should be followed, particularly when one is mounting and connecting the solar collector panels on the roof. Roof leaks can be caused by improper mounting. Flexing of loose panels caused by insecure mounting can break connections at the manifold. Improper soldering can also result in joint leaks. Particular attention should be given to installing the insulation on connecting tubing. As with the internal plumbing system in the building, the solar system is expected to have a life at least equal to the collector panels themselves, usually 15 to 25 years.

REVIEW QUESTIONS AND PROBLEMS

1. What is meant by *azimuth angle* and *altitude angle?*

2. What are the three components of a solar system?

3. What is the difference between direct and diffuse radiation?

4. What are some differences between flat plate and concentrator collectors?

5. Why is the plumber involved in solar systems?

6. How are domestic hot water needs (load) determined?

7. Approximately how much heat is required to raise the temperature of 50 gal (189 1) of water 100°F (55°C)?

8. What is the monthly (30 days) value of 90 000 Btu/day collected by using a solar system, if the cost of energy is 4¢ per KWH?

9. Why is a solar system considered to be a "supplemental" rather than the main system?

10. What fixed direction and angle should a flat plate collector have to collect the most radiation for the entire year?

11. What is the collector tilt factor F_b, and what use does it have in computing the solar harvest?

12. What is the collector panel efficiency factor, and how does it affect the efficiency of the collector?

13. Plot the load vs. harvest (insolation) from Examples 12-1 and 12-4, using months for the x-axis and Btu load/output for the y-axis.

14. From climatic sources and local weather data, assemble the following data for one specific geographic location for each month of the year:
a. latitude
b. collector tilt for hot water systems
c. ground water temperature
d. ambient temperature
e. horizontal solar insolation
f. percentage diffuse radiation
g. tilt factor
h. solar daytime hours

15. From the specifications of one of the many domestic hot water solar packages available and using the optimum number of panels for your area, estimate the performance (percentage of hot water supplied) and utility savings for one year. Use a load of 75 gal/day heated to 135°F (57.2°C) and a utility rate of 4.5¢/KWH.

REFERENCES

[1] Sheldon H. Butt (President of Solar Energy Industries Association), "The Solar Future, 1978," *Solar Engineering Magazine,* March, 1978.

[2] Ibid., pp. 33–43.

[3] *Solar Energy Systems,* Copper Development Association Inc., 405 Lexington Avenue, New York, N.Y. 10017.

[4] Scotch Programs Incorporated, 8107 72nd Avenue, Box 430734, Miami, Florida 33143.

[5] "1976 Systems," *ASHRAE Handbook and Product Directory, Chapter 43, ASHRAE, Inc. New York, N.Y. 1976.*

[6] ASHRAE Standard 93-77 (also ANSI Standard B 198.1-1977), "Methods of Testing to Determine the Thermal Performance of Solar Collectors," ASHRAE, Inc., 345 E. 47th Street, New York, N.Y. 10017.

[7] H. McKinley Conway, Jr., *The Weather Handbook* (Atlanta, Ga.: Conway Publications, Inc., 1963).

[8] Charles Hill, "Gauging Hydraulic Performance For Solar Hot Water Systems," *Solar Engineering Magazine,* March, 1978.

[9] Ibid.

[10] See Federal Tax Credit Form Number 5695, available through local IRS Offices.

[11] For example, Solar Energy Products, Inc. 1208 NW 8th Avenue, Gainesville, Fl. 32601.

Appendices

A. Plumbing Definitions
B. Graphic Symbols for Plumbing Fixtures, Pipe Fittings and Valves, and Piping
C. Mathematical Symbols and Plumbing Abbreviations
D. Conversion Factors
E. Capacities of Cylinders and Tanks
F. Properties of Fluids
G. Friction Losses in Pipes and Fittings
H. Powers and Roots of Numbers
I. Natural Trigonometric Functions
J. Natural Trigonometric Functions Table
K. Plumbing Standards
L. Solar Insolation Maps
M. Answers to Even-Numbered Problems

Appendix A

PLUMBING DEFINITIONS

The following terms shall have the meaning indicated in this list. No attempt is made to define ordinary words which are used in accordance with their established dictionary meaning except where it is necessary to define their meaning as used in the National Standard Plumbing Code to avoid misunderstanding.

Accessible: Accessible means having access thereto but which first may require the removal of an access panel, door or similar obstruction. "Readily accessible" means direct access without the necessity of removing or moving any panel, door or similar obstruction.

Acid Waste: See "**Special Wastes.**"

Administrative Authority: The individual official, board, department, or agency established and authorized by a state, county, city or other political subdivision created by law to administer and enforce the provisions of the plumbing code as adopted or amended.

Air Break (Drainage System): A piping arrangement in which a drain from a fixture, appliance, or device discharges indirectly into a fixture, receptacle, or interceptor at a point below the flood level rim of the receptacle so installed as to prevent backflow or siphonage.

Air Gap (Drainage System): The unobstructed vertical distance through the free atmosphere between the outlet of waste pipe and the flood level rim of the receptacle into which it is discharging.

Air Gap (Water Distribution System): The unobstructed vertical distance through the free atmosphere between the lowest opening from any pipe or faucet supplying water to a tank, plumbing fixture or other device and the flood level rim of the receptacle.

Anchors: See **Supports.**

Approved: Accepted or acceptable under an applicable standard stated or cited in the National Standard Plumbing Code, or accepted as suitable for the proposed use under procedures and powers of the Administrative Authority.

Area Drain: A receptacle designed to collect surface or storm water from an open area.

Aspirator: A fitting or device supplied with water or other fluid under positive pressure which passes through an integral orifice or "constriction" causing a vacuum.

Autopsy Table: A fixture or table used for the postmortem examination of a body.

Backflow: The flow of water or other liquids, mixtures, or substances into the distributing pipes of a potable supply of water from any source or sources other than its intended source. Backsiphonage is one type of backflow.

Backflow Connection: Any arrangement whereby backflow can occur.

Reprinted with permission from the National Standard Plumbing Code, 1978.

Backflow Preventer: A device or means to prevent backflow.

Backflow Preventer, Reduced Pressure Zone Type: An assembly of differential valves and check valves including an automatically opened spillage port of the atmosphere.

Back-Siphonage: The flowing back of used, contaminated, or polluted water from a plumbing fixture or vessel or other sources into a potable water supply due to a negative pressure in such pipe.

Back Water Valve: A device installed in a drain or pipe to prevent backflow.

Barometric Loop: A loop of pipe rising approximately 35 feet, at its topmost point, above the highest fixture it supplies.

Battery of Fixtures: Any group of two or more similar adjacent fixtures which discharge into a common horizontal waste or soil branch.

Bedpan Steamer: A fixture used for scalding bedpans or urinals by direct application of steam.

Bedpan Washer: A fixture designed to wash bedpans and to flush the contents into the soil drainage system. It may also provide for steaming the utensils with steam or hot water.

Bedpan Washer Hose: A device supplied with hot and cold water and located adjacent to a water closet or clinic sink to be used for cleansing bedpans.

Boiler Blow-Off: An outlet on a boiler to permit emptying or discharge of sediment.

Boiler Blow-Off Tank: A vessel designed to receive the discharge from a boiler blow-off outlet and to cool the discharge to a temperature which permits its safe discharge to the drainage system.

Branch: Any part of the piping system other than a riser, main or stack.

Branch, Fixture: See **Fixture Branch.**

Branch, Horizontal: See **Horizontal Branch.**

Branch Interval: A distance along a soil or waste stack corresponding in general to a story height, but in no case less than 8 feet, within which the horizontal branches from one floor or story of a building are connected to the stack.

Branch Vent: A vent connecting one or more individual vents with a vent stack or stack vent.

Building: A structure having walls and a roof designed and used for the housing, shelter, enclosure or support of persons, animals or property.

Building Classification: The arrangement adopted by the Administrative Authority for the designation of buildings in classes according to occupancy.

Building Drain: That part of the lowest piping of a drainage system which receives the discharge from soil, waste and other drainage pipes inside the walls of the building and conveys it to the building sewer beginning 3 feet outside the building wall.

Building Drain—Combined: A building drain which conveys both sewage and storm water or other drainage.

Building Drain—Sanitary: A building drain which conveys sewage only.

Building Drain—Storm: A building drain which conveys storm water or other drainage but no sewage.

Building Sewer: That part of the drainage system which extends from the end of the building drain and conveys its discharge to a public sewer, private sewer, individual sewage-disposal system, or other point of disposal.

Building Sewer—Combined: A building sewer which conveys both sewage and storm water or other drainage.

Building Sewer—Sanitary: A building sewer which conveys sewage only.

Building Sewer—Storm: A building sewer which conveys storm water or other drainage but no sewage.

Building Subdrain: That portion of a drainage system which does not drain by gravity into the building sewer.

Building Trap: A device, fitting, or assembly of fittings installed in the building drain to prevent circulation of air between the drainage system of the building and the building sewer.

Cesspool: A lined and covered excavation in the ground which receives the discharge of domestic sewage or other organic wastes from a drainage system, so designed as to retain the organic matter and solids, but permitting the liquids to seep through the bottom and sides.

Chemical Waste: See "**Special Wastes.**"

Circuit Vent: A branch vent that serves two or more traps and extends from the downstream side of the highest fixture connection of a horizontal branch to the vent stack.

Clear Water Waste: Cooling water and condensate drainage from refrigeration, and air conditioning equipment; cooled condensate from steam heating systems; cooled boiler blowdown water; waste water drainage from equipment rooms and other areas where water is used without an appreciable addition of oil, gasoline, solvent, acid, etc., and treated effluent in which impurities have been reduced below a minimum concentration considered harmful.

Clinic Sink (Bedpan Hopper): A sink designed primarily to receive wastes from bedpans provided with a flush rim, integral trap with a visible trap seal, having the same flushing and cleansing characteristic as a water closet.

Code: Regulations, subsequent amendments thereto, or any emergency rule or regulation which the Administrative Authority having jurisdiction may lawfully adopt.

Combination Fixture: A fixture combining one sink and laundry tray or a two- or three-compartment sink or laundry tray in one unit.

Combined Building Drain: See **Building Drain—Combined.**

Combined Building Sewer: See **Building Sewer—Combined.**

Combination Waste and Vent System: A specially designed system of waste piping embodying the horizontal wet venting of one or more sinks or floor drains by means of a common waste and vent pipe adequately sized to provide free movement of air above the flow line of the drain.

Common Vent: A vent connected at a common connection of two fixture drains and serving as a vent for both fixtures.

Conductor: A conductor is the water conductor from the room to the building storm drain, combined building sewer, or other means of disposal and located inside of the building.

Continuous Vent: A vertical vent that is a continuation of the drain to which it connects.

Continuous Waste: A drain from two or more fixtures connected to a single trap.

Critical Level: The critical level marking on a backflow prevention device or vacuum breaker is a point established by the manufacturer and usually stamped on the device by the manufacturer which determines the minimum elevation above the floor level rim of the fixture or receptacle served at which the device may be installed. When a backflow prevention device does not bear a critical level marking, the bottom of the vacuum breaker, combination valve, or the bottom of any approved device shall constitute the critical level.

Cross Connection: Any connection or arrangement between two otherwise separate piping systems, one of which contains potable water and the other either water of questionable safety, steam, gas, or chemical whereby there may be a flow from one system to the other, the direction of flow depending on the pressure differential between the two systems. (See **Backflow** and **Back-Siphonage.**)

Dead End: A branch leading from a soil, waste, or vent pipe, building drain, or building sewer, and terminating at a developed length of 2 feet or more by means of a plug, cap, or other closed fitting.

Department Having Jurisdiction: See **Administrative Authority.**

Developed Length: The length of a pipe line measured along the center line of the pipe and fittings.

Diameter: The nominal diameter as designated commercially.

Double Check Valve Assembly: A backflow prevention device consisting of two independently acting check valves, internally force loaded to a normally closed position between two tightly closing shut-off valves, and with means of testing for tightness.

Double Offset: Two changes of direction installed in succession or series in a continuous pipe.

Downspout: A downspout is the rainleader from the roof to the building storm drain, combined building sewer, or other means of disposal and located outside of the building.

Domestic Sewage: The water-borne wastes derived from ordinary living processes.

Drain: Any pipe which carries waste water or water-borne wastes in a building drainage system.

Drainage Pipe: See **Drainage System.**

Drainage System: Includes all the piping, within public or private premises, which conveys sewage, rainwater, or other liquid wastes to a point of disposal. It does not include the mains of a public sewer system or private or public sewage-treatment or disposal plant.

Drainage System, Building Gravity: A drainage system which drains by gravity into the building sewer.

Drainage System, Sub-Building: See **Building Subdrain.**

Dry Well: See **Leaching Well.**

Dual Vent: See **Common Vent.**

Dwelling Unit—Multiple: A room or group of rooms forming a single habitable unit with facilities which are used or intended to be used for living, sleeping, cooking and eating; and whose sewer connections and water supply within its own premise are shared with one or more other dwelling units.

Dwelling Unit—Single: A room or group of rooms forming a single habitable unit with facilities which are used or intended to be used for living, sleeping, cooking and eating; and whose sewer connections and water supply are within its own premise separate from and completely independent of any other dwelling.

Effective Opening: The minimum cross-sectional area at the point of water supply discharge, measured or expressed in terms of (1) diameter of a circle, or (2) if the opening is not circular, the diameter of a circle of equivalent cross-sectional area.

Existing Work: A plumbing system or any part thereof installed prior to the effective date of this Code.

Family: One or more individuals living together and sharing the same facilities.

Fixture: See **Plumbing Fixture.**

Fixture Branch: A pipe connecting several fixtures.

Fixture Drain: The drain from the trap of a fixture to the junction of that drain with any other drain pipe.

Fixture Supply: The water supply pipe connecting a fixture to a branch water supply pipe or directly to a main water supply pipe.

Fixture Unit (Drainage—d.f.u.): A measure of the probable discharge into the drainage system by various types of plumbing fixtures. The drainage fixture-unit value for a particular fixture depends on its volume rate of drainage discharge, on the time duration of a single drainage operation, and on the average time between successive operations.

Fixture Unit (Supply—s.f.u.): A measure of the probable hydraulic demand on the water supply by various types of plumbing fixtures. The supply fixture-unit value for a particular fixture depends on its volume rate of supply, on the time duration of a single supply operation, and on the average time between successive operations.

Flood Level: See **Flood Level Rim.**

Flood Level Rim: The edge of the receptacle from which water overflows.

Flooded: The condition which results when the liquid in a container or receptacle rises to the flood-level rim.

Flow Pressure: The pressure in the water supply pipe near the faucet or water outlet while the faucet or water outlet is wide-open and flowing.

Flushing-Type Floor Drain: A floor drain which is equipped with an integral water supply, enabling flushing of the drain receptor and trap.

Flush Valve: A device located at the bottom of a tank for flushing water closets and similar fixtures.

Flushometer Valve: A device which discharges a predetermined quantity of water to fixtures for flushing purposes and is closed by direct water pressure.

Frostproof Closet: A hopper with no water in the bowl and with the trap and water supply control valve located below frost line.

Grade: The fall (slope) of a line of pipe in reference to a horizontal plane. In drainage it is usually expressed as the fall in a fraction of an inch per foot length of pipe.

Grease Interceptor: See **Interceptor.**

Grease Trap: See **Interceptor.**

Ground Water: Subsurface water occupying the zone of saturation.

A. Confined Ground Water—A body of ground water overlain by material sufficiently impervious to sever free hydraulic connection with overlying ground water.
B. Free Ground Water—Ground water in the zone of saturation extending down to the first impervious barrier.

Hangers: See **Supports.**

Horizontal Branch Drain: A drain branch pipe extending laterally from a soil or waste stack or building drain, with or without vertical sections or branches, which receives the discharge from one or more fixture drains and conducts it to the soil or waste stack or to the building drain.

Horizontal Pipe: Any pipe or fitting which makes an angle of less than 45° with the horizontal.

Hot Water: Hot water is supplied to plumbing fixtures at a temperature of not less than 120 degrees Fahrenheit.

House Drain: See **Building Drain.**

House Sewer: See **Building Sewer.**

House Trap: See **Building Trap.**

Individual Sewage Disposal System: A system for disposal of domestic sewage by means of a septic tank, cesspool or mechanical treatment, designed for use apart from a public sewer to serve a single establishment or building.

Indirect Waste Pipe: A waste pipe which does not connect directly with the drainage system, but which discharges into the drainage system through an air break or air gap into a trap, fixture, receptor or interceptor.

Individual Vent: A pipe installed to vent a fixture drain. It connects with the vent system above the fixture served or terminates outside the building into the open air.

Individual Water Supply: A supply other than an approved public water supply which serves one or more families.

Industrial Wastes: Liquid or liquid borne wastes resulting from the processes employed in industrial and commercial establishments.

Insanitary: Contrary to sanitary principles—injurious to health.

Interceptor: A device designed and installed so as to separate and retain deleterious, hazardous, or undesirable matter from normal wastes while permitting normal sewage or liquid wastes to discharge into the drainage system by gravity.

Installed: An altered, changed or new installation.

Leaching Well or Pit: A pit or receptacle having porous walls which permit the contents to seep into the ground.

Leader: An exterior vertical drainage pipe for conveying storm water from roof or gutter drains.

Liquid Waste: The discharge from any fixture, appliance, area or appurtenance, which does not contain human or animal waste matter.

Load Factor: The percentage of the total connected fixture unit flow which is likely to occur at any point in the drainage system.

Local Ventilating Pipe: A pipe on the fixture side of the trap through which vapor or foul air is removed from a fixture.

Loop Vent: A circuit vent which loops back to connect with a stack vent instead of a vent stack.

Main: The principal pipe artery to which branches may be connected.

Main Sewer: See **Public Sewer.**

Main Vent: The principal artery of the venting system to which vent branches may be connected.

Multiple Dwelling: A building containing two or more dwelling units.

Nonpotable Water: Water not safe for drinking or for personal or culinary use.

Nuisance: Public nuisance at common law or in equity jurisprudence; whatever is dangerous to human life or detrimental to health; whatever building, structure, or premise is not sufficiently ventilated, sewered, drained, cleaned, or lighted, in reference to its intended or actual use; and whatever renders the air or human food or drink or water supply unwholesome.

Offset: A combination of elbows or bends which brings one section of the pipe out of line but into a line parallel with the other section.

Oil Interceptor: See **Interceptor.**

Pitch: See **Grade.**

Plumbing: The practice, materials and fixtures used in the installation, maintenance, extension, alteration and removal of all piping, plumbing fixtures, plumbing appliances and plumbing appurtenances in connection with any of the following: Sanitary drainage or storm facilities and venting system and the public or private water supply systems within or adjacent to any building, structure or conveyance; also the pratice and materials used in the installation, maintenance, extension, alteration or removal of storm water, refrigeration and air conditioning drains, liquid waste or sewage, and water supply systems of any premises to their connection with the public water supply system or to an acceptable disposal facility.

Except for the initial connection to a potable water supply and the final connection that discharges indirectly into a public or private disposal system, the following are excluded from this definition: All piping, equipment or material used exclusively for environmental control, for incorporation of liquids or gases into any product or process for use in the manufacturing or storage of any product, including product development, or for the installation, alteration, repair or removal of automatic sprinkler systems installed for fire protection only or their related appurtenances or standpipes connected to automatic sprinkler systems or overhead or underground fire lines beginning at a point where water is used exclusively for fire protection.

Plumbing Appliance: Any one of a special class of plumbing fixture which is intended to perform a special plumbing function. Its operation and/or control may be dependent upon one or more energized components, such as motors, controls, heating elements, or pressure or temperature-sensing elements. Such fixtures may operate automatically through one or more of the following actions: a time cycle, a temperature range, a pressure range, a measured volume or weight; or the fixture may be manually adjusted or controlled by the user or operator.

Plumbing Appurtenance: A manufactured device, or a prefabricated assembly, or an on-the-job assembly of component parts, and which is an adjunct to the basic piping system and plumbing fixtures. An appurtenance demands no additional water supply, nor does it add any discharge load to a fixture or the drainage system. It is presumed that it performs some useful function in the operation, maintenance, servicing, economy, or safety of the plumbing system.

Plumbing Fixture: A receptacle or device which is either permanently or temporarily connected to the water distibution system of the premises, and demands a supply of water therefrom, or it discharges used water, liquid-borne waste materials, or sewage either directly or indirectly to the drainage system of the premises, or which requires both a water supply connection and a discharge to the drainage system of the premises. Plumbing appliances as a special class of fixture are further defined.

Plumbing Inspector: See **Administrative Authority.**

Plumbing System: Includes the water supply and distribution pipes, plumbing fixtures and traps; soil, waste and vent pipes; sanitary and storm drains and building sewers, including their respective connections, devices and appurtenances to an approved point of disposal.

Pollution: The addition of sewage, industrial wastes, or other harmful or objectionable material to water. Sources of sewage pollution may be privies, septic tanks, subsurface irrigation fields, seepage pits, sink drains, barnyard wastes, etc.

Pool: See **Swimming Pool.**

Potable Water: Water free from impurities present in amounts sufficient to cause disease or harmful physiological effects and conforming in its bacteriological and chemical quality to the requirements of the Public Health Service Drinking Water Standards or the regulations of the public health authority having jurisdiction.

Private or Private Use: In the classification of plumbing fixtures, private applies to fixtures in residences and apartments and similar installations.

Private Sewer: A sewer not directly controlled by public authority.

Public or Public Use: In the classification of plumbing fixtures, public applies to every fixture not defined under private use and public shall include all installations where a number of fixtures are installed and their use may be restricted or unrestricted.

Public Sewer: A common sewer directly controlled by public authority.

Public Water Main: A water supply pipe for public use controlled by public authority.

Receptor: A fixture or device which receives the discharge from indirect waste pipes.

Reduced Pressure Backflow Preventer: A backflow prevention device consisting of two independently acting check valves, internally force loaded to a normally closed position and separated by an intermediate chamber (or zone) in which there is an automatic relief means of venting to atmosphere internally loaded to a normally open position between two tightly closing shut-off valves and with means for testing for tightness of the checks and opening of relief means.

Relief Vent: An auxiliary vent which permits additional circulation of air in or between drainage and vent systems.

Return Offset: A double offset installed so as to return the pipe to its original alignment.

Revent Pipe: See **Individual Vent.**

Rim: An unobstructed open edge of a fixture.

Rise: A water supply pipe which extends vertically one full story or more to convey water to branches or to a group of fixtures.

Roof Drain: A drain installed to receive water collecting on the surface of a roof and to discharge it into a leader or a conductor.

Roughing-In: The installation of all parts of the plumbing system which can be completed prior to the installation of fixtures. This includes drainage, water supply, and vent piping, and the necessary fixture supports, or any fixtures that are built into the structure.

Safe Waste: See **Indirect Waste.**

Sand Filter: A treatment device or structure, constructed above or below the surface of the ground, for removing solid or colloidal material of a type that cannot be removed by sedimentation, from septic tank effluent.

Sand Interceptor: See **Interceptor.**

Sand Trap: See **Interceptor.**

Sanitary Sewer: A sewer which carries sewage and excludes storm, surface and ground water.

Scavenger: Any person engaged in the business of cleaning and emptying septic tanks, seepage pits, privies or any other sewage disposal facility.

Seepage Well or Pit: See **Leaching Well.**

Separator: See **Interceptor.**

Septic Tank: A water-tight receptable which receives the discharge of a building sanitary drainage system or part thereof, and is designed and constructed so as to separate solids from the liquid, digest organic matter through a period of detention, and allow the liquids to discharge into the soil outside of the tank through a system of open joint or perforated piping, or a seepage pit.

Sewage: Any liquid waste containing animal or vegetable matter in suspension or solution, and may include liquids containing chemicals in solution.

Sewage Ejectors: A device for lifting sewage by entraining it in a high velocity jet of steam, air or water.

Sewage Pump: A permanently installed mechanical device other than an ejector for removing sewage or liquid waste from a sump.

Side Vent: A vent connecting to the drain pipe through a fitting at an angle not greater than 45° to the vertical.

Size of Pipe and Tubing: See **Diameter.**

Slope: See **Grade.**

Soil Pipe: A pipe which conveys sewage containing human or animal waste to the building drain or building sewer.

Soil Vent: See **Stack Vent.**

Special Wastes: Wastes which require special treatment before entry into the normal plumbing system.

Special Waste Pipe: Pipes which convey special wastes.

Stack: A general term for any vertical line of soil, waste, vent or inside conductor piping. This does not include vertical fixture and vent branches that do not extend through the roof or that pass through not more than two stories before being reconnected to the vent stack or stack vent.

Stack Group: A group of fixtures located adjacent to the stack so that by means of proper fittings, vents may be reduced to a minimum.

Stack Vent: The extension of a soil or waste stack above the highest horizontal drain connected to the stack.

Stack Venting: A method of venting a fixture or fixtures through the soil or waste stack.

Sterilizer, Boiling Type: A fixture (nonpressure type) used for boiling instruments, utensils, and/or other equipment (used for disinfection) and may be portable or connected to the plumbing system.

Sterilizer Instrument: See **Sterilizer, Boiling Type.**

Sterilizer, Pressure, Instrument Washer: A fixture (pressure vessel) designed to both wash and sterilize instruments during the operating cycle of the fixture.

Sterilizer, Pressure (Autoclave): A fixture (pressure vessel) designed to use steam under pressure for sterilizing. See **Sterilizer, Boiling Type.**

Sterilizer Vent: A separate pipe or stack, indirectly connected to the building drainage system at the lower terminal, which receives the vapors from nonpressure sterilizers, or the exhaust vapors from pressure sterilizers, and conducts the vapors directly to the outer air. Sometimes called vapor, steam, atmosphere or exhaust vent.

Sterilizer, Water: A device for sterilizing water and storing sterile water.

Still: A device used in distilling liquids.

Storm Drain: See **Building Storm Drain.**

Storm Sewer: A sewer used for conveying rainwater, surface water, condensate, cooling water, or similar liquid wastes.

Subsoil Drain: A drain which collects subsurface or seepage water and conveys it to a place of disposal.

Sump: A tank or pit, which receives sewage or liquid waste, located below the normal grade of the gravity system and which must be emptied by mechanical means.

Sump Pump: A permanently installed mechanical device other than an ejector for removing sewage or liquid waste from a sump.

Supports: Devices for supporting and securing pipe, fixtures and equipment.

Swimming Pool: Any structure, basin, chamber or tank containing an artificial body of water for swimming, diving, or recreational bathing.

Tempered Water: Water at a temperature of not less than 90° and not more than 105°.

Trap: A fitting or device which provides a liquid seal to prevent the emission of sewer gases without materially affecting the flow of sewage or waste water through it.

Trap Arm: A trap arm is that portion of a fixture drain between a trap and its vent.

Trap Primer: A trap primer is a device or system of piping to maintain a water seal in a trap.

Trap Seal: The maximum vertical depth of liquid that a trap will retain, measured between the crown weir and the top of the dip of the trap.

Vacuum: Any pressure less than that exerted by the atmosphere.

Vacuum Breaker: See **Backflow Preventer.**

Vacuum Breaker, Nonpressure Type (Atmospheric): A vacuum breaker which is not designed to be subject to static line pressure.

Vacuum Breaker, Pressure Type: A vacuum breaker designed to operate under conditions of static line pressure.

Vacuum Relief Valve: A device to prevent excessive vacuum in a pressure vessel.

Vent Pipe: Part of the vent system.

Vent Stack: A vertical vent pipe installed to provide circulation of air to and from the drainage system and which extends through one or more stories.

Vent System: A pipe or pipes installed to provide a flow of air to or from a drainage system or to provide a circulation of air within such system to protect trap seals from siphonage and back pressure.

Vertical Pipe: Any pipe or fitting which makes an angle of 45° or less with the vertical.

Wall Hung Water Closet: A water closet installed in such a way that no part of the water closet touches the floor.

Waste: Liquid waste and industrial waste.

Waste Pipe: A pipe which conveys only waste.

Water-Distributing Pipe: A pipe within the building or on the premises which conveys water from the water-service pipe to the point of usage.

Water Lifts: See **Sewage Ejector.**

Water Main: A water supply pipe for public use.

Water Outlet: A discharge opening through which water is supplied to a fixture, into the atmosphere (except into an open tank which is part of the water supply system), to a boiler or heating system, to any devices or equipment requiring water to operate but which are not part of the plumbing system.

Water Riser Pipe: See **Riser.**

Water Service Pipe: The pipe from the water main or other source of potable water supply to the water-distributing system of the building served.

Water Supply System: The water service pipe, the water-distributing pipes, and the necessary connecting pipes, fittings, control valves, and all appurtenances in or adjacent to the building or premises.

Wet Vent: A vent which receives the discharge of wastes other than from water closets and kitchen sinks.

Yoke Vent: A pipe connecting upward from a soil or waste stack to a vent stack for the purpose of preventing pressure changes in the stack.

Appendix B

GRAPHIC SYMBOLS FOR PLUMBING FIXTURES, PIPE FITTINGS AND VALVES, AND PIPING

Fixtures

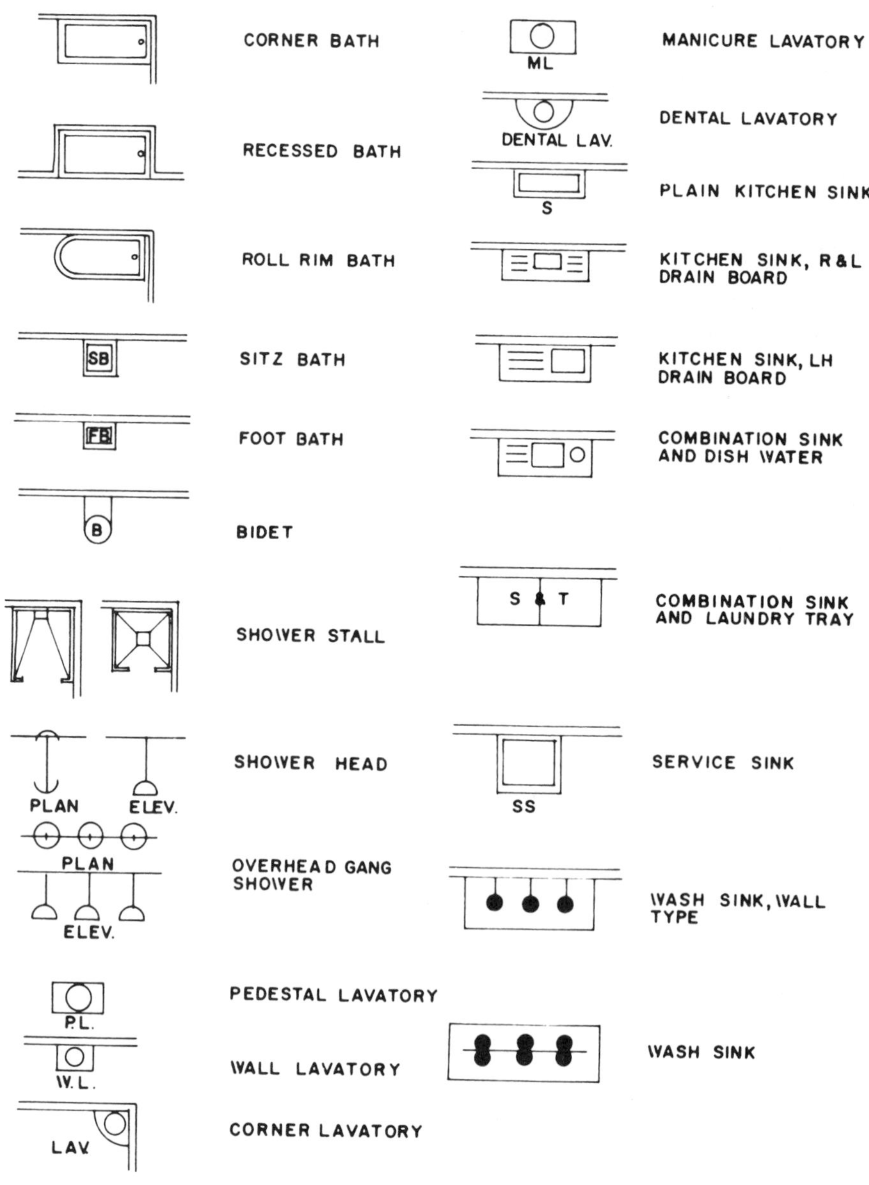

Fixtures

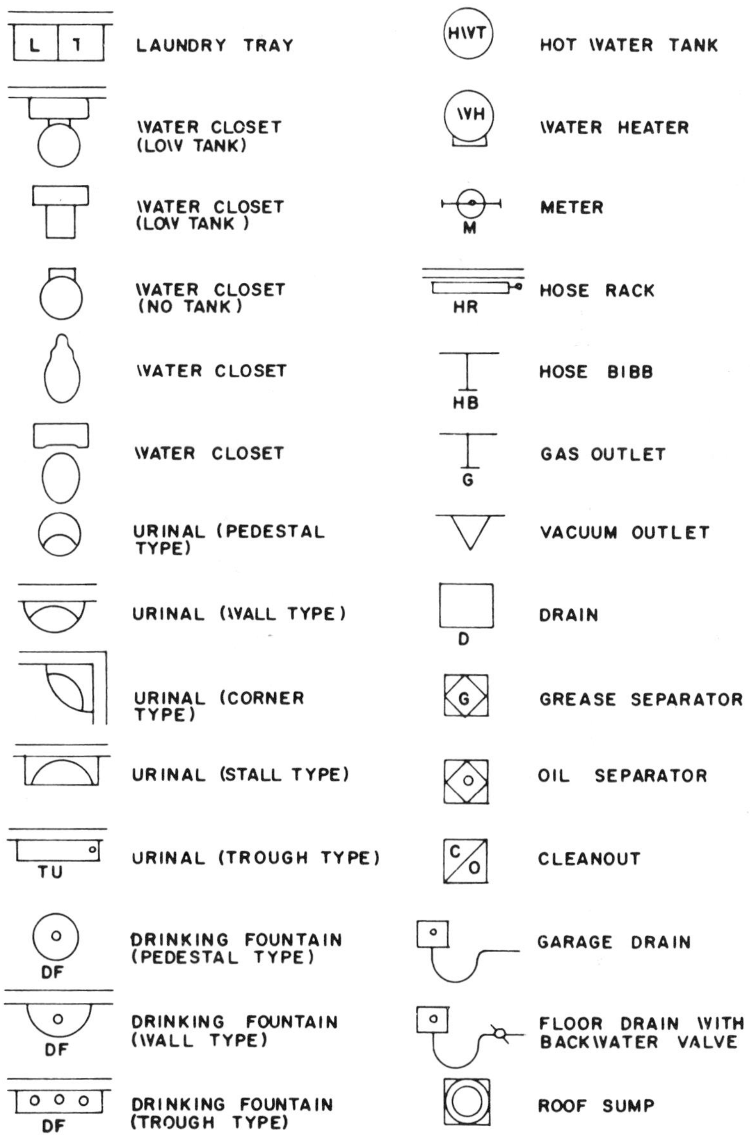

LAUNDRY TRAY

WATER CLOSET (LOW TANK)

WATER CLOSET (LOW TANK)

WATER CLOSET (NO TANK)

WATER CLOSET

WATER CLOSET

URINAL (PEDESTAL TYPE)

URINAL (WALL TYPE)

URINAL (CORNER TYPE)

URINAL (STALL TYPE)

URINAL (TROUGH TYPE)

DRINKING FOUNTAIN (PEDESTAL TYPE)

DRINKING FOUNTAIN (WALL TYPE)

DRINKING FOUNTAIN (TROUGH TYPE)

HOT WATER TANK

WATER HEATER

METER

HOSE RACK

HOSE BIBB

GAS OUTLET

VACUUM OUTLET

DRAIN

GREASE SEPARATOR

OIL SEPARATOR

CLEANOUT

GARAGE DRAIN

FLOOR DRAIN WITH BACKWATER VALVE

ROOF SUMP

Pipe Fittings and Valves

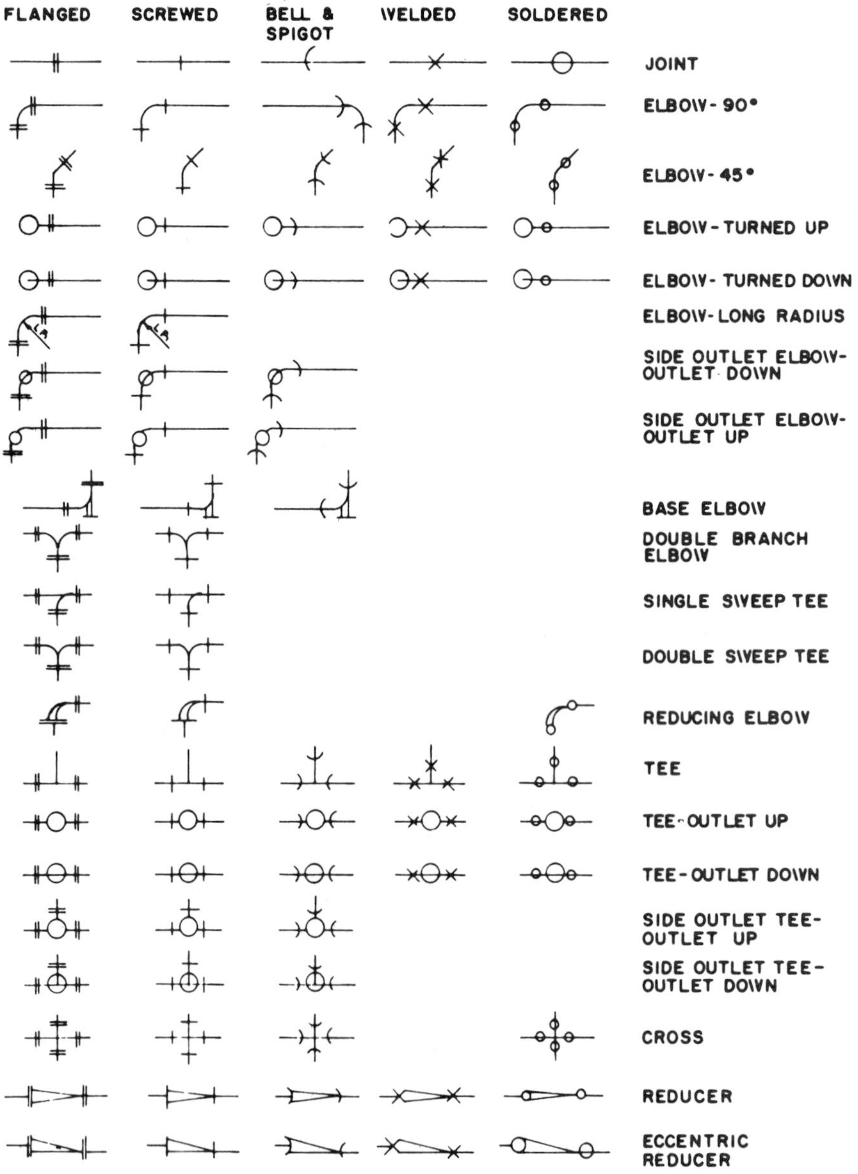

FLANGED	SCREWED	BELL & SPIGOT	WELDED	SOLDERED	
					JOINT
					ELBOW - 90°
					ELBOW - 45°
					ELBOW - TURNED UP
					ELBOW - TURNED DOWN
					ELBOW - LONG RADIUS
					SIDE OUTLET ELBOW-OUTLET DOWN
					SIDE OUTLET ELBOW-OUTLET UP
					BASE ELBOW
					DOUBLE BRANCH ELBOW
					SINGLE SWEEP TEE
					DOUBLE SWEEP TEE
					REDUCING ELBOW
					TEE
					TEE - OUTLET UP
					TEE - OUTLET DOWN
					SIDE OUTLET TEE-OUTLET UP
					SIDE OUTLET TEE-OUTLET DOWN
					CROSS
					REDUCER
					ECCENTRIC REDUCER

Pipe Fittings and Valves

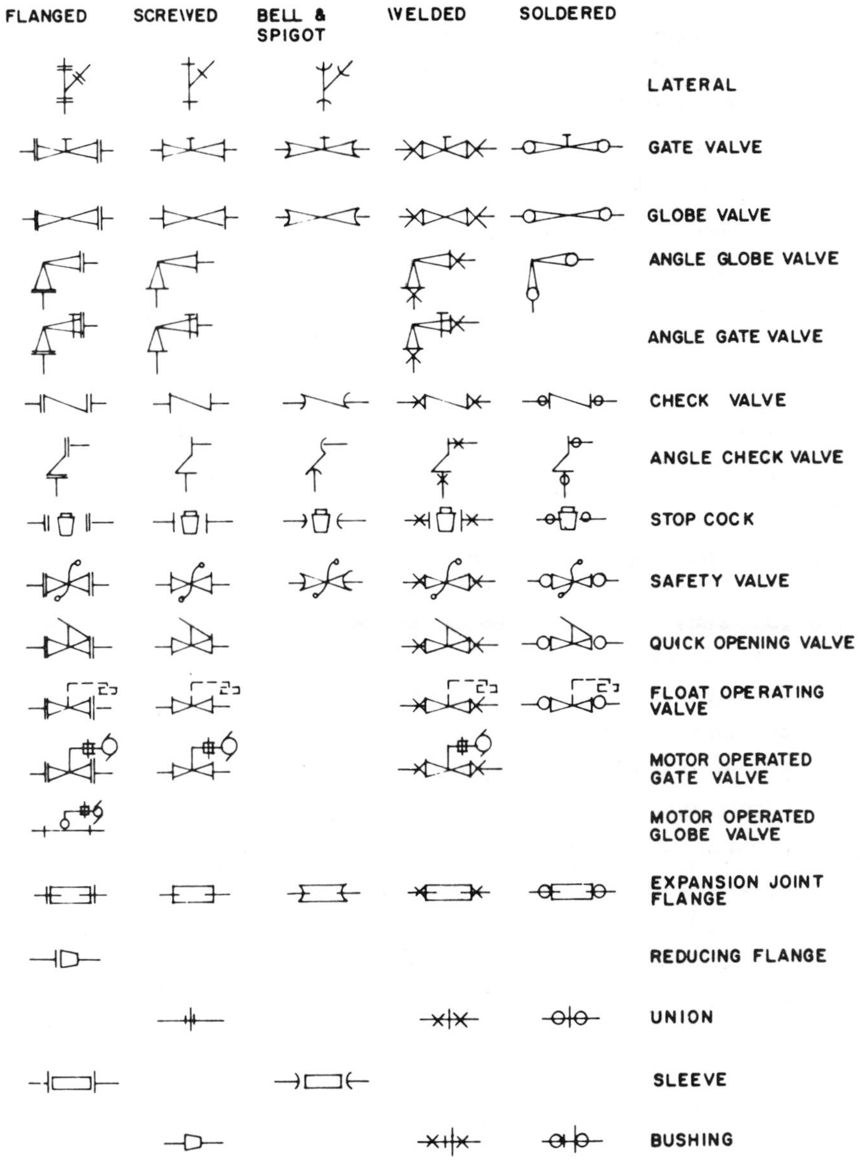

FLANGED SCREWED BELL & SPIGOT WELDED SOLDERED

LATERAL

GATE VALVE

GLOBE VALVE

ANGLE GLOBE VALVE

ANGLE GATE VALVE

CHECK VALVE

ANGLE CHECK VALVE

STOP COCK

SAFETY VALVE

QUICK OPENING VALVE

FLOAT OPERATING VALVE

MOTOR OPERATED GATE VALVE

MOTOR OPERATED GLOBE VALVE

EXPANSION JOINT FLANGE

REDUCING FLANGE

UNION

SLEEVE

BUSHING

Piping

CHARACTER	PLAN LINE	OR
CIRC. HOT CITY WATER		
CHILLED DRINK. WATER		
FIRE LINE		F
COLD INDUSTRIAL WATER		
HOT INDUSTRIAL WATER		
CIRC. HOT INDUS. WATER		
AIR		A
GAS		G
OIL		O
VACUUM CLEANER		V
LOCAL OR SURFACE VENT		

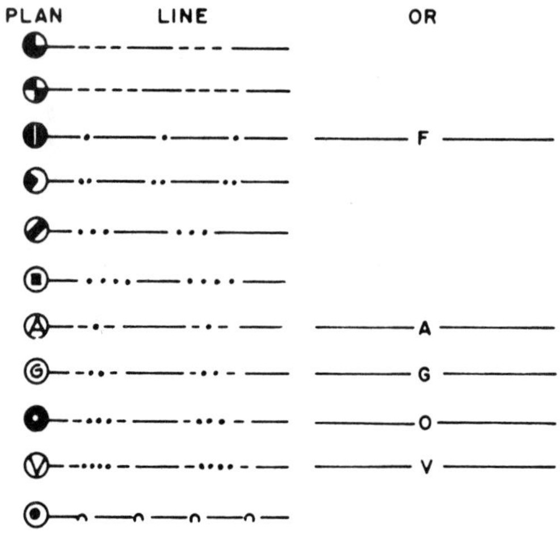

CHARACTER	PLAN LINE
SANITARY SEWAGE	
SOIL STACK	24
WASTE STACK	17
VENT STACK	18
COMBINED SEWAGE	
STORM SEWAGE	
ROOF LEADER	
INDIRECT WASTE	
INDUSTRIAL SEWAGE	
ACID OR CHEMICAL WASTE	
COLD CITY WATER	
HOT CITY WATER	

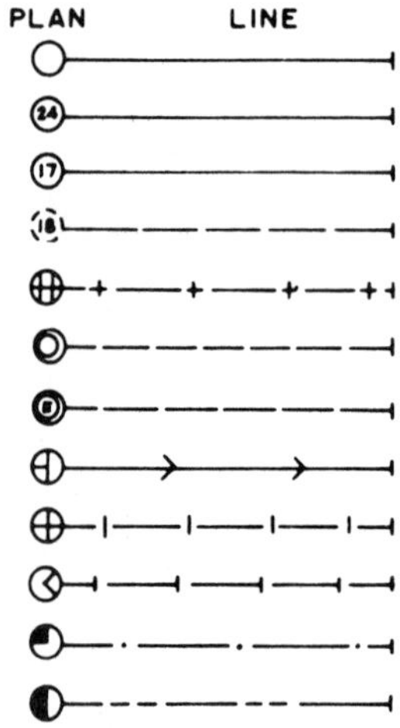

Appendix C

MATHEMATICAL SYMBOLS AND PLUMBING ABBREVIATIONS

Letter Symbol	*Quantity*	*English Units*	*SI Units*	*SI Symbol*
A	Area	ft^2, in^2	m^2, cm^2	
D, d	Diameter	ft, in	m, cm, mm	
F	Force	lbf	newtons	N
F_b	Solar panel factor			
h	Head energy	ft, in	m, cm, mm	
I_d	Average incident-radiation	Btu/ft^2	joule/m^2	
I_h	Average radiation falling on a horizontal surface	Btu	joule/m^2	
J	Joule constant	778 ft-lbf/Btu		
K, k	Various constants			
L, l	Distance, length	ft, in	m, cm, mm	m
M	Mass	slugs	kg($N \cdot s^2/m$)	M
N	Speed	rpm	rpm	
P	Power	J/s	J/s	
Q, q	Liquid volume flow rate	ft^3/sec, gal/min	M^3/s, l/m	
R	Hydraulic radius	ft	m	
S	Slope			
T	Temperature	°F	°C	
V	Volume	ft^3, in^3	m^3	
W	Work	ft-lbf	m·N	
a	Acceleration	ft/sec^2	m/s^2	
g	Acceleration due to gravity	32.2 ft/sec^2	9.806 m/s^2	
f	Friction factor			
h	Head energy	ft	m	
h_f	Head loss due to friction			
p	Pressure	lbf/in^2	N/m^2	Pa
q_u	Energy	Btu/ft^2	joule/m^2	
r	Radius	ft, in	m, cm, mm	
s	Distance	ft, in	m, cm	
t	Time	sec	sec	s
v	Velocity	ft/sec	m/s	
w	Weight, Load	lbf	N	
x	Distance	ft	m	
z	Potential energy from elevation	ft	m	

Greek Symbols	Quantity	English Units	SI Units	SI Symbol
γ (gamma)	Specific weight	lbf/ft^3	N/m^3	
η (eta)	Solar panel efficiency			
θ (theta)	Angle			
ϕ (phi)	Angle			
π (pi)	Constant	3.14	3.14	
ϱ (rho)	Density	slugs/ft^3	kg/m^3	

Abbreviations	Quantity
abs	Absolute pressure
AGA	American Gas Association
AISI	American Iron and Steel Institute
ANSI	American National Standards Institute
API	American Petroleum Institute
ASA	American Standards Association (Now ANSI)
ASCE	American Society of Civil Engineering
ASHRAE	American Society of Heating, Refrigeration and Air Conditioning
ASME	American Society of Mechanical Engineers
ASPE	American Society of Plumbing Engineers
ASSE	American Society of Sanitation Engineers
ASTM	American Society for Testing and Materials
AWWA	American Water Works Association
B & S	Bell and spigot
B.O.C.A.	Building Officials and Code Administrators International
Btu	British thermal unit
C–C	Center-to-center measurement
C–F	Center-to-face measurement
C I	Cast iron
CISP	Cast iron soil pipe
CISPI	Cast Iron Soil Pipe Institute
C.O.	Cleanout
CWFU	Cold water fixture unit
C.W.	Cold water
dfu	Drainage fixture unit
DWV	Drain, waste and vent
Fe	Iron (chemical abbreviation)
FF	Finish floor
F–F	Face-to-face measurement
F.G.	Finish grade
gal	Gallon
GPD	Gallons per day
GPM	Gallons per minute
H_2O	Water (chemical abbreviation)

Abbreviations	*Quantity*
H.B.	Hose bib
H.W.	Hot water
hwfu	Hot water fixture unit
I.A.P.M.O.	International Association of Plumbing and Mechanical Officials
I.C.B.O.A.	International Conference of Building Officials Association
I.D.	Internal diameter
IPS	Iron pipe size
lav	Lavatory
MCA	Mechanical Contractors Association
Mfr.	Manufacturer
MGD	Million gallons per day
M.H.	Manhole
M.U.L.	Make-up length
NAPHCC	National Association of Plumbing-Heating-Cooling Contractors
NBFU	National Board of Fire Underwriters
NBS	National Bureau of Standards
NCPI	National Clay Pipe Institute
NFPA	National Fire Protective Association
NPS	Nominal pipe size
O.D.	Outside diameter
Pb	Lead (chemical abbreviation)
PDI	Plumbing Drainage Institute
pH	Hydrogen-ion concentration
ppm	Parts per million
psi	lbf/in^2 pressure (usually gauge)
psig	lbf/in^2 gauge pressure
psia	lbf/in^2 absolute pressure
qt	Quart
R.D.	Roof drain
R.L.	Roof leader
S.C.	Sill cock
Sn	Tin (chemical abbreviation
S.S.	Service sink
Std	Standard
SV	Service (service weight)
S & D	Soil and drain
S & W	Soil and waste
U or Urn	Urinal
UPC	Uniform Plumbing Code
vtr	Vent through roof
W	Waste
W.C.	Water closet
W.H.	Wall hydrant
W.L.	Water level
Wt	Weight
XH	Extra heavy

Appendix D

CONVERSION FACTORS

Acceleration

ft/sec^2 × 0.3048 = m/s^2
in/sec^2 × 0.0254 = m/s^2

Area

ft^2 × 0.0929 = m^2
in^2 × 0.000 645 = m^2
ft^2 × 144 = in^2
in^2 × 0.006 94 = ft^2
in^2 × 6.452 = cm^2
in^2 × 645.2 = mm^2

Density

slug/ft^3 × 515.38 = kg/m^3
gm/cm^3 × 1000 = kg/m^3

Energy, Work

Btu × 1055.87 = joule (J)
Calorie × 4.19 = joule (J)
kilowatt hour (KWH)
× 3.6 × 10^6 = joule (J)
ft-lbf × 1.356 = joule (J)

Flow Rate

ft^3/min × 7.48 = gal/min
ft^3/min × 0.028 = m^3/min
gal/min × 3.785 = liters/min (l/min)

Force

kilogram force (kgf) × 9.806 = newton N
pound force (lbf) × 4.448 = newton N

Length

ft × 0.3048 = meter (m)
in × 0.0254 = meter (m)
ft × 0.000 189 = mile
mile × 1609.34 = meter (m)
yard × 0.914 = meter (m)

Mass

gram (gm) × 0.001 = kg
slug × 14.59 = kg

Power

Btu/sec × 1054.35 = watt
Btu/min × 17.57 = watt
Btu/hr × 0.293 = watt
hp (550 ft-lbf/sec) × 746 = watt
ft-lbf/min × 0.02259 = watt

Pressure

lbf/in^2 (psi) × 6894.76 = Pa (N/m^2)
bar × 100 000 = Pa
atmosphere × 101 133 = Pa
ft of H_2O (at 32°F) × 2988.98 = Pa
in of Hg (at 32°F) × 3386.39 = Pa
in of H_2O (at 39.2°F) × 249.08 = Pa
lbf/ft^2 × 47.88 = Pa
kgf/m^2 × 9.806 = Pa
kgf/cm^2 × 9 8066.5 = Pa

Temperature

Celsius to kelvin $t_k = t_c + 273.15$
Fahrenheit to kelvin $t_k = (5/9)(t_f + 459.67)$
Fahrenheit to Celsius $t_c = (5/9)(t_f - 32)$
Celsius to Fahrenheit $t_f = (9/5 \times t_c) + 32$

Velocity

ft/min × 0.005 = m/s (meter/second)
ft/hr × 8.47 × 10^{-5} = m/s
ft/sec × 0.3048 = m/s
in/sec = 0.0254 m/s

Volume

ft^3 × 0.0283 = m^3
gal × 0.0038 = m^3
in^3 × 1.638 × 10^{-5} = m^3
ounce × 2.957 × 10^{-5} = m^3
pint × 4.731 × 10^{-4} = m^3
quart × 9.463 × 10^{-4} = m^3
yard × 0.7645 = m^3

Appendix E CAPACITIES OF CYLINDERS AND TANKS

Capacities, in U.S. Gallons, of Cylinders and Tanks of Various Diameters and Lengths

Diam inches	Length of cylinder																		Diam inches
	1″	1′	5′	6′	7′	8′	9′	10′	11′	12′	13′	14′	15′	16′	17′	18′	20′	22′	24′	
1	....	0.04	0.20	0.24	0.28	0.32	0.36	0.40	0.44	0.48	0.52	0.56	0.60	0.64	0.68	0.72	0.80	0.88	0.96	1
2	0.01	0.16	0.80	0.96	1.12	1.28	1.44	1.60	1.76	1.92	2.08	2.24	2.40	2.56	2.72	2.88	3.20	3.52	3.84	2
3	0.03	0.37	1.84	2.20	2.56	2.92	3.30	3.68	4.04	4.40	4.76	5.12	5.48	5.84	6.22	6.60	7.36	8.08	8.80	3
4	0.05	0.65	3.26	3.92	4.58	5.24	5.88	6.52	7.18	7.84	8.50	9.16	9.82	10.5	11.1	11.8	13.0	14.4	15.7	4
5	0.08	1.02	5.10	6.12	7.14	8.16	9.18	10.2	11.2	12.2	13.3	14.3	15.3	16.3	17.3	18.4	20.4	22.4	24.4	5
6	0.12	1.47	7.34	8.80	10.3	11.8	13.2	14.7	16.1	17.6	19.1	20.6	22.0	23.6	25.0	26.4	29.4	32.2	35.2	6
7	0.17	2.00	10.0	12.0	14.0	16.0	18.0	20.0	22.0	24.0	26.0	28.0	30.0	32.0	34.0	36.0	40.0	44.0	48.0	7
8	0.22	2.61	13.0	15.6	18.2	20.8	23.4	26.0	28.6	31.2	33.8	36.4	39.0	41.6	44.2	46.8	52.0	57.2	62.4	8
9	0.28	3.31	16.5	19.8	23.1	26.4	29.8	33.0	36.4	39.6	43.0	46.2	49.6	52.8	56.2	60.0	66.0	72.4	79.2	9
10	0.34	4.08	20.4	24.4	28.4	32.6	36.8	40.8	44.8	48.8	52.8	56.8	61.0	65.2	69.4	73.6	81.6	89.6	97.6	10
11	0.41	4.94	24.6	29.6	34.6	39.4	44.4	49.2	54.2	59.2	64.2	69.2	74.0	78.8	83.8	88.8	98.4	104.	118.	11
12	0.49	5.88	29.4	35.2	41.0	46.8	52.8	58.8	64.6	70.4	76.2	82.0	87.8	93.6	99.6	106.	118.	129.	141.	12
13	0.57	6.90	34.6	41.6	48.6	55.2	62.2	69.2	76.2	83.2	90.2	97.2	104.	110.	117.	124.	138.	152.	166.	13
14	0.67	8.00	40.0	48.0	56.0	64.0	72.0	80.0	88.0	96.0	104.	112.	120.	128.	136.	144.	160.	176.	192.	14
15	0.77	9.18	46.0	55.2	64.4	73.6	82.8	92.0	101.	110.	120.	129.	138.	147.	156.	166.	184.	202.	220.	15
16	0.87	10.4	52.0	62.4	72.8	83.2	93.6	104.	114.	125.	135.	146.	156.	166.	177.	187.	208.	229.	250.	16
17	0.98	11.8	59.0	70.8	81.6	94.4	106.	118.	130.	142.	153.	163.	177.	189.	201.	212.	236.	260.	283.	17
18	1.10	13.2	66.0	79.2	92.4	106.	119.	132.	145.	158.	172.	185.	198.	211.	224.	240.	264.	290.	317.	18
19	1.23	14.7	73.6	88.4	103.	118.	132.	147.	162.	177.	192.	206.	221.	235.	250.	265.	294.	324.	354.	19
20	1.36	16.3	81.6	98.0	114.	130.	147.	163.	180.	196.	212.	229.	245.	261.	277.	294.	326.	359.	392.	20
21	1.50	18.0	90.0	108.	126.	144.	162.	180.	198.	216.	238.	252.	270.	288.	306.	324.	360.	396.	432.	21
22	1.65	19.8	99.0	119.	139.	158.	178.	198.	218.	238.	257.	277.	297.	317.	337.	356.	396.	436.	476.	22
23	1.80	21.6	108.	130.	151.	173.	194.	216.	238.	259.	281.	302.	324.	346.	367.	389.	432.	476.	518.	23
24	1.96	23.5	118.	141.	165.	188.	212.	235.	259.	282.	306.	330.	353.	376.	400.	424.	470.	518.	564.	24
25	2.12	25.5	128.	153.	179.	204.	230.	255.	281.	306.	332.	358.	383.	408.	434.	460.	510.	562.	612.	25
26	2.30	27.6	138.	166.	193.	221.	248.	276.	304.	331.	359.	386.	414.	442.	470.	496.	552.	608.	662.	26
27	2.48	29.7	148.	178.	208.	238.	267.	297.	326.	356.	386.	416.	426.	476.	504.	534.	594.	652.	712.	27
28	2.67	32.0	160.	192.	224.	256.	288.	320.	352.	384.	416.	448.	480.	512.	544.	576.	640.	704.	768.	28
29	2.86	34.3	171.	206.	240.	274.	309.	343.	377.	412.	446.	480.	514.	548.	584.	618.	686.	754.	824.	29
30	3.06	36.7	183.	220.	257.	294.	330.	367.	404.	440.	476.	514.	550.	588.	624.	660.	734.	808.	880.	30
32	3.48	41.8	209.	251.	293.	334.	376.	418.	460.	502.	544.	586.	628.	668.	710.	752.	836.	920.	1004.	32
34	3.93	47.2	236.	283.	330.	378.	424.	472.	520.	566.	614.	660.	708.	756.	802.	848.	944.	1040.	1132.	34
36	4.41	52.9	264.	317.	370.	422.	476.	528.	582.	634.	688.	740.	792.	844.	898.	952.	1056.	1164.	1268.	36

Reprinted from Cameron Hydraulic Data, *1977 with permission of the Ingersoll-Rand Company.*

Appendix F PROPERTIES OF FLUIDS

Physical Properties of Water

Temperature (t)		Specific Weight (γ)		Density (ϱ)		Absolute Viscosity (μ)		Vapor Pressure Head	
°F	°C	lbf/ft³	N/m³	Slugs/ft³	kg/m³	lb•s/ft²	N•s/m²	ft	m
	0		9805		999.9		1.792×10^{-3}		0.06
32	0.01	62.42		1.940		3.746×10^{-5}		0.20	
40	4.44	62.43		1.940		3.229×10^{-5}		0.28	
	5		9806		1000		1.519×10^{-3}		0.09
50	10	62.41	9803	1.940	999.7	2.735×10^{-5}	1.308×10^{-3}	0.41	0.12
	15		9798		999.1		1.140×10^{-3}		0.17
60	15.6	62.37		1.938		2.359×10^{-5}		0.59	
	20		9789		998.2		1.005×10^{-3}		0.25
70	21.1	62.30		1.936		2.050×10^{-5}		0.84	
	25		9779		997.1		0.894×10^{-3}		0.33
80	26.7	62.22		1.934		1.799×10^{-5}		1.17	
	30		9767		995.7		0.801×10^{-3}		0.44
90	32.2	62.11		1.931		1.595×10^{-5}		1.61	
	35		9752		994.1		0.723×10^{-3}		0.58
100	37.8	62.00		1.927		1.424×10^{-5}		2.19	
	40		9737		992.3		0.656×10^{-3}		0.76
110	43.3	61.86		1.923		1.284×10^{-5}		2.95	
	45		9720		990.2		0.599×10^{-3}		0.98
120	48.9	61.71		1.918		1.168×10^{-5}		3.91	
	50		9697		988.1		0.549×10^{-3}		1.26
130	54.4	61.55		1.913		1.069×10^{-5}		5.13	
	55		9679		985.7		0.506×10^{-3}		1.61

(Continued on following page.)

Temperature (t) °F	Temperature (t) °C	Specific Weight (γ) lbf/ft³	Specific Weight (γ) N/m³	Density (ϱ) Slugs/ft³	Density (ϱ) kg/m³	Absolute Viscosity (μ) lb•s/ft²	Absolute Viscosity (μ) N•s/m²	Vapor Pressure Head ft	Vapor Pressure Head m
140	60	61.38	9658	1.908	983.2	0.981×10^{-5}	0.469×10^{-3}	6.67	2.03
	65		9635		980.6		0.436×10^{-3}		2.56
150	65.5	61.21		1.902		0.905×10^{-5}		8.58	
	70		9600		977.8		0.406×10^{-3}		3.20
160	71.1	61.00		1.896		0.838×10^{-5}		10.95	
	75		9589		974.9		0.380×10^{-3}		3.96
170	76.6	60.80		1.890		0.780×10^{-5}		13.83	
	80		9557		971.8		0.357×10^{-3}		4.86
180	82.2	60.58		1.883		0.726×10^{-5}		17.33	
	85		9529		968.6		0.336×10^{-3}		5.93
190	87.8	60.36		1.876		0.678×10^{-5}		21.55	
	90		9499		965.3		0.317×10^{-3}		7.18
200	93.3	60.12		1.868		0.637×10^{-5}		26.59	
	95		9469		961.9		0.299×10^{-3}		8.62
212	100	59.83	9438	1.860	958.4	0.593×10^{-5}	0.284×10^{-3}	33.90	10.33

Density and Specific Gravity of Various Fluids

	Temperature (t) °F	Temperature (t) °C	Specific Weight (γ) lbf/ft³	Specific Weight (γ) N/m³	Density (ϱ) Slugs/ft³	Density (ϱ) kg/m³	Specific Gravity (Sg)
Fuel Oil #3	60	15.6	56.02	8800	1.740	897.4	0.898
Gasoline	60	15.6	46.81	7353	1.454	749.8	0.751
Kerosene	60	15.6	50.85	7987	1.579	814.5	0.815
Mercury	60	15.6	846.32	132 940	26.283	13 557	13.568
SAE 10 lube	60	15.6	54.64	8583	1.697	875.3	0.876
SAE 30 lube	60	15.6	56.02	8800	1.740	897.4	0.898

Values Based on *Flow of Fluids* (410 & 410M), Appendix A-6, Crane Company; and *Fluid Mechanics*, Streeter & Wylie, McGraw-Hill Book Company, Sixth Edition, Appendix C.1 and C.2.

Appendix G

FRICTION LOSSES IN PIPES AND FITTINGS

Flow of Water Through Schedule 40 Steel Pipe

Discharge		Pressure Drop per 100 feet and Velocity in Schedule 40 Pipe for Water at 60 F.															
Gallons per Minute	Cubic Ft. per Second	Velocity Feet per Second	Press. Drop Lbf per Sq. In.	Velocity Feet per Second	Press. Drop Lbf per Sq. In.	Velocity Feet per Second	Press. Drop Lbf per Sq. In.	Velocity Feet per Second	Press. Drop Lbf per Sq. In.	Velocity Feet per Second	Press. Drop Lbf per Sq. In.	Velocity Feet per Second	Press. Drop Lbf per Sq. In.	Velocity Feet per Second	Press. Drop Lbf per Sq. In.	Velocity Feet per Second	Press. Drop Lbf per Sq. In.
		1/8"		**1/4"**		**3/8"**		**1/2"**		**3/4"**							
.2	0.000446	1.13	1.86	0.616	0.359												
.3	0.000668	1.69	4.22	0.924	0.903	0.504	0.159	0.317	0.061								
.4	0.000891	2.26	6.98	1.23	1.61	0.672	0.345	0.422	0.086								
.5	0.00111	2.82	10.5	1.54	2.39	0.840	0.539	0.528	0.167	0.301	0.033						
.6	0.00134	3.39	14.7	1.85	3.29	1.01	0.751	0.633	0.240	0.361	0.041						
.8	0.00178	4.52	25.0	2.46	5.44	1.34	1.25	0.844	0.408	0.481	0.102	**1"**		**1¼"**		**1½"**	
1	0.00223	5.65	37.2	3.08	8.28	1.68	1.85	1.06	0.600	0.602	0.155	0.371	0.048				
2	0.00446	11.29	134.4	6.16	30.1	3.36	6.58	2.11	2.10	1.20	0.526	0.743	0.164	0.429	0.044		
3	0.00668			9.25	64.1	5.04	13.9	3.17	4.33	1.81	1.09	1.114	0.336	0.644	0.090	0.473	0.043
4	0.00891			12.33	111.2	6.72	23.9	4.22	7.42	2.41	1.83	1.49	0.565	0.858	0.150	0.630	0.071
5	0.01114	**2"**				8.40	36.7	5.28	11.2	3.01	2.75	1.86	0.835	1.073	0.223	0.788	0.104
				2½"													
6	0.01337	0.574	0.044			10.08	51.9	6.33	15.8	3.61	3.84	2.23	1.17	1.29	0.309	0.946	0.145
8	0.01782	0.765	0.073			13.44	91.1	8.45	27.7	4.81	6.60	2.97	1.99	1.72	0.518	1.26	0.241
10	0.02228	0.956	0.108	0.670	0.046	**3"**		10.56	42.4	6.02	9.99	3.71	2.99	2.15	0.774	1.58	0.361
15	0.03342	1.43	0.224	1.01	0.094			**3½"**		9.03	21.6	5.57	6.36	3.22	1.63	2.37	0.755
20	0.04456	1.91	0.375	1.34	0.158	0.868	0.056			12.03	37.8	7.43	10.9	4.29	2.78	3.16	1.28
										4"							
25	0.05570	2.39	0.561	1.68	0.234	1.09	0.083	0.812	0.041			9.28	16.7	5.37	4.22	3.94	1.93
30	0.06684	2.87	0.786	2.01	0.327	1.30	0.114	0.974	0.056			11.14	23.8	6.44	5.92	4.73	2.72
35	0.07798	3.35	1.05	2.35	0.436	1.52	0.151	1.14	0.704	0.882	0.041	12.99	32.2	7.51	7.90	5.52	3.64
40	0.08912	3.83	1.35	2.68	0.556	1.74	0.192	1.30	0.095	1.01	0.052	14.85	41.5	8.59	10.24	6.30	4.65
45	0.1003	4.30	1.67	3.02	0.668	1.95	0.239	1.46	0.117	1.13	0.064	**5"**		9.67	12.80	7.09	5.85
50	0.1114	4.78	2.03	3.35	0.839	2.17	0.288	1.62	0.142	1.26	0.076			10.74	15.66	7.88	7.15
60	0.1337	5.74	2.87	4.02	1.18	2.60	0.406	1.95	0.204	1.51	0.107			12.89	22.2	9.47	10.21
70	0.1560	6.70	3.84	4.69	1.59	3.04	0.540	2.27	0.261	1.76	0.143	1.12	0.047			11.05	13.71
80	0.1782	7.65	4.97	5.36	2.03	3.47	0.687	2.60	0.334	2.02	0.180	1.28	0.060	**6"**		12.62	17.59
90	0.2005	8.60	6.20	6.03	2.53	3.91	0.861	2.92	0.416	2.27	0.224	1.44	0.074			14.20	22.0
100	0.2228	9.56	7.59	6.70	3.09	4.34	1.05	3.25	0.509	2.52	0.272	1.60	0.090	1.11	0.036	15.78	26.9
125	0.2785	11.97	11.76	8.38	4.71	5.43	1.61	4.06	0.769	3.15	0.415	2.01	0.135	1.39	0.055	19.72	41.4
150	0.3342	14.36	16.70	10.05	6.69	6.51	2.24	4.87	1.08	3.78	0.580	2.41	0.190	1.67	0.077		
175	0.3899	16.75	22.3	11.73	8.97	7.60	3.00	5.68	1.44	4.41	0.774	2.81	0.253	1.94	0.102	**8"**	
200	0.4456	19.14	28.8	13.42	11.68	8.68	3.87	6.49	1.85	5.04	0.985	3.21	0.323	2.22	0.130		
225	0.5013	...	...	15.09	14.63	9.77	4.83	7.30	2.32	5.67	1.23	3.61	0.401	2.50	0.162	1.44	0.043
250	0.557	...	...	...	...	10.85	5.93	8.12	2.84	6.30	1.46	4.01	0.495	2.78	0.195	1.60	0.051
275	0.6127	...	...	...	...	11.94	7.14	8.93	3.40	6.93	1.79	4.41	0.583	3.05	0.234	1.76	0.061
300	0.6684	...	...	...	...	13.00	8.36	9.74	4.02	7.56	2.11	4.81	0.683	3.33	0.275	1.92	0.072
325	0.7241	...	...	...	...	14.12	9.89	10.53	4.09	8.19	2.47	5.21	0.797	3.61	0.320	2.08	0.083
350	0.7798			...	...	...	...	11.36	5.41	8.82	2.84	5.62	0.919	3.89	0.367	2.24	0.095
375	0.8355			...	...	...	...	12.17	6.18	9.45	3.25	6.02	1.05	4.16	0.416	2.40	0.108
400	0.8912			...	...	...	...	12.98	7.03	10.08	3.68	6.42	1.19	4.44	0.471	2.56	0.121
425	0.9469			...	...	...	...	13.80	7.89	10.71	4.12	6.82	1.33	4.72	0.529	2.73	0.136
450	1.003	**10"**		...	...	...	...	14.61	8.80	11.34	4.60	7.22	1.48	5.00	0.590	2.89	0.151
475	1.059	1.93	0.054			...	...	...	...	11.97	5.12	7.62	1.64	5.27	0.653	3.04	0.166
500	1.114	2.03	0.059			...	...	...	...	12.60	5.65	8.02	1.81	5.55	0.720	3.21	0.182
550	1.225	2.24	0.071			...	...	...	...	13.85	6.79	8.82	2.17	6.11	0.861	3.53	0.219
600	1.337	2.44	0.083			...	...	...	...	15.12	8.04	9.63	2.55	6.66	1.02	3.85	0.258
650	1.448	2.64	0.097	**12"**		...	...	...	...	...	...	10.43	2.98	7.22	1.18	4.17	0.301
						14"											
700	1.560	2.85	0.112	2.01	0.047			...	...	...	...	11.23	3.43	7.78	1.35	4.49	0.343
750	1.671	3.05	0.127	2.15	0.054			...	...	...	...	12.03	3.92	8.33	1.55	4.81	0.392
800	1.782	3.25	0.143	2.29	0.061			...	...	...	...	12.83	4.43	8.88	1.75	5.13	0.443
850	1.894	3.46	0.160	2.44	0.068	2.02	0.042	...	...	...	...	13.64	5.00	9.44	1.96	5.45	0.497
900	2.005	3.66	0.179	2.58	0.075	2.13	0.047	...	...	...	...	14.44	5.58	9.99	2.18	5.77	0.554
950	2.117	3.86	0.198	2.72	0.083	2.25	0.052			...	...	15.24	6.21	10.55	2.42	6.09	0.613
1 000	2.228	4.07	0.218	2.87	0.091	2.37	0.057	**16"**		...	...	16.04	6.84	11.10	2.68	6.41	0.675
1 100	2.451	4.48	0.260	3.15	0.110	2.61	0.068			...	...	17.65	8.23	12.22	3.22	7.05	0.807
1 200	2.674	4.88	0.306	3.44	0.128	2.85	0.080	2.18	0.042	...	...	...	...	13.33	3.81	7.70	0.948
1 300	2.896	5.29	0.355	3.73	0.150	3.08	0.093	2.36	0.048	...	...	...	...	14.43	4.45	8.33	1.11
1 400	3.119	5.70	0.409	4.01	0.171	3.32	0.107	2.54	0.055					15.55	5.13	8.98	1.28
1 500	3.342	6.10	0.466	4.30	0.195	3.56	0.122	2.72	0.063	**18"**				16.66	5.85	9.62	1.46
1 600	3.565	6.51	0.527	4.59	0.219	3.79	0.138	2.90	0.071					17.77	6.61	10.26	1.65
1 800	4.010	7.32	0.663	5.16	0.276	4.27	0.172	3.27	0.088	2.58	0.050			19.99	8.37	11.54	2.08
2 000	4.456	8.14	0.808	5.73	0.339	4.74	0.209	3.63	0.107	2.87	0.060	**20"**		22.21	10.3	12.82	2.55
2 500	5.570	10.17	1.24	7.17	0.515	5.93	0.321	4.54	0.163	3.59	0.091			**24"**		16.03	3.94
3 000	6.684	12.20	1.76	8.60	0.731	7.11	0.451	5.45	0.232	4.30	0.129	3.46	0.075			19.24	5.59
3 500	7.798	14.24	2.38	10.03	0.982	8.30	0.607	6.35	0.312	5.02	0.173	4.04	0.101			22.44	7.56
4 000	8.912	16.27	3.08	11.47	1.27	9.48	0.787	7.26	0.401	5.74	0.222	4.62	0.129	3.19	0.052	25.65	9.80
4 500	10.03	18.31	3.87	12.90	1.60	10.67	0.990	8.17	0.503	6.46	0.280	5.20	0.162	3.59	0.065	28.87	12.2
5 000	11.14	20.35	4.71	14.33	1.95	11.85	1.21	9.08	0.617	7.17	0.340	5.77	0.199	3.99	0.079	...	...
6 000	13.37	24.41	6.74	17.20	2.77	14.23	1.71	10.89	0.877	8.61	0.483	6.93	0.280	4.79	0.111	...	...
7 000	15.60	28.49	9.11	20.07	3.74	16.60	2.31	12.71	1.18	10.04	0.652	8.08	0.376	5.59	0.150	...	...
8 000	17.82	...	...	22.93	4.84	18.96	2.99	14.52	1.51	11.47	0.839	9.23	0.488	6.38	0.192	...	...
9 000	20.05	...	...	25.79	6.09	21.34	3.76	16.34	1.90	12.91	1.05	10.39	0.608	7.18	0.242	...	...
10 000	22.28	...	...	28.66	7.46	23.71	4.61	18.15	2.34	14.34	1.28	11.54	0.739	7.98	0.294	...	...
12 000	26.74	...	...	34.40	10.7	28.45	6.59	21.79	3.33	17.21	1.83	13.85	1.06	9.58	0.416	...	...
14 000	31.19	...	...	...	...	33.19	8.89	25.42	4.49	20.08	2.45	16.16	1.43	11.17	0.562	...	...
16 000	35.65	...	...	...	...	...	...	29.05	5.83	22.95	3.18	18.47	1.85	12.77	0.723	...	...
18 000	40.10	...	...	...	...	...	...	32.68	7.31	25.82	4.03	20.77	2.32	14.36	0.907	...	...
20 000	44.56	...	...	...	...	...	...	36.31	9.03	28.69	4.93	23.08	2.86	15.96	1.12	...	...

For pipe lengths other than 100 feet, the pressure drop is proportional to the length. Thus, for 50 feet of pipe, the pressure drop is approximately one-half the value given in the table . . . for 300 feet, three times the given value, etc.

Velocity is a function of the cross sectional flow area; thus, it is constant for a given flow rate and is independent of pipe length.

Flow of Water Through Schedule 40 Steel Pipe

Discharge	Pressure Drop per 100 metres and Velocity in Schedule 40 Pipe for Water at 15 C															
Litres per Minute	Velocity Metres per Second	Press. Drop 100 kPa (bars)	Velocity Metres per Second	Press. Drop 100 kPa (bars)	Velocity Metres per Second	Press. Drop 100 kPa (bars)	Velocity Metres per Second	Press. Drop 100 kPa (bars)	Velocity Metres per Second	Press. Drop 100 kPa (bars)	Velocity Metres per Second	Press. Drop 100 kPa (bars)	Velocity Metres per Second	Press. Drop 100 kPa (bars)	Velocity Metres per Second	Press. Drop 100 kPa (bars)
	1/8"		1/4"		3/8"		1/2"		3/4"		1"					
1	0.459	0.726	0.251	0.17												
2	0.918	2.59	0.501	0.60	0.272	0.136	0.170	0.044								
3	1.38	5.59	0.752	1.22	0.407	0.29	0.255	0.091	0.144	0.023						
4	1.84	9.57	1.00	2.09	0.543	0.48	0.340	0.151	0.192	0.038	0.120	0.012				
5	2.29	14.45	1.25	3.18	0.679	0.70	0.425	0.223	0.241	0.057	0.150	0.017				
													1 1/4"		1 1/2"	
6	2.75	20.29	1.50	4.46	0.815	0.98	0.510	0.309	0.289	0.077	0.180	0.024				
8	3.67	35.16	2.01	7.36	1.09	1.69	0.680	0.524	0.385	0.129	0.240	0.041	0.138	0.011		
10			2.51	11.81	1.36	2.52	0.850	0.798	0.481	0.193	0.300	0.061	0.172	0.015	0.127	0.008
15			3.76	25.67	2.04	5.37	1.28	1.69	0.722	0.403	0.450	0.124	0.258	0.032	0.190	0.015
20	2"				2.72	9.24	1.70	2.84	0.962	0.683	0.600	0.210	0.344	0.054	0.254	0.026
			2 1/2"													
30	0.231	0.016					2.55	6.17	1.44	1.45	0.900	0.442	0.517	0.114	0.380	0.053
40	0.308	0.027	0.216	0.010			3.40	10.72	1.92	2.50	1.20	0.758	0.689	0.193	0.507	0.091
50	0.385	0.039	0.270	0.017					2.41	3.83	1.50	1.14	0.861	0.290	0.634	0.135
60	0.462	0.055	0.324	0.023					2.89	5.41	1.80	1.61	1.03	0.400	0.761	0.187
70	0.539	0.098	0.378	0.031	3"				3.37	7.27	2.10	2.15	1.21	0.541	0.888	0.248
							3 1/2"									
80	0.616	0.092	0.432	0.039	0.280	0.014			3.85	9.27	2.40	2.76	1.38	0.690	1.01	0.315
90	0.693	0.115	0.486	0.048	0.315	0.017	0.235	0.008			2.70	3.47	1.55	0.862	1.14	0.397
									4"							
100	0.770	0.141	0.540	0.059	0.350	0.020	0.261	0.010			3.00	4.25	1.72	1.05	1.27	0.488
150	1.15	0.295	0.810	0.125	0.524	0.042	0.392	0.021	0.304	0.011	4.50	9.30	2.58	2.26	1.90	1.03
200	1.54	0.512	1.08	0.212	0.699	0.072	0.523	0.036	0.405	0.019			3.44	3.91	2.54	1.81
											5"		6"			
250	1.92	0.773	1.35	0.322	0.874	0.108	0.653	0.053	0.507	0.028					3.17	2.74
300	2.31	1.10	1.62	0.449	1.05	0.152	0.784	0.074	0.608	0.040	0.387	0.014			3.80	3.82
350	2.69	1.47	1.89	0.606	1.22	0.203	0.915	0.099	0.710	0.053	0.452	0.018			4.44	5.18
400	3.08	1.92	2 16	0.780	1.40	0.264	1.05	0.128	0.811	0.068	0.516	0.023	0.357	0.009	5.07	6.69
450	3.46	2.39	2.43	0.979	1.57	0.329	1.18	0.161	0.912	0.084	0.581	0.028	0.402	0.012	5.71	8.45
500	3.85	2.95	2.70	1.20	1.75	0.403	1.31	0.196	1.01	0.101	0.646	0.034	0.447	0.014		
550	4.23	3.55	2.97	1.44	1.92	0.479	1.44	0.232	1.11	0.122	0.710	0.041	0.491	0.016		
600	4.62	4.20	3.24	1.69	2.10	0.566	1.57	0.273	1.22	0.146	0.775	0.047	0.536	0.019		
650	5.00	6.88	3.51	1.97	2.27	0.658	1.70	0.319	1.32	0.169	0.839	0.055	0.581	0.022		
700	5.39	5.63	3.78	2.28	2.45	0.759	1.83	0.368	1.42	0.194	0.904	0.063	0.625	0.025		
															8"	
750	5.77	6.44	4.05	2.60	2.62	0.863	1.96	0.420	1.52	0.218	0.968	0.072	0.670	0.029		
800			4.32	2.95	2.80	0.977	2.09	0.473	1.62	0.246	1.03	0.081	0.715	0.032		
850			4.59	3.31	2.97	1.09	2.22	0.528	1.72	0.277	1.10	0.091	0.760	0.036	0.439	0.009
900					3.15	1.22	2.35	0.585	1.82	0.308	1.16	0.010	0.804	0.041	0.465	0.010
950					3.32	1.35	2.48	0.649	1.93	0.342	1.23	0.111	0.849	0.045	0.491	0.012
1000					3.5	1.50	2.61	0.714	2.03	0.377	1.29	0.122	0.894	0.049	0.516	0.013
1100					3.85	1.75	2.87	0.860	2.23	0.452	1.42	0.147	0.983	0.059	0.568	0.015
1200					4.20	2.14	3.14	1.02	2.43	0.534	1.55	0.172	1.07	0.069	0.620	0.018
1300							3.40	1.19	2.64	0.627	1.68	0.200	1.16	0.080	0.671	0.021
1400							3.66	1.37	2.84	0.722	1.81	0.232	1.25	0.091	0.723	0.024
1500							3.92	1.56	3.04	0.818	1.94	0.264	1.34	0.105	0.775	0.027
1600	10"						4.18	1.78	3.24	0.924	2.07	0.297	1.43	0.118	0.826	0.031
1700							4.44	1.99	3.45	1.04	2.19	0.331	1.52	0.132	0.878	0.035
1800	0.590	0.012							3.65	1.16	2.32	0.369	1.61	0.147	0.930	0.039
1900	0.622	0.014							3.85	1.28	2.45	0.410	1.70	0.163	0.981	0.042
2000	0.655	0.015							4.05	1.41	2.58	0.452	1.79	0.181	1.03	0.046
2200	0.721	0.018	12"						4.46	1.70	2.84	0.545	1.97	0.217	1.14	0.056
2400	0.786	0.021									3.10	0.645	2.14	0.253	1.24	0.065
2600	0.852	0.025	0.600	0.010							3.36	0.749	2.32	0.296	1.34	0.076
2800	0.917	0.028	0.646	0.012	14"						3.61	0.859	2.50	0.339	1.45	0.087
3000	0.983	0.032	0.692	0.013	0.573	0.008					3.87	0.982	2.68	0.387	1.55	0.099
3500	1.15	0.043	0.810	0.018	0.668	0.011	16"				4.52	1.33	3.13	0.526	1.81	0.134
4000	1.31	0.055	0.923	0.023	0.764	0.014					5.16	1.72	3.57	0.673	2.07	0.172
4500	1.47	0.068	1.04	0.029	0.860	0.018	0.658	0.009					4.02	0.853	2.32	0.214
5000	1.64	0.084	1.15	0.034	0.955	0.022	0.731	0.011					4.47	1.04	2.58	0.262
									18"							
6000	1.96	0.118	1.38	0.049	1.15	0.031	0.877	0.016					5.36	1.47	3.10	0.373
7000	2.29	0.158	1.61	0.065	1.34	0.042	1.02	0.021	0.808	0.012			6.25	2.0	3.61	0.499
8000	2.62	0.204	1.84	0.085	1.53	0.054	1.17	0.027	0.924	0.015			7.15	2.59	4.13	0.650
9000	2.95	-0.256	2.08	0.107	1.72	0.067	1.31	0.033	1.04	0.019					4.65	0.816
10 000	3.28	0.313	2.31	0.130	1.91	0.081	1.46	0.041	1.15	0.023	20"				5.16	0.992
													24"			
12 000	3.93	0.447	2.77	0.184	2.29	0.114	1.75	0.057	1.38	0.032	1.11	0.019			6.20	1.41
14 000	4.59	0.600	3.23	0.246	2.67	0.153	2.05	0.077	1.62	0.044	1.30	0.025			7.23	1.91
16 000	5.24	0.776	3.69	0.317	3.06	0.198	2.34	0.099	1.85	0.056	1.49	0.032	1.03	0.013	8.26	2.48
18 000	5.90	0.975	4.15	0.398	3.44	0.246	2.63	0.124	2.08	0.069	1.67	0.040	1.16	0.016		
20 000	6.55	1.19	4.61	0.487	3.82	0.302	2.92	0.152	2.31	0.084	1.86	0.049	1.28	0.020		
25 000	8.19	1.83	5.77	0.758	4.77	0.469	3.65	0.234	2.89	0.130	2.32	0.076	1.61	0.030		
30 000			6.92	1.08	5.73	0.669	4.38	0.332	3.46	0.183	2.79	0.108	1.93	0.043		
35 000			8.07	1.46	6.68	0.903	5.12	0.446	4.04	0.248	3.25	0.144	2.25	0.057		
40 000			9.23	1.90	7.64	1.17	5.85	0.578	4.62	0.319	3.72	0.186	2.57	0.074		
45 000			10.38	2.39	8.59	1.47	6.58	0.726	5.19	0.400	4.18	0.233	2.89	0.092		
50 000					9.55	1.81	7.31	0.888	5.77	0.491	4.64	0.284	3.21	0.113		
55 000							8.04	1.07	6.35	0.594	5.11	0.343	3.53	0.136		
60 000							8.77	1.27	6.93	0.708	5.58	0.411	3.86	0.161		
65 000							9.5	1.49	7.50	0.822	6.04	0.475	4.18	0.189		
70 000							10.2	1.70	8.08	0.955	6.51	0.552	4.50	0.216		
75 000							11.0	1.98	8.66	1.10	6.97	0.628	4.82	0.246		

1 cubic meter = 1000 liters

Reprinted with permission from Flow of Fluids Through Valves, Fittings and Pipe, *Publication 410M2(Metric), Crane Company, Chicago, Illinois.*

Pressure Loss Due to Friction in Type M Copper Tube

Flow, gpm	Pressure Loss per 100 Feet of Tube in lbf/in² (kPa) — Standard Type M Tube Size -- inches: 3/8	1/2	3/4	1	1 1/4	1 1/2	2	2 1/2	3	4	5	6
1	2.5(17.2)	0.8(5.5)	0.2(1.4)									
2	8.5(58.6)	2.8(19.3)	0.5(3.4)	0.2(1.4)								
3	17.3(119.3)	5.7(39.3)	1.0(6.9)	0.3(2.1)	0.1(6.9)							
4	28.6(197.2)	9.4(64.8)	1.8(12.4)	0.5(3.4)	0.2(1.4)							
5	42.2(291)	13.8(95.2)	2.6(17.9)	0.7(4.8)	0.3(2.1)	0.1(0.7)						
10		**46.6(321.3)**	8.6(59.3)	2.5(17.2)	0.9(6.2)	0.4(2.8)	0.1(0.7)					
15			17.6(121.4)	5.0(34.5)	1.9(13.1)	0.9(6.2)	0.2(1.4)					
20			**29.1(200.6)**	8.4(57.9)	3.2(22.1)	1.4(9.7)	0.4(2.8)	0.1(0.7)				
25				12.3(84.8)	4.7(32.4)	2.1(14.5)	0.6(4.1)	0.2(1.4)				
30				**17.0(117.2)**	6.5(44.8)	2.9(20.0)	0.8(5.5)	0.3(2.1)	0.1(0.7)			
35					8.5(58.6)	3.8(26.2)	1.0(6.9)	0.4(2.8)	0.2(1.4)			
40					11.0(75.8)	4.9(33.8)	1.3(9.0)	0.5(3.4)	0.2(1.4)			
45					**13.6(93.8)**	6.1(42.1)	1.6(11.0)	0.6(4.1)	0.2(1.4)			
50						7.3(50.3)	2.0(13.8)	0.7(4.8)	0.3(2.1)			
60						10.2(70.3)	2.7(18.6)	1.0(6.9)	0.4(2.8)			
70						**13.5(93.1)**	3.6(24.8)	1.2(8.3)	0.5(3.4)	0.1(0.7)		
80							4.6(31.7)	1.6(11.0)	0.7(4.8)	0.2(1.4)		
90							5.7(39.3)	2.0(13.8)	0.9(6.2)	0.2(1.4)		
100							**7.5(51.7)**	2.7(18.6)	1.0(6.9)	0.3(2.1)	0.1(0.7)	
200								**8.5(58.6)**	3.6(24.8)	1.0(6.9)	0.3(2.1)	0.1(0.7)
300									**8.0(55.2)**	2.0(13.8)	0.7(4.8)	0.3(2.1)
400										**3.3(22.8)**	1.2(8.3)	0.5(3.4)
500											1.7(11.7)	0.7(4.8)
750											**3.6(24.8)**	1.5(10.3)
1000												**2.5(17.2)**

NOTE: Numbers in bold face correspond to flow velocities of just over 10 ft. per sec.

Derived from Copper Tube Handbook with permission from the Copper Development Association, Inc.

Allowance for Friction Loss in Valves and Fittings Expressed as Equivalent Length of Tube

Fitting Size, inches	Equivalent Length of Tube in Feet (Meters)						
	Standard Ells		90° Tee		Coupling	Gate Valve	Globe Valve
	90°	45°	Side Branch	Straight Run			
3/8	0.5(0.15)	0.3(0.09)	0.75(0.23)	0.15(0.05)	0.15(0.05)	0.1(0.30)	4(0.12)
1/2	1(0.30)	0.6(0.18)	1.5(0.46)	0.3(0.09)	0.3(0.09)	0.2(0.06)	7.5(2.29)
3/4	1.25(0.38)	0.75(0.22)	2(0.61)	0.4(0.12)	0.4(0.12)	0.25(0.08)	10(3.05)
1	1.5(0.46)	1.0(0.30)	2.5(0.76)	0.45(0.14)	0.45(0.14)	0.3(0.09)	12.5(3.81)
1 1/4	2(0.61)	1.2(0.37)	3(0.91)	0.6(0.18)	0.6(0.18)	0.4(0.12)	18(5.49)
1 1/2	2.5(0.76)	1.5(0.46)	3.5(1.07)	0.8(0.24)	0.8(0.24)	0.5(0.15)	23(7.01)
2	3.5(1.06)	2(0.61)	5(1.52)	1(0.30)	1(0.30)	0.7(0.21)	28(8.53)
2 1/2	4(1.22)	2.5(0.76)	6(1.83)	1.3(0.40)	1.3(0.40)	0.8(0.24)	33(10.06)
3	5(1.52)	3(0.91)	7.5(2.29)	1.5(0.46)	1.5(0.46)	1(0.30)	40(12.19)
3 1/2	6(1.83)	3.5(1.07)	9(2.74)	1.8(0.55)	1.8(0.55)	1.2(0.37)	50(15.24)
4	7(2.13)	4(1.22)	10.5(3.20)	2(0.61)	2(0.61)	1.4(0.43)	63(19.20)
5	9(2.74)	5(1.52)	13(3.96)	2.5(0.76)	2.5(0.76)	1.7(0.52)	70(21.34)
6	10(3.05)	6(1.83)	15(4.57)	3(0.91)	3(0.91)	2(0.61)	84(25.60)

NOTE: Allowances are for streamlined soldered fittings and recessed threaded fittings. For threaded fittings, double the allowances shown in the table.

Derived from Copper Tube Handbook with permission from the Copper Development Association Inc.

Appendix H

POWERS AND ROOTS OF NUMBERS

No.	Square	Cube	Square Root	Cube Root	No.	Square	Cube	Square Root	Cube Root
1	1	1	1.000	1.000	51	2,601	132,651	7.141	3.708
2	4	8	1.414	1.260	52	2,704	140,608	7.211	3.732
3	9	27	1.732	1.442	53	2,809	148,877	7.280	3.756
4	16	64	2.000	1.587	54	2,916	157,464	7.348	3.780
5	25	125	2.236	1.710	55	3,025	166,375	7.416	3.803
6	36	216	2.449	1.817	56	3,136	175,616	7.483	3.826
7	49	343	2.646	1.913	57	3,249	185,193	7.550	3.848
8	64	512	2.828	2.000	58	3,364	195,112	7.616	3.871
9	81	729	3.000	2.080	59	3,481	205,379	7.681	3.893
10	100	1,000	3.162	2.154	60	3,600	216,000	7.746	3.915
11	121	1,331	3.317	2.224	61	3,721	226,981	7.810	3.936
12	144	1,728	3.464	2.289	62	3,844	238,328	7.874	3.958
13	169	2,197	3.606	2.351	63	3,969	250,047	7.937	3.979
14	196	2,744	3.742	2.410	64	4,096	262,144	8.000	4.000
15	225	3,375	3.873	2.466	65	4,225	274,625	8.062	4.021
16	256	4.096	4.000	2.520	66	4,356	287,496	8.124	4.041
17	289	4,913	4.123	2.571	67	4,489	300,763	8.185	4.062
18	324	5,832	4.243	2.621	68	4,624	314,432	8.246	4.082
19	361	6,859	4.359	2.668	69	4,761	328,509	8.307	4.102
20	400	8,000	4.472	2.714	70	4,900	343,000	8.367	4.121
21	441	9.261	4.583	2.759	71	5,041	357,911	8.426	4.141
22	484	10,648	4.690	2.802	72	5,184	373,248	8.485	4.160
23	529	12,167	4.796	2.844	73	5,329	389,017	8.544	4.179
24	576	13,824	4.899	2.884	74	5,476	405,224	8.602	4.198
25	625	15,625	5.000	2.924	75	5,625	421,875	8.660	4.217
26	676	17,576	5.099	2.962	76	5,776	438,976	8.718	4.236
27	729	19,683	5.196	3.000	77	5,929	456,533	8.775	4.254
28	784	21,952	5.292	3.037	78	6,084	474,552	8.832	4.273
29	841	24,389	5.385	3.072	79	6,241	493,039	8.888	4.291
30	900	27,000	5.477	3.107	80	6,400	512,000	8,944	4.309
31	961	29,791	5.568	3.141	81	6,561	531,441	9.000	4.327
32	1,024	32,768	5.657	3.175	82	6,724	551,368	9.055	4.344
33	1,089	35,937	5.745	3.208	83	6,889	571,787	9.110	4.362
34	1,156	39,304	5.831	3.240	84	7,056	592,704	9.165	4.380
35	1,225	42,875	5.916	3.271	85	7,225	614,125	9.220	4.397
36	1,296	46,656	6.000	3.302	86	7,396	636,056	9.274	4.414
37	1,369	50,653	6.083	3.332	87	7,569	658,503	9.327	4.431
38	1,444	54,872	6.164	3.362	88	7,744	681,472	9.381	4.448
39	1,521	59,319	6.245	3.391	89	7.921	704,969	9.434	4.465
40	1,600	64,000	6.325	3.420	90	8,100	729,000	9.487	4.481
41	1,681	68,921	6.403	3.448	91	8,281	753,571	9.539	4.498
42	1,764	74,088	6.481	3.476	92	8,464	778,688	9.592	4.514
43	1,849	79,507	6.557	3.503	93	8,649	804,357	9.644	4.531
44	1,936	85,184	6.633	3.530	94	8,836	830,584	9.695	4.547
45	2,025	91,125	6.708	3.577	95	9,025	857,375	9.747	4.563
46	2,116	97,336	6.782	3.583	96	9,216	884,736	9.798	4.579
47	2,209	103,823	6.856	3.609	97	9,409	912,673	9.849	4.595
48	2,304	110,592	6.928	3.634	98	9.604	941,192	9.899	4.610
49	2,401	117,649	7.000	3.659	99	9,801	970,299	9.950	4.626
50	2,500	125,000	7.071	3.684	100	10,000	1,000,000	10.000	4.642

Appendix I

NATURAL TRIGONOMETRIC FUNCTIONS

TRIGONOMETRIC FUNCTIONS

$$\sin\theta = \frac{y}{r} \qquad \csc\theta = \frac{r}{y}$$

$$\cos\theta = \frac{x}{r} \qquad \sec\theta = \frac{r}{x}$$

$$\tan\theta = \frac{y}{x} \qquad \cot\theta = \frac{x}{y}$$

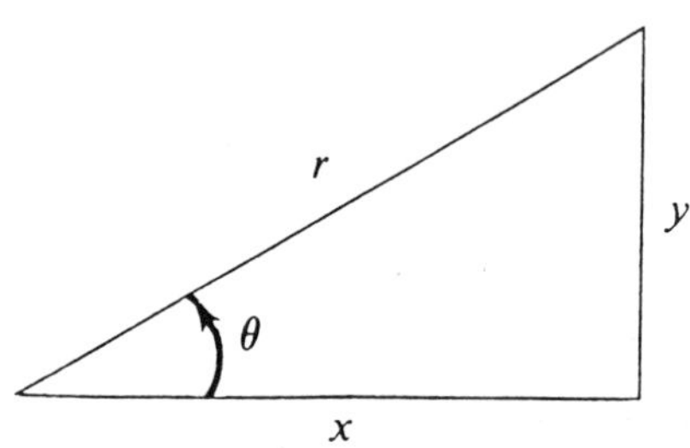

Appendix J

NATURAL TRIGONOMETRIC FUNCTIONS TABLE

Degrees	Sin	Cos	Tan	Cot	Sec	Cosec	
0	.0000	1.0000	.0000		1.0000		90
1	.0175	.9998	.0175	57.29	1.0000	57.30	89
2	.0349	.9994	.0349	28.64	1.001	28.65	88
3	.0523	.9986	.0524	19.08	1.001	19.11	87
4	.0698	.9976	.0699	14.30	1.002	14.34	86
5	.0872	.9962	.0875	11.43	1.004	11.47	85
6	.1045	.9945	.1051	9.514	1.006	9.567	84
7	.1219	.9925	.1228	8.144	1.008	8.206	83
8	.1392	.9903	.1405	7.115	1.010	7.185	82
9	.1564	.9877	.1584	6.314	1.012	6.392	81
10	.1736	.9848	.1763	5.671	1.015	5.759	80
11	.1908	.9816	.1944	5.145	1.019	5.241	79
12	.2079	.9781	.2126	4.705	1.022	4.810	78
13	.2250	.9744	.2309	4.331	1.026	4.445	77
14	.2419	.9703	.2493	4.011	1.031	4.134	76
15	.2588	.9659	.2679	3.732	1.035	3.864	75
16	.2756	.9613	.2867	3.487	1.040	3.628	74
17	.2924	.9563	.3057	3.271	1.046	3.420	73
18	.3090	.9511	.3249	3.078	1.051	3.236	72
19	.3256	.9455	.3443	2.904	1.058	3.072	71
20	.3420	.9397	.3640	2.747	1.064	2.924	70
21	.3584	.9336	.3839	2.605	1.071	2.790	69
22	.3746	.9272	.4040	2.475	1.079	2.669	68
23	.3907	.9205	.4245	2.356	1.086	2.559	67
24	.4067	.9135	.4452	2.246	1.095	2.459	66
25	.4226	.9063	.4663	2.145	1.103	2.366	65
26	.4384	.8988	.4877	2.050	1.113	2.281	64
27	.4540	.8910	.5095	1.963	1.122	2.203	63
28	.4695	.8829	.5317	1.881	1.133	2.130	62
29	.4848	.8746	.5543	1.804	1.143	2.063	61
30	.5000	.8660	.5774	1.732	1.155	2.000	60
31	.5150	.8572	.6009	1.664	1.167	1.942	59
32	.5299	.8480	.6249	1.600	1.179	1.887	58
33	.5446	.8387	.6494	1.540	1.192	1.836	57
34	.5592	.8290	.6745	1.483	1.206	1.788	56
35	.5736	.8192	.7002	1.428	1.221	1.743	55
36	.5878	.8090	.7265	1.376	1.236	1.701	54
37	.6018	.7986	.7536	1.327	1.252	1.662	53
38	.6157	.7880	.7813	1.280	1.269	1.624	52
39	.6293	.7771	.8098	1.235	1.287	1.589	51
40	.6428	.7660	.8391	1.192	1.305	1.556	50
41	.6561	.7547	.8693	1.150	1.325	1.524	49
42	.6691	.7431	.9004	1.111	1.346	1.494	48
43	.6820	.7314	.9325	1.072	1.367	1.466	47
44	.6947	.7193	.9657	1.036	1.390	1.440	46
45	.7071	.7071	1.0000	1.000	1.414	1.414	45
	Cos	Sin	Cot	Tan	Cosec	Sec	Degrees

Appendix K

PLUMBING STANDARDS

Organizations

ANSI—American National Standards Institute
ASTM—American Society for Testing and Materials
CISPI—Cast Iron Soil Pipe Institute

Iron and Steel Pipe

ANSI B 36.1	Welded and Seamless Steel Pipe (also ASTM A 53)
ANSI B 36.20	Black and Hot Dipped Zinc Coated (Galvanized) Welded and Seamless Steel for Ordinary Uses (also ASTM A 120)

Iron and Steel Fittings

ANSI B 16.12	Cast Iron Screwed Drainage Fittings
ANSI B 16.19	Malleable-Iron Screwed Fittings, 300 lb
ANSI B 16.3	Malleable-Iron Screwed Fittings, 150 lb
ANSI B 16.4	Cast Iron Screwed Fittings, 125 lb and 250 lb

Cast Iron Pipe

ANSI A 21.50	Thickness Design of Ductile-Iron Pipe
ANSI A 21.51	Ductile-Iron Pipe, Centrifugally Cast in Metal Molds or Sand-Lined Molds, for Water or Other Liquids
ANSI A 40.1	Cast Iron Soil Pipe and Fittings
ASTM A 74	Cast Iron Soil Pipe and Fittings
ASTM A 126	Gray Iron Castings for Valves, Flanges and Pipe Fittings
CISPI 301	Specification Data for Hubless Cast Iron Sanitary System with No-Hub Pipe and Fittings

Copper Tube

ASTM B 31	Corrosion Control for ANSI B 31.1 Power Piping Systems-Guide
ASTM B 68	Seamless Copper Tube, Bright Annealed
ASTM B 75	Seamless Copper Tube
ASTM B 88	Seamless Copper Water Tube—Types K, L & M (Also ANSI H 23.1)
ASTM B 306	Streamline Copper Tube, Type DWV (Also ANSI 23.6)

Copper and Brass Fittings

ANSI B 16.15	Cast Bronze Solder-Joint Pressure Fittings
ANSI B 16.18	Cast-Brass Solder-Joint Fittings
ANSI B 16.22	Wrought Copper and Bronze Solder-Joint Fittings
ANSI B 16.23	Cast-Brass Solder-Joint Drainage Fittings
ANSI B 16.24	Brass or Bronze Flanges and Flange Fittings
ANSI B 16.26	Brass Fittings for Flared Copper Tubes
ANSI B 16.29	Wrought Solder-Joint DWV Fittings
ANSI B 16.32	Cast Bronze Solder-Joint Fittings for SOLVENT Drainage Fittings

Clay Pipe

ASTM C 12	Installing Vitrified Clay Pipe Lines
ASTM C 13	Standard Strength Clay Sewer Pipe
ASTM C 200	Extra Strength Clay Pipe
ASTM C 261	Ceramic Glazed Standard Strength Sewer Pipe
ASTM C 278	Ceramic Glazed Extra Strength Sewer Pipe
ASTM C 301	Testing Vitrified Clay Pipe
ASTM C 425	Compression Joints for Vitrified Clay Pipe and Fittings
ASTM C 700	Vitrified Clay Pipe, Extra Strength, Standard Strength and Perforated

Plastic Pipe

(see Table 4–5)

Appendix L SOLAR INSOLATION MAPS

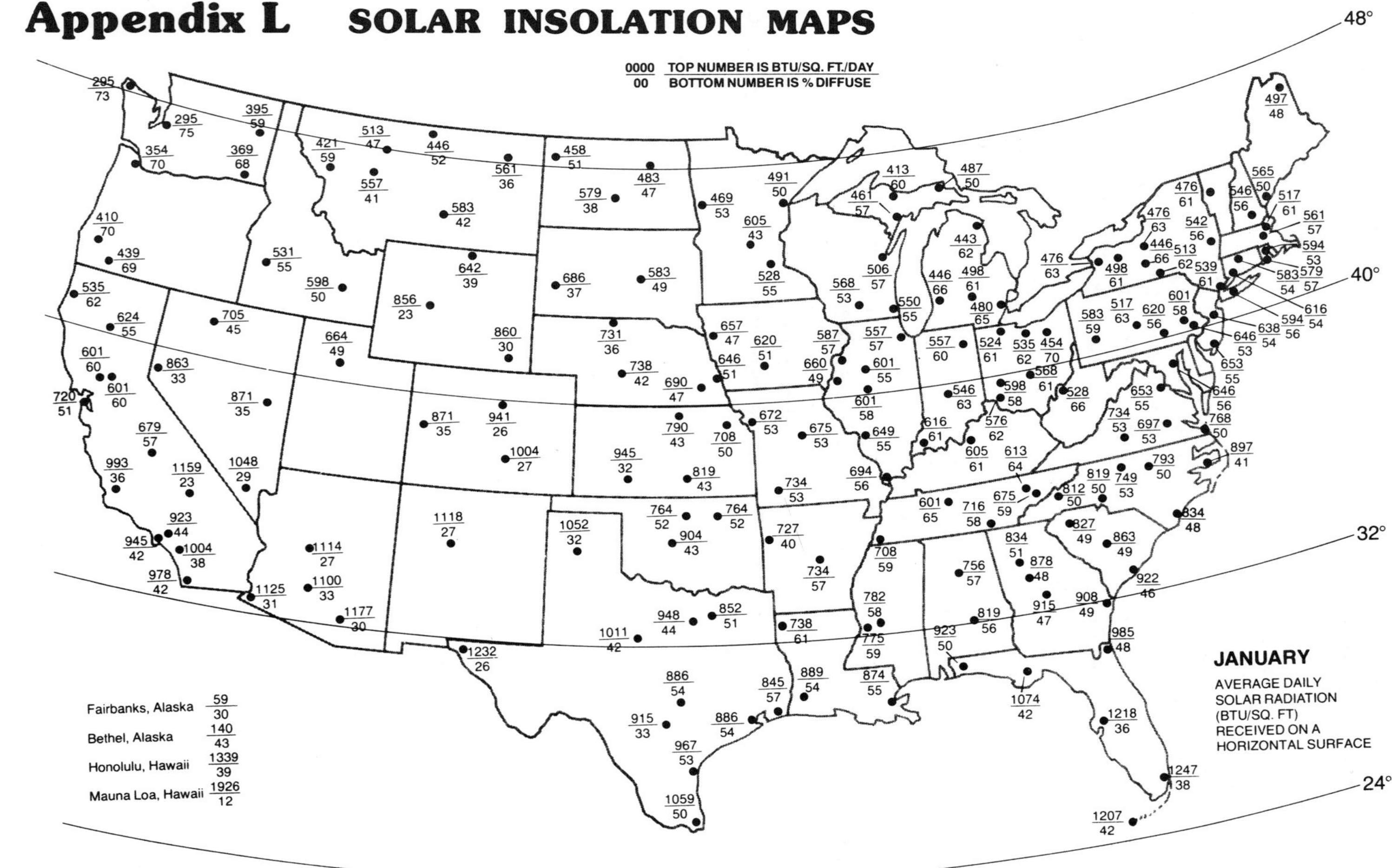

Reproduced from Copper Brass Bronze Design Handbook, Solar Energy Systems *with permission of the Copper Development Association Inc., New York.*

FEBRUARY
AVERAGE DAILY
SOLAR RADIATION
(BTU/SQ. FT)
RECEIVED ON A
HORIZONTAL SURFACE
48°
40°
32°
24°
Fairbanks, Alaska 262/39
Bethel, Alaska 399/39
Honolulu, Hawaii 1557/38
Mauna Loa, Hawaii 2125/15

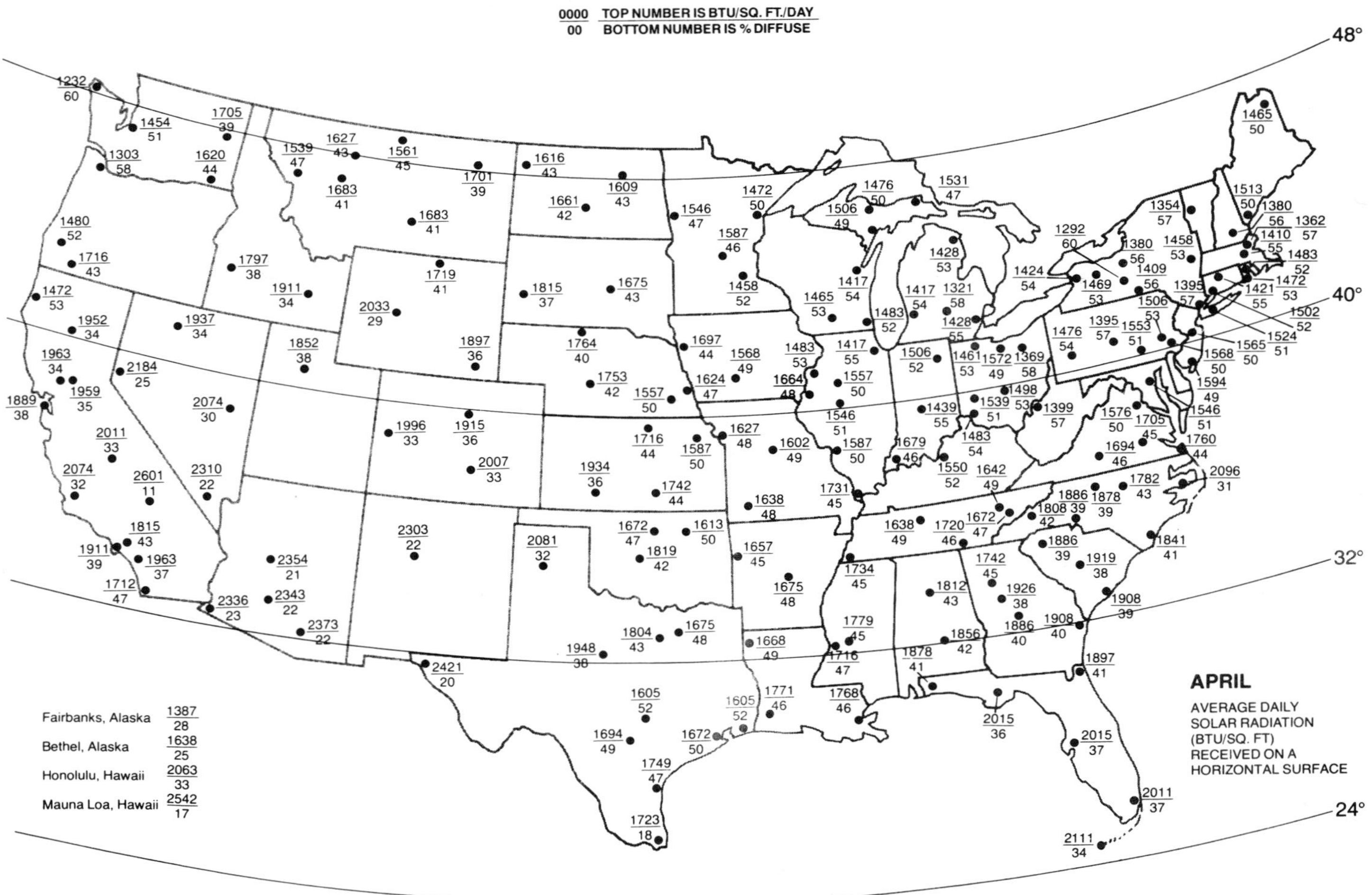
0000 TOP NUMBER IS BTU/SQ. FT./DAY
00 BOTTOM NUMBER IS % DIFFUSE
48°
40°
32°
24°
APRIL
AVERAGE DAILY
SOLAR RADIATION
(BTU/SQ. FT)
RECEIVED ON A
HORIZONTAL SURFACE
Fairbanks, Alaska 1387/28
Bethel, Alaska 1638/25
Honolulu, Hawaii 2063/33
Mauna Loa, Hawaii 2542/17

MARCH
AVERAGE DAILY SOLAR RADIATION (BTU/SQ. FT) RECEIVED ON A HORIZONTAL SURFACE
48°
40°
32°
24°
Fairbanks, Alaska 786 / 25
Bethel, Alaska 1041 / 22
Honolulu, Hawaii 1904 / 32
Mauna Loa, Hawaii 2509 / 14

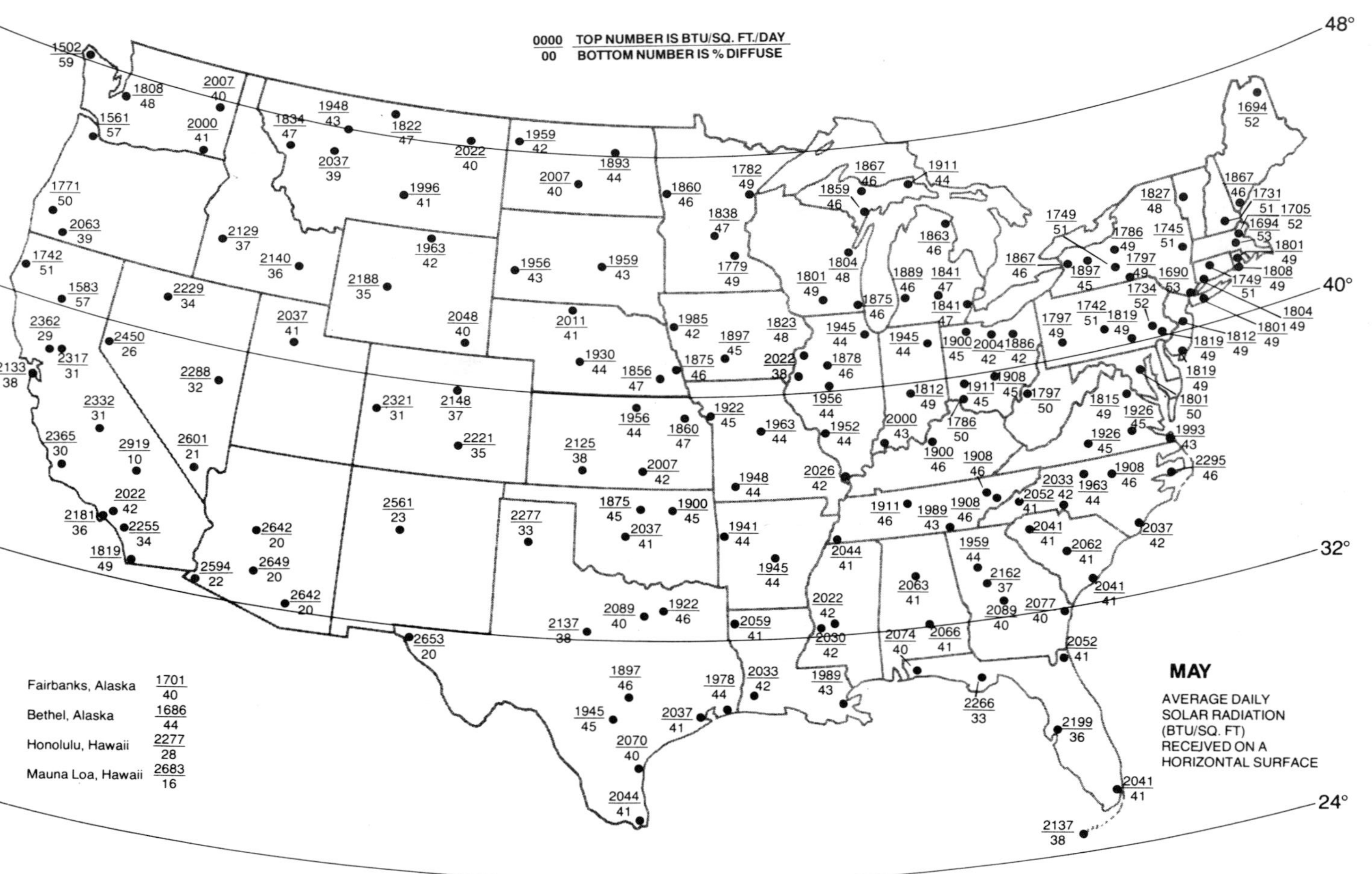
0000 TOP NUMBER IS BTU/SQ. FT./DAY
00 BOTTOM NUMBER IS % DIFFUSE
48°
40°
32°
24°
Fairbanks, Alaska 1701 40
Bethel, Alaska 1686 44
Honolulu, Hawaii 2277 28
Mauna Loa, Hawaii 2683 16
MAY
AVERAGE DAILY
SOLAR RADIATION
(BTU/SQ. FT)
RECEIVED ON A
HORIZONTAL SURFACE

JUNE
AVERAGE DAILY
SOLAR RADIATION
(BTU/SQ. FT)
RECEIVED ON A
HORIZONTAL SURFACE
48°
40°
32°
24°
Fairbanks, Alaska 1860/44
Bethel, Alaska 1675/53
Honolulu, Hawaii 2269/30
Mauna Loa, Hawaii N/A

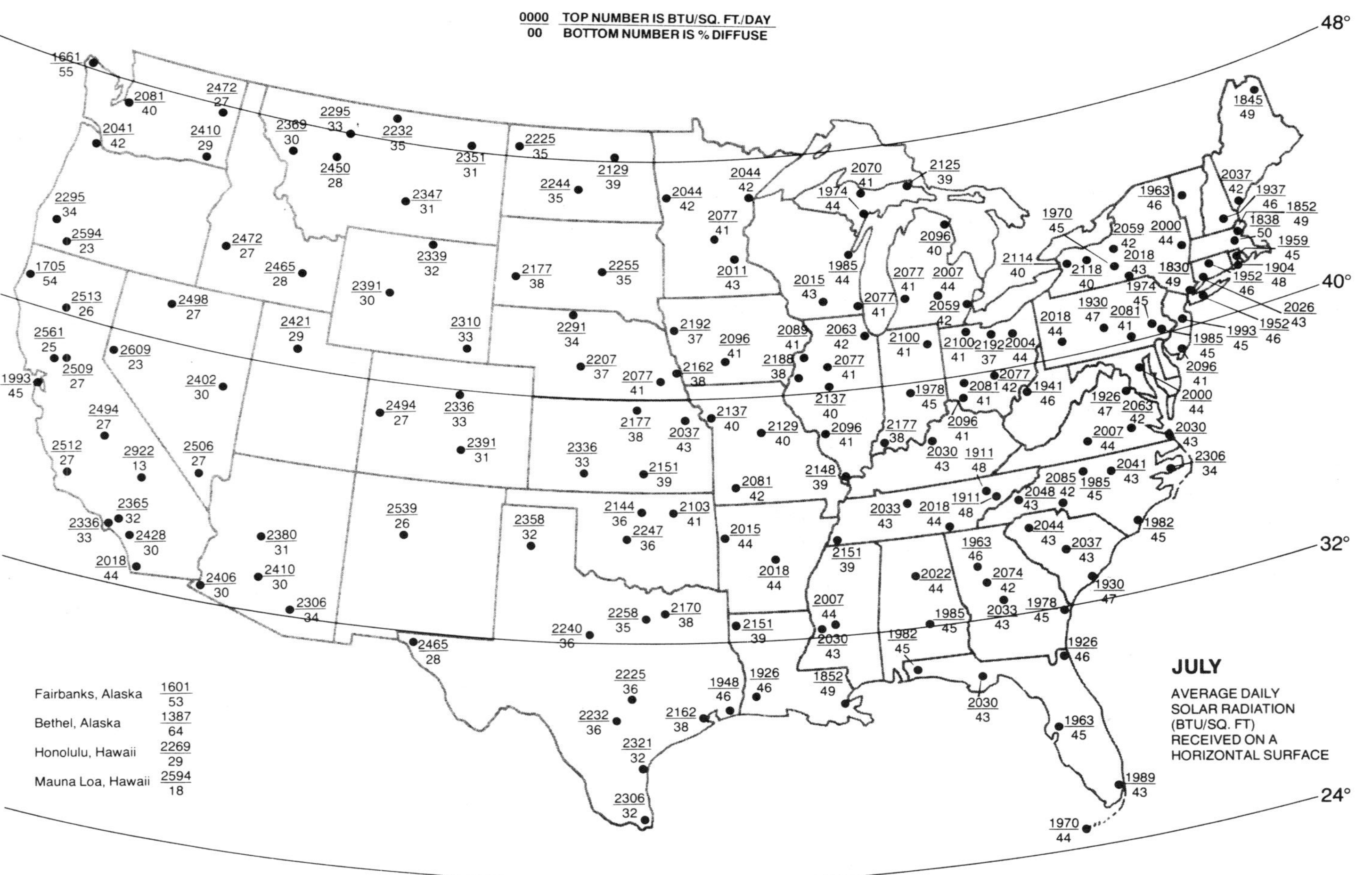

0000 TOP NUMBER IS BTU/SQ. FT./DAY
00 BOTTOM NUMBER IS % DIFFUSE
48°
40°
32°
24°
JULY
AVERAGE DAILY SOLAR RADIATION (BTU/SQ. FT) RECEIVED ON A HORIZONTAL SURFACE
Fairbanks, Alaska 1601/53
Bethel, Alaska 1387/64
Honolulu, Hawaii 2269/29
Mauna Loa, Hawaii 2594/18

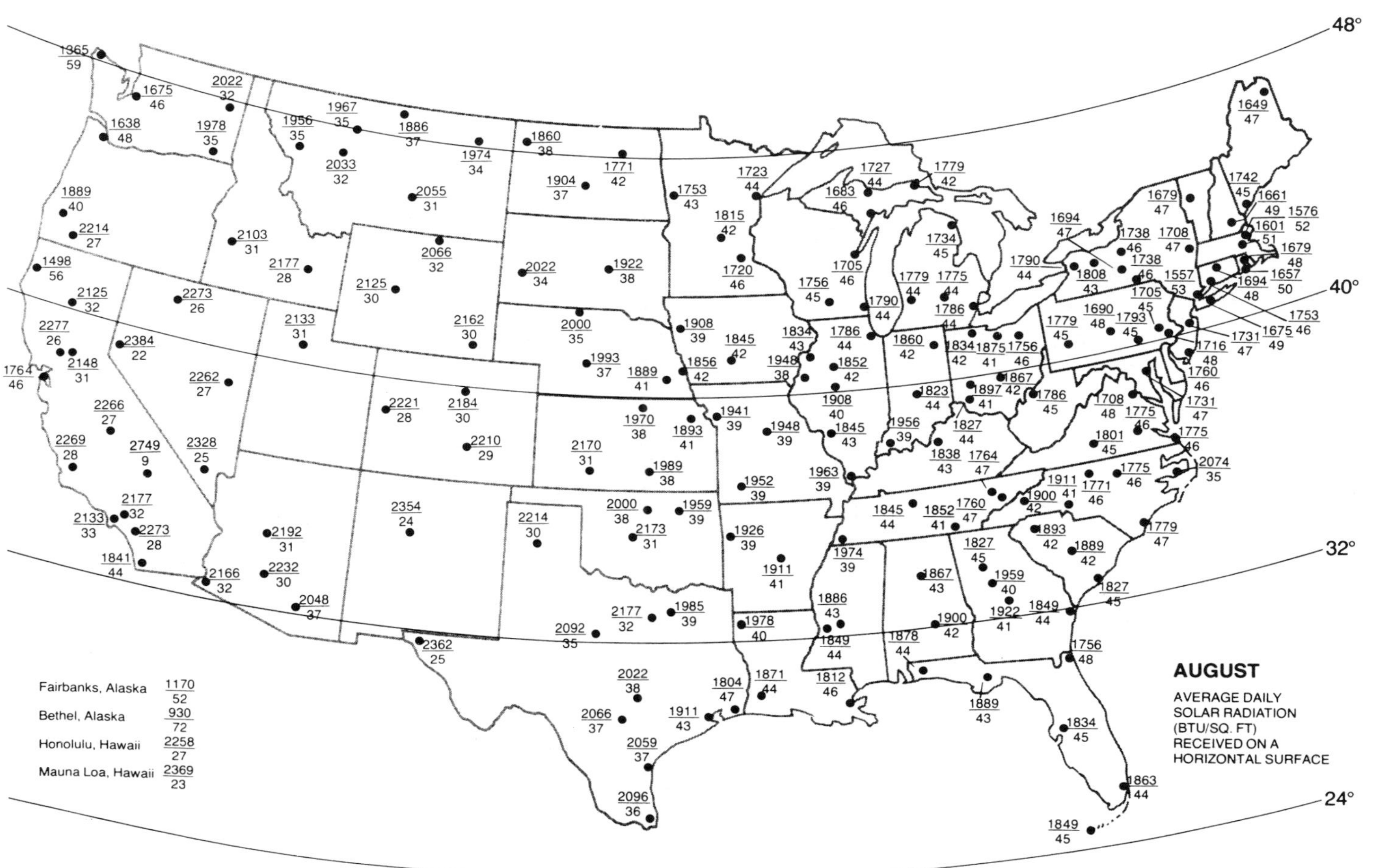
AUGUST
AVERAGE DAILY
SOLAR RADIATION
(BTU/SQ. FT)
RECEIVED ON A
HORIZONTAL SURFACE
48°
40°
32°
24°
Fairbanks, Alaska 1170 52
Bethel, Alaska 930 72
Honolulu, Hawaii 2258 27
Mauna Loa, Hawaii 2369 23

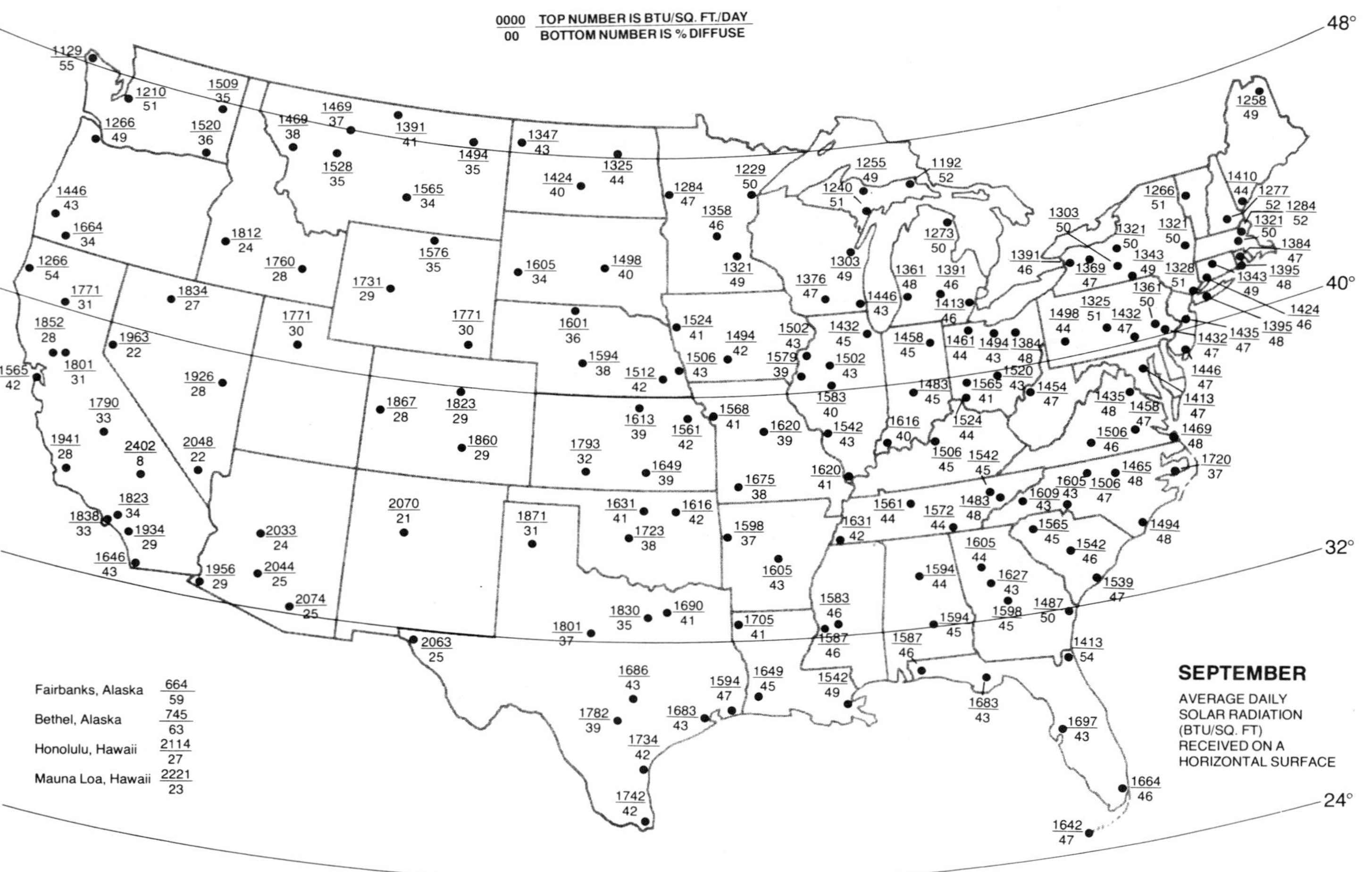
0000 TOP NUMBER IS BTU/SQ. FT./DAY
00 BOTTOM NUMBER IS % DIFFUSE
48°
40°
32°
24°
SEPTEMBER
AVERAGE DAILY
SOLAR RADIATION
(BTU/SQ. FT)
RECEIVED ON A
HORIZONTAL SURFACE
Fairbanks, Alaska 664/59
Bethel, Alaska 745/63
Honolulu, Hawaii 2114/27
Mauna Loa, Hawaii 2221/23

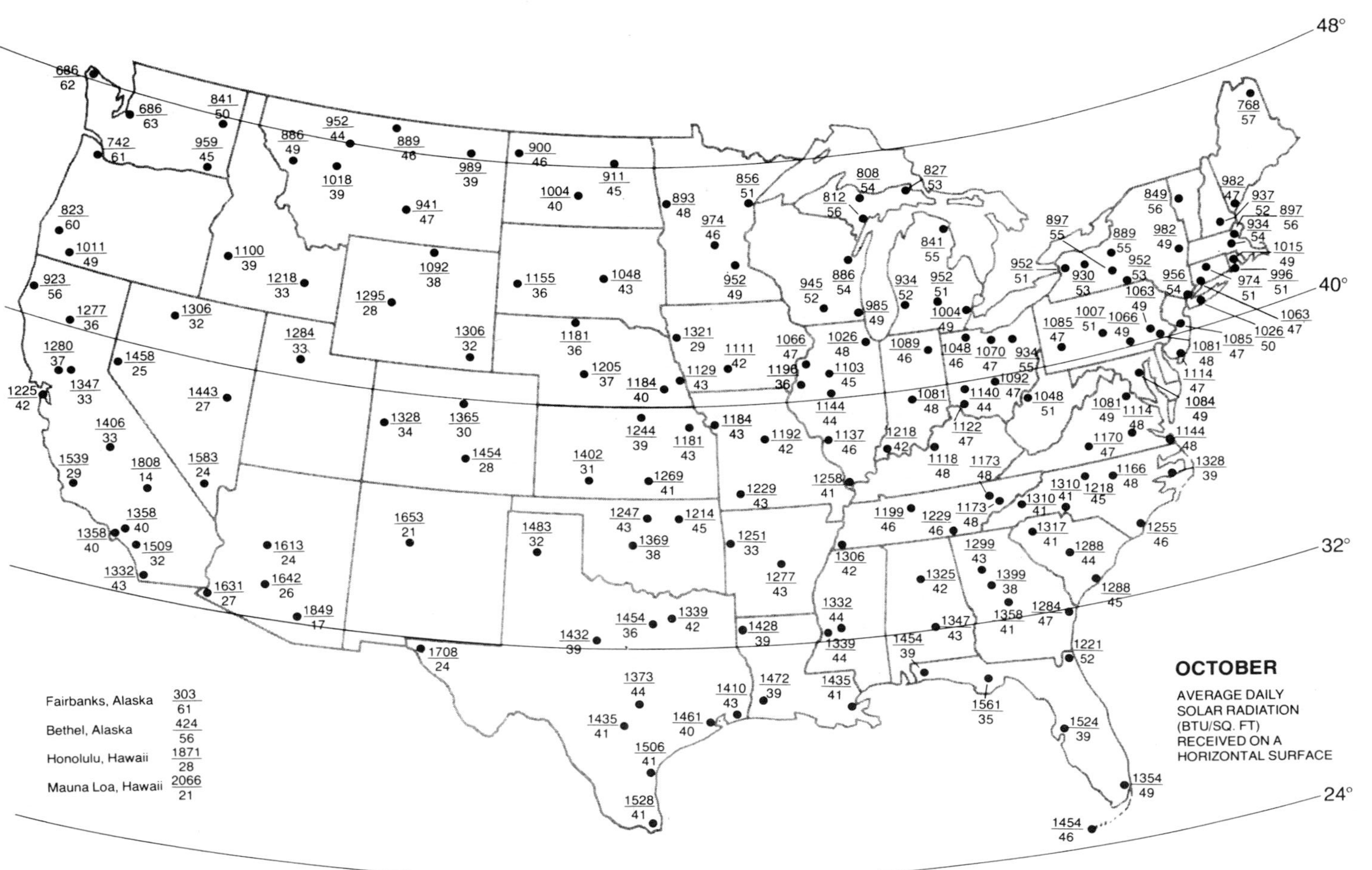
OCTOBER
AVERAGE DAILY SOLAR RADIATION (BTU/SQ. FT) RECEIVED ON A HORIZONTAL SURFACE
48°
40°
32°
24°
Fairbanks, Alaska 303 61
Bethel, Alaska 424 56
Honolulu, Hawaii 1871 28
Mauna Loa, Hawaii 2066 21

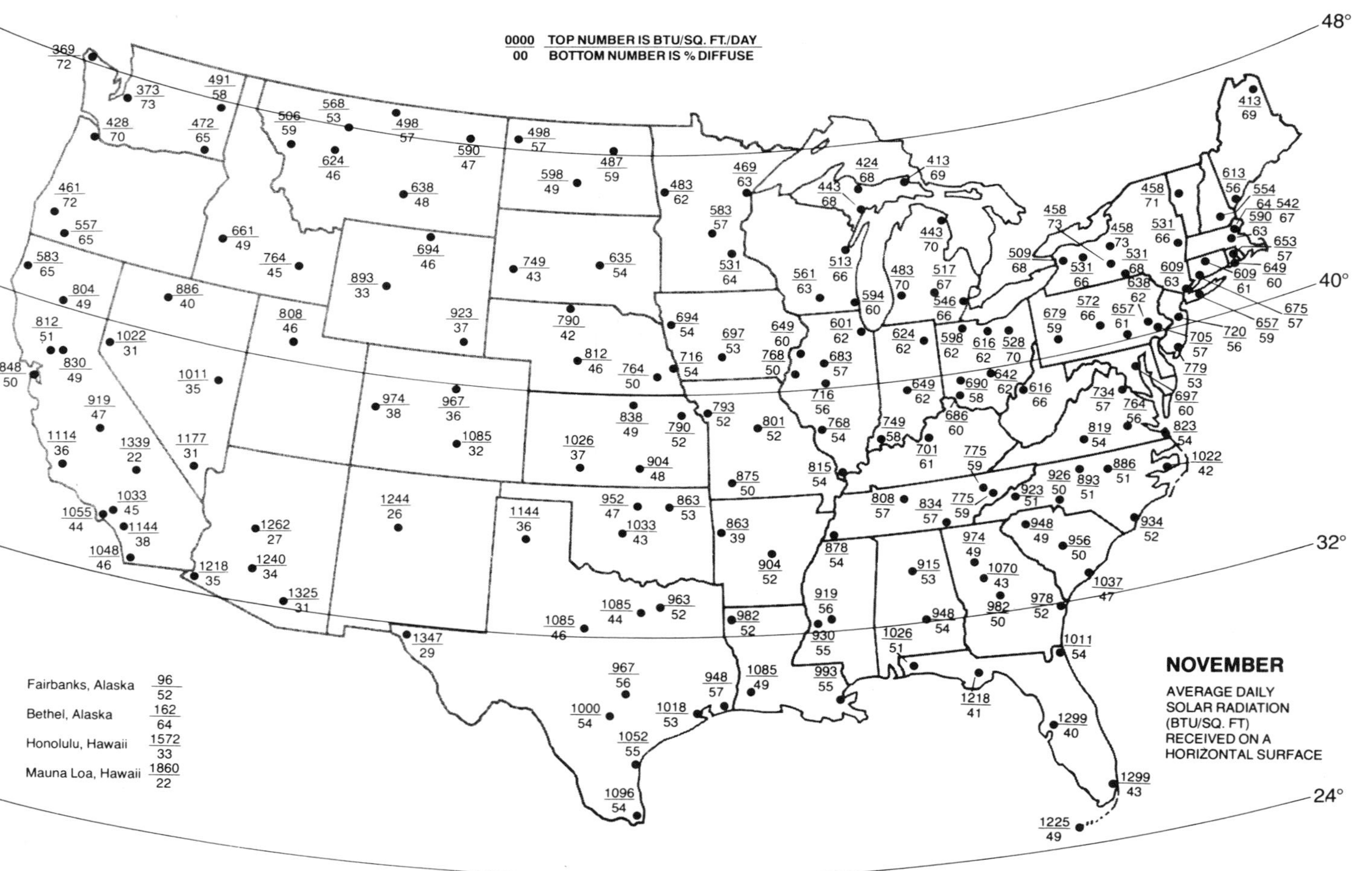

0000 TOP NUMBER IS BTU/SQ. FT./DAY
00 BOTTOM NUMBER IS % DIFFUSE
48°
40°
32°
24°
NOVEMBER
AVERAGE DAILY SOLAR RADIATION (BTU/SQ. FT) RECEIVED ON A HORIZONTAL SURFACE
Fairbanks, Alaska 96/52
Bethel, Alaska 162/64
Honolulu, Hawaii 1572/33
Mauna Loa, Hawaii 1860/22

48°
40°
32°
24°
DECEMBER
AVERAGE DAILY
SOLAR RADIATION
(BTU/SQ. FT)
RECEIVED ON A
HORIZONTAL SURFACE
Fairbanks, Alaska 22/53
Bethel, Alaska 81/53
Honolulu, Hawaii 1369/33
Mauna Loa, Hawaii 1775/15

Appendix M

ANSWERS TO EVEN-NUMBERED PROBLEMS

Chapter 3

4. a. 11–3/8
   b. 6–1/8
   c. 735/32 = 22–31–32
   d. 3–1/3

6.

Sixteenths (16ths)	*Millimeters (mm)*
1/16	1.588
1/8	3.175
3/16	4.763
1/4	6.350
5/16	7.938
3/8	9.525
7/16	11.113
1/2	12.700
9/16	14.288
5/8	15.875
11/16	17.463
3/4	19.050
13/16	20.638
7/8	22.225
15/16	23.813
1 in	25.400

8. 22.97

10. a. 17.06
    b. 18.09
    c. 23.98
    d. 13.24
    e. 39.26

12. a. Offset = 30 in (76 cm)
    b. Run =30 in (76 cm)
    c. Travel = 42–3/8 in (108 cm)

14. $108

Chapter 6

4. 80.3 lbf/in^2 (209 kPa)

6. 1.07 in^2 (6.93 cm^2)

8. 1101 gal (4168 liters)

Chapter 7

2.

Nominal Size	*Schedule 40 Steel Pipe*	*Schedule 40 PVC Plastic*	*Type M Copper Tube*
1 in	0.864	0.864	0.874
2 in	3.354	3.354	3.459
4 in	12.724	12.724	12.155
8 in	50.002	50.002	45.576

4. 18 min, 38 sec

8. 12.5 ft/sec

10. 62 lbf/in^2 approx.

Chapter 9

14. Gutter: 6 in @ 1/8-in pitch per ft
Downspout: 2-1/2 in
Drain: 3 in @ 1/4-in pitch per ft

Chapter 11

6. 1174 gal/hr

Chapter 12

8. $31.64

10. South at an angle equal to the latitude

Index

D

E

F

G

H

I

J

T